HYDROLOGY OF DISASTERS

Water Science and Technology Library

VOLUME 24

Editor-in-Chief
V. P. Singh, *Louisiana State University,
Baton Rouge, U.S.A.*

Editorial Advisory Board

M. Anderson, *Bristol, U.K.*
L. Bengtsson, *Lund, Sweden*
A. G. Bobba, *Burlington, Ontario, Canada*
S. Chandra, *New Delhi, India*
M. Fiorentino, *Potenza, Italy*
W. H. Hager, *Zürich, Switzerland*
N. Harmancioglu, *Izmir, Turkey*
A. R. Rao, *West Lafayette, Indiana, U.S.A.*
M. M. Sherif, *Giza, Egypt*
Shan Xu Wang, *Wuhan, Hubei, P.R. China*
D. Stephenson, *Johannesburg, South Africa*

HYDROLOGY OF DISASTERS

edited by

VIJAY P. SINGH

*Water Resources Program,
Department of Civil and Environmental Engineering,
Louisiana State University, Baton Rouge, U.S.A.*

KLUWER ACADEMIC PUBLISHERS
DORDRECHT / BOSTON / LONDON

Library of Congress Cataloging-in-Publication Data

```
Hydrology of disasters / edited by V.P. Singh.
      p.    cm. -- (Water science and technology library ; v. 24)
   Includes index.
   ISBN 0-7923-4092-2 (hb : alk. paper)
   1. Floods. 2. Natural disasters.    I. Singh, V. P. (Vijay P.)
 II. Series.
 GB1399.H88   1996
 363'.492--dc20                                              96-18911
```

ISBN 0-7923-4092-2

Published by Kluwer Academic Publishers,
P.O. Box 17, 3300 AA Dordrecht, The Netherlands.

Kluwer Academic Publishers incorporates
the publishing programmes of
D. Reidel, Martinus Nijhoff, Dr W. Junk and MTP Press.

Sold and distributed in the U.S.A. and Canada
by Kluwer Academic Publishers,
101 Philip Drive, Norwell, MA 02061, U.S.A.

In all other countries, sold and distributed
by Kluwer Academic Publishers Group,
P.O. Box 322, 3300 AH Dordrecht, The Netherlands.

Printed on acid-free paper

All Rights Reserved
© 1996 Kluwer Academic Publishers
No part of the material protected by this copyright notice may be reproduced or
utilized in any form or by any means, electronic or mechanical,
including photocopying, recording or by any information storage and
retrieval system, without written permission from the copyright owner.

Printed in the Netherlands

to: ANITA, VINAY AND ARTI

Table of contents

Preface   **xiii**

1 Disasters: Natural or Man-made
V.P. Singh   **1**
- 1.1 Types of Disasters 1
  - 1.1.1 Hurricanes and Tornadoes 5
- 1.2 Environmental and Hydrologic Consequences 13
  - 1.2.1 Hurricanes 13
  - 1.2.2 Floods 14
  - 1.2.3 Earthquakes 14
- 1.3 Mitigation of Disasters 14
  - 1.3.1 Hurricanes 15
  - 1.3.2 Floods 16
  - 1.3.3 Earthquakes 16

2 Representativity of Extreme Wind Data
J. Wieringa   **19**
- 2.1 Introduction 19
- 2.2 Wind phenomena Classified by the Wind Sspectrum 20
- 2.3 What Are Representative Wind Data? 22
- 2.4 Representativity of "Potential Wind" 23
- 2.5 Roughness Determination at Ordinary Wind Stations 25
- 2.6 Scale of Application of Station-Observed Wind 28
- 2.7 Getting and Using Potential Wind Data 30
- 2.8 Distribution of Average Wind Speeds 31
- 2.9 Shelter Problems in Analysis of Extreme Stormwind speeds 33
- 2.10 Some Available Studies of Extreme wind Speeds 35

3 Climate Change and Hydrological Disasters
M.A. Beran and N.W. Arnell   **41**
- 3.1 Introduction 41
  - 3.1.1 Questions 41
  - 3.1.2 Concepts and definitions 41
  - 3.1.3 Is climate change a hydrological disaster? 42
- 3.2 The Greenhouse Effect, Climate Change and Hydrological Regimes 43
- 3.3 Estimating the Impacts of Climate Change on Hydrological Characteristics 45
  - 3.3.1 Introduction: Climate Change Scenarios 45
  - 3.3.2 Creating Climate Change Scenarios 45
  - 3.3.3 Equilibrium and Transient Scenarios 48
- 3.4 Impacts of Climate Change on Floods 48
  - 3.4.1 Introduction 48
  - 3.4.2 Factors Responsible for Floods 49
- 3.5 Impact of Climate Change on Drought 52

	3.5.1	Introduction	52
	3.5.2	Drought Definition	52
	3.5.3	Factors Responsible for Drought	54
	3.5.4	Climatic Drought: A Shortage of Rainfall and Soil Water	55
	3.5.5	Hydrological Drought: A Shortage of Runoff and Recharge	56
	3.5.6	Water Resources Drought	57
	3.5.7	The Case of Sahel Drought	58
	3.5.8	Direct Effect of Carbon Dioxide	59
3.6	Conclusions		59

4 Extreme Floods
F. Ashkar   **63**

4.1	Introduction	63
4.2	Limitations of Statistical Methods for Estimating Design Floods	64
4.3	Data Series Used in the Estimation of Design Floods	64
	4.3.1 Brief Description of the POT Procedure	65
	4.3.2 Brief Comparison between AMF and POT Modelling	67
4.4	Flood "Quantiles" as Design Events	67
4.5	Statistical Hypotheses Used for Estimating Design Floods	68
4.6	Some Remarks Pertaining to Flood Data	69
4.7	Choice of "D/E Procedure" for Flood Frequency Analysis	70
	4.7.1 Choice of Statistical Distribution (D)	71
	4.7.2 Choice of Parameter-Estimation Method (e)	75
	4.7.3 Sources of Uncertainty in Estimating Design Floods	76
	4.7.4 Statistical Tools for Selecting D/E Procedures	77
4.8	Regional Flood Frequency Estimation	77
4.9	Conclusions	79

5 Dam-Breach Floods
D.L. Fread   **85**

5.1	Introduction	85
5.2	Breach Description	87
	5.2.1 Mathematical Description of Breach	87
	5.2.2 Concrete Dams	89
	5.2.3 Earthen Dams	89
	5.2.4 Assessment of Breach Parameters	94
5.3	Dam-Breach Flood Routing	96
	5.3.1 Dynamic Routing	96
5.4	Dam-Breach Flood Routing Data	112
	5.4.1 Cross-Sectional Properties	112
	5.4.2 Sinuosity Factors	113
	5.4.3 Manning n Friction Coefficients	113
	5.4.4 Levee Properties	115
	5.4.5 Lateral Flows	116
5.5	Teton Dam-Breach Flood Case Study	116
5.6	Uncertainties of Dam-Breach Flood Modeling	121
	5.6.1 Two-Dimensional Effects	121
	5.6.2 Cross-Sectional Degradation	121
	5.6.3 Manning n	122
	5.6.4 Debris Effects	123
	5.6.5 Breach Properties	123
	5.6.6 Flow Losses	123

6 Extreme Droughts
M.L. Kavvas and M.L. Anderson   **127**

  6.1   Introduction 127
  6.2   Physical Systems Involved in a Drought 129
        6.2.1   Atmospheric System 129
  6.3   The Evolution of Extreme Drought 135
        6.3.1   Initiation 135
        6.3.2   Growth 138
        6.3.3   Recovery 140
        6.3.4   Example 140
  6.4   Modeling an Extreme Drought 145
        6.4.1   General Circulation Models and Energy Balance Models 146
        6.4.2   Energy Balance Model Parameterization 147
        6.4.3   Simulation of an Extreme Drought by an Energy Balance Climate Model 151
  6.5   Discussion and Conclusions 156

7 Mud and Debris Flows
P.A. Johnson and R.H. McCuen   **161**

  7.1   Introduction 161
  7.2   Physical Processes 162
        7.2.1   Factors Affecting Debris Flow Initiation 162
        7.2.2   Laboratory and Mathematical Studies 166
  7.3   Methods of Prediction: When, Where, and How Much 167
        7.3.1   Magnitude 167
        7.3.2   Frequency of Occurrence 169
  7.4   Debris Flow Mitigation 171
        7.4.1   Warning Systems 171
        7.4.2   Passive Mitigation Measures 171
        7.4.3   Active Mitigation Measures 172
  7.5   Statistical Modeling of Debris Flows 173
        7.5.1   The Regionalization Process 174
        7.5.2   Regionalization of Debris Volumes 175
        7.5.3   Data Requirements for Regionalization 176
  7.6   Conclusions 177

8 Landslides
T.P. Gostelow   **183**

  8.1   Introduction 183
  8.2   Hydrological Triggering Mechanisms 183
  8.3   Rainfall and Landslide Disasters 185
  8.4   Regional Groundwater Flow 187
        8.4.1   General 187
        8.4.2   Groundwater in Mountainous Settings 188
  8.5   First-time and reactivated landslides 188
  8.6   First-time Translational Slides 188
        8.6.1   Examples 188
        8.6.2   Physical Models of Rainfall Infiltration and Mechanisms of Translational Failure 191
        8.6.3   First-Time Translational Slides: Rainfall Triggers 193
  8.7   First-Time Rotational and Complex Deep-Seated Pre-Existing Landslide Movements 195

		8.7.1 Hydrogeology of Pre-Existing Landslides 195

        8.7.1   Hydrogeology of Pre-Existing Landslides 195
        8.7.2   Mass Movement Associated with Depression and Perched Springs 196
        8.7.3   Geological Susceptibility to Hydrological Landslide Disasters . . . 197
  8.8 Landslide Caprocks and Their Response to Rainfall 206
        8.8.1   General 206
        8.8.2   Landslide and Aquifer Response to Rainfall 206
        8.8.3   Spring Discharge from Aquifers and Landslide Caprocks: Monitoring a Potential Disaster? 210
        8.8.4   Geological Structure, Aquifers, Valley Sides and Landslides 212
        8.8.5   Development of Aquifer Properties over Time 216
  8.9 Groundwater Models, Pre-Existing Landslide Complexes and Regional Planning 216
        8.9.1   Models, Planning and Geographic Information Systems (GIS) . . . 216
        8.9.2   Recognition and Mapping Groundwater Discharge and Recharge Areas for Hazard Assessment 217
  8.10 Landslides Associated with Snowmelt, Permafrost and Glaciers 218
  8.11 Erosion, Rivers and Landslides 219
  8.12 Storm-Induced Submarine Landslides 220
  8.13 Conclusions 221

9 Land Subsidence
G. Gambolati, M. Putti and P. Teatini   **231**

  9.1 Introduction 231
  9.2 Review of Mathematical Theory of Land Subsidence due to Fluid Withdrawal 235
        9.2.1   Coupled (Biot) Model of Land Subsidence 236
        9.2.2   Uncoupled Model of Land Subsidence 238
        9.2.3   Comparison of Coupled and Uncoupled Land Subsidence Predictions 240
  9.3 Illustrative Case Studies 243
        9.3.1   Land Subsidence at Ravenna due to Groundwater Withdrawal . . . 243
        9.3.2   Land Subsidence in the Ravenna Area Caused by Gas Production . 256
        9.3.3   Land Subsidence Prediction at Mexico City 258

10 Saltwater Intrusion
M.M. Sherif and V.P. Singh   **264**

  10.1 Hydrological Aspects 269
  10.2 Sharp Interface and Density Dependent Approaches 270
        10.2.1 Sharp Interface Approach 270
        10.2.2 Density Dependent Approach 278
  10.3 Dispersion in Porous Media 279
        10.3.1 Experimental Investigations on Longitudinal and Lateral Dispersion 281
        10.3.2 Hydrodynamic Dispersion Equation 284
  10.4 Mechanism of Saltwater Intrusion into Coastal Aquifers 287
  10.5 Governing Equations 287
        10.5.1 Sharp Interface Approach 287
        10.5.2 Density Dependent Approach 289
  10.6 Initial and Boundary Conditions 290
        10.6.1 Initial Conditions 291
        10.6.2 Boundary Conditions 291
  10.7 Numerical Methods 295
        10.7.1 Finite Difference Method 295
        10.7.2 Finite Element Method 296
  10.8 Finite Element Formulation for Density Dependent Problems 297

Table of Contents

    10.9 Study Cases 299
        10.9.1 Hypothetical Case 299
        10.9.2 The Madras Aquifer 302
        10.9.3 The Nile Delta Aquifer in Egypt 307
    10.10 Concluding Remarks 312

11 Avalanche Dynamics
K. Hutter   **317**

    11.1 Introduction 319
        11.1.1 Some Historical Notes 319
        11.1.2 Physical behaviour 320
        11.1.3 Laws of Similitude 323

I. Dynamics of Granular Avalanches   **325**

    11.2 Some Distinctive Characteristics of Granular Flows 325
        11.2.1 Dilatancy, Internal Friction, Rate Dependence of Stress, Large
               Energy Dissipation 325
        11.2.2 Large Travelled Distances, Size Effects 327
        11.2.3 Three Different Flow Regimes 329
    11.3 One-Dimensional Model 332
        11.3.1 Governing Equations 332
        11.3.2 Comparison with Experiments 338
        11.3.3 Similarity Solutions 340
    11.4 Two-Dimensional Unconfined Flow 344
        11.4.1 Equations 344
        11.4.2 Experiments and First Results 346

II. Powder Avalanches   **351**

    11.5 Density and Turbidity Current Concept 352
        11.5.1 Long Gravity or Turbidity Currents 353
        11.5.2 Short gravity currents. "Thermals" on Inclined Boundaries . . . 363
        11.5.3 Other Mixture Models and Critique 374
    11.6 Two-phase Flow Models 376
        11.6.1 Treatment of Boundary Conditions 376
        11.6.2 Experimental and Computational Results 378
    11.7 Concluding remarks 386

12 Hydrological Disasters Associated with Volcanoes
V.E. Neall   **395**

    12.1 Introduction 395
    12.2 Steam (Phreatic) Explosions 396
    12.3 Eruptions Through a Crater Lake 398
    12.4 Pyroclastic Flows Interacting with Water 403
    12.5 Volcanic Melting of Snow and Ice 403
    12.6 Volcanogenic Tsunamis 409
    12.7 Release of Gases from a Crater Lake 411
    12.8 Non-Volcanic Initiated Collapse of a Crater Lake 413
    12.9 Heavy Rains on Recently Erupted Materials 414
    12.10 Conclusion 419

13 Earthquakes
A. Terakawa and O. Matsuo **427**

    13.1 Introduction 427
    13.2 Example of Hydrologic Consequences of Earthquakes 427
        13.2.1 The Zenkouji Earthquake 427
        13.2.2 The Naganoken Seibu Earthquake 430
        13.2.3 The Matsushiro Earthquake 431
        13.2.4 The Izu Oshima Kinkai Earthquake 433

List of Contributors **435**

Index **437**

PREFACE

The General Assembly of the United Nations passed a resolution on December 11, 1987, designating the 1990s as the International Decade for Natural Disaster Reduction. This resolution has served as a catalyst in promotion of international cooperation in the field of natural disaster reduction; in initiation of wide-ranging research activities on natural and man-made disasters; in development of technologies for assessment, prediction, prevention, and mitigation through technical assistance, technology transfer, demonstration projects, and education and training; and in dissemination of information related to measures for assessment, prediction, prevention, and mitigation of natural disasters.

Disasters are manifestations of environmental extremes. Depending upon the type of disasters, their occurrence may have short-term and/or long-term detrimental environmental consequences. Disasters cannot be prevented altogether, but their impact can be mitigated. This book is an attempt to provide a discussion of hydrological aspects of the various types of natural disasters. It is hoped that others will be stimulated to write more comprehensive texts on this subject of enormous importance.

The subject matter of this book is divided into 13 chapters. A brief perspective of natural disasters is provided in the introductory Chapter 1. Also included in the chapter are the consequences of these disasters and their mitigation. Representativity of extreme wind data is discussed in Chapter 2. It discusses the concept of representative wind information, classification of wind phenomena, requirements for various wind-data users, methods for conforming wind stations with WMO specifications, roughness determination, distribution of average wind speeds, and practical ways to obtain design wind speeds. Climate change and its relation to hydrological disasters constitute the subject matter of Chapter 3. Climate change by itself may not be a disaster but may alter the exposure to hydrological disasters. Therefore, it must be taken into account when planning and designing disaster mitigation schemes.

Floods occur every year throughout the world and cause loss of life and property. Design of flood-control works is based on statistical methods which are discussed in Chapter 4. Floods are also caused by breaching of dams and cause death and destruction of people and their property in the downstream valley. Such floods can be catastrophic. Chapter 5 discusses dam-breach floods, with particular attention to breach modeling, flood routing, and uncertainties associated with dam-breach flood models. On the other end of the hydrological spectrum are droughts which visit one or the other part of the world virtually every year, causing loss of property and disruption of socio-economic infrastructure. Chapter 6 focuses on the physical processes present in and modeling of droughts.

Debris flows occur in many parts of the world, killing people and destroying roadways, bridges, and homes. Such flows are typical in steep, mountainous areas. Chapter 7 reviews the physical processes initiating a debris flow, the models proposed to predict the magnitude and frequency of debris flows, and the mitigation methods. Chapter 8 discusses land slides. Some of the world's worst landslide disasters have been caused by hydrological factors, especially rainfall. Reviewing some of the factors responsible for the distribution of hydrologically induced slides, the chapter goes on to discuss triggering mechanisms and hydrogeological models, and is concluded with suggestions for identifying and mapping the geological, topographical, and climatic conditions leading to landslides.

Land subsidence is another disaster, caused primarily by over-exploitation of subsurface fluids, such as water, oil, or natural gas. It constitutes the subject matter of Chapter 9. Presenting a summary of the most well-known subsiding sites in the world, it goes on to discuss the basic linear theory employed to build mathematical models for prediction of land subsidence due to fluid withdrawal, a comparison of coupled and uncoupled approaches, and three well-known examples of land subsidence due to water and gas extraction. On the other hand, extensive pumping of fresh groundwater and the consequent lowering of the water table or piezometric head causes salt water intrusion in coastal areas, and may degrade the water quality in the aquifer or may even destroy the freshwater resource. Salt water intrusion is the focus of Chapter 10. Presenting the hydrology of salt water intrusion, it discusses the mechanisms, mathematical models, and the initial and boundary conditions, and concludes with a discussion of two case studies.

Avalanche dynamics is the subject of Chapter 11. Introducing the topic with a discussion of some historical notes, physical behavior, and laws of similitude, it provides an extensive review of the dynamics of granular as well as powder avalanches with particular emphasis on their characteristics and modeling. Hydrological disasters associated with volcanoes constitute the subject matter of Chapter 12. Volcanoes are triggered by a variety of mechanisms. The chapter reviews each of these mechanisms and emphasizes the need for volcanic monitoring and civilian response in order to help reduce the burgeoning number of casualties. The concluding Chapter 13 discusses the hydrological consequences resulting from earthquakes, with examples from Japan.

The editor would like to express his deep gratitude to the chapter contributors who, despite their numerous engagements and hectic schedule, were generous to complete their contributions. He also acknowledges the support and cooperation of his wife, Anita, and his children, Vinay and Arti, without which this book would not have come to fruition.

V.P. SINGH
Baton Rouge, Louisiana, USA

CHAPTER 1

Disasters: Natural or Man-made

V.P. Singh

The term 'disaster' has the connotation of an event capable of inflicting damage or causing danger to human and animal life and/or property. A disaster can be natural or manmade. Regardless of their origin, disasters have occurred to mankind from the time immemorial and will continue to occur in the years ahead. We cannot eliminate disasters, but certainly can mitigate their impact.

1.1 Types of Disasters

There is a wide range of natural and man-induced disasters (Starosolszky and Melder, 1989), including storms (hurricanes, typhoons, and cyclones), floods, droughts, extreme heat, extreme cold, volcanoes, earthquakes, tsunamis, landslides, dam breaching, land subsidence, saltwater encroachment, chemical spills, wild fires, snow avalanches, mud and debris flow, viral epidemic, climate change, sea-level rise, tidal waves, famine, accidents (train accidents, air crashes, etc.), bomb explosion, nuclear explosion as exemplified by Chernobyle nuclear power plant, hazardous contamination as exemplified by Love Canal and Superfund sites in the United States, chemical explosion as exemplified by Bhopal accident in India, and so on. To this litany of disasters, one can arguably add such man-triggered disasters as unchecked population growth, increasing dependence on chemicals, depletion of natural resources, and rising tide of social and religious fanaticism. Due to different mechanisms causing them, these disasters have different characteristics in terms of their spatial and temporal scales; time, place, and frequency of their occurrence; and their duration and intensity. Consequently, their impact vary in time, space and cost. According to the U.S. National Committee for the Decade for Natural Disaster Reduction, the 1989 estimated insurance payments for catastrophic losses in the United States due to high winds, tornadoes, floods, tropical storms, hurricanes, hail, severe winter storms, and earthquakes exceeded $7.6 billion. More than 15 catastrophic events, with insured losses exceeding $5 million each, occurred in 35 states, Puerto Rico, and Virgin Islands. National Academy

Press (1987) catalogued selected natural disasters of this century as shown in Table 1.1.

Some disasters may last a long time, while others a few seconds or a few minutes. For example, climate change may last a long time, but earthquakes a few seconds. Similarly, droughts occur over a long time whereas floods, by comparison, a very short time. Some of the disasters have local impact, whereas others have impact over large areas. For example, hurricanes affect large areas, whereas a chemical spill has a local effect. Volcanic eruptions, on the other hand, are intermediate to regional in their impact. As an example, Alaska's Mount Spurr erupted on September 17, 1992, for the third time in 12 weeks. The erupted cloud drifted east and south across Canada and re-entered U.S. airspace over the upper Midwest. By the morning of September 20, 1992, the cloud had moved northeast out of U.S. domestic air corridors and had started to disperse. The eruption cloud is a hazard to jet aircraft. Mount Rainier in Washington, U.S.A., was designated by the United Nations in 1987 as one of the fifteen volcanoes targetted for hazard reduction. St. Helens erupted in 1980 and killed 69 people. Like St. Helens, Rainier is part of the cascade range but Rainier is bigger and more dangerous than St. Helens because it is covered with 2.9 billion m^3 of ice and snow.

Some of the disasters occur more frequently than others. For example, hurricanes occur annually along the Atlantic Coast and the Gulf Coast region in the United States. Similarly, extreme winds are quite common in the Rocky Mountains. On the other hand, drought and dam breaching are not as frequent. Some disasters are confined to certain areas, while others are more ubiquitous. As an example, earthquakes occur in areas that are seismically active, as in the case in California in the U.S.A. Similarly, droughts are common in low rainfall areas of the world, as is the case with certain countries in Africa and Asia. Snow avalanches occur in snow-capped mountains, and land subsidence in areas having indiscriminate withdrawal of groundwater.

In certain years or decades, too many disasters seem to occur. For example, in 1981, a tidal wave submerged the Maldives in the month of April; the drought lingered on in the African continent; the serious shortage of monsoon rains caused severe drought in India; in Bangladesh 50% of the land was flooded and twenty-one million people were affected by extreme floods; Greece was struck by a heat wave, whereas the neighboring European countries struggled against floods and mudslides due to unusually heavy rains and wind storms; the Philippines and Vietnam were hit hard by cyclones and typhoons; and an earthquake hit Mexico and El Salvador. Similarly, in 1985, cyclones ravaged Bangladesh and the Philippines, and volcanic mudflows caused destruction in Columbia. In 1993, major flooding occurred in many parts of world, including Bangladesh, China, India, Nepal, Pakistan, and U.S.A. The Mississippi River flood that occurred during the spring and summer of 1993 was a major natural disaster for the United States, taking many lives and causing significant loss of property.

TABLE 1.1
Selected* natural disasters of this century (after National Academy Press, 1987)

Year	Event	Location	Approximate Death Toll
1900	Hurricane	USA	6,000
1902	Volcanic Eruption	Martinique	29,000
1902	Volcanic Eruption	Guatemala	6,000
1906	Typhoon	Hong Kong	10,000
1906	Earthquake	Taiwan	6,000
1906	Earthquake/Fire	USA	1,500
1908	Earthquake	Italy	75,000
1911	Volcanic Eruption	Philippines	1,300
1915	Earthquake	Italy	30,000
1916	Landslide	Italy, Austria	10,000
1919	Volcanic Eruption	Indonesia	5,200
1920	Earthquake/Landslide	China	200,000
1923	Earthquake/Fire	Japan	143,000
1928	Hurricane/Flood	USA	2,000
1930	Volcanic Eruption	Indonesia	1,400
1932	Earthquake	China	70,000
1933	Tsunami	Japan	3,000
1935	Earthquake	India	60,000
1938	Hurricane	USA	600
1939	Earthquake/Tsunami	Chile	30,000
1945	Floods/Landslides	Japan	1,200
1946	Tsunami	Japan	1,400
1948	Earthquake	USSR	100,000
1949	Floods	China	57,000
1949	Earthquake/Landslide	USSR	12,000-20,000
1951	Volcanic Eruption	Papua New Guinea	2,900
1953	Floods	North Sea coast (Europe)	1,800
1954	Landslide	Austria	200
1954	Floods	China	40,000
1959	Typhoon	Japan	4,600
1960	Earthquake	Morocco	12,000
1961	Typhoon	Hong Kong	400
1962	Landslide	Peru	4,000-5,000
1962	Earthquake	Iran	12,000
1963	Tropical Cyclone	Bangladesh	22,000
1963	Volcanic Eruption	Indonesia	1,200
1963	Landslide	Italy	2,000
1965	Tropical Cyclone	Bangladesh	17,000
1965	Tropical Cyclone	Bangladesh	30,000
1965	Tropical Cyclone	Bangladesh	10,000

continued on the next page

TABLE 1.1
continued

Year	Event	Location	Approximate Death Toll
1968	Earthquake	Iran	12,000
1970	Earthquake/Landslide	Peru	70,000
1970	Tropical Cyclone	Bangladesh	300,000-500,000
1971	Tropical Cyclone	India	10,000-25,000
1976	Earthquake	China	250,000
1976	Earthquake	Guatemala	24,000
1976	Earthquake	Italy	900
1977	Tropical Cyclone	India	20,000
1978	Earthquake	Iran	25,000
1982	Volcanic Eruption	Mexico	1,700
1985	Tropical Cyclone	Bangladesh	10,000
1985	Earthquake	Mexico	10,000
1985	Volcanic Eruption	Columbia	22,000
1987	Wildfire	China	200

* Disasters selected to represent global vulnerability to rapid-onset natural disasters.
Sources: Compiled from (a) Office of U.S. Foreign Disaster Assistance (1987); (b) National Geographic Society (1986); (c) K. Toki, Disaster Prevention Research Institute (Japan), personal communication (1987); (d) R.L. Schuster, U.S. Geological Survey, personal communication (1987); (e) C. Newhall, U.S. Geological Survey, personal communication (1987).

TABLE 1.2
The States in the United States most frequently hit by hurricanes since 1990.

State	Number of Hurricanes
Florida	55
Texas	36
Louisiana	25
North Carolina	22
South Carolina	14
Alabama	10
New York	9
Mississippi	8
Connecticut	8
Massachusetts	6

1.1.1 HURRICANES AND TORNADOES

Hurricanes and tornadoes are highly destructive storms. They are spatially confined, and the forces that drive them are highly concentrated (Davis and Dolan, 1993). Hurricanes occur every year along the Atlantic Coast and the Gulf Coast. Table 1.2 shows the states most hit by hurricanes since 1990. Florida, Texas, and Louisiana are the three most affected states. To better grasp the consequences, hydrologic and otherwise, of such storms, it may be worthwhile to recount the occurrence of Hurricane Andrew and Hurricane Iniki in the United States.

Florida's Dade County, where Miami is located, was the site of one of the most costly natural disasters ever to occur in the United States. Hurricane Andrew, one of the most powerful storms of the century, struck in late August, 1992. It devastated Dade County and caused damage in widespread areas of Florida. It had gusts reportedly reaching 177 mph in some areas and was far more brutal to the built environment. Damage estimates were $25 to $35 billion, and more than 80,000 homes felt Andrew's wrath. More than 5,000 mobile homes were destroyed. Areas as far as Louisiana could not escape its fury and suffered considerable damage. Indeed, it has been stated (Tarricone, 1994) that even after adjustment for inflation, Hurricane Andrew caused more damage than the sum of all earthquake damage in the U.S. in this century. In the wake of such a storm, the massive clean-up had to be organized. Andrew is alleged to have generated 20 million cubic yards (about 5 million tons) of debris. An effective clean-up required painstaking separation, processing, management, and environmentally sensitive disposal.

On September 11, 1992, a major tropical storm, Hurricane Iniki, hit Hawaii. The word "Iniki" is translated as "piercing wind", and Hurricane Iniki lived up to its name. It had sustained winds of 150 mph and gusts up to 225 mph, and stormed through all the Hawaiian islands. However, it unleashed its worst fury on Kanai, where damages totalled $1.5 billion. Virtually every home on Kanai suffered some damage, and the majority (roughly 70% of the 18,000 homes) sustained major damage. Electrical and telephone lines were down all over the island, wiping out power and communication systems. Debris resulting from wreckage of buildings and vegetation blocked roads and disrupted transportation. Iniki uprooted trees, resulting in shifted power lines, pulled-up pipes, and breaking of service lines.

There are other equally destructive storms as exemplified by nor'easters that cause significant damage along the east coast of North America. While hurricanes typically menace a relatively small stretch of coastline, roughly 100 to 150 kilometers, strong nor'easters can affect stretches of coast over 1500 kilometers long. Their high waves can cause damage comparable to or even occasionally exceeding that of a hurricane (Davis and Dolan, 1993).

1.1.1.1 Floods

The summer of 1993 witnessed catastrophic flooding in the Mississippi and Missouri River basins. The flooding wrought havoc in the midwestern United States, caused damage totalling over $12 billion (about $7 billion in agricultural loss and $5 billion in property damage), and claimed 48 lives in nine states, including Iowa, Illinois, Kansas, Minnesota, Missouri, Nebraska, North Dakota, South Dakota, and Wisconsin. More than 20 million acres were affected by the floods. Crops on more than 9 million acres were damaged and planting on more than 8 million acres was prevented. The floods damaged nearly 38,000 homes and displaced more than 30,000 people. Other damaging consequences included destruction of municipal water supply systems for many cities, disruption of river navigation and automobile and truck traffic, damaging of many buildings, and affecting of parks, preserves and recreational areas.

Beginning with April, 1993, the midwestern region experienced almost four months of rain as a persistent pressure system sat off the Southeastern U.S., pulling wet air up from the Gulf of Mexico, and cold air from a lower-than-normal-latitude jet stream turned that moisture into record-breaking precipitation over the aforementioned nine states (Denning, 1994). Compounding the precipitated water was snow-melt. The result was the worst flood the area had experienced this century, possibly even bigger than the flood of 1844 — the biggest on record. Almost 1000 miles of the Mississippi and Missouri Rivers rose beyond their normal flood levels, deluging 16,000 square miles of land. Nearly 800 of the region's 1300 levees were overtopped or damaged (Denning, 1994). This flood was so gigantic that its effect was witnessed some 1500 miles away in the Gulf Stream of the Atlantic Ocean off the east coast of Florida. Parts of the Gulf Stream, normally deep blue, were turned dark green as the flood gush of the Mississippi River dispensed in the Gulf of Mexico. Several months after the flooding, there were signs of the river's fresh water in an area 10 to 12 miles wide and more than 60 feet deep. Although this freshwater reduces salinity, it has its own detrimental effect on marine life. Because floodwater is known to carry pollutants such as pesticides and oils, it can grow organisms such as red tides which can then cause fish kills and produce toxins that get into food chains and into shellfish. It can affect lobster and fish. Therefore, coastal fishermen had reported having to head south in order to get good catches.

The flood debilitated both residential and commercial buildings, ravaged roadways and bridges, damaged water supplies and sewer systems, and contaminated the river with raw sewage and industrial wastes. Estimates of crop losses totalled over $7 billion, and miles of farmland were severely harmed. In Missouri alone, 450,000 acres of soil were blanketed with sand up to 8 feet deep. Nearly 60% of the aforementioned nine states, totalling 10% of the land area of the contiguous United States, were declared federal disaster areas. In Illinois, 1,871,928 acres of farmland were affected by high water tables. Twenty-four percent of this area's harvest was destroyed and 36% of the damaged acres could not be planted the following spring. The total crop loss in Illinois alone for the 1993 year was estimated to be $183 million. Similar damage estimates were likely throughout the area. Based on the 1973 flood, some of these farmlands would not be in production for the next 2 to 3 years. In the state of Iowa, nearly 3,000 people lost their jobs. Significant problems caused by floods include significant erosion and siltation (primarily affecting agricultural areas), water pollution, flooding of hazardous waste sites, and wash out of industrial and agricultural chemicals.

In early July of 1994, torrential rains stemming from a tropical storm that came ashore from the Gulf of Mexico caused massive flooding in Georgia in the U.S.

TABLE 1.3
Flood Damages Exceeding $50 Million in the United States

Year	Stream or Place	Damage ($ millions)		Cause
		Contemporary Dollars	1966 Dollars	
1844	Upper Mississippi River	N.A.	1161	Rainfall-river flood
1889	Johnstown, Pennsylvania	20	84	Dam failure
1900	Galveston, Texas	25	100	Hurricane tidal floods
1903	Passaic and Delaware Rivers	25	273	Rainfall and dam failure
1903	Missouri River basin	50	N.A.	Rainfall-river flood
1913	Ohio River basin	150	516	Rainfall-river flood
1913	Brazos and Colorado Rivers, Texas	128	349	Hurricane rainfall-river floods
1921	Arkansas River	13	64	Rainfall-river flood
1926	Miami and Clewiston, Florida	70	130	Hurricane-tidal and river floods
1926	Illinois River	N.A.	51	Rainfall-river floods
1927	New England	50	178	Rainfall-river flood
1927	Lower Mississippi	284	N.A.	Rainfall-river flood
1928	Puerto Rico	50	90	Hurricane tide and waves
1935	Susquehanna-Delaware Rivers	36	185	Rainfall-river flood
1936	Northeastern United States	221	374	Rainfall-river flood
1936	Ohio River basin	150	371	Rainfall snowmelt flood
1937	Ohio River basin	418	996	Rainfall-river flood
1938	New England streams	125	376	Hurricane tidal and river floods
1938	California streams	100	294	Rainfall-river floods
1942	Mid-Atlantic coastal streams	28	103	Rainfall-river floods
1943	Central States	172	N.A.	Rainfall-river floods
1944	South Florida	63	117	Hurricane tidal and river floods
1944	Missouri River basin	52	N.A.	Rainfall-river floods
1945	Hudson River basin	24	75	Rainfall-river floods
1945	South Florida	54	98	Hurricane-tidal and river floods
1945	Ohio River basin	34	61	Rainfall-river floods

TABLE 1.3
Flood Damages Exceeding $50 Million in the United States (continued)

Year	Stream or Place	Damage ($ millions)		Cause
		Contemporary Dollars	1966 Dollars	
1947	South Florida	60	88	Hurricane tidal and river floods
1947	Missouri River basin	178	N.A.	Rainfall-river floods
1948	Columbia River basin	102	226	Rainfall-river floods
1950	San Joaquin River, California	32	57	Rainfall-river floods
1951	Kansas River basin	883	N.A.	Rainfall-river floods
1952	Missouri River basin	180	N.A.	Snowmelt floods
1952	Upper Mississippi River	198	N.A.	Rainfall-river floods
1954	New England streams	180	216	Hurricane tidal floods
1955	Northeastern United States	684	879	Hurricane tidal and river floods
1955	California and Oregon streams	271	405	Rainfall-river floods
1957	Ohio River basin	65	72	Rainfall-river floods
1957	Texas rivers	144	188	Rainfall-river floods
1959	Ohio River basin	114	120	Rainfall-river floods
1960	South Florida	78	86	Hurricane tidal and river floods
1961	Texas coast	300	336	Hurricane tidal floods
1964	Florida	325	342	Hurricane tidal and river floods
1964	Ohio River basin	106	112	Rainfall-river floods
1964	California streams	173	183	Rainfall-river floods
1964	Columbia River - N. Pacific	289	311	Rainfall-river floods
1965	South Florida	139	144	Hurricane tidal and river floods
1965	Upper Mississippi River	158	162	Rainfall snowmelt river flood
1965	Platte River, Colorado - Nebraska	191	N.A.	Rainfall-river flood
1965	Arkansas River, Colorado - Kansas	61	65	Rainfall-river flood
1965	New Orleans and vicinity	322	338	Hurricane tidal flood

N.A. = Not Available. Source: U.S. Water Resources Council, 1968.

and hit hard many of the state's dams. Prior to this disaster, the last time Georgia suffered devastating losses due to flooding was when the Kelly Barnes Dam in Taccoa Falls failed in 1977. That flooding was caused by a period of intense rain creating a 25-foot flood wave, killing 39 people, damaging a bible college, and causing $2.5 million in damages. The July, 1994 flood, according to state dam safety officials, was responsible for failure of two category 1 dams, 26 category 2 dams and more than 100 smaller structures, as well as for dozens of drownings and washing out of roads and bridges.

In many parts of the world, flood is an annual phenomenon. Even in the United States, virtually every year flooding occurs in one part or the other, causing damage of all sorts — loss of life and property. Table 1.3 gives a glance of flood damage in the United States during the period 1844 to 1965. Projected annual damages are reflected in Table 1.4, and loss of human life is given in Table 1.5. Bangladesh perhaps is the most flood-prone nation in the world. The country has generally a flat topography, with the northern part being slightly steeper and the southern part flatter. On an average year, one-fifth of its area suffers inundation during monsoon months. In case of an extreme flood, two-third of the country is flooded. Approximately 9.35 million ha of agricultural land are susceptible to flooding to varying degrees, as shown in Table 1.6. In the flood of 1988, three-quarters of the country the size of Wisconsin was submerged. More than 2,000 people died and the national economy was devastated. Every year there is flooding of varying intensity and approximately every 5 years there is severe flooding (see Table 1.7) causing damage worth millions of dollars and claiming numerous lives.

1.1.1.2 Earthquakes

The word 'earthquake' sends shivers among people. This can be appreciated by a quick survey of two recent earthquakes that occurred in California. At 4:30 a.m. Pacific time on January 17, 1994, a 6.7 (on the Richter scale) magnitude earthquake rocked the Los Angeles area, crippling much of the local infrastructure, claiming 61 lives, and causing more than 9,000 injuries. Damage estimates totalled over $30 billion, making it the costliest natural disaster in the U.S. history. From its epicenter in the Northridge section of the San Fernando Valley, about 20 miles northwest of downtown Los Angeles, the earthquake, located on a previously unmapped fault, was felt from San Diego to Las Vegas. The fault plane dipped toward the south into densely populated valley. Nearly 2,500 aftershocks in the 10 days following the quake, with several measuring more than 5.0 on the Richter scale, were felt. The resulting damage was so widespread that it affected every aspect of the built environment. It damaged thousands of buildings to the extent of being demolished, collapsed six bridges on major freeways, crippling highway traffic in auto-dependent Los Angeles; broke naturalgas and water pipelines; and damaged or destroyed 45,000 residences. However, the 6.7 quake and the aftershocks might be considered as just another rehearsal for the fabled "big one" in Los Angeles, expected to be an 8.0 magnitude earthquake (on Richter scale) on the San Andreas fault, releasing 139 times more energy than Northridge. Another earthquake of 6.5 magnitude occurred nearly 23 years earlier in 1971, and about 10 miles apart from

TABLE 1.4

Projected Annual Flood Damages in the United States (Values in $ millions. Upstream refers to those streams above a point where the total area drained is 250,000 acres or less; downstream refers to the stream pattern below that point)

Region	1957 down-stream	1957 up-stream	1966 down-stream	1966 up-stream	1980[a] down-stream	1980[a] up-stream	2000[a] down-stream	2000[a] up-stream	2020[a] down-stream	2020[a] up-stream
North Atlantic	64.3	62.6	63.1	70.7	75.6	91.2	89.8	120.9	116.2	163.3
South Atlantic-Gulf	46.7	109.6	44.1	123.8	55.8	183.2	74.8	267.3	90.4	383.7
Great Lakes	12.3	29.8	13.0	33.7	15.8	43.8	21.0	57.2	27.7	76.1
Ohio	78.7	49.2	73.9	55.6	99.5	68.3	151.0	90.9	237.0	116.6
Tennessee	3.5	27.3	4.9	30.9	7.6	42.6	8.3	58.3	[b]	80.2
Upper Mississippi	60.0	52.0	64.5	68.5	96.0	101.9	151.0	143.2	218.0	197.2
Lower Mississippi	66.0	38.5	86.8	43.5	117.2	55.3	164.2	73.5	224.5	100.1
Souris-Red-Rainy	5.8	13.5	5.6	15.3	6.4	18.4	7.5	24.1	8.6	32.2
Missouri	101.4	148.1	44.0	167.3	69.0	222.3	118.0	302.7	221.0	430.1
Arkansas-White-Red	48.6	129.2	50.0	146.0	61.6	184.0	90.6	245.3	127.0	330.0
Texas-Gulf	32.5	49.5	28.2	55.9	39.5	86.1	59.3	125.3	86.4	178.4
Rio Grande	12.2	10.4	14.7	11.8	14.8	19.5	15.8	30.9	18.8	44.9
Upper Colorado	0.9	16.9	13.0	19.1	19.0	27.4	30.3	42.1	57.0	62.1
Lower Colorado	5.4	25.8	10.0	29.1	20.2	59.3	42.2	93.3	96.7	141.3
Great Basin	3.0	8.4	4.1	9.5	6.7	17.5	10.2	27.5	14.1	42.0
Columbia-North Pacific	52.2	106.7	52.1	120.6	73.6	170.1	120.6	235.3	197.7	325.8
California	36.9	67.2	61.6	75.9	102.1	134.3	185.9	211.0	262.6	311.2
Alaska	3.2	[c]	4.3	[c]	5.6	[c]	8.4	[c]	12.4	[c]
Hawaii	1.2	10.6	1.8	12.2	2.2	16.8	2.8	23.7	3.6	34.0
Puerto Rico-Virgin Islands	2.6	3.8	2.9	4.3	3.2	6.0	3.5	8.5	4.0	12.0
Total[d]	637	959	643	1094	891	1548	1355	2181	2024	3061

[a] Projected damages based on 1968 flood control works. [b] Not reported. [c] Not available. [d] Rounded.

Source: U.S. Water Resources Council, 1968.

TABLE 1.5
Floods Causing 100 or More Deaths in the United States

Year	Stream or Place	Lost Lives	Cause
1831	Barataria Isle, LA	150	Hurricane tidal flood
1856	Isle Derniere, LA	320	Hurricane tidal flood
1874	Connecticut River tributary	143	Dam failure
1875	Indianola, Texas	176	Hurricane tidal flood
1886	Sabine, Texas	150	Hurricane tidal flood
1889	Johnstown, PA	2100	Dam failure
1893	Vic. Grand Isle, LA	2000	Hurricane tidal flood
1899	Puerto Rico	3000	Hurricane tide and waves
1900	Galveston, Texas	6000+	Hurricane tidal flood
1903	Central States	100+	Rainfall-river floods
1903	Heppner, Oregon	247	Rainfall-river floods
1906	Gulf Coast	151	Hurricane tidal flood
1909	Gulf Coast-New Orleans, LA	700	Hurricane tidal flood
1913	Miami, Muskingham, and Ohio Rivers	467	Rainfall-river floods
1913	Brazos River, Texas	177	Rainfall-river floods
1915	Louisiana and Texas Gulf Coast	550	Hurricane tidal flood
1919	Louisiana and Texas Gulf Coast	284	Hurricane tidal flood
1921	Upper Arkansas River	120	Rainfall-river flood
1926	Miami and Clewiston, Florida	350	Hurricane tidal and river flood
1927	Lower Mississippi River	100+	Rainfall-river flood
1927	Vermont	120	Rainfall-river flood
1928	Puerto Rico	300	Hurricane tide and waves
1928	Lake Okeechobee, Florida	2400	Hurricane tidal flood
1928	San Francisco, California	350	Dam failure
1932	Puerto Rico	225	Hurricane tide and waves
1935	Florida Keys	400	Hurricane tidal flood
1935	Republican R., Kansas, Nebraska	110	Rainfall-river flood
1936	Northeastern United States	107	Rainfall, snow melt-river floods
1937	Ohio River	137	Rainfall-river flood
1938	New England Coast	200	Hurricane tidal and river flood
1955	Northeastern United States	115	Hurricane rainfall-river floods
1957	West Coast, Louisiana	556	Hurricane tide and river floods
1960	Puerto Rico	107	Hurricane rainfall-river floods

Source: U.S. Water Resources Council, 1968.

TABLE 1.6
Flooding of agricultural land in Bangladesh

Flood Depth (cm)	Flooded Area (10^6 ha)	Flooded Area (in %)	Nature of Flooding
<30	2.35	27	Intermittent
30 to 90	3.68	39	Seasonal
90 to 180	1.66	18	Seasonal
>180	1.46	16	Seasonal/Perennial
	9.15	100	

TABLE 1.7
Number of flood events and extreme flood events in Bangladesh during the 1954 to 1988 period

Period	Number of Floods	Number of Extreme Floods
1954 to 1960	4	2
1961 to 1970	10	2
1971 to 1980	10	1
1981 to 1984	8	2

the Northridge in the San Fernando Valley. At that time, the fault plane dipped toward the north, away from the population, under the mountains. It claimed 64 lives in one building and caused damage over $511 million.

The Loma Prieta earthquake struck northern California just after 5 p.m. on October 17, 1989. It had a magnitude of 7.1, not quite the fabled "big one" but close enough. Its epicenter was in Southern Santa Cruz Mountains segment of the San Andreas fault in a mostly rural area. It killed 63 people and hospitalized 350 more. In less than 10 seconds of strong shaking, it resulted in collapses at the I-880 Cypress Viaduct and on I-80 on the Oakland side of the San Francisco-Oakland Bay bridge as well as in major localized damage to elevated viaducts (Prendergast, 1994). The water main break cut off supply to a pipeline designated for fire fighting. The most damage occurred in San Francisco and Oakland, 50-60 miles away from the quake's epicenter. The hardest hit were the double-deck structures, especially the Southern, Central, and Embarcadero freeways. Hundreds of unreinforced masonry buildings were damaged in the quake-affected area and a major fire in San Francisco's Marina District had to be controlled using a potable water supply system.

The twentieth century's most lethal earthquakes (to 1990) are given by Coburn and Spence (1992) as shown in Table 1.8. The five worst earthquakes caused over half of the total fatalities this century.

Disasters: Natural or Man-made

TABLE 1.8
The twentieth century's most lethal earthquakes (to 1990) (after Coburn and Spence, 1992); Total: 1,042,005 fatalities

	Fatalities	Earthquake			Magnitude
1	242,469	1976	Tangshan	China	7.7
2	200,000	1920	Kansu	China	8.5
3	99,331	1923	Kanto	Japan	8.3
4	66,794	1970	Ancash	Peru	7.8
5	58,000	1908	Messina	Italy	7.5
6	40,912	1927	Tsinghai	China	8.0
7	40,000	1990	Manjil	Iran	7.3
8	32,700	1939	Erzincan	Turkey	8.0
9	32,610	1915	Avezzano	Italy	7.5
10	28,000	1939	Chillan	Chile	8.3
11	25,000	1935	Quetta	Pakistan	7.5
12	24,944	1988	Armenia	USSR	6.9
13	23,000	1976	Guatemala	Guatemala	7.5
14	20,000	1974	China	China	6.8
15	19,800	1948	Ashkhabad	USSR	7.3
16	19,000	1905	Kangra	India	8.6
17	18,220	1978	Tabas	Iran	7.7
18	15,000	1917	Indonesia	Indonesia	NA
19	12,225	1962	Buyin Zhara	Iran	7.3
20	12,000	1968	Dasht-i Biyaz	Iran	7.3
21	12,000	1960	Agadir	Morocco	5.9

1.2 Environmental and Hydrologic Consequences

Environmental consequences of natural disasters depend upon the type of disaster. Nevertheless, they can be summarized as hydrological, environmental, public health, and property damage.

1.2.1 HURRICANES

In the aftermath of a hurricane, the massive clean-up job has to be undertaken. The hurricane generates huge amounts of debris, produces hazardous wastes from household solvents, cleaners, batteries, and tanks of propane or natural gas from destroyed mobile homes. These items have to be sorted out and collected as they are discovered.

Hurricanes may damage water supply systems or render usable water unusable, destroy other utilities such as light and power, heat, and communication systems, and disrupt transportation systems. Repairing these systems within as short a time as possible becomes an enormous challenge.

1.2.2 FLOODS

The hydrological consequences of a flood are many, including increased infiltration and contaminants, rising groundwater levels, reduced subsurface storage capacity, increased groundwater discharge to streams, erosion as well as deposition of soil, loss of vegetation, etc. If much of the area affected is agricultural, nitrates and pesticides are carried into the surface and groundwater bodies. The water systems are disrupted. The flood of 1993 in the midwestern U.S. either flooded or affected 12,000 water systems and 1,000,000 private wells through rising water tables. More than 75% of the population of the nine states and a large portion of the population in the flooded valley use groundwater as their drinking water source. Septic tanks are displaced by rising groundwater pressure, flooded, and get hydraulically connected with the wells. Private wells are bacteriologically contaminated. High water tables and wet soils contribute to disease outbreaks amongst people and animals. Flooded basements contribute to mold, fungal, and spare growth impacting human health and damaging property.

The 1993 Mississippi flood is believed to have caused low-salinity water in the northeastern Gulf of Mexico, south Florida region, and along the U.S. east cost during the 1993 summer (Walker et al., 1994). Oxygen concentrations below 2 mg/l were elevated along the Louisiana coastline west of the delta by introduction of abnormal amounts of agricultural nutrients and the widespread low-salinity plume of river water. The nutrients fueled phytoplankton growth.

1.2.3 EARTHQUAKES

In addition to the damages to the built environment, the public utilities in general and the water systems and gas pipelines in particular may be severely destroyed. Communication systems may be damaged and transportation systems disrupted. Earth dams may be undermined through cracking and may run the risk of breaking.

1.3 Mitigation of Disasters

Natural and man-induced disasters cannot be eliminated entirely. However, through a comprehensive and integrated planning approach and preventive measures, the impact of disasters can be minimized. Such an approach involves coordinated participation by the government, the industry, and the society, as shown in Figure 1.1. When a disaster strikes, through coordinated work, human suffrage and property loss can be mitigated. This may involve evacuation, provision of shelter, food, health and medical facilities, and post-disaster clean-up. This is a short-term approach. On the other hand, a long-term approach may involve warning and communication systems, mitigation and control measures, and post-disaster recovery. These elements, of course, depend upon the type of disaster.

Disasters: Natural or Man-made 15

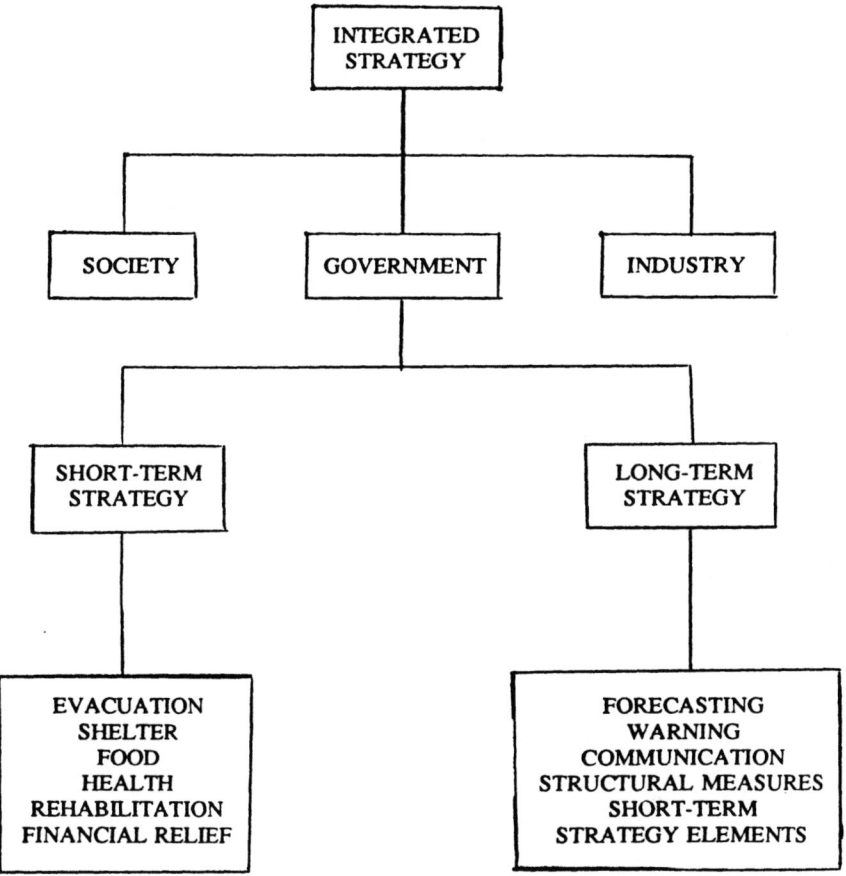

Fig. 1.1. Integrated approach to disaster mitigation.

1.3.1 HURRICANES

One of the most difficult challenges in the aftermath of a hurricane is the collection and disposal of debris. An effective clean-up may require setting up a long-term waste management plan, including separation, processing, management, and environmentally sensitive disposal. The debris may be sorted into medical (or biohazardous) wastes, hazardous or toxic wastes, household garbage, burnable roadside debris, and construction and demolition debris. Pre-disaster planning and coordination with waste generators can play a significant role in effective waste management.

Disposal may entail landfilling, air-curtain burning, and chipping/mulching. Hurricane Andrew generated an estimated 20 million cubic yards (about 5 million tons) of debris. Landfilling this much rubble is not feasible. Thus, all methods of disposal may have to be used.

Temporary storage and processing of debris is another key element. A good debris site may be relatively close to residential areas, resulting in reduced haul times, increased productivity, and decreased costs. Several sites are usually needed that permit multiple disposal operations, storage, and sorting of debris.

Environmental sensitivity is an important consideration in waste management. A testing and monitoring program should be in place to check for possible contamination both during and after operations. Recycling hurricane debris can be important. The island of Kanai is currently processing 70% of Iniki's debris, from steel reinforcing bars to biofuels. After Hurricane Andrew, more than 300,000 tons of mulch was processed and redistributed to the farmers in South Dade County for use in top soil.

1.3.2 Floods

There are a multitude of approaches to flood management, including modification of floods, reduction in susceptibility to flood damages, minimization of the loss, and bearing of the loss. The flood modification may be conducted in the atmospheric phase, land phase, and/or channel phase. The atmospheric phase involves weather modification, whereas the land phase includes watershed management through afforestation, land use, etc. The channel phase involves embankment, reservoirs, channel improvement, diversion works, etc.

The vulnerability to flood damage can be reduced by flood plain management, structural modifications, flood proofing, flood forecasting and warning, disaster preparedness and response planning, development policies, etc. The minimization of loss during and after the flood may be effected by evacuation, flood fighting, public health measures, disaster relief, tax remission, and flood insurance. A schematic of flood management is shown in Figure 1.2.

1.3.3 Earthquakes

Many of the hundreds of earth dams in the United States are in need of seismic remediation to operate safely (Marcuson, et al., 1993). Rehabilitation of an earth dam to prevent seismic instability requires modifying the engineering properties of the dam and/or the foundation, modifying the geometry of the dam, or both. To that end, various options are presented by Marcuson, et al. (1993): (1) use of berms, buttresses and blanks to decrease the liquefaction potential and increase the length of the failure surface; (2) excavation and replacement of liquefaction-prone material to remediate shallowing cracking; (3) in-situ densification to decrease potential for liquefaction by decreasing the void ratio; (4) in-situ strengthening through soil nailing, stone columns, and deep soil mixing; (5) increasing the freeboard; (6) drainage through installation of strip drains, stone columns, or gravel walls or trenches; and (7) various combinations of (1) to (5).

There is a need to develop a realistic assessment of existing buildings and other built environments. This will involve knowing how buildings were actually built. Buildings may need retrofit work and building design has to be upgraded. Seismic

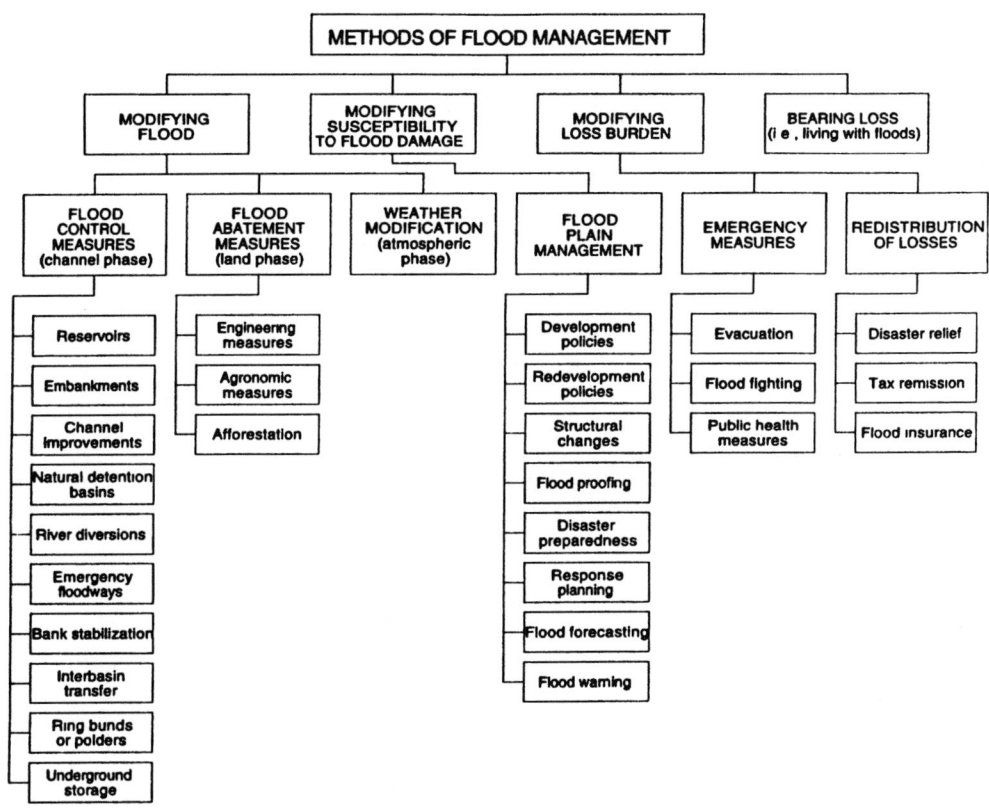

Fig. 1.2. Schematic diagram for Flood Management.

isolation is a promising technology for construction of buildings that can sustain earthquakes.

References

Coburn, A. and Spence, R.J.S., 1992. Earthquake Protection. Wiley, New York.
Davis, R.E. and Dolan R., 1993. Nor'easters. American Scientist, Vol. 81, pp. 428-439.
Denning, J., 1994. When the levee breaks. Civil Engineering, January, pp. 38-41.
Marcuson, W.F., III., Hadala, P.F., and Ledbetter, R.H., 1993. Seismic remediation for earth dams. Civil Engineering, December, pp. 76-78.
National Academy Press, 1987. Confronting natural disasters. Washington, D.C.
Predergast, J., 1994. Aftershocks: The legacy of Loma Prieta. Civil Engineering, January, pp. 46-49.
Starosolszky, O. and Melder, O.M., editors, 1989. Hydrology of Disasters. World Meteorological Disasters, Geneva, Switzerland.
Tarricone, P., 1994. The winds of change? Civil Engineering, January, pp. 42-45.

Walker, N.D., Fargion, G.S., Rouse, L.J. and Biggs, D.C., 1994. The great flood of summer, 1993: Mississippi River discharge studied. Transactions, American Geophysical Union, Vol. 75, No. 36, pp. 403, 414-415.

CHAPTER 2

Representativity of Extreme Wind Data

J. Wieringa

2.1 Introduction

Each time that some region is struck by disastrously strong winds, news media surpass each other in nervous guesses about the reason for this disaster. Some will blame the greenhouse effect, others will invoke divine interference. More practically minded people inquire about the degree of risk of a repetition, in particular how often they can expect to suffer wind speeds which are higher than their trade can withstand.

This survey provides a guide to methods for such enquiries. For more extensive information on various aspects, e.g. existing climatological studies, accessible references are given. Particular attention is given here to spatial representativity of wind information, a subject which is not very well covered in most standard texts on storminess.

Climate change is *not* discussed here. All existing storm investigation methods are based on the assumption of the long-term constancy of the wind climate, irrespective of any "greenhouse" climate changes. People may not believe this, but at present there are no solid reasons to reject this assumption. Existing climate models are still too coarse to model storms adequately, and actual theories underlying such models do not indicate clearly to what extent any changes in global average temperature would generate regional changes in average storminess. Therefore IPCC climate change conclusions [26] do not even mention wind or storms. If global temperature were to rise by some degrees, local windiness might remain the same, it might increase or decrease — that depends on location. Models cannot yet provide reliable information on local wind climate development.

Good wind observations are only available since Dines developed the first reliable anemometer about a hundred years ago [42], and only a few observation sites have been retained in acceptable, constantly unsheltered surroundings all the time. Exposure-corrected normalized annual wind speed averages for 20 stations in the Netherlands [69] show over the period 1951–1980 a trend of 0.036% per year, insignificant compared with their annual uncertainty $\sigma = 4.6\%$. Also, climatological studies on available data [33,77] have not shown any significant trends in *occurrences* of heavy storms over this century.

However, the *risks and damages* of storms have been rising significantly in the last decennia for sociological reasons. There are more people on this world, particularly in coastal regions, they possess more, they are increasingly lax in windproofing their buildings, and they insure more heavily. Moreover, probable increases in global temperature will certainly raise the sea level and lead to more damage during coastal storms. Recent large premium increases by wind-insuring firms are therefore economically defensible [6], but are not based on meteorological proof of significant wind climate change.

2.2 Wind Phenomena Classified by the Wind Spectrum

Requirements of users of maximum-wind information vary widely. Airfield controllers need continuous information on the probable largest 3-second gust in the next 30 minutes. Builders of sea dikes want to estimate for the next hundred years the occurrence risk of violent storms which last long enough (say ten hours) to drive the sea high against the dikes. It seems sensible to classify existing wind events along time-and-space scales, by using the entire wind spectrum as guidance.

The first attempt to make a full-range wind spectrum was published in 1957 by Van der Hoven. It is still faithfully reproduced in many handbooks, because most people do not know that his spectrum is quite unrepresentative. First, his high-frequency turbulence peak is too large, being derived from one hurricane occurrence. Second, his diurnal wind variability peak is almost absent, since he analysed only data observed on a mast at 100 m height, where diurnal wind variability has an average minimum. At 80 to 100 m height the season-averaged diurnal course of wind reverses its phase, changing over from a daytime maximum in the surface layer to a nighttime maximum in the upper boundary layer [71].

Many better spectra have been produced since 1957 [e.g. 20,38]. Looking at those spectra, we find large wind variability at the following frequencies n (in cycles/hour):

(a) The boundary-layer turbulence peak, consisting of:

(a1) High-frequency micrometeorological turbulence ($3000 > n > 30$), caused by interaction at the earth's surface of average airflow with obstacles and surface roughness, on a horizontal spatial scale of a kilometer or less. This produces random *gustiness*, which can be quite damaging in rough terrain where large gusts are formed by obstacle wakes.

(a2) Convective turbulence ($30 > n > 3$), caused by thermal circulation at horizontal scales between 1 km and 10 km. Extreme gusts at this scale are often caused by *thunderstorms*; these may be organized in *squall lines*, and may also produce a *tornado*.

(b) *Diurnal course* wind variability ($0.3 > n > 0.03$) gives the largest spectral peak in the lowest fifty meters of the atmosphere. Most of this variability is gradual; but regional changes in the terrain at spatial scales of 10 km to 100 km (coasts, hills, mountains) can occasionally give rise to local *topographical winds*, which can become quite violent. Usually these wind events have local names, e.g. bora, chinook, mistral, and so on.

(c) Large-scale weather systems ($0.03 > n > 0.003$), which can cause occasional storms with local durations of about one day. The latitude-dependent alternatives are:

(c1) Tropical storms, most often called *hurricanes* (in America) or typhoons (in Asia), can be formed occasionally over a warm ocean at latitudes between 5° and 25°.

(c2) Extratropical *gales* are the most violent members of the families of low-pressure areas, wandering along the polar front at $\approx 50°$ latitude during westerly circulations, especially during the winter season.

The above review indicates that a wind variability minimum occurs around $3 > n > 0.3$, the so-called spectral gap. Therefore hourly wind averages are relatively steady data. This is a major reason why generally averaging periods between 10 minutes and 3 hours are used in statistical calculation of extreme winds — both for gusts and for storms. Wind data, which have been averaged over 1 or 2 minutes only (e.g. aviation data, or U.S. "fastest miles"), have their sampling period located in the middle of the spectral turbulence peak and show therefore an excessive variability [16,25,51]. Such very-short-period data are unreliable for climatological application. On the other hand, wind averages over 24 hours cannot document the important wind changes due to the diurnal course. Therefore long-period averages are of little use for purposes requiring information on the exceedance of given thresholds during some hours (e.g. wind energy, wind risk studies).

For estimation of largest possible design gusts, which is of great importance in other applications than hydrology, there is a physical reason for using hourly averages as basic data in wind statistics. There are *two* sources of gust variability: first, the variable meteorological phenomena which generate the wind, and second, the local aerodynamic effects which cause the turbulence. Sometimes, maybe for the sake of convenience, the effect of the two sources is combined into a single "largest gust in 50 years". However, that approach requires the unrealistic assumption, that all local wind statistics are determined adequately by the aerodynamics of wind flow at the nearest meteorological station [22]. This assumption is incorrect: gustiness is very terrain-dependent, also its magnitude varies with duration, so gust observations depend on instrument sensitivity [64,67].

More sound procedures for gust estimation exist: ratios of median largest gusts u_{mx} to the simultaneous average wind speeds U depend on local height z and local surface roughness length z_0. For the hourly expected gust u_{mx} of duration t we get [64,67]:

$$u_{mx}/U = 1.1(1 + \{1.42 + 0.3\ln[(10^3/[Ut]) - 4]\}/\ln[z/z_0]) \qquad (2.1)$$

for near-neutral stability conditions. So local values of maximum gusts can be derived very well from statistics of extreme local hourly-averaged wind [22,47,56]. On the other hand, data series of largest gusts are unsuitable material for reliable extreme value analysis of wind, unless we know hourly-averaged simultaneous winds and also local observation circumstances.

Concluding, both for gust estimation and for estimating 50-year design winds, the best material for analysis are representative observations of hourly wind.

2.3 What Are Representative Wind Data?

We may define "representativity" of an observation as the degree, to which it represents the character of the measured parameter usefully towards a specific purpose. In other words, representativity depends both on the properties of the parameter (e.g. wind) and on the application, and is therefore *not* a universal quality. Representative observation requires good and well-placed instruments, but is not the same as measurement accuracy [19]. Important other aspects are observation location, and sampling frequency [41], and for hasty applications also signal conversion and presentation. It is easy to optimize the observations towards one application so much, that other applications become impossible.

Each application of wind data requires its own matched space and time conditions. In the time domain, airport controllers require actual two-minute-average wind, while building climatologists need reliable historical series over periods of 20 years at least. In the space domain, weather forecasters and global climatologists need gridpoint data, averaged over areas of >50 km diameter, *'regionally representative'*. At the other end, pollution monitors, hydrologists and agricultural meteorologists are interested in turbulent atmospheric transport properties at scales less than integral scales of turbulence, between centimeters and kilometers. They strive to obtain a *'locally representative'* picture of the wind at scales so small, that forecasters would eliminate them because they are 'subgrid'.

A third form of spatial representativity is required by construction engineers and wind energy developers. Because wind pressure is proportional to the square of wind speed U, and wind energy even to a higher power of U, they require accurate wind knowledge at the actual place of their building, or wind turbine. Good station observations should be *'point-to-point representative'*, i.e. be able to provide wind knowledge elsewhere by extrapolation, or interpolation [37].

Finally, climatologists require *'long-term representativity'*. Wind observations are very sensitive to changes in the surroundings, much more than observations of rainfall, temperature, or radiation. Therefore most station wind series show gradual or instant changes because of nearby vegetation growth and building activity, or relocation of the wind mast. An interesting example is given in Figure 2.1, showing how inhomogeneous a long series of wind observations can become due to shelter changes in the surroundings. Another example of the systematic shelter change problem is given by Tamura and Suda [57], who showed that in Japan the average station observations of largest annual hourly wind have decreased from ≈ 21 m/s to ≈ 15 m/s between ≈ 1955 and ≈ 1985. This decrease was caused by growth of cities around the weather stations. Circumstantial facts are not climate, but can make weather data look as if the climate has changed. Such data are not reliable for computing of wind energy, or of 50-year design wind for constructions.

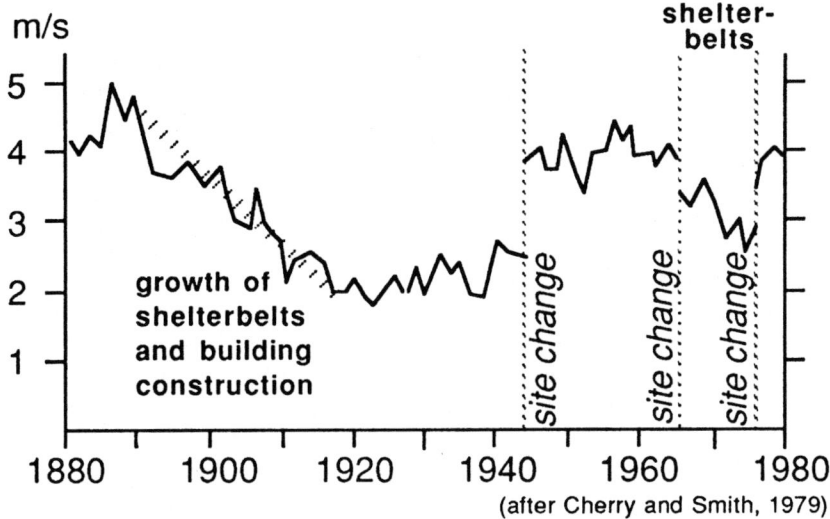

Fig. 2.1. A century of wind observed at a station in changing surroundings [10].

Summarizing, representativeness requirements are user-dependent functions of scale [37]. The scales vary widely. For the global network of regular weather stations, organized by the World Meteorological Organization to support weather forecasting, the preferred *time* scale is hourly sampling of wind, averaged over at least 10 minutes. In view of existence of the spectral gap (discussed above) this is an optimal compromise choice. It includes, for typical wind speeds of ≈5 m/s, use of a *space* averaging scale of ≈10 km.

The practical representativity problem is now how, using these data, we can serve the various applications best, while their spatial requirements vary so widely. A general solution is to shift our attention from the application to the source. For wind, the source is the pressure-gradient-related airflow in the atmosphere, well above retarding influences of the cluttered surface. *Wind data should be primarily representative of the wind itself*, rather than be tied to the spatial requirements of any particular application at ground level. It is shown below that by this approach, from above, we can also deal satisfactorily with purely-local shelter problems in wind observations. The purpose is to obtain generalized wind data which are most useful for *all* mentioned applications, not just for a few.

2.4 Representativity of "Potential Wind"

The World Meteorological Organization requires: "Measurements of wind for synoptic purposes should *refer to* a height of ten metres in an unobstructed area, and should consist of the means of values taken over a period of about ten minutes." Though WMO suggests that a free fetch of 10 obstruction heights, a few hundred metres, is a sufficient amount of open space, anemometers at that distance are not outside the obstruction wake. A fetch of 12 to 15 obstruction heights is needed to be free of the wake. Also, if the size of the represented area is to match with a 10 minute sampling period, we need for winds of 4 to 5 m/s a similar upwind fetch of ≈2 km. If the far terrain is not hilly, the surface roughness length of such a fetch of unobstructed terrain will be ≈0.03 m [72].

At first sight the WMO rule provides observations, which are only representative of the regional wind climate in homogeneously unobstructed flat landscapes. Moreover, unobstructed surroundings are seldom available; e.g. safety rules do not allow wind masts in the middle of airports, only at the (obstructed) edges. Meteorological services with limited budgets must use any station locations which they can get, if only reliable observers are available. So unfortunately there are not many stations, which really do make their observations according to the WMO rule, at 10 m above really open terrain.

However, assume that we have such a station. There an anemometer records a windspeed U_p, driven by large-scale air motion U_b at height z_b higher in the atmosphere. Within the region we take another nearby surface station, where the wind speed at height z_s is U_s. If we can relate the driving flow U_b *both* to U_p observed at the WMO-station *and* to U_s which is observed nearby, then the WMO-station and the nearby station are mutually 'point-to-point representative', as will be shown.

It is well known, that flow over obstacles regains a reasonable degree of horizontal homogeneity at levels above two or three times the obstacle height. For defining a regionally driving airflow we need a reference level, where presence or absence of individual obstructions of ≈20 m size causes no localized retardation or speedup. The lowest useful height choice for this level then is a *blending height* $z_b \approx 60$ m [65]. Its matching horizontal region size is ≈5 km [73]. Below $z_b = 60$ m we assume, that the wind profile is logarithmic and that it can be specified by a local roughness length z_0 and the average wind speed at a single height z_s. (Use of logarithmic profiles requires that $z_s \geq 20z_0$.)

This leads to the exposure correction model shown in Figure 2.2. Here the measured wind U_s is observed at height z_s at a station with *local* roughness z_{os}, and upward transformation gives a regional 'mesowind' $U(z_b)$ at blending height z_b. Then exposure-corrected U_p is obtained from $U(z_b)$ by downward transformation over open-terrain roughness.

In other words, we can estimate from observations of U_s at height z_s at a non-ideal station with upwind roughness z_{os}, what at that location the wind speed would be at 10 m above open terrain. This wind speed for a hypothetical local WMO-station situation is called the local *potential wind speed* U_p. The transformation formula [69] is:

$$U_p = 0.76(\ln(60/z_{os})/\ln(z_s/z_{os}))U_s \equiv FU_s, \qquad (2.2)$$

where $F \equiv U_p/U_s$ is called '*exposure correction factor*' and may depend on azimuth.

Assume that we know U_p at some station. By inversion of eq. (2.2) we can then derive for an arbitrary nearby location (within the influence region of U_b), what the

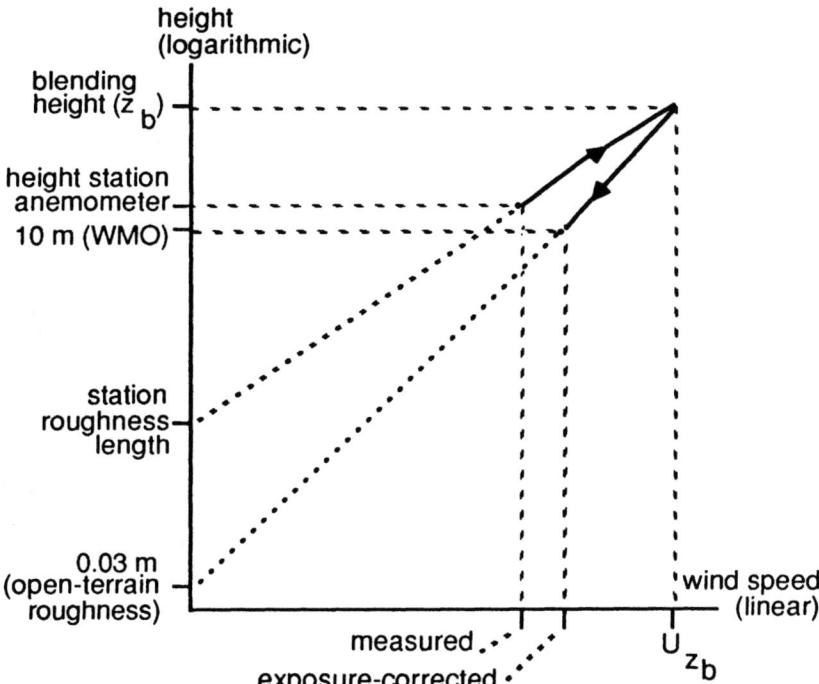

Fig. 2.2. Logarithmic model for transformation from measured to exposure-corrected wind [65].

wind speed U_r at height z_r will be, if only we know the local upwind roughness z_{or}. In other words, a WMO-station has *point-to-point* representativity, and therefore for the region within ≈5 km radius also *local* representativity. Potential wind data are also excellent material for interpolation with distant wind stations, or for large-scale mapping of wind.

Summarizing, wind measurements made according to WMO-rule, or else potential wind speeds *referring to* 10 m over really open terrain, are representative wind speeds for a region of at least 5 km radius. Obviously, it is practical to obtain potential wind speed values from all the wind stations which we are using. For this purpose we need methods to determine local roughness lengths at existing non-ideal stations.

2.5 Roughness Determination at Ordinary Wind Stations

When strong wind speeds blow over rough terrain, the surface layer of the atmosphere is nearly neutrally stratified because of mixing. The increase of wind speed U_x at height z_x with increasing height then agrees with the logarithmic wind profile

$$U_1/U_2 = \ln(z_1/z_o)/(\ln(z_2/z_o)), \tag{2.3}$$

where z_0 is the roughness length of the terrain. This relation is valid at heights between $\approx 20z_0$ and ≈ 100 m. Defining z_0 as "the height at which U becomes zero" is utter nonsense, since at heights $\approx z_0$ the logarithmic profile is no longer valid.

There are five practicable ways to determine z_0 at some location:

(a) *Wind profile* measurements and application of Eq. (2.3). Formally this is correct, but operationally it does not work for two reasons. First, usually in practice only a single anemometer is available. Second, good profile analysis needs some research experience, its results are highly sensitive to nearby upwind terrain inhomogeneities [3,29,63,73], and it requires also a temperature profile or other forms of stability determination.

(b) *Turbulence* measurements, making use of the fact that standard deviations of wind speed, σ_u, increase with increasing roughness. In neutral stability is valid

$$\sigma_U/U = 1/\ln(z/z_0). \qquad (2.4)$$

This can be objectively handled with a single anemometer, but recording of standard deviations requires more instrumentation and recording facilities than are usually available at simple weather stations. For details about this method see Beljaars [4].

(c) *Gustiness* measurements, actually a poor-man's version of method (b). Its practical advantage is that records of maximum gusts are often available at simple weather stations. The method is objective, and will be summarized below.

(d) Use of a *terrain classification* for visual estimation of roughness. For ordinary terrain situations, the best choice is the updated Davenport classification. Other classification tables, e.g. ESDU, have been shown to underestimate z_0 of typical land surfaces by a factor two at least [72]. Table 2.1 gives the abbreviated classification description.

The method is subjective, but very fast with some experience. The possibility of wrong class assignment results in a random z_0-uncertainty factor ≈ 2 for this classification table. This gives wind analysis errors of $\pm 6\%$ at 10 m height. Tables with fewer classes give higher random errors (apart from the systematic underestimation in some tables).

(e) *IBL terrain analysis*. For this method we need a very detailed description of the surrounding terrain within a radius of ≈ 3 km, with sizes, directions and distances of major obstacles, and roughness estimates of homogeneous terrain areas. (Due to the latter, the method is subjective.) Internal Boundary Layer analysis [18] is then applied to the entire information set. This requires much expertise and is very tiresome work, though it can be made easier by use of existing computer programs such as WAsP [35]. The IBL method was used for making the European Wind Atlas, because for application of the objective gustiness method (c) no gust data were available from some of the countries which participated in the atlas.

A practical advantage of using gustiness for determining roughness is the fact, that data on extreme gusts u_{mx} are part of many routine weather reports and can be extracted from wind speed records on analog charts. The ratio between u_{mx} and the average U over the period T in which u_{mx} occurred is the gust factor $G \equiv u_{mx}/U$. In strong winds the wind variation around the average is nearly Gaussian. This

TABLE 2.1
Revised Davenport terrain roughness classification (summarized from [72])

Class Number	Name	Roughness length (m)	Description of landscape within ≈3 km radius
1	Sea	0.0002	Open water, tidal flat, snow, with free fetch >3 km
2	Smooth	0.005	Featureless land with negligible cover, or ice
3	Open	0.03	Flat terrain with grass or very low vegetation, and widely separated low obstacles; airport runway
4	Roughly open	0.10	Cultivated area, low crops, occasional obstacles separated by more than 20 obstacle heights H
5	Rough	0.25	Open landscape, crops of varying height, scattered shelterbelts etcetera, separation distance ≈$15H$
6	Very rough	0.5	Heavily used landscape with open spaces ≈$10H$; bushes, low orchards, young dense forest
7	Closed	1.0	Full obstacle coverage with open spaces ≈H, e.g. mature forests, low-rise built-up areas
8	Chaotic	≥2	Irregular distribution of very large elements: city centre, big forest with large clearings

means that the median gust factor $\langle G \rangle$ is statistically related to the average wind standard deviation σ_u by

$$(\langle G \rangle / f_T) - 1 = E_{Ut}(\sigma_u/U) \equiv [1.42 + 0.30 \ln(-4 + 10^3/Ut)](\sigma_u/U). \quad (2.5)$$

Factor f_T is 1.00 for $T = 10$ minutes, and 1.10 for $T = 1$ hour [64,65]. Median gust factor averaging (with 50% exceedance probability) is used, because it is less sensitive than arithmetical averaging to the occurrence of excessively extreme gusts from thunderstorms.

The value of the excentricity E_{Ut} depends on gust wavelength Ut, product of the duration t of the extreme gust u_{mx} and the average wind U. Short gusts are relatively large, and the duration of u_{mx} is so short that it is attenuated by common wind instrumentation. The attenuation A of u_{mx} depends on Ut, on the anemometer response distance l and on the first-order response time t_{RC} of the registration chain by way of

$$\frac{1}{A} = \sqrt{1 + (2\pi t_{RC} U/Ut)^2} \sqrt{1 + (2\pi\lambda/Ut)^2}. \tag{2.6}$$

The gust response of some standard instrument combinations is for $6 < U < 12$ m/s:

Pitot tube windvane with Dines manometer:	$Ut \approx 70$ m,	$A \approx 0.90$,
Heavy cup anemometer with galvanometric recorder:	$Ut \approx 110$ m,	$A \approx 0.86$,
Light cup anemometer or propeller with servo-recorder:	$Ut \approx 30$ m,	$A \approx 0.93$.

More gust response information, like graphs for deriving Ut from λ and t_{RC}, is given elsewhere [3,4,65,66,67,68]. Because of the attenuation, recorded gust factors G_m are smaller than the gust factors G in the real wind: $\langle G_m \rangle - 1 = A(\langle G \rangle - 1)$.

In strong winds ($U > 6$ m/s) gustiness is mainly determined by the roughness of the nearest few km of upwind terrain for any azimuth sector a (width $\approx 30°$). This leads, in combination with equations (2.4) and (2.5), to a relation between station-measured gust factors and roughness z_{os} of the terrain upwind of the station:

$$z_{os}(\alpha) = z_s \exp\left[\frac{-Af_T E_{Ut}}{\langle G_m(\alpha)\rangle - 1 + A - f_T A}\right], \tag{2.7}$$

where z_s is the observation height. For reliable results we need per sector $\geq 20 G_m$-values with two-decimal accuracy. Because dynamic instrument specifications are often difficult to get, it is advisable to apply gustiness analysis first to some upwind terrain sector which is rather homogeneous, so that for it a reliable Davenport roughness estimate is possible. In this way, tentatively chosen values of Ut and A can be "calibrated", and can after minor correction be used for azimuth-dependent roughness analysis of other directions.

The validity of these roughness determination methods was investigated for inhomogeneous landscape by Barthelmie et al. [2], checked against profile-derived z_0-values; both gustiness (c) and classification (d) gave similar z_0-values, while (b), using s_u, gave unexplainable underestimation of roughness. Another study, by Palutikof et al. [40], found methods (c) and (d) to be both satisfactory, while IBL-methods (e) gave unrealistic obstacle effects. IBL-models, which are used in WAsP-type programs, usually assume obstacles to be 2-dimensional and perpendicular to the wind [59].

In Germany the gustiness method proved to be an excellent method to gauge actual exposure of stations and homogeneity of wind data series [12]. E.g., some unreported wind mast relocations were discovered by analysis of gustiness in station data.

2.6 Scale of Application of Station-Observed Wind

The applicability of the terrain- and height-dependent exposure correction eq. (2.2) seems limited by its dependence on the neutral-stability wind profile, eq. (2.3).

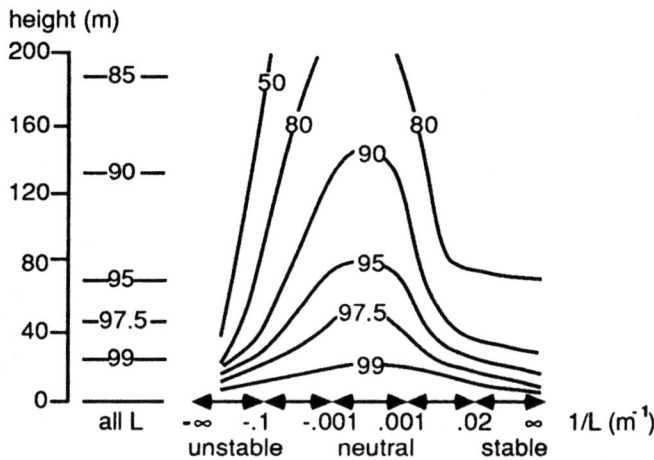

Fig. 2.3. Correlation coefficients (%) between mast-observed wind speeds and stability-corrected wind estimates from observations at 10 m height (Cats, [8]). L = Obukhov length.

At levels >30 m, influence of thermal stability on the wind profile becomes practically important if the wind is not strong. But, surprisingly, changes in wind profile shape with varying stability do not spoil the validity of the transformation given in Figure 2.2 and eq. (2.2), because the deviation in the upward part of the transformation is compensated in the downward part by an opposite deviation — if only the lengths of these two profile legs are comparable. So transformations by eq. (2.2) between heights lower than 30 m are not significantly affected by thermal stability. Field comparison of two 10m-masts at 2 km relative distance showed that roughness-caused wind differences of 9% could be reduced to 2%, even for U-values as low as 3 m/s in quite stable or unstable conditions [67].

It has been asked, why the blending height z_b is not taken to lie at the "geostrophic" height ($\approx$500 m). One reason is that the wind at such height is influenced by the surface roughness and orography over a horizontal distance of 20 to 50 km [28]. It is unlikely, that surface characteristics are sufficiently homogeneous over such a distance.

Equally or more important is the fact that observed surface wind and the wind at z_b would not be very closely related, if z_b were chosen very high above the surface layer. Formally, surface winds can be related to geostrophic wind by application of stability-dependent Rossby-number similarity profiles [29]. However, above $\approx$100 m the climatological wind behavior differs from the wind climate in the surface layer in various ways. For example, the average diurnal course of the wind reverses its phase at $\approx$80 m height [71]. Therefore downward-transformed upper wind data (e.g. pressure-gradient wind or radiosonde wind distributions) will not be equivalent to real surface wind information.

Figure 2.3 shows the lack of correlation between observed surface winds and 'upper' winds, using one year of hourly Cabauw mast wind observations. Holtslag [24] proves that, with careful use of stability-dependent wind profiles, U_{80} can still be estimated well enough from surface winds, but U_{200} only very poorly. So there are obvious reasons why for transformation eq. (2.2) it is *not* wise to take z_b at "geostrophic" height.

For surface layer climatology, we need surface station wind data — but we should be aware, that they primarily represent the wind in a surface layer of ≈ 60 m height. The consequence is, that their direct full representativity is restricted to ≈ 5 km distance from the station anemometer.

Usefulness of station data at larger distance depends on the geographical situation. In non-mountainous country, use of map-extracted estimates of "meso"-roughness at a larger scale (5×5 km^2) allows calculation of a larger-scale "macro-wind" (a model quantity, not an actual flow speed) from the station mesowind $U(z_b)$. Then from the macrowind we can derive surface-layer winds anywhere in the analysed region within 10% [69,70]. In mountainous country, gridpoint modelling has to be applied [2,35,58,61,63].

2.7 Getting and Using Potential Wind Data

If we are to obtain potential wind data from a station, logarithmic profiles must be applicable at its anemometry site. An appendix below describes necessary location requirements. Moreover, the station must be well-run and its data quality-controlled, it must measure wind azimuth as well as speed, and preferably also 3-sec-gusts. For further requirements and methods see various reviews [4,11,12, 63,66,68,71].

Use of potential wind U_p instead of original wind observations U_s has the very great practical advantage of normalization. The user does not need to worry about the imperfect surroundings of the actual station (with corrections of the type "De Bilt underestimates westerly winds by 20%") since the resulting anomalies in the data have been removed. Instead, he can restrict himself to solving his local wind problems, and from the station-interpolated U_p-values he can get an objective estimate of his actual local wind. For that he only has to specify his upwind roughness (Table 2.2) and his application height z (which must be $\geq 20z_0$). Then he can apply eq. (2.2), of which Table 2.2 is a summary.

Assume that at some location we know upwind roughness classes for all sectors, and potential wind data U_p are available from a nearby weather station which is in a similar climatological location (do not use coastal stations for inland sites!). Then Table 2.2 gives per z_0-class, by way of eq. (2.2), factors F_z with which U_p must be multiplied to get the local wind speeds U_z at various heights z:

$$U_z = F_z U_p.$$

(Over dense high roughness, we must relate z to a "displacement level" at $\approx 2/3$ of the roughness element height.)

Using this objective method, we can for example achieve the following results:
(1) We can make data series homogeneous in time. *Relocations* are automatically

TABLE 2.2
Ratios F_z of local terrain-dependent wind speed to potential wind speed, and gust factors

Roughness class →	1	2	3	4	5	6
Class name →	sea	smooth	open	roughly open	rough	very rough
$F_{30} \equiv U_{(z=30)}/U_p$	1,24	1,21	1,19	1,17	1,14	1,12
$F_{10} \equiv U_{(z=10)}/U_p$	1,12	1,06	1,00	0,94	0,88	0,82
$F_2 \equiv U_{(z=2)}/U_p$	0,96	0,83	0,72	0,61	[0,5]	[0,4]
$(u_{mx} \cdot hr/U)_{(z=10)}$	*1,3*	*1,4*	*1,5*	*1,6*	*1,7*	*1,8*

Italic figures are gust factors for gust wavelength $Ut \approx 100$ m.
Bracketed figures are below model limit height ($20z_0$).

"repaired", if data from the old site and the new site are both transformed into potential winds. In the same way we can get reliable design winds from data series obtained in gradually changing surroundings, like the Japanese series mentioned earlier [57], provided that good historic descriptions of those terrain changes are obtainable.

(2) We can use potential wind data as input to wind models for *complex terrain*. Some useful models exist, in particular for rather smooth topography. However, in field trials it has been shown that the quality of model output is increased, if local shelter problems at the observation site have been eliminated from the input wind data [40,43,60].

(3) We can make better *wind maps* of non-complex regions from the available station data. For interpolation, the Davenport classification can be used to make roughness maps [69]. Engineers have been universally content with the maps and tables of potential wind speeds, which have been published in the Dutch wind atlas [69,74] instead of the customary original station-measured wind speeds. A similar approach, giving wind distributions for normalized terrain types, is used in the European Wind Atlas for wind energy [61].

2.8 Distribution of Average Wind Speeds

Investigations of the occurrence probability of extreme wind speeds require for a start the parent frequency distributions of average wind. Wind speeds are described by positive numbers, and the ubiquitous availability of Gaussian graph paper made initially use of the lognormal distribution quite popular. Though the lognormal function fits tolerably well to many wind distributions, there is no particular reason why it should do so. Better fits and some theoretical foundations for hourly-averaged wind are available when using the Weibull distribution function:

$$F(U) = 1 - \exp\left[-\left(\frac{U}{a}\right)^k\right], \qquad (2.8)$$

where $F(U)$ is the cumulative probability of occurrence of wind speeds $<U$. The scale parameter a has dimensions of wind speed, increases regularly with increasing height, and is proportional to the average wind speed calculated from the entire distribution: for $k = 1.7$ we have $a = 1.123\,U$, and for $k = 2.2$ we have $a = 1.129\,U$.

The shape parameter $k\,(\geq 1)$ describes the skewness of the distribution function. For $k \approx 3.5$, as is sometimes observed in the tropics, the function is nearly symmetrical. For decreasing values of k, the distribution modus shifts to lower wind speeds and the probability of high wind speeds increases simultaneously. For mid-latitude wind speed distributions the k-parameter has typically values between 1.6 and 2.2; the so-called Rayleigh distribution has $k = 2$. In the surface boundary layer, k increases approximately linearly with height [71]: if k_s is the shape parameter value at the observation level z_s, then

$$k_z = k_s + 0.0084(z - z_s), \tag{2.9}$$

but k decreases again with height above ≈ 80 m. For distributions of 10-minute wind averages, the shape parameter is ≈ 0.1 smaller than for distributions of hourly averages.

The Weibull distribution will give a fully adequate representation of a circular normal frequency distribution of wind [62]. This requires that the orthogonal components of the wind velocity, transformed by raising to the power $k/2$, have the following properties: (a) a normal distribution; (b) equal variances; (c) zero means; (d) they are uncorrelated. These conditions will be approximately met for wind climates, where the frequency distribution does not vary strongly with wind direction and has a low frequency of calms. In other words, the Weibull distribution should be a good variability descriptor for hourly-averaged wind speeds at rather windy locations without strong topographical influences. In practice, the Weibull formula is used to describe wind distributions as long as a reasonable fit can be obtained, because its parameters are rather simply related to statistical moments and to the average nth power of wind speed [29,74].

When fitting a Weibull function to (potential) wind data [11,29,30,55], one should be aware that many observed wind distributions contain an unrealistically large zero-wind class, either because the station anemometer has a start-up speed of 2 or 3 m/s, or because the station is too sheltered. For fitting a distribution to wind data with too many calms, a useful method is calculation of frequencies for classes of ≈ 1 m/s width, then using only classes with $U > 3$ m/s for linear regression calculation of a and k from eq. (2.8), rewritten

$$\ln\{-\ln[1 - F(U)]\} = k(\ln U) - k \ln a. \tag{2.10}$$

Weibull-distribution graph paper has been devised to assist in such a calculation [55,74].

The calms-frequency, representing the unused classes, should be retained as separate information. In distribution fitting by the method of moments, calms-correction is cumbersome, since it requires elimination of low-speed hours from the full original data set [29].

Direct use of Weibull parent distributions for the estimation of extreme wind risks is feasible in "well-behaved" wind climate [53,56]. However, they also are used to estimate extremes by direction, a difficult matter. Extrapolation of partial data sets can give higher extreme values than the value obtained from extrapolating the complete series [33,45,56].

2.9 Shelter Problems in Analysis of Extreme Stormwind speeds

Statistical estimation of extreme events can proceed in two alternative ways: either the tail of the total parent distribution is analyzed, or the extreme events are analyzed separately. The latter method, which is more convenient because it involves less data to handle, was pioneered by Fisher and Tippett [17]. They showed that, if a set of independent data is divided into sufficiently large samples, the extreme values of these samples have three possible alternative distributions, FT-I, FT-II and FT-III. Full mathematical descriptions are given e.g. by Mayne [33], and in many handbooks.

Of these extreme-value distributions, the first one (also called Gumbel distribution) occurs, if the parent distribution consists of a single, identically distributed population with an exponential distribution function, like eq. (2.8). Physically, such a population will be obtained in two cases: either in case of a single dominant source of variability, or else when out of a large number of variability sources none is dominant [56]. Those wind climates, in which a single weather mechanism generates all really important storms, are called "well-behaved". Such a climate occurs e.g. in Western Europe, where the largest observed hourly wind speeds occur almost always in polar front depressions.

Problems in storm statistics arise, when several weather mechanisms compete with each other for causing the largest storms. This is for example the case in the hurricane areas of the world, such as the Caribbean and the Western Pacific. Since hurricanes are infrequent and geographically relatively small, the great majority of annual wind maxima at any chosen place in that region will *not* have been caused by a hurricane — a 'mixed' stormwind climate. On Gumbel-type statistical graph paper, which linearizes the FT-I distribution, the annual wind extremes then make up two intersecting lines, one for hurricanes and one for 'ordinary' storms, merging into a mock FT-II distribution curve [56].

Good studies of this mixed-population problem are given by Gomes and Vickery [21], Riera et al. [44], Simiu et al. [52], and recently Tabony [56]. Of course, there are different possibilities of a mix. In some regions, violent mountain winds (e.g. katabatic winds, mistral) may cause major storms, and e.g. in the Mississippi valley (nicknamed 'Tornado Alley') thunderstorms should be included in any extreme wind investigation.

The probabilities of exceedance of specified hourly stormwind speeds are usually given by statisticians as return periods, inverse values of the *average* frequency of such speeds. The word "return period" causes much misunderstanding, since the majority of the public thinks deterministically and reads into it a non-existing regularity. Mayne [33] reports an enlightening statistical experiment on an event with a 50-year return period: a 5000-year random series contained one 150-year

TABLE 2.3
Influence of anemometer height change on estimates of extreme wind

	Measured annual extremes	50-y.-return wind
10 years at 10 m	14,15,15,16,17, 14,15,15,16,17	18.5 m/s
10 years at 25 m	18,19,19,20,21, 18,19,19,20,21	22.5 m/s
5 years at 10 m, 5 years at 25 m	14,15,15,16,17; 18,19,19,20,21	24.4 m/s

period *without* the event and also a 50-year period in which the event occurred *four* times. On the average, in any specified 50-year period the occurrence probability of one such event would be 63%, and of two events still 27%.

Developing methods of extreme wind speed analysis, much attention has been given to the choice of the best frequency distribution for extrapolation in a particular type of wind climate. A second much-debated issue is the extrapolation problem, i.e. how to estimate from short data runs the occurrence of very high (and infrequent) winds with long return periods. Can we sensibly estimate a 100-year-return-period wind from a 10-year series?

An attempted solution to this problem is to maximize the number of extremes by taking the shortest sample periods, which might still be meteorologically uncorrelated and therefore statistically independent. Cook [14] claims that in the British climate the sample period can be shortened to 3 to 4 days, giving 100 extremes per year for analysis. However, strong depressions come in pairs, strung like beads on the jetstream, and in Western Europe there is a significant chance of a storm being followed by a similar storm two days later [46]. Moreover, this approach enhances the mixed-population problem, because storminess varies by season. As a result, separate analysis of stormy months will produce extremes, which are significantly higher than the extremes from analysis of the entire year — that is of course nonsense [56]. If series are too short, a better way out is to use terrain-normalized U_p-series, which allows joint analysis of series from neighbouring stations into a statistically more reliable *regional* estimate of extreme winds [47].

A third problem gets much less attention, namely errors due to shelter inhomogeneities in the series itself. No design winds should be derived from e.g. the 100 annual extremes of the variably-sheltered station illustrated in fig. 2.1. It can be shown by a simple example, that inhomogeneities of any type will *always increase* the wind speeds estimated for a given return period. Take a station in rough surroundings ($z_o = 0.3$ m), where the wind was measured at 10 m and at 25 m height, with in the two 5-year periods identical annual maxima: 14, 15, 15, 16, 17 m/s at 10 m. Applying Gumbel-extrapolation gives the results shown in Table 2.3.

Exaggeration of expected wind speeds results from the omission to homogenize — in the case of Table 2.3 for a height change, in most cases for shelter changes. Inhomogeneous series always will produce an excessive slope of the Gumbel graph, quite different from the graph slope for nearby stations with unchanged local situations.

Figure 2.4 is an actual example of the exaggeration due to series inhomogeneity. The Gumbel-graphs of the three illustrated neighbouring stations are numbered in the expected order of their windiness. However, observations at station 2 were first made at an inland airport on a very short mast above a Nissen hut (data series 2 b) and were incorrigible. Subsequent observations at station 2 (data series 2 a) were made on the coastline, first at 10 m height for $\approx$10 years, then for a similar period at 24 m height. The inhomogeneity, combined with an open-water fetch in SW directions at station 2, gave the implausible result that *original* estimates from station 2 exceeded largely the estimates for the (more windy) station 1, which had an overland fetch in SW directions. After objective data correction, the Gumbel-graphs for potential wind speeds at the three stations are similar in slope, as they should be in this "well-behaved" region, and in proper order of magnitude.

Extrapolation of those extreme potential winds to a 1000-year return period can be done with some confidence, while for the original measured winds the extrapolations do not look very reliable for 50-year return periods.

Concluding, published estimates of extreme wind speeds can only be trusted if two conditions are satisfied. Primarily, the wind study should have assessed carefully if the wind climate of the investigated region is "well-behaved", i.e. if locally the occurrence of excessively high wind speeds results from just one major meteorological mechanism. If this is not the case and the wind climate is "mixed", the wind study should recognize this and have attempted to deal with it explicitly. Secondarily, used wind data series should have been checked on homogeneity. Inhomogeneous series should either have been disregarded, or else have been corrected, e.g. by transforming wind data to potential values.

2.10 Some Available Studies of Extreme Wind Speeds

For estimating return periods of "ordinary" hourly-averaged wind speed, pioneer reviews by Shellard [50] and by Jensen and Franck [27] are still worth reading. Useful extreme wind studies exist for Britain [15], for Denmark [1], for France [9], for Germany [49], for India [23], for Ireland [32], for Japan [36], for the Netherlands [45,74], for Poland [76], and for the contiguous United States [51]. Next to these, there are many climatological studies of tornadoes, e.g. for Australia [13,75], for Britain [34], for India [54], for Italy [39], for the Netherlands [74], for the United States [31,48], and for central Europe [5,7]. This listing is not exhaustive. Even if your region of interest *is* mentioned, do inquire at national weather or climate institutes about available local climate studies. A basic problem is that so many climatological studies are written in the local language, for obvious reasons, and often have not found their way to major meteorological libraries of the world. Additional extreme wind information sources are building codes.

However, when working with such available material, regard it critically. See if the author(s) took care in selecting their wind stations, in homogenizing their data series in some way, and in their treatment of various sources of stormy wind. Look at the age of their references: knowledge has increased in the last twenty years. And if you have to analyze raw data yourself, using methods mentioned above, stick to two golden rules:

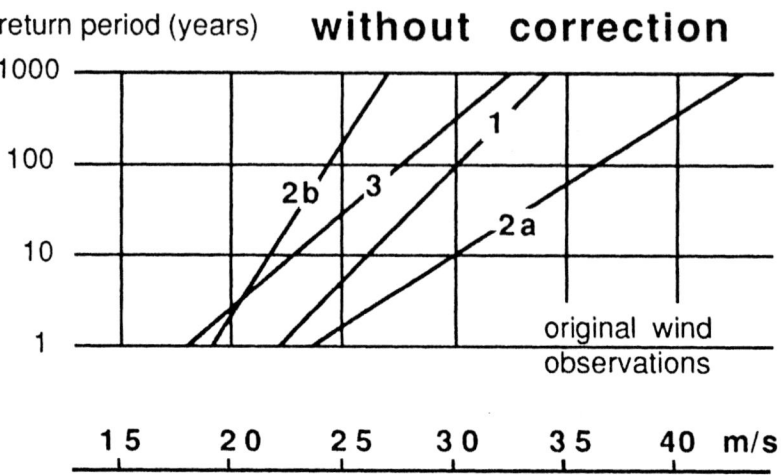

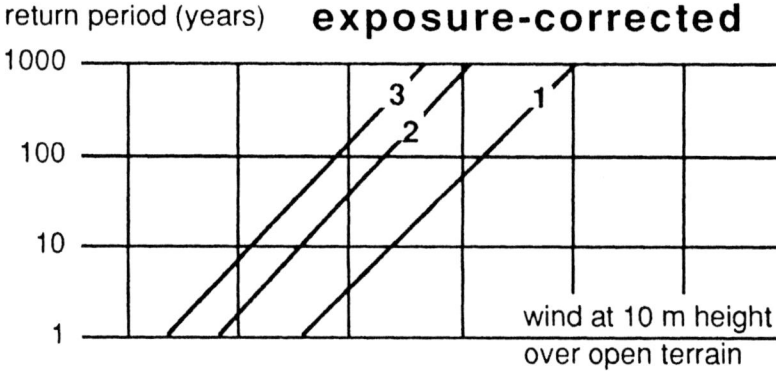

Fig. 2.4. Effect of exposure correction on extreme value estimates of hourly-averaged wind speed, statistically expected to occur once within the indicated return period at three stations in the Dutch coastal zone. Station 1 lies at the windy northern end of the coast, station 2 at the calmer southern end of the coast, 200 km distant, and station 3 lies between 1 and 2 at ≈80 km distance from the coast. Basic data sets (length >20 years) are the same in both graphs [69].

(1) *Never* use wind data from unspecified stations in unknown surroundings.
(2) Too many arrogant climatology amateurs and model-playboys have never checked up their data. If you have checked yours, do not be cowed if your results differ from theirs.

Appendix. Wind station siting requirements, agreed at the Wind Standards Workshop, Rockville 1992

The anemometer shall be located at 10 m height in level or gently sloping terrain with an open fetch of at least 150 m in all directions, with the largest fetch possible in the prevailing wind direction. Obstacles in the vicinity shall be at least 10 times their own height distant from the wind sensors.

The wind sensors should preferably be located on top of a solitary mast. If side mounting is necessary, the boom length should be at least 3 times the mast width. In the undesirable case that locally no open terrain is available and the measurement is made above some building, the wind sensor height above the roof top should not be less than the largest width of the building. In this case the station description file should indicate the height above ground level (AGL) of the highest part of the building and the height AGL of the wind sensors.

References

[1] Abild, J., Nielsen, B.: Extreme values of wind speeds in Denmark. Risø-M-2842 (1991).
[2] Barnard, J. C.: An evaluation of three models designed for siting windturbines in areas of complex terrain. Solar Energy 46 (1991), 283-294.
[3] Barthelmie, R. J., Palutikof, J. P., Davies, T. D.: Estimation of sector roughness lengths and the effect on prediction of the vertical wind speed profile. Bound. -Layer Meteor. 66 (1993), 19-47.
[4] Beljaars, A. C. M.: The measurement of gusts at routine wind stations–a review. WMO/CIMO Instr. Obs. Meth. Rep. 31 (1987) (= Roy. Neth. Meteor. Inst. WR-87-11).
[5] Berz, G.: Das Trombenrisiko in Europa nach Untersuchungen Alfred Wegeners. Ann. Meteor. 15 (1980), 74-76.
[6] Berz, G.: The increasing significance of windstorms and the INDR. J. Wind Engin. Industr. Aerodyn. 41 (1992), 23-25.
[7] Böllmann, G., Jurksch, G.: Ein Beitrag zur Festlegung der Grundwind- und Nennböen-Geschwindigkeit im Binnenland der BRD für die DIN-Norm 1055. Meteor. Rundschau 37 (1984), 1-10
[8] Cats, G. J.: Wind profile estimates up to 200 m from synoptic observations. Roy. Neth. Meteor. Inst. V-326 (1979).
[9] Chémery, L., Duchène-Marullaz, P.: Atlas climatique de la construction. Centre Sc. Techn. Bâtiment (Nantes, 1987), 182 pp.
[10] Cherry, N. J., Smyth, V. G.: Wind energy resource survey of Canterbury. WER-11 (1979), Lincoln College, New Zealand.
[11] Cherry, N. J.: Wind energy resource survey methodology. J. Industr. Aerodyn. 5 (1980), 281-296.
[12] Christoffer, J., Jurksch, G.: Untersuchung von Anemometerstandorten mittels sektorabhängiger Böenfaktoren. Meteor. Rundschau 38 (1985), 43-48.
[13] Clarke, R. H.: Severe local wind storms in Australia. CSIRO Div. Meteor. TP-13 (1962).
[14] Cook, N. J.: Towards better estimation of extreme winds. J. Wind Engin. Industr. Aerodyn. 9 (1982), 295-323
[15] Cook, N. J., Prior, M. J. (1987): Extreme wind climate of the United Kingdom. J. Wind Engin. Industr. Aerodyn. 26, 371-389.

[16] Dutton, M. J. O.: Optimum averaging times for reports of along-runway and across-runway wind components to aviation. Meteor. Mag. 105 (1976), 343-350.
[17] Fisher, R. A., Tippett, L. H. C.: Limiting forms of the frequency distribution of the largest or smallest member of a sample. Proc. Cambridge Philos. Soc. 24 (1928), 180-190.
[18] Garratt, J. R.: The internal boundary layer–a review. Bound. -Layer Meteor. 50 (1990), 171-203
[19] Giraytys, J.: Can instruments be designed to be both accurate and representative? Am. Meteor. Soc. Meteor. Monogr. 11-33 (1970), 437-443.
[20] Gomes, L., Vickery, B. J.: On the prediction of extreme wind speeds from the parent distribution. J. Industr. Aerodyn. 2 (1977), 21-36.
[21] Gomes, L., Vickery, B. J.: Extreme wind speeds in mixed wind climates. J. Industr. Aerodyn. 2 (1978), 331-334.
[22] Gumley, S. J., Wood, C. J.: A discussion of extreme wind-loading probabilities. J. Wind Engin. Industr. Aerodyn. 10 (1982), 31-45.
[23] Harihara Ayyar, P. S., Goyal, S. C.: Extreme wind speeds over India. Indian J. Meteor. Geophys. 23 (1972), 67-70.
[24] Holtslag, A. A. M.: Estimates of diabatic wind speed profiles from near-surface weather observations. Bound. -Layer Meteor. 29 (1984), 225-250.
[25] Huss, P. O.: Estimation of distributions and maximum values of horizontal wind speeds. J. Appl. Meteor. 13 (1974), 647-653.
[26] IPCC (Intergovernmental Panel on Climate Change): Policymakers Summary of the Scientific Assessment of Climate Change. WMO-UNEP-WG-1 (1990).
[27] Jensen, M., Franck, N.: The climate of strong winds in Denmark. Danish Techn. Press, Copenhagen, 1970.
[28] Jensen, N. O.: Change of surface roughness and the planetary boundary layer. Quart. J. Roy. Meteor. Soc. 104 (1978), 351-356.
[29] Jensen, N. O., Petersen, E. L., Troen, I.: Extrapolation of mean wind statistics with special regard to wind energy applications. WMO-WCP- 86 (1984).
[30] Justus, C. G., Hargraves, W. R., Mikhail, A., Graber, D. G.: Methods for estimating wind speed frequency distributions. J. Appl. Meteor. 17 (1978), 350-353.
[31] Kessler, E., Lee, J. T.: Distribution of the tornado threat in the United States. Bull. Am. Meteor. Soc. 59 (1978), 61-62.
[32] Logue, J. J.: Extreme wind speeds in Ireland for periods ending in 1974. Meteor. Serv. Dublin Techn. Note 41 (1975).
[33] Mayne, J. R.: The estimation of extreme winds. J. Industr. Aerodyn. 5 (1979), 109-137.
[34] Meaden, G. T.: Tornadoes in Britain: their intensities and distribution in space and time. J. Meteor. 1 (1976), 242-251.
[35] Mortensen, N. G., Landberg, L., Troen, I., Petersen, E. L.: Wind atlas analysis and application program (WAsP). Risø Nation. Lab. (Denmark) I-666 (1993).
[36] Murakami, S.: Extreme wind speeds for various return periods during rainfall. J. Wind Engin. Industr. Aerodyn. 26 (1987), 105-125.
[37] Nappo, C. J. et al.: Workshop on the representativeness of meteorological observations. Bull. Am. Meteor. Soc. 63 (1982), 761-764.
[38] Oort, A. H., Taylor, A.: On the kinetic energy spectrum near the ground. Month. Weath. Rev. 97 (1969), 623-636.
[39] Palmieri, S., Pulcini, A.: Trombe d'aria sull'Italia. Riv. Meteor. Aeron. 39 (1979), 263-277.
[40] Palutikof, J. P., Davies, T. D., Watkins, C. P., Bass, J. H., Halliday, J. A.: A methodology for the prediction of windspeeds at a candidate wind turbine site. Proc. 10th British Wind Energy Ass. Conf. (London, 1988) 95-103.

[41] Petersen, D. P., Middleton, D.: On representative observations. Tellus 15 (1963), 387-405.
[42] Pike, W. S.: One hundred years of the Dines pressure-tube anemometer. Meteor. Mag. 118 (1989), 209-214.
[43] Reid, S. J.: The accuracy of mountain-top winds estimated from free air data. J. Wind Engin. Industr. Aerodyn. 26 (1987), 179-193.
[44] Riera, J. D., Viollaz, A. J., Reimundin, J. C.: Some recent results on probabilistic models of extreme wind speeds. J. Industr. Aerodyn. 2 (1977), 271-287.
[45] Rijkoort, P. J.: A compound Weibull model for the description of surface wind velocity distributions. KNMI-WR- 83-13 (1983).
[46] Rijkoort, P. J., Hemelrijk, J.: The occurrence of "twin" storms from the North West on the Dutch coast. Statistica Neerl. 11 (1957), 121-130.
[47] Rijkoort, P. J., Wieringa, J.: Extreme wind speeds by compound Weibull analysis of exposure-corrected data. J. Wind Eng. Industr. Aerodyn. 13 (1983), 93-104.
[48] Schaefer, J. T.: Tornadoes – when, where, how often. Weatherwise 33 (1980), 52-59.
[49] Schmidt, H.: Zur Extrapolation empirische Verteilungen der Windgeschwindigkeit für Standorte in Flachland und auf freier See. Meteor. Rundschau 33 (1980), 129-137.
[50] Shellard, H. C.: Extreme wind speeds over Great Britain and Northern Ireland. Meteor. Mag. 87 (1958), 257-265.
[51] Simiu, E., Filliben, J. J.: Probability distributions of extreme wind speeds. ASCE J. Struct. Div. 102 (1976), 1861-1877.
[52] Simiu, E., Biétry, J. J., Filliben, J. J.: Sampling errors in estimation of extreme winds. ASCE J. Struct. Div. 104 (1978), 491-501.
[53] Simiu, E., Filliben, J. J.: Weibull distributions and extreme wind speeds. ASCE J. Struct. Div. 106 (1980), 2365-2374.
[54] Singh, R.: Occurrence and distribution of tornadoes in India. Mausam 32 (1982), 307-314.
[55] Stevens, M. J. M., Smulders, P. T.: The estimation of the parameters of the Weibull wind speed distribution for wind energy utilization purposes. Wind Engin. 3 (1979), 132-145.
[56] Tabony, R. C.: Extreme value analysis in meteorology. Meteor. Mag. 112 (1983), 77-98.
[57] Tamura, Y., Suda, K.: Correction of annual maximum windspeed considering yearly variation of the ground roughness in Japan. Proc. 7th Internat. Conf. Wind Eng. (Aachen, 1987) Vol. 1, 31-40.
[58] Taylor, P. A., Mason, P. J., Bradley, E. F.: Boundary-layer flow over low hills (a review). Bound. -Layer Meteor. 39 (1987), 107-132.
[59] Taylor, P. A., Salmon, J. R.: A model for the correction of surface wind data for sheltering by upwind obstacles. J. Appl. Meteor. 32 (1993), 1683-1694.
[60] Tieleman, H. W.: Wind characteristics in the surface layer over heterogeneous terrain. J. Wind Engin. Industr. Aerodyn. 41 (1992), 329-340.
[61] Troen, I., Petersen, E. L. (eds.): European Wind Atlas. Risø Nation. Lab., Roskilde, 1989.
[62] Tuller, S. E., Brett, A. C.: The characteristics of wind velocity that favor the fitting of a Weibull distribution in wind speed analysis. J. Clim. Appl. Meteor. 23 (1984), 124-134.
[63] Walmsley, J. L.: Assessing the wind resource. In: R. Hunter, G. Elliot (eds.), Wind-diesel systems, Cambridge Univ. Press, U. K. (1994), pp. 54-94.
[64] Wieringa, J.: Gust factors over open water and built-up country. Bound. -Layer Meteor. 3 (1973), 424-441.

[65] Wieringa, J.: An objective exposure correction method for average wind speeds measured at a sheltered location. Quart. J. Roy. Meteor. Soc. 102 (1976), 241-253.
[66] Wieringa, J.: Wind representativity increase due to an exposure correction, obtainable from past analog station wind records. In: Proc. TECIMO Conf., WMO-No. 480 (1977), 39-44.
[67] Wieringa, J.: Representativeness of wind observations at airports. Bull. Am. Meteor. Soc. 61 (1980), 962-971.
[68] Wieringa, J.: Description requirements for assessment of non-ideal wind stations – for example Aachen. J. Wind Eng. Industr. Aerodyn. 11 (1983), 121-131.
[69] Wieringa, J.: Roughness-dependent geographical interpolation of surface wind speed averages. Quart. J. Roy. Meteor. Soc. 112 (1986), 867-889.
[70] Wieringa, J.: Kartering van Nederland's windklimaat boven 40 m hoogte. Proc. Nation. Windenergie Conf., publ. Energie Anders, Rotterdam (1988), 102-106.
[71] Wieringa, J.: Shapes of annual frequency distributions of wind speed observed on high meteorological masts. Bound. -Layer Meteor. 47 (1989), 85-110.
[72] Wieringa, J.: Updating the Davenport roughness classification. J. Wind Eng. Industr. Aerodyn. 41 (1992), 357-368.
[73] Wieringa, J.: Representative roughness parameters for homogeneous terrain. Bound. -Layer Meteor. 63 (1993), 323-363.
[74] Wieringa, J., Rijkoort, P. J.: Windklimaat van Nederland. Staatsuitgeverij, Den Haag (1983), 263 pp. (ISBN 90 12 04466 9).
[75] Whittingham, H. E.: Extreme wind gusts in Australia. Bur. Meteor. Austr. Bull. 46 (1964).
[76] Zuranski, J. A.: Analiza charakterystycznych predkosci wiatru w Polsce. In: Obciazenia wiatrem budowli i konstrukciji, Arkady Publ., Warszawa (1978), 49-52.
[77] Zwart, B., Hatch, D. J.: Is the number of severe gales increasing? J. Meteor. 10 (1985), 40-45.

CHAPTER 3

Climate Change and Hydrological Disasters

Max A. Beran and Nigel W. Arnell

3.1 Introduction

3.1.1 QUESTIONS

Many regard climate change as a disaster-in-waiting, a view reinforced by society's evident dependence on systems to protect itself against the vagaries of weather (Ausubel, 1991), and in some instances to exploit them. Professor Obasi, Secretary General of the World Meteorological Organization, opened the 1988 conference on the Hydrology of Disasters with the words, *"losses from natural disasters appear to be increasing every year with the growth of population, and we must now consider the complications introduced by climate change"* (Starosolsky and Melder, 1989). In this chapter we attempt to answer the embedded questions — what will be the impact of climate change on hydrological disasters? Will they become more or less disastrous, more or less frequent, of greater or lesser magnitude?

3.1.2 CONCEPTS AND DEFINITIONS

The term "climate" is best understood through its relationship to "weather", the latter referring to the specific amounts of rainfall, sunshine, radiation, etc. received at a given time and place. Climate is a statistical statement about the assemblage of weather; its description includes information on averages, variations and extreme values of the various elements. It is not the weather which changes when climate changes, but the relative frequency of the different types of weather. A useful analogue for climate is as a multifaceted dice marked up with weather types. A daily throw of the dice gives a pattern of weather whose frequency is determined by the frequency of its appearance on the dice. Some individual faces, or runs of faces constitute disasters, for example when dry weather is indicated many days in succession. Climate change, on this scheme, is represented by a relabelling of the faces such that the various weather types occur with different frequencies.

When we speak of the "climate system" we refer to the interacting processes in the atmosphere, ocean and land surface that control the mixture of weathers that a region can experience. In the current context we are most concerned with the role of the water cycle in the climate system. However it has become increasingly apparent that the climate system is inextricably bound up with biological and chemical processes which determine the regional energy and water budget, and the radiative properties of the atmosphere.

Climate has changed throughout geological and historical time due to mechanisms internal to the climate system as well as to rapid and slow evolution of external factors. In this chapter we concentrate on anthropogenic changes superimposed on this natural background. The important change currently in prospect is global warming due to man's emissions of greenhouse gases such as carbon dioxide, methane, ozone, CFCs and nitrous oxide. Their effect is to increase the intensity of the exchange of long wave radiation between the earth's surface and the lower atmosphere. Such a change will have immmediate consequences through raising the near-surface temperature and indirect consequences on rainfall, evaporation and the hydrological cycle. It may also have indirect effects acting via ecosystem changes. Some of these consequences feed back into the climate system and may further aggravate the initial forcing due to the grenhouse gas emissions. Here we are concerned with hydrological disasters, largely to be equated with floods and droughts.

Floods and droughts have many forms. Overbank flooding from rivers is the simplest and has received most attention. Hydrological droughts likewise are most readily analysed in terms of their effect on river flows and the use made by society of that resource. A more complex example of a flood-related hydrological disaster is slope instability giving rise to landslides. Mudflows and ice bursts are other forms of hydrological disaster which can occur only under rather particular circumstances. Sea levels are expected to rise as a consequence of global warming so coastal and estuary flooding will increase. Agricultural drought is felt when there is a severe and sustained soil moisture deficit which may have different origins to a drought that is responsible for low river discharges. Both floods and drought may initiate other processes such as pollution incidents or ice flows. Actions taken to mitigate risks can themselves sow the seed of future enhanced disaster, for example by dam-break, by concentrating populations immediately beyond a currently protected zones, or by increasing food-supply vulnerability.

In the space available, and given the current state of knowledge, it is necessary to take a broadbrush approach to reviewing possible effects of climate change on hydrological disasters. This chapter therefore focuses on flood and drought disasters in rivers and how they may change under an altered climate regime.

3.1.3 Is Climate Change a Hydrological Disaster?

There is a popular vision, reinforced by some sections of the media and environmental lobby, that portrays the greenhouse effect as a series of catastrophes in-

Climate Change and Hydrological Disasters

volving submerging islands, devastating floods and hurricanes, threatening famine, outbreaks of pest and disease, and enforced migrations that destabilize entire countries. Such an extreme view is untenable in the light of accumulating knowledge, and it is critical that one holds on to a scientific appreciation of the real causes and processes. In fact what will be learned is how little certainty there is about the future course of events, especially in relation to extreme events.

A disaster is an interaction between an environmental event and a vulnerable society. The environmental event acts as the trigger; vulnerable society experiences the consequences. In hydrological terms, the potentially disastrous aspects of the environment to society are an *excess* of water, and a *lack* of water. Climate change itself should not be regarded as a disaster, but rather as something which may change the *risk* of hydrological disaster.

The following sections of this chapter describe how climate change might affect exposure to hydrological disasters, focusing on floods and droughts in rivers. It examines how specific impacts in particular catchments may be assessed, and reviews current views on the impact of climate change on hydrological disasters.

3.2 The Greenhouse Effect, Climate Change and Hydrological Regimes

The climatic system of planet earth is driven by short-wave energy received from the sun. A portion of this radiation is reflected from clouds and the earth's surface and the balance heats the ocean and land surface thus powering the climate system. Warm surfaces re-radiate energy at longer wavelength and it is this long-wave radiation which is absorbed by certain radiatively-active trace gases in the atmosphere. Their action is to intensify the exchange of energy between the lower atmosphere and the surface causing their temperatures to rise. Without this natural "greenhouse effect" — so called because the radiatively-active gases act in a similar way to the panes of glass in a greenhouse — the temperature of the earth would be around 33°C cooler than it actually is, and unsuitable for life as we know it. The most important of the radiatively-active trace gases are water vapour and carbon dioxide.

Human activity has led to increases in the concentration of these gases — particularly of carbon dioxide — and has also introduced new man-made greenhouse gases into the atmosphere. Carbon dioxide comes from deforestation and combustion of forest fuels, methane is emitted from grazing animals and plants grown in marshy conditions (such as rice), and nitrous oxides derive from several agricultural and industrial practices. Chlorofluorocarbons (CFCs) are very powerful man-made greenhouse gases, and did not exist before the 1920s (although their net contribution to the enhanced greenhouse effect is offset by their impact on stratospheric ozone: IPCC, 1992).

Water vapour, the most important greenhouse gas, is not directly affected to any appreciable degree by human activity, at least not on the global scale, but would increase due to the higher temperatures resulting from higher concentrations of the other greenhouse gases. This is an example of the feedbacks which characterise the climate system, atmospheric chemistry, and the sources and sinks of greenhouse gases. Other feedbacks also involve the hydrological cycle. Another such example in which the hydrological cycle is implicated is the ice-albedo feedback in which an initial warming causes shelf-ice to melt with consequent lowering of the albedo in circumpolar regions and enhanced regional heating.

An increasing concentration of greenhouse gases in the atmosphere must mean increased radiative forcing at the bottom of the atmosphere, and higher temperatures. Estimates of the global warming effect of past and future human activities are, however, uncertain. The magnitude of future warming depends on assumptions about the rate of population and economic growth, and there are major uncertainties in (i) the effects of terrestrial and marine ecosystems on sources and sinks of greenhouse gases in a warmer world, (ii) the operation of chemical processes within the atmosphere, (iii) the precise radiative effects of a change in greenhouse gas concentrations (particularly due to changes in cloud cover), (iv) the counteracting effects of sulphate aerosols in emitted into the atmosphere by pollution, and (v) the climatic effects of a change in radiation received at the ground.

Nevertheless, the United Nations Intergovernmental Panel on Climate Change (IPCC) estimated that, under a "business-as-usual" growth scenario, global mean temperature would be expected to rise by 0.3°C per decade (IPCC, 1990; 1992). This would result in a global mean temperature by 2030 1.2°C higher than at present, higher than at any time in human history. The rate of change in temperature too is very high, and may be faster than the rate at which natural and human systems can adapt.

Hydrological processes play a fundamental role in the climate system. Some globally significant feedbacks have been mentioned above. Others are more local in their operation. In arid regions an initial drying, due to differential temperature rise between ocean and continental areas, reduces vegetation cover and increases reflection and sensible heat loss, gives rise to sinking air mass and reduced production of cloud condensation nuclei, and provides the conditions for increased aridity. However the aspects of hydrological disaster that we focus on here are those which affect society.

Changes in temperature will result in changes in rainfall and evaporative demand, and will therefore produce changes in river runoff and groundwater recharge. The precise impacts of climate change on hydrological regimes in a catchment — and hence floods and droughts — will depend on the regional change in climate, on the current climatic and physical characteristics of the catchment, and also on its soil and vegetation. Other things being equal, a given change in temperature and rainfall will have a greater proportional effect in a drier catchment; in other words, where rainfall input and evaporation are most closely in balance.

The effects of a change in flood and drought magnitude and frequency on human water-use systems depend also on the characteristics of those systems. The more stressed a water resources system, in the sense of near complete exploitation of the available resource, the more sensitive it will tend to be to any change in climatic inputs. It is important to emphasise that the sequence running from a change in climate through a change in hydrology to an impact on potential for disaster is not linear. In particular it is characterised by the presence of critical levels, beyond which there may be a qualitative change in impact.

Climate Change and Hydrological Disasters 45

Whilst it is possible to estimate in general terms the global effect of an increased concentration of greenhouse gases on the radiative balance and global temperature, it is currently not possible to forecast future changes in catchment hydrological regimes and hence the potential for flood and drought in similar general terms. It is necessary to consider the individual features of the catchment and water use or flood protection systems. The next section therefore reviews the methods that can be used to estimate possible impacts of climate change in a catchment.

3.3 Estimating the Impacts of Climate Change on Hydrological Characteristics

3.3.1 INTRODUCTION: CLIMATE CHANGE SCENARIOS

A climate change scenario is a feasible, internally-consistent description of a possible future climate. It is not a forecast, nor indeed a prediction. Its purpose is to guide high-level policy rather than to intiate a practical response such as construction of an engineering scheme. Climate change impact studies are driven by estimates of change based on climate change scenarios, and commonly use a range of alternative scenarios.

The basic methodology adopted by most climate change impact studies is a two-stage process. The first stage traces the consequence of the assumed change in climatic characteristics to changes in hydrological regime making use of some representation of the transfer between climate and catchment hydrology. The subsequent stage follows through to impacts on the operation of a specific water resources system (the impact approach: Carter et al., 1992). It is possible in principle to work the other way round, and determine the changes in climate that would be needed to push the system of interest across some critical threshold, but this approach has not been used in any published hydrological impact studies.

The rest of this section examines ways of creating climate change scenarios.

3.3.2 CREATING CLIMATE CHANGE SCENARIOS

Four basic methods have been used for creating climate change scenarios.

Arbitrary changes in climate characteristics. This most basic approach involves perturbing, in an arbitrary way, either the statistical properties of an observed hydrological series, or the properties of the temperature, rainfall and evaporation inputs to a hydrological model. This approach has the merit of simplicity, but should be seen more as a sensitivity analysis than a climate change impact assessment; there is no attempt to ensure that the changes imposed are feasible or internally-consistent. Nevertheless, the approach does provide useful information about the sensitivity of a hydrological or water resources system to changes in inputs.

Temporal analogues. This approach assumes that the future will be like some defined period in the past. There are two variants. The first takes advantage of historical or instrumental data, collected over the past 100 years or so. Certain periods within these records are taken to be representative of warm conditions, and others cool; the effect of global warming is assumed to be represented by the difference between the cool and warm periods (Palutikof, 1987; Krasovskaia and Gottschalk, 1993). The method can be criticized on the grounds that the past decade-to-decade contrasts between a cool and a warm period do not provide an analogue for the effects of global warming due to a progressive change in radiative forcing. The results are also sensitive to the data used to define warm and cool periods. Also, the difference between cool and warm periods in the recent instrumental past may be smaller than the possible effects of global warming. However, by providing information on the performance of a water resources system during extreme events of the past, the approach is likely to be of benefit in determining how to manage the system in the face of possible increased extreme events in the future.

The second variant uses periods in the geological past as analogues for the future. Past climates (palaeoclimates) and environments are reconstructed from geological deposits, geomorphological forms and vegetation histories (based on pollen analyses, for example). There are three major assumptions. First, it is assumed that the relationships between form and process operating today are the same as those operating in the past, and that it is therefore possible to deduce past process from past form. Second, it is assumed that data from the past represent equilibrium conditions, but in practice environmental systems are rarely in equilibrium with climatic conditions. Third, and most importantly, it is assumed that the same pattern of consequential change occurs regardless of the cause of change in climate. Again, this is unlikely to be true in practice. The post-glacial climatic optimum, for example, was associated with a change in solar radiation receipts due to a change in the earth's orbit; the spatial patterns in these radiation changes are different to the patterns in radiation change due to an increasing concentration of greenhouse gases, and their effects on climate are therefore different.

Nevertheless this variant of the approach has achieved great prominence largely through the work of the Russian school (IPCC, 1990) which identified three palaeoclimatic analogues. The Holocene climatic optimum (6.2 to 5.3 ka BP) is assumed to represent an increase in temperature of 1°C, and the last interglacial (125 ka BP) is taken as an analogue for a 2°C increase in temperature. A 4°C increase is represented by the Pliocene (3 to 4 ma BP). In practice, however, it is difficult to extract reliable *quantitative* information on hydrological characteristics from palaeoclimatic information. The best prospects exist for past catastrophic floods which leave their mark in geomorphological and sedimentological evidence (Knox, 1993).

Spatial analogues. This approach assumes that the future climate of one region can be represented by the current climate of a second region; the future climate of the Canadian mid-west, for example, may be regarded as similar to the current climate of the US mid-west (Parry et al., 1989). In actuality, however, the weather and climate of a particular region are a result of the interaction between large-scale circulation features, local topography, and the local configuration of land cover and water. The greater the relative importance of large-scale features, the more feasible it will be to transfer the climate of one area to another.

Even where an adequate climate spatial analogue can be identified difficulties will remain in equating hydrological regimes which are controlled not just by climate but also by catchment physical characteristics, and particularly geology which is the critical factor governing drought response. Differences in physical characteristics can be taken into account through the use of suitable models when transferring hydrological data from one region to another, though at the cost of additional uncertainty.

Statistical relationships predicting hydrological parameters such as flood or low flow quantiles from catchment and climatic characteristics can be viewed as a mathematically formal approach to quantifying spatial analogues. It is tempting to use such a relationship to estimate the effects in a catchment of a given change in the values of climatic parameters input to the relationship. For example, it would be possible to estimate changes in the mean annual flood in a catchment associated with a change in annual rainfall, if an empirical relationship existed which related those two parameters. In practice, however, the estimated sensitivity of the mean annual flood to change would be determined entirely by the value of the coefficients of the statistical relationship, which is itself dependent on the other predictor variables present and the calibration data used. Such empirical relationships internalise the current *causative* relationships between processes involved in creating hydrological response, and current relationships may not hold in the future.

Use of climate model simulations. The majority of climate change impact studies have used scenarios based on the output from general circulation models (GCMs). However, it is not possible to use directly the output from a GCM in a hydrological impact study, for two main reasons. First, GCMs are designed to simulate large-scale circulation patterns, rather than detailed climate in specific regions. Second, the spatial resolution of current GCMs is very coarse. The highest resolution GCM used routinely in greenhouse-gas studies uses a grid spacing of around 250×350 km, far larger than typical scales employed by hydrologists for the study of floods and droughts.

Several methods have therefore been proposed to express GCM output at a scale and in a form suitable for hydrological impact studies. The most popular method uses the GCM simulations to determine *change* in climatic characteristics (such as temperature and rainfall), which are then used to adjust the climatic data of the catchment under study. These capitalise on the belief that, even if the GCM does not simulate the actual climate correctly, it will simulate the change in climate with greater reliability. It also assumes that such changes, evaluated at the GCM grid points, are reasonably spatially conservative so can be applied to adjacent regions on the ground in one of the ways enumerated below.

(i) *Simple interpolation.* The first method interpolates GCM output down to the catchment of interest, either statistically or subjectively. It has been very widely used.

(ii) *Use empirical relationships between large-scale climatic characteristics and catchment scale variability.* This approach assumes that it is possible to develop empirical relationships between GCM-scale climate and local climate, and that these relationships will continue to hold in the future. One variation of this approach (Wigley et al., 1990) developed regression relationships to predict temperature and precipitation at selected points from large-scale average climate, and applied these relationships to the output from GCM climate estimates. Another variation uses relationships between some index of large-scale circulation (such as a pressure gradient or a weather type) and point climate (Bardossy and Plate, 1992; Hay et al., 1990, von Storch et al., 1991). The major assumption is that these empirical relationships hold in the future.

(iii) *Use of nested meso-scale atmospheric models.* It is possible to run a high-resolution meso-scale atmospheric model for a region using inputs from a GCM, and thus produce high-resolution climate simulations (Giorgi and Mearns, 1990). The regional simulations are limited by the accuracy of the inputs derived from the GCM, but have the advantage of providing spatially and temporally coherent fields of inputs to hydrological models. The method has not yet been widely applied to create climate change scenarios, but has considerable potential.

3.3.3 Equilibrium and Transient Scenarios

Most climate change impact studies have examined the impacts of a new climate stabilised around doubled-CO_2 concentrations; ie they have assumed that the climate has reached a new equilibrium. In actuality greenhouse gas concentrations are increasing over time, and the change in climate will be evolutionary. Beyond the point in time in the middle of the next century when equivalent CO_2 concentrations have doubled, climate will continue to change as concentrations increase further, and the climate and ocean system catches up. There will not therefore be a step change in vulnerability to disaster, but rather a transient change. This transient change may take the form of a gradual underlying increase, or may occur as a series of sudden jumps. Too few transient climate change experiments have yet been run with GCMs to determine the characteristics of change, but historical and geological evidence suggest that change may occur by a series of short shocks (see for example Taylor et al., 1993).

Transient climate change scenarios can be derived from the results of equilibrium climate change experiments (Viner and Hulme, 1993), essentially by rescaling equilibrium model output according to global temperature change simulated for the year of interest using a one-dimensional global energy balance model. Transient scenarios can also be developed from transient climate simulation runs. Few hydrological studies, however, have yet exploited transient scenarios to investigate the evolution of change in vulnerability to disaster over time.

3.4 Impacts of Climate Change on Floods

3.4.1 Introduction

Although there have been many generalized statements made about the impacts of climate change on hydrological regimes and water resources, few studies address explicitly the issue of hydrological disasters. Drought incidence is considered by some, fewer have looked at possible changes in flood occurrence. This and the

Climate Change and Hydrological Disasters

next section attempt to assess the potential impacts of climate change on flood and drought occurrence, based on speculation and information gleaned from a few published studies, but at the same time reviewing why it is currently difficult to determine changes in disaster occurrence.

3.4.2 Factors Responsible for Floods

To the hydrologist a flood is an incidence of high runoff. To society a flood disaster consists of massive out-of-bank flow with large areas inundated, loss of life, and damage to property, livestock and crops. When considering the consequences of climate change one needs to distinguish rain-generated from snowmelt-generated floods.

Rain-generated floods. Rain-generated flash floods are triggered by intense rainfall over the catchment. However, the intensity of rainfall required to generate a flood depends significantly on catchment characteristics, and particularly on land cover, soil properties and the state of the catchment. The wetter the catchment, the lower the rainfall which may be needed to cause a flood. In some catchment types, antecedent conditions are very important in controlling flood occurrence — particularly in larger catchments with permeable soils and large amounts of storage to fill — but in others, such as very responsive clay catchments or catchments prone to soil crusting, antecedent conditions may be unimportant and flood magnitudes are very closely related to the amount of storm rainfall.

On the world scale the most notable flood disasters are those which occur periodically on the major rivers which drain south from the Himalayas, run through eastern China, and in the central USA. In these cases details of catchment characteristics diminish in importance and the driving force is the incidence of widespread rainfall immediately prior to the flood occurring on a catchment that has experienced generally wet conditions in the previous season. Geomorphological considerations often apply where such rivers have created natural levees in their middle reaches which man might have brought into management.

Globally speaking the expected warming can be expected to intensify the hydrological cycle. This expectation arises from the first order effect where higher temperatures leads to an increase in the moisture holding capacity of the atmosphere. Such a line of argument may lead one to suppose that conditions giving rise to flooding would be more frequently experienced on the large basins, either because of seasonal humidification, or an increase in catchment rainfall.

Looking at the process of flood formation in greater detail, global warming can be expected to produce changes in the frequency of occurrence of intense rainfall in a catchment, for two main reasons. First, there may be a change in the *paths and intensities of depressions and storms*, and second there will probably be an *increase in convective activity*.

The latitudinal position and intensity of rain-producing depressions across the North Atlantic and North Pacific are influenced by the precise position of the jetstream, which is itself a characteristic of large-scale atmospheric circulation. A change in the position of the jetstream would mean a change in the route of depressions. Some areas would experience them, and their associated rainfall, more often; other areas would be visited less frequently and would thus enjoy a reduction in flood-producing rainfall. Stephens and Held (1993) found few dramatic changes in storm tracks, using the GFDL coupled ocean–atmosphere model, but noted that storm tracks in the North Atlantic became slightly weaker whilst those in the North Pacific became slightly stronger. However, there have so far been too few studies with other models to lend support to this finding.

The lower reaches of low-latitude rivers and coastal basins experience flooding due to cyclone activity. Tropical cyclones can develop only when sea surface temperatures exceed 27°C, and beyond that threshold storm intensity is related to sea surface temperature (Emanuel, 1987). An increase in sea surface temperatures can be expected to increase the intensity of tropical cyclones *and* to expand the area over which they may develop, exposing new regions to the cyclone threat. Gutowski et al. (1994) calculated an increase in surface pressure gradients off Florida of 40% due to global warming, and assumed that rainfall intensities would increase by a corresponding percentage. This altered the return period of the current 1000-year event in southern Florida to one in a 100 years. The IPCC (1990; 1992) found some evidence that the frequency, intensity and area of occurrence of tropical storms may increase, but noted that the evidence was "not compelling" (IPCC, 1990: p154). However, sea surface temperatures are not the only control on cyclone intensity (Evans, 1993), making changes in occurrence harder to determine.

The El Nino-Southern Oscillation (ENSO) phenomenon has a significant effect on climatic anomalies and weather extremes across a large part of the globe. In most years, currents transport water from east to west across the Pacific Ocean, and this water is held in place off Indonesia by trade winds blowing in the same direction. Eventually, however, so much water is accumulated that any random weakening in the trade winds causes a surge of warm water eastwards across the Pacific. Sea surface temperature in the mid-Pacific then rises, warming the atmosphere, changing atmospheric circulation patterns and causing the Indonesian low pressure system to move eastwards. During an ENSO event, some parts of the world experience more rainfall, whilst other parts receive less, but the anomaly pattern varies from event to event. Meehl et al. (1993) were able to simulate the ENSO phenomenon using the NCAR climate model, and investigated the possible effects of global warming. They found little change in the anomaly patterns, but found that the anomalies intensified: dry areas became drier and wet areas wetter. They also noted a change in mid-latitude circulation patterns with the more intense ENSO events.

As already mentioned, meteorological theory leads to the expectation that global warming should produce an increase in the intensity of rainfall for several reasons. The amount of water held in the atmosphere increases as air temperature rises. The volume of water held in air over the sea also rises, non-linearly, as sea surface temperature rises (Stephens, 1990). Higher temperatures can also be expected to lead to greater convective activity, and the condensation of the additional water vapour in convective systems releases latent heat which leads to even greater convective activity.

It is, however, difficult to extract from GCMs reliable information on changes in extreme rainfall. Intense rain-storms are very localised features which are not explicitly simulated by current GCMs with their coarse spatial resolution. Although GCMs work on a very short time step, the simulated rainfall is an areal total distributed over several tens of thousands of square kilometres, and it is difficult to infer local point or catchment rainfall from a large-scale regional average. Nevertheless, Mearns et al. (1990) and Whetton et al. (1993) found, for the United States and Australia respectively, a reduction in the number of days on which rain fell somewhere in a GCM grid-cell and an increase in mean rain-day rainfall, following a doubling of atmospheric CO_2. Similar results have been found in other regions (IPCC, 1992). Whetton et al. (1993) also found an increase in the total volume of convective rainfall over the rainfall season.

Very few studies have attempted to model changes in river flood characteristics following changes in rainfall, largely because of the difficulties in defining reasonable scenarios for change in rainfall characteristics, but also partly because of the difficulties in modelling the translation of rainfall into flood. Gellens (1991) used a daily rainfall-runoff model to investigate changes in flood occurrence in three catchments in Belgium. He assumed an increase in mean daily rainfall in winter — the flood-generating season — and found more frequent floods and flows remaining above high thresholds for longer. The mean flood peak increased by between 2 and 10 per cent, depending on catchment. Bultot et al. (1992) used the same daily rainfall-runoff model and approach to estimate possible changes in flood characteristics in Switzerland, and found an increase in the mean annual flood of around 10 per cent.

Snow-melt floods. Two main factors affect the magnitude of a flood caused by the melting of snow: the volume of water and the rate of melt. The rate at which snow melts depends on the change in temperature and also on the amount of rain that falls. The greatest snow-melt floods occur when a rise in temperature is associated with heavy rainfall.

A general increase in air temperature can initially be expected to result in fewer and smaller snowmelt floods, simply because less precipitation would fall as snow, and indeed many studies have shown that global warming would reduce snow cover (Gleick, 1987; Bultot et al., 1992, van Katwijk et al., 1993, for example). In their Swiss catchment, Bultot et al. (1992) found that an increase in temperature of 2.8°C would reduce the number of days with snow lying by over 50 per cent. However, increased precipitation might compensate for the higher temperatures, resulting in a greater snowfall during precipitation events and hence a greater snowmelt flood potential. It is also possible that the reduction of snow cover would increase the frequency of rain-generated floods during winter: precipitation would no longer be stored in the snowpack, and would run off rapidly. Global warming may not therefore necessarily reduce snow-melt flooding or flooding during what is at present the snow season. Lettenmaier and Gan (1990), for example, found a reduction in snowmelt floods but an increase in the frequency of rain-generated floods in the Sacramento-San Joaquin basin in California, with a consequent increase in the risk of occurrence of a particular discharge being exceeded.

Sensitivity analysis. The potential impacts of climate change on flood risk can be evaluated for a catchment using a simple sensitivity analysis using a probability distribution. Figures 3.1a and 3.1b show the return period of the current 10-year and 100-year floods, given changes in the mean and coefficient of variation (CV) of annual flood magnitudes. The Figures assume an Extreme Value Type 1 (EV1 or Gumbel) distribution with a slope typical of frequency curves in the UK. The same sensitivity analysis can be performed with a three-parameter distribution either by holding the third parameter (such as skewness) constant or by attempting a three-dimensional plot.

From Figure 3.1a it can be seen that a 10% increase in the mean would result in the current 10-year flood occurring on average approximately once in every 7 years; the 100-year flood would be exceeded once in around 50 years (Figure 3.1b). Note that Bultot et al. (1992) simulated a 10% increase in the mean and a 17% increase in CV in their Swiss study catchment. The effect of a change in mean and CV on the return period of a particular magnitude event depends on the CV: the lower the CV (and the flatter the frequency curve) the greater the change.

Uncertainty in flood risk assessment. Global warming may lead to changes in the estimated risk of flooding. It is important to remember, however, that estimation of flood risk under current climatic conditions is fraught with uncertainty. The most appropriate form and parameterisation of a flood frequency distribution are unknown, and considerable statistical uncertainties arise due to the estimation of model parameters from small samples. Finally, the estimated risk of flooding may depend considerably on the period of record available. Whilst reviewing some of the implications of global warming for flooding, Smith (1993) showed how very different estimates of flood risk and flood damage potential could be derived from different periods of record at Lismore, New South Wales, Australia.

3.5   Impact of Climate Change on Drought

3.5.1   INTRODUCTION

In many respects drought has been at the heart of the debate about climate change, the greenhouse effect and its propensity to cause disasters. In the mind of the populace, who have been introduced to the issue through the prospect of disaster, the basic sequence of events is that the greenhouse effect induces global warming, the warming increases evaporation, there is less water in the atmosphere as a consequence to fall as rain, plants wilt, crops fail, water supplies dry up, and finally there is widespread famine and drought.

Is such a scenario realistic? In this section we analyse more carefully what is known about mechanisms of drought and review what evidence there is on its possible trend under a warming climate.

3.5.2   DROUGHT DEFINITION

From the point of view of scientific analysis, drought differs from flood in a rather essential respect. Whereas floods can be thought of as the presence (to excess) of something, drought is the absence of something, often for an extended period. This makes drought intrinsically the more "slippery" concept capable of a large number of interpretations. As a disaster, drought is certainly no less important,

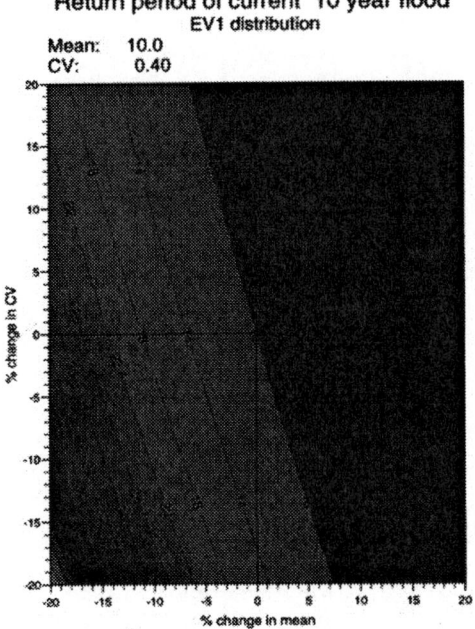

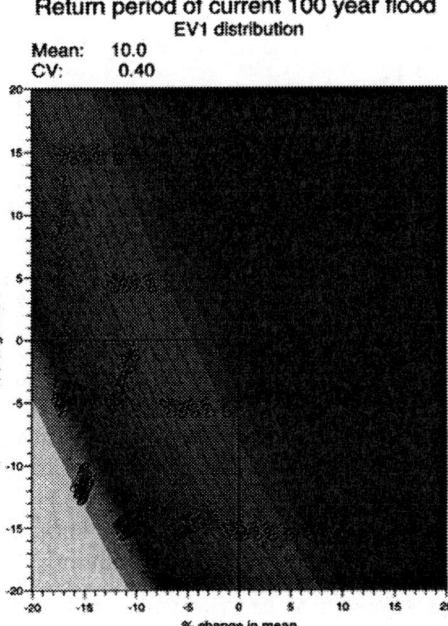

Fig. 3.1. Effect of changes in the mean and coefficient of variation of annual maximum floods on the return period of the current 10-year flood (a, *top*) and 100-year flood (b, *bottom*). Extreme Value Type 1 distribution with parameters typical of annual maximum floods in the UK.

affecting approximately double the number of people as do floods. Data compiled by the Red Cross quoted in Wijkman and Timberlake (1984) place drought as the number one disaster measured in terms of the number of people affected, though less catastrophic than earthquake, hurricane or flood in terms of human deaths caused.

Hydrologists view drought as a sustained and regionally extensive occurrence of below normal natural water availability (Beran and Rodier, 1985). Though satisfactory as a starting point this statement requires further specification for quantified studies of drought.

Many solutions have been devised by hydrologists and others to the problem of how to attach numerical indices to drought severity. Two elements are normally necessary: the duration of the drought event and, within the duration, its magnitude. Textbook definitions of hydrological drought magnitude commonly express it as a departure below some rainfall or streamflow threshold — e.g. a quantile or proportion of the mean. The duration element of the definition may enter through the use of a period-integrated value of rainfall or runoff. Sometimes a third element is included in the definition, the area affected. It is in the nature of drought and its causative agents that it influences an appreciable area.

Many definitions take such a form as "drought occurs when the annual rainfall falls below 75 per cent of the long-term average". Variants on the basic theme may substitute other percentages or quantiles, or may consider other measures of normality. All reinforce the notion of drought as a relative phenomenon and imply that drought afflicts well watered areas as frequently as more dessicated areas. This conflicts with the more popular perception of drought, certainly those that can be labelled a disasters, which would tend to be located in semi-arid regions. This is because the effect of drought that is of most concern is famine. In a semi-arid region where water shortage is in any case endemic, any reduction in water availability is liable to lower it below the point where food may be produced and water for drinking and hygiene is available. In a well-watered region the consequences of an equivalent relative reduction may disrupt economic life but seldom threaten life in a direct sense.

3.5.3 Factors Responsible for Drought

Drought is a "climatic" phenomenon in the sense that it requires a major modifcation of the normal weather patterns sustained over a lengthy period and affecting a wide region (Beran and Rodier, 1985). Only by understanding its climatic origins can we arrive rationally at a forecast of the effect of climate change.

A drought's most characteristic climate signature is warm dry air in the troposphere. A stable system develops in which slow subsidence inhibits raincloud formation and any local moisture is entrained. Balancing divergent airflows within the boundary layer remove soil mosture from the affected region. Resistance to humidifying conditions such as the importation of moist tongues, e.g. of oceanic air, can be induced by a blocking pattern of high pressure cells that divert such air from the drought region.

Conditions like those outlined in the previous paragraph are largely a matter of chance occurrence, and can be thought of as responsible for the onset of drought. Whether a drought is self sustaining has been a question of overriding research interest over decades. Several self-generating mechanisms have been postulated. Hot dry surfaces heat the air above them and can enhance the high pressure cell; airborne dust particles can inhibit droplet formation. Parched vegetation and dry soil has high albedo and this contributes to a greater radiative heat loss from the affected region than the surrounding zone. The horizontal temperature gradient resulting from this so-called Charney mechanism induces a frictionally controlled circulation pattern that imports heat and maintains a sinking motion. One external factor that is responsible for regional circulation is sea surface temperature, and this has the necessary high inertia to prolong a weather pattern (Namias, 1983).

The discussion on drought causes has so far focused on weather anomalies during the drought period and this is what is important for most types of drought. On the other hand, in many circumstances of hydrological importance it is the weather preceding the water resource drought that is most relevant. For example a region's summer water supply may depend on meltwater from a snow pack laid down the previous winter. Similarly water levels in aquifers and water supply reservoirs are replenished by winter rainfall. In these circumstances it is the weather of the preceding winter that is critical, though the greatest impact of drought will be felt when this is coupled with drought during the supply season that can deplete alternative sources and elevate demand.

3.5.4 Climatic Drought: A Shortage of Rainfall and Soil Water

Most GCM simulations of a future "$2 \times CO_2$" world predict an increase in summer dryness in midlatitude regions (IPCC, 1990; 1992). This represents a convergence from results reported as little as five years previous which revealed strong intermodel disagreements at the zonal and seasonal level. The basic mechanism is for higher temperatures to lead to higher evaporation, and hence a more rapid decline in soil moisture content from spring into summer. The reduced evaporation also means that there is less evaporative cooling, so warming the land surface still further. A second positive feedback may arise because of reduced low cloud cover due to reduced evaporation. Snow also may melt earlier, thus reducing albedo earlier in the year and leading to increased surface temperatures in spring (Manabe and Wetherald, 1987).

In some regions a reduction in summer rainfall reinforces summer dryness, but in others soil moisture deficits decrease despite an increase in rainfall (Rind et al., 1990). Most climate models simulate saturated soils in winter, so any extra winter precipitation runs off and does not delay the onset of drier conditions in spring; in some models (eg Mitchell and Warrilow, 1987) the extra winter rainfall does lead to wetter soils, but even here the effects of the increased evaporation are strong enough to result in lower soil moisture contents by summer.

Despite the apparent consistency between different GCMs, the reliability of these estimated changes is low because of the simplified representation of land-surface processes in climate models. Vinnikov and Yeserkopova (1991) showed,

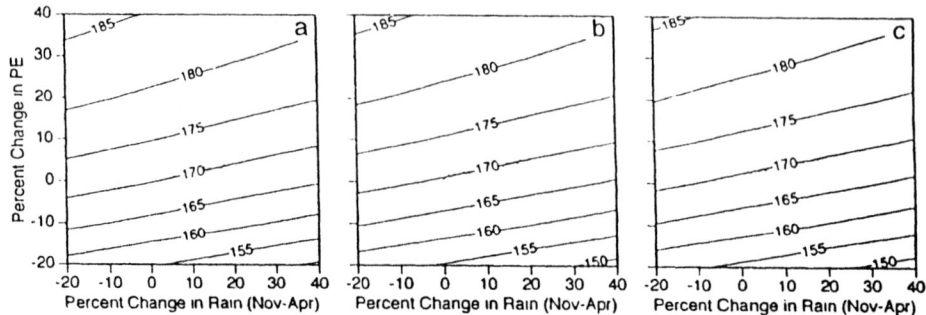

Fig. 3.2. Effect of changes in summer (November to April) rainfall and potential evaporation on mean summer soil water deficit at Perth, Australia (Whetton et al., 1993). Panels a, b and c represent −10, 0 and +10% changes in winter (May to October) rainfall. (Taken from Whetton et al. (1993)).

using observed soil moisture data from the USSR, that most climate models overestimated the seasonal amplitude in soil moisture: the increased intensity in soil moisture deficits may therefore be overstated.

Potential changes in point or catchment average soil moisture contents have been assessed in many studies, using soil water balance models run with various climate change scenarios. Whetton et al. (1993), for example, used a simple soil water balance model to investigate potential changes in soil moisture content at several locations in Australia. Figure 3.2 shows the mean summer soil water deficit at Perth to changes in rainfall and potential evaporation. Change in winter precipitation has little effect, and changes in potential evaporation have a greater impact than changes in summer rainfall. In the Sacramento-San Joaquin basin in California, Lettenmaier and Gan (1990) simulated an increase in winter and early spring soil moisture contents — due to increased rainfall and less snow — but large reductions in soil water in late spring and summer.

3.5.5 Hydrological Drought: A Shortage of Runoff and Recharge

Many studies have assessed the potential effects of global warming on river flows, and most have considered changes in monthly and seasonal runoff. The results depend both on the scenarios considered and the catchment environment, but in all cases except those with a large increase in summer rainfall, higher evaporation has led to lower flows during summer and hence a greater risk of drought. Few studies, however, have presented explicitly information about changes in measures of extreme low flow or drought.

Figure 3.3 shows the change in one low-flow measure — the flow exceeded 95 per cent of the time — in several UK catchments under a range of change scenarios (Arnell and Reynard, 1993). Each scenario assumes the same change in rainfall (an increase in winter rainfall and either little change or a small decrease in summer, depending on the catchment), but a different change in potential evaporation. Under all scenarios the low flow value is reduced, and hence the frequency of occurrence of the current value increased. In comparison, Bultot et al. (1992) simulated a

Climate Change and Hydrological Disasters

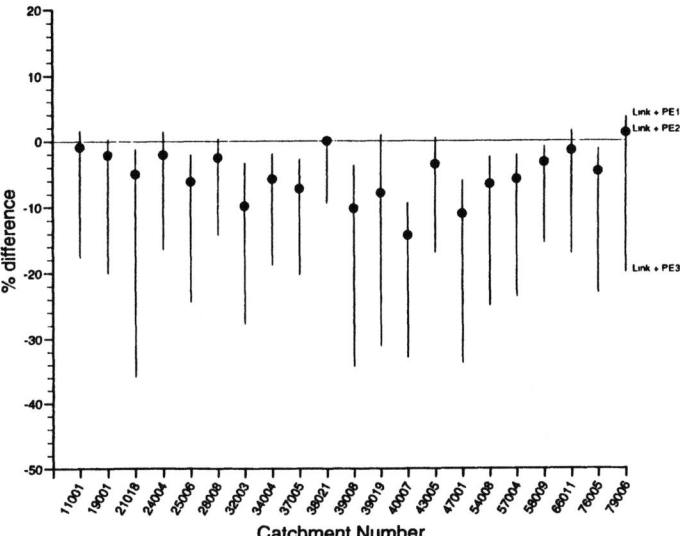

Fig. 3.3. Percentage change in the flow exceeded 95% of the time, under a range of change scenarios, in 21 UK catchments (Arnell and Reynard, 1993). The differences between the scenarios essentially reflect different assumed changes in potential evaporation.

decrease in the magnitude of the 95 percentile flow of around 10 per cent in their Swiss study catchment.

3.5.6 WATER RESOURCES DROUGHT

An even smaller number of studies have considered explicitly the impacts of global warming on water supply during droughts. The storage provided by reservoirs may mitigate to a certain extent any change in dry season water availability, depending on the amount of extra precipitation and runoff that may occur at other times of the year and the capacity of the reservoir system to carry over excess water. It is, however, quite possible that a reservoir may not be able to store any of the extra water, and therefore would have little mitigating impact. It is obviously very difficult to generalise.

However, some studies have indicated the potential magnitude of change in system reliability. Cole et al. (1990) concluded that the capacity of a hypothetical reservoir in southern England would need to be increased by between 10 and 21 per cent (depending on the magnitude of current yield) in order to maintain yields with the same degree of reliability: these percentage increases in storage were larger than the percentage reductions in annual runoff. Table 3.1 shows the percentage of time the storage reservoirs in the Delaware River Basin supplying New York

TABLE 3.1
Percentage of time in which the New York City reservoirs in the Delaware River Basin are in a "crisis" condition (Wolock et al, 1993)

Change in temperature	Change in precipitation		
	−20%	0	20%
0°	27	4	3
+2°C	36	7	3
+4°C	42	11	3

City would be in "crisis" conditions, assuming no change in operating rules, under a range of scenarios for change in precipitation and temperature (Wolock et al., 1993). The percentage change in risk is greater than the percentage change in runoff, and under extreme change scenarios the change is very large.

3.5.7 THE CASE OF SAHEL DROUGHT

The Sahel region is one of intense interest because of the scale of the human tragedy arising out of the recent and continuing drought event, and to the unique long-run characteristics of its climate. According to Lare and Nicholson (1994) rainfall has not exceeded the long-term average since 1969, and exceeded it in every year during the 1950s. There is good understanding of the meterorological and dynamic mechanisms that control the short (July to September) monsoon season. The source of moist air is from the southern tade winds which can penetrate as far north as the Inter-tropical Convergence Zone (ITCZ). The relative strengths and locations of the Saint Helena "high" and the Saharan trough govern the strength of the moist air transport. The strength of the Hadley cell governs the northward shift of the ITCZ.

However this understanding does little to elucidate the reasons for the long-run behaviour which would in turn provide a basis for forecasting whether the situation would be affected by global warming. Clues have been provided by some observed teleconnections with sea surface temperature (SST) and remote pressure centres which have led to seasonal rainfall forecast. However recent successful seasonal forecasts have been based on a GCM driven by *global* SST and the relative importance for interannual persistence of SST and positive feedback mechanisms through albedo changes and dust remains unknown.

The prospect of changes in ocean circulation is one of the more frightening aspects of human interventions in global systems such as climate change, and of course is not limited to the Sahel (Palmer and Brankovic, 1989).

3.5.8 DIRECT EFFECT OF CARBON DIOXIDE

A specific feature of the greenhouse-induced climate change is the simultaneous change in the carbon dioxide content of the atmosphere. This impacts directly on vegetation by increasing productivity and water use efficiency (WUE = biomass gain per unit water evaporated). These phenomena are clearly positive factors in terms of reducing drought frequency and increasing the resistance of vegetation to drought. A useful entry point to a large literature on this area from a hydrological standpoint is Rosenberg et al. (1990).

Increases of WUE in the range 25 to over 100 per cent have been recorded though in most cases from short-term experiments conducted under ideal circumstances for plant growth. Major uncertainties surround the magnitude of the effect in field conditions such that few impact studies currently take the WUE increase into account. Downward regulation through acclimation certainly occurs in productivity increase, though the WUE effect seems more robust. However the influences of limitations in other resources such as light, nutrient and space may nullify the effect. Phenological changes are also expected: an early season boost in water use or end-of-season waterlogging may have detrimental impacts on local water resources. Over the long term changes in soil structure may occur which may have a negative impact on water holding capacity. This possibility arises both as a direct effect of climate, and also through the higher ratio of carbon to nitrogen in litter derived from vegetation growing in a high CO_2 atmosphere and hence an increased fraction of poorly decomposable material.

3.6 Conclusions

Climate change is not of itself a disaster. It may, however, change the exposure to hydrological disaster. In the above analysis, which has focused on floods and drought, three main points have emerged:

- There is considerable uncertainty, due to difficulties in estimating changes in extremes and inadequate knowledge of driving processes and their likely change.

- There are differences between catchments and their response to changes in climate.

- Climatic change is superimposed on climatic variability: there is a considerable variation in risk from decade to decade.

Remembering these points, it is perhaps not possible to conclude definitively that global warming will increase exposure to drought and flood; in any event change in exposure will not be uniform and some areas will enjoy a reduction in exposure. Uncertainty carries its own penalties and steps to resolve these need to be strengthened. International programmes such as IGBP-BAHC[1], WMO-GEWEX[2] and the HDP[3] already address the scientific and socio-economic aspects and have

---

[1] International Geosphere-Biosphere Programme: Biospheric Aspects of the Hydrological Cycle
[2] World Meteorological Organisation Global Energy and Water Cycle Experiment
[3] International Social Science Council Human Dimensions of Global Environmental Change Programme

their national counterparts in many countries of the world which require further encouragement and participation.

What are the implications of global warming for the management of hydrological disasters? Global warming is not yet conclusively proven - although the evidence is increasingly convincing - and if it does occur, will show its impact over the next few decades. The impact *may* be felt within the design life of flood defence and drought mitigation systems. However, it is too early to build in specific adjustments to global warming now, partly because of uncertainty and partly because impacts are expected some years into the future. Nevertheless, it is appropriate to consider global warming when planning and designing disaster mitigation schemes:

- Assess the sensitivity of the proposed scheme to global warming. Would feasible changes threaten the integrity of the scheme? Could the proposed scheme be progressively adapted to cope with changed circumstances?

- Should robustness to an uncertain future influence the selection of a disaster mitigation scheme? "The best projects are those that make sense even when the future turns out worse than forecast" (Kelly and Kelly, 1986, p. 115).

- Follow the precautionary principle: do not necessarily respond to global warming now, but do not take any actions which will make response more difficult in the future.

- Build in flexibility into new schemes, where economically-justifiable, so that the scheme can be upgraded in response to new information.

References

Arnell, N.W. and Reynard, N.S. (1993) Impact of climate change on river flow regimes in the United Kingdom. Report to the Department of the Environment. Institute of Hydrology, Wallingford, UK 130pp.

Ausubel, J.H. (1991) Does climate still matter? Nature, 350, 649-652

Bardossy, A. and Plate, E.J. (1992) Space-time model for daily rainfall using atmospheric circulation patterns. Water Resources Research 28, 1247-1259

Beran, M.A. and Rodier, J. (1985) Hydrological aspects of drought. Unesco: Paris

Bultot, F., Gellens, D., Spreafico, M. and Schädler, B. (1992) Repercussions of a CO_2 doubling on the water balance - a case study in Switzerland. J. Hydrology 137, 199-208.

Carter, T.R., Parry, M.L., Nishioka, S., and Harasawa, H. (1992) Preliminary Guidelines for Assessing Impacts of Climate Change. IPCC/WMO/UNEP, 28 pp.

Cole, J.A., Slade, S., Jones, P.D. and Gregory, J.M. (1991) Reliable yield of reservoirs and possible effects of climate change. Hydrol. Sci. J. 36, 579-598.

Emanuel, K.A. (1987) The dependence of hurricane intensity on climate. Nature 326, 483-485.

Evans, J.L. (1993) Sensitivity of tropical cyclone intensity to sea surface temperatures. J. Climate 6, 1133-1140.

Gellens, D. (1991)Impact of a CO_2-induced climate change on river flow variability in three rivers in Belgium. Earth Surface Processes and Landforms 16, 619-625.

Giorgi, F. and Mearns, L.O. (1991)Approaches to the simulation of regional climate change: a review. Reviews of Geophysics 29, 191-216
Gleick, P.H. (1987) Regional hydrologic consequences of increases in atmospheric CO_2 and other trace gases. Climatic Change 10, 137-161.
Gutowski, W.J., McMahon, G.F., Schluchter, S. and Kirshen, P.H. (1994) Effects of global warming on hurricane-induced flooding. J. Water Resources Planning and Management 120, 176185.
Hay, L.E., McCabe, G.J., Wolock, D.M. and Ayers, M.A. (1992) The use of weather types to disaggregate general circulation model predictions. J. Geophys. Res. 97, 2781-2790.
IPCC (Intergovernmental Panel on Climate Change) (1990) Climate Change. The IPCC Scientific Assessment. Houghton, J.T., Jenkins, G.J. and Ephraums, J.J. (eds.) Cambridge University Press: Cambridge.
IPCC (Intergovernmental Panel on Climate Change) (1992) Climate Change 1992. The Supplementary Report to the IPCC Scientific Assessment. Houghton, J.T., Callander, B.A. and Varney, S.K. (eds.) Cambridge University Press: Cambridge.
Kelly, F.J. and Kelly, H.M. (1986) What they really teach you at the Harvard Business School. Piatkus: London
Knox, J.C. (1993) Large increases in flood magnitude in response to modest changes in climate. Nature 361, 430-432.
Krasovskaia, I. and Gottschalk, L. (1993) Frequency of extremes and its relation to climatic fluctuations. Nordic Hydrology 24, 1-12.
Lare, A.R. and Nicholson, S.E. (1994) Contrasting conditions of surface water balance in wet and dry years as a possible land surface-atmosphere feedback mechanism in the West African Sahel. J. Climate, 7, 653-668.
Lettenmaier, D.P. and Gan, T.Y. (1990) Hydrologic sensitivities of the Sacramento-San Joaquin river basin, California, to global warming. Water Resources Research 26, 69-86.
Manabe, S. and Wetherald, R. (1987) Large-scale changes of soil wetness induced by an increase in atmospheric carbon dioxide. J. Atmos. Sci. 44, 1211-1236.
Mearns, L.O., Schneider, S.H., Thompson, S.L. and McDaniel, L.R. (1990) Analysis of climate variability in general circulation models: comparison with observations and changes in variability in $2 \times CO_2$ experiments. J. Geophys. Res. 95. 20469-20490.
Meehl, G.A., Branstator, G.W. and Washington, W.M. (1993) Tropical Pacific interannual variability and CO_2 climate change. J. Climate 6, 42-63.
Mitchell, J.F.B. and Warrilow, D.A. (1987) Summer dryness in northern mid-latitudes due to increased CO_2. Nature 330, 238-240.
Namias, J. (1983) Some causes of United States drought. J. Climate and Applied Meteorology. 22, 30-39.
Palmer, T.N. and Brankovic, C. (1989) The 1988 US drought linked to anomalous sea surface temperature. Nature, 339, 54-57.
Palutikof, J.P. (1987) Some possible impacts of greenhouse gas induced climatic change on water resources in England and Wales. Impacts of climatic variability and climate change on hydrological regimes and water resources. Int. Ass. Hydrol. Sci. Publ. 168, 585-596.
Parry, M.L., Carter, T.R. and Konijn, N.T. (eds.) (1989) The Impact of Climatic Variations on Agriculture. Volume 1: Assessment in Cool Temperate and Cold Regions. Kluwer: Dordrecht.
Rind, D., Goldberg, R., Hansen, J., Rosensweig, C. and Ruedy, R. (1990) Potential evapotranspiration and the likelihood of future drought. J. Geophys. Res. 95, 9983-10004.

Rosenberg, N.J., Kimball, B.A., Martin, Ph., and Cooper, C.F. (1990) From climate and CO_2 enrichment to evapotranspiration. in P.E. Waggoner (ed) Climate change and US water resources. Wiley, New York. pp151-175

Smith, D.I. (1993) Greenhouse climatic change and flood damages: the implications. Climatic Change 25, 319-333.

Starosolsky, O and Melder, O.M. eds. (1989) Hydrology of disasters. Proceedings of WMO Technical Conference held in Geneva, November 1988. James and James: London. 319p.

Stephens, D.B. and Held, I.M. (1993) GCM response of northern winter stationary waves and storm tracks to increasing amounts of carbon dioxide. J. Climate 6, 1859-1870.

Stephens, G.L. (1990) On the relationship between water vapour over the oceans and sea surface temperature. J. Climate 3, 634-645.

van Katwijk, V.F., Rango, A. and Childress, A.E. (1993) Effect of simulated climate change on snowmelt runoff modeling in selected basins. Water Resources Bulletin 29, 755-766.

Taylor, K.C., Lamorey, G.W., Doyle, G.A. et al. (1993) The 'flickering switch' of late Pleistocene climate change. Nature 361, 432-434.

Viner, D. and Hulme, M. (1993) Construction of climate change scenarios by linking GCM and STUGE output. Technical Note 2, Climatic Research Unit, University of East Anglia, Norwich, UK.

Vinnikov, K.Y. and Yeserkepova, I.B. (1991) Soil moisture: empirical data and model results. J. Climate 4, 66-79.

von Storch, H., Zorita, E. and Cubasch, U. (1991) Downscaling of global climate change estimates to regional scales: an application to Iberian rainfall in wintertime. Max-Planck-Institut für Meteorologie. Report 64. 36pp.

Whetton, P.H., Fowler, A.M. Haylock, M.R. and Pittock, B. (1993) Implications of climate change due to the enhanced greenhouse effect on floods and droughts in Australia. Climatic Change 25, 289-317.

Wigley, T.M.L., Jones, P.D., Briffa, K.R. and Smith, G. (1990) Obtaining sub-grid scale information from coarse resolution GCM output. J. Geophys. Res. 95, 1943-1953.

Wijkman, A. and Timberlake, L. (1984) Natural disasters; acts of God or acts of man? Earthscan, London. 145p.

Wolock, D.M., McCabe, G.J., Tasker, G.D. and Moss, M.E. (1993) Effects of climate change on water resources in the Delaware River basin. Water Resources Bulletin 29, 475-486.

CHAPTER 4

Extreme Floods

Fahim Ashkar

ABSTRACT. Floods occur every year throughout the world causing loss of life and property damage, and necessitating large investments into flood engineering projects in public works, water resources, and land development. In the planning, design, and operation of flood related hydrotechnical works, a primary design variable known as the "design flood", needs to be estimated. Statistical methods have been used for estimating design flood values by two main types of models: (1) "annual maximum flood" (AMF) series models, and (2) "peaks over threshold" (POT) series models. Many "D/E procedures" involving the choice of a probability distribution, D, coupled with a method of parameter estimation, E, for analyzing AMF and POT series have been proposed. We briefly review and discuss some of these classical statistical procedures and identify some of the important difficulties that have been involved in choosing among them for the purpose of estimating design flood values. We also briefly discuss regional flood frequency estimation methods which have been used: (1) as a means of improving the at-site estimation of flood quantiles in the case where there is a lack of sufficient quantity of hydrometric data at a specific flood site for single-site flood frequency analysis (by jointly using atsite and regional data, i.e. by substituting spacial data in a region for inadequate temporal data at the site); and (2) as a way of obtaining flood estimates from regional information when there is no local hydrometric record at all (ungauged sites).

4.1 Introduction

Flood conditions usually occur as a result of extreme hydrological events such as high rainfall and/or snowmelt. Large investments are made into flood related engineering projects such as floodplain delineation, design of flood protection works, urban stormwater management, building of bridges and culverts, projects in channel improvement, and river diversions. In the planning, design, and operation of hydrotechnical works, a primary design variable, commonly known as the "design flood", needs to be estimated. This could be peak flow, maximum water level, flood volume, or the entire flood hydrograph. Although flood characteristics such as duration or volume are of practical interest in many applications, most research and published work on flood frequency estimation have focused on peak flows. The calculation of a design flood value is one of the key elements on which the overall design of the engineering project depends.

After having identified the required results of the engineering project to be undertaken and analyzed the general nature of the data base, the hydrologist might decide to use statistical methods for estimating a design flood. Statistical techniques are likely to be the hydrologist's first choice if the project location has a sufficiently long streamflow record, or if there are sufficient streamflow data in the locality or region.

In the present chapter, we shall focus on statistical methods for analyzing extreme floods for the purpose of estimating design flood values. These methods have been used extensively in many areas of hydrology and water resources for analyzing hydrological data sets. Much of the discussion will concentrate on statistical flood frequency analysis through the use of probability distributions ("parametric approach"), which has been the most widely employed approach for estimating the relationship between the magnitude and frequency of occurrence of flood events for use in design. However, many other important aspects of flood analysis and management will not be treated. Such aspects include flood forecasting and warning systems, operation of flood control systems, and the environmental implications of flood related projects.

4.2 Limitations of Statistical Methods for Estimating Design Floods

It should be stressed that a wide variety of methods other than statistical ones are usually available to the hydrologist for estimating design flood values. Reliance on any one single method, is not recommended. According to Yevjevich (1972, p. 3) future progress of hydrology may depend to a large extent on how hydrologists combine deterministic and statistical approaches for "discovering, understanding, and generalizing hydrologic regularities of nature".

In calculating an appropriate design flood value, the geographic and jurisdictional setting of the engineering structure, or flood related project, need to be considered, and any regulatory requirements or standards that might be relevant to the project should be consulted. The use of professional judgement, within limits, should also be applied, by investigating the regional technical and historical data base (including hydrological as well as meteorological data records), published regional analyses of flood data, and any other general guideline documents pertaining to the engineering project that might be available. Where results from applying different methods or judgemental factors differ substantially, it is unwise to discount such differences without further investigation.

4.3 Data Series Used in the Estimation of Design Floods

The most frequently used statistical frequency analysis procedure for estimating design floods involves: (1) selecting a sample of flood values that satisfies certain statistical criteria (to be outlined later); (2) fitting a probability distribution to this sample using the best fitting technique available; and (3) using this fitted distribution to make statistical inferences about the underlying flood "population" (a hypothetical, infinite data series), such as estimating an extreme flood event to be used as a design value, and assessing the error of estimation.

The typical sample considered in a statistical flood frequency analysis contains n elements corresponding to the maximum water level, river discharge, or any set of hydraulic conditions that a project facility is designed to withstand, with these maxima being taken over time intervals of equal length (e.g., $\{X_i; i = 1, \ldots, n\}$ = monthly, seasonal or annual maximum discharge values). These maxima define the random variable, X, to be analyzed statistically.

We shall focus considerable attention on a case very commonly employed in hydrological practice, where the sample $\{X_i\}$, of size n, is a series of *annual flood maxima*. These are observed over an n-year period of hydrometric record at a given gauging station. This is the *annual-floodseries* (AFS) or *annual-maximum-flood* (AMF) type of flood frequency analysis, where only one flood event per time interval (the year), is considered in the modelling.

Another, equally important, but less commonly employed approach for estimating design floods, uses an average of m *flood values* per year of streamflow record. These are often taken to be those nxm *peak* values of the hydrograph that exceed a certain "truncation", "threshold", or "base" level, Q_B. These peak values (or some other variable of interest related to them), are called flood *exceedances* (of the threshold). This is the *peaks over threshold* (POT), or *partial duration series* (PDS) approach to flood frequency analysis.

4.3.1 BRIEF DESCRIPTION OF THE POT PROCEDURE

The POT procedure can be useful for analyzing a hydrological process such as river flow or precipitation intensity that varies randomly as a function of time (t), as shown in Figure 4.1. In hydrology, the POT methodology was developed mainly by Bernier (1967), Shane and Lynn (1964) and Todorovic (1970), although a number of authors have subsequently improved upon it. If we consider only those values of the hydrological process $Q = Q(t)$ that exceed a threshold or base level Q_B (Figure 4.1), i.e.

$$\xi = \xi(t) = \begin{cases} 0; & Q \leq Q_B, \\ Q - Q_B; & Q > Q_B \end{cases}$$

then the separate peak magnitudes ξ_i; $i = 1, 2, \ldots$ of the process Q exceeding Q_B ("exceedances"), define what is known as a "marked point process", as depicted in Figure 4.2. Two important variables of interest in this process are (1) the times of occurrence $\tau(i)$ of the exceedances (these time arrivals or "point events" define a "point process", when considered separately in the time domain), and (2) the ecxeedance magnitudes ξ_i in the space domain (the ith exceedance magnitude ξ_i constitutes the "mark" of the ith point event occurring at time $\tau(i)$; therefore the couples $(\xi_i, \tau(i))$; $i = 1, 2, \ldots$ define a "marked" point process in the time-space domain). Within an arbitrarily fixed time interval such as the k-th season or the k-th year of observation (T_{k-1}, T_k), as in Figures 4.1 and 4.2, a random number (v) of flood events occur. The main objective of the POT model is to study simultaneously

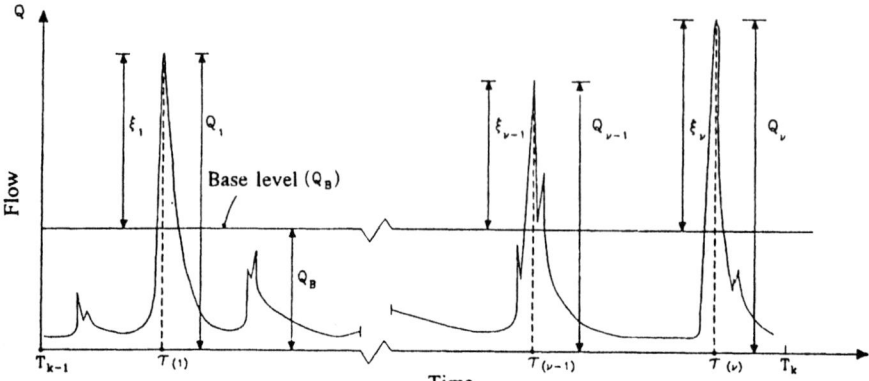

Fig. 4.1. Instantaneous hydrologic process identifying the base level and flood exceedances.

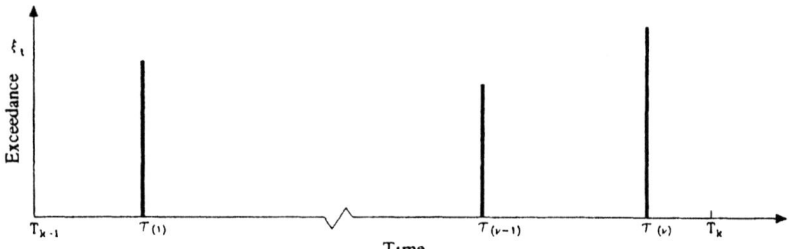

Fig. 4.2. Stochastic process representing exceedances in the time interval (T_{k-1}, T_k).

the two sequences ξ_i and $\tau(i)$, i.e. the points $(\xi_i, \tau(i))$ in two-dimensional space. Todorovic and Zelenhasic (1970), among others, studied the case where all the exceedance magnitudes are assumed to be independent and identically distributed (iid), irrespective of their times of occurrence (i.e., all flood magnitudes within the year, as well as from year to year, are assumed to be independent, and to belong to the same population). It is often assumed that flood events arrive according to a Poisson process, which implies that the number of flood events within an arbitrarily fixed time interval $(0, t)$ follows a Poisson distribution, but more general models have also been proposed and used (see, for example, Rasmussen et al., 1994 for more detail).

Todorovic and Rousselle (1971), among others, considered a more general POT model than the "iid" one mentioned above. In this more general "seasonal model", which has also been extensively used in hydrology, an arbitrary time interval $(0, t)$ is decomposed into k sub-intervals or "seasons" and all flood magnitudes are again considered to be independent (as in the iid case), but exceedance magnitudes are considered to follow the same probability density function (i.e. to belong to the same population) only if they belong to the same season.

4.3.2 Brief Comparison between AMF and POT Modelling

One readily notes that POT flood modelling is a compromise between annual-maximum-flood (AMF) analysis and a more detailed analysis of, for example, the complete daily-discharge series using classical time series procedures. The more detailed time series analysis of daily flow data focuses on modelling the nonstationarity and dependence structure of *all* the daily values, but POT modelling concentrates only on the high flows, which contain most of the information about the extreme-value process, and which are in certain respects easier to model than the complete daily series. Whereas AMF modelling requires that floods be modeled only in the magnitude domain (i.e., main interest is in the *magnitude* of the n observed AMF values), it was seen in the previous section that POT modelling requires that flood peaks be modeled both in the time and magnitude (space) domains. In comparison to AMF modelling, POT models consider a wider and more selective spectrum of events than only one event per year (noting that in certain years the annual maximum might not qualify as a flood at all). Probably the main strength of POT models is that by appropriate selection of the threshold, they allow for a better inclusion/exclusion of events in the analysis to be considered as "floods", as compared to AMF modelling.

Due to this added flexibility of POT models in comparison to their AMF counterparts (but which implies also an added analytical complexity), some authors have argued that POT modelling provides a better tool for describing seasonality in flood data (Todorovic and Rousselle, 1971; North, 1980; Waylen and Woo, 1982; Cruise and Arora, 1990; Rasmussen and Rosbjerg, 1991). For example, it has been argued that POT models are more adapted to combining rainfall and snowmelt floods [which often (but not always) occur in different parts of the year, and may have quite different characteristics], into one single statistical model. However, the added mathematical complexity of POT models has made them less popular among practitioners in comparison to AMF models. Some of the areas of research that need to be pursued further with regard to POT modelling are presented by Ashkar (1995).

4.4 Flood "Quantiles" as Design Events

In the planning, design and operation of water resource systems, it is useful to base decisions on some conveniently defined design quantity or quantities. Among the design quantities most commonly used are flood events X_T of "return period" T (or "recurrence interval" T). In AMF analysis, if $f(x)$ denotes the probability density function (p.d.f.), and $F(x)$ is the cumulative distribution function (c.d.f.) of the random variable X of interest (the maximum annual discharge, for example), then X_T and T (in years) are associated to each other according to the probabilistic relationship:

$$p = \text{Prob}(X \geq X_T) = 1 - F(X_T) = 1/T,$$

e.g. when $p = 0.01$, $F(X_T) = 0.99$, and $T = 100$ years, where p is the probability ("risk") that the T-year event X_T (also often denoted by X_p), is exceeded in any particular year. In other words, the T-year flood X_T is the $(1 - p)$th "quantile" in the distribution of annual flood maxima whose p.d.f. is $f(x)$. Because they are often specified as design events, certain flood quantiles such as X_T; $T = 2, 10, 100, \ldots$ years, are computed more frequently than others. It can be shown that the return period T, which is the reciprocal of the exceedance probability p, is the average number of years separating two annual floods equalling or exceeding X_T ($X \geq X_T$). This means that, on the average, in 1,000 years there will be ten 100-year floods but there is no implication that these will be equally spaced.

In the more general POT modelling context, it can be shown that if m is the annual average number of exceedances of the threshold, then for relatively large return periods (which are important in design), the T-year flood X_T is the $(1 - 1/mT)$th quantile in the distribution of exceedances. In other words, if $h(y)$ is the p.d.f. and $H(y)$ is the c.d.f. of the distribution of exceedances, Y, then the T-year flood X_T, will be associated to the return period T (in years), according to the probabilistic relationship (ACH, 1989):

$$p = \text{Prob}(Y \geq X_T) = 1 - H(X_T) = 1/mT \quad (\text{for large } T).$$

4.5 Statistical Hypotheses Used for Estimating Design Floods

In estimating design flood values from AMF or POT series, four statistical hypotheses are commonly used. These are briefly discussed below.

(1) *Randomness*. This condition means that the fluctuations in the streamflow process arise from natural causes. Flood flows appreciably altered by dam operation, for example, are not random. No general statistical tests exist to check for randomness in streamflow data sets.

(2) *Independence*. Prior to fitting a statistical distribution to an AMF or POT flood sample, it should be verified that the sample observations are independent from each other. This means that the occurrence and magnitude of one flood observation should not influence the occurrence or magnitude of any other flood observation. A possible non-parametric statistical test to check for independence in AMF or POT series is the Wald–Wolfowitz test (1943) (Bobée and Ashkar, 1991). In the POT approach, the serial correlation between flood exceedances increases with the average number of exceedances per year; therefore a sufficiently high threshold value, Q_B, has to be chosen if the assumption of independent exceedances is to be justified.

(3) *Homogeneity*. In fitting a statistical distribution to an AMF or POT flood sample, the sample's observations must all be drawn from a common statistical population (hypothetical infinite data series). A change in the location of a streamflow gauge, for example, may significantly influence streamflow measurements, so that data collected before and after the change might be coming from two significantly different populations. In this case, it is better to analyze the sub-samples separately, and then combine them in order to provide the needed overall design flood estimate.

The non-parametric Mann–Whitney (1947) test can be used to check for non-homogeneity in two flood sub-samples (Bobée and Ashkar, 1991). In the case where more than two seasons are believed to have significant influence on the flood process (case of several nonhomogeneous sub-populations), the Kruskal–Wallis test for homogeneity (Noether, 1976) could be used. It should be emphasized, however, that to determine the existence, type and amount of nonhomogeneity in AMF/POT flood series, reliance should not be placed entirely on statistical tests, but all the relevant physical factors that are judged to influence the flood process, should also be investigated.

In the case where two independent sub-populations of floods, a and b, have been identified, the return period of an annual flood of magnitude x can be calculated from the relation (ACH, 1989):

$$T = T_a T_b / (T_a + T_b - 1),$$

where T_a is the return period of a flood of magnitude x belonging to population a, and T_b is its return period if it belonged to population b.

(4) *Stationarity.* The condition of stationarity means that, excluding random fluctuations, the flood series is invariant with respect to time. The two most important forms of nonstationarity that might be present in the streamflow series are: (1) "jumps" and (2) "trends". Possible causes of such nonstationarities could be the construction of a dam or a levee, the occurrence of a forest fire in the basin, forest clearing, gradual extension of agricultural drainage, alterations in agricultural cropping and cultivation practices, or similar causes. Statistical tests for detecting nonstationarities include the Mann–Whitney (1947) test for jumps, and the Wald–Wolfowitz (1943) test for trend (Bobée and Ashkar, 1991). On the other hand, nonstationarity caused by longterm changes in hydrological or meteorological variables, such as global changes in weather or very slow evolution in nature, cannot be detected by statistical tests because of the general shortness of available records of these variables.

If nonstationarities are not taken into account in calculating design flood values, it means that an assumption is made that historic data are applicable to future considerations. However, as is cautioned by the Canadian flood Guide (ACH, 1989): "a tendency to overlook potential future changes is particularly strong when reliance is placed on a statistical approach to flood analysis".

When a flood sample satisfies all the foregoing four conditions of randomness, independence, homogeneity, and stationarity, it can be considered a random sample of independent identically distributed (iid) variates. However, certain floods are caused by special kinds of conditions that make them non-iid, so they cannot be statistically treated by the general procedures outlined in the present chapter. Examples of such phenomena are given by urban floods, and ice or debris jam floods.

4.6 Some Remarks Pertaining to Flood Data

Random and systematic errors are often present in river flow data, and these are usually greatest during high-flow periods. It is therefore important to use professional judgement to take account of these errors in estimating a design flood value.

Also, in certain cases data for certain years may be missing within the period of record. Attempts have been made by government agencies to classify these cases and to recommend procedures for their treatment (see U.S. Interagency Advisory Committee on Water Data (1982), for example).

Statistical analyses of river discharge by AMF or POT methods should ideally be carried on maximum instantaneous hydrometric data (peak instantaneous annual maxima, in the case of AMF models, for example), but in many cases the hydrometric station may not be of the sophisticated type to furnish such "instantaneous" measurements. In such cases, analyses have tended to concentrate on maximum daily discharges, but this implies a need to consider corrections to be made to the estimated design flood values (e.g., based on ratios of instantaneous to daily maxima, if available).

Values that are far from the bulk of the data are also quite frequently present in AMF or POT flood series. These "outlying" observations, or "outliers", as they have often been called, can cause significant difficulties in fitting a statistical distribution to the sample. In particular, high- or low-valued "outliers" in flood data can lead to design-flood estimates that are significantly under- or over-estimated.

Two fundamental questions have yet to be clearly resolved from both the hydrological and statistical viewpoints with regard to the problem of "handling outliers": (1) how is an outlier defined? and (2) what should be done when an outlier exists? In particular, it is not yet clear what weights should be given to any outlying or "exceptional" values in the data when estimating a design flood value. It is clear, however, that all the physical and statistical evidence should be gathered and evaluated before any final decision about what to do with "outliers" is taken. It is important to keep in mind that exceptional data values are in many instances observations that have actually occurred as a result of a rare (unique) combination of hydrological and/or meteorological events, although they could also be caused by errors in data reporting.

To be able to handle flood samples with exceptional values more adequately, it should be helpful to be able to "detect" these values (maybe only crudely) by a quantitative statistical test. The Grubbs–Beck (1972) test for outlier detection (Bobée and Ashkar, 1991) is one possible statistical test to use.

There are also situations where approximate values of some large floods predating the systematic flow record are known. These are commonly known as "historical floods". It is strongly recommended that such historical flood information, when available, be incorporated into the flood frequency analysis, and methods for doing so have been proposed by several authors (e.g., NERC, 1975; Tasker and Thomas, 1978; U.S. Interagency Advisory Committee on Water Data, 1982; Condie and Lee, 1982; Hosking and Wallis, 1986; Stedinger and Cohn, 1986; Hirsch and Stedinger, 1987).

4.7 Choice of "D/E Procedure" for Flood Frequency Analysis

After having: (1) verified that the flood sample satisfies the conditions of randomness, independence, homogeneity, and stationarity, and (2) decided upon how to handle any exceptional values that might be present in the data set, the usual next step is to fit the sample by an appropriate distribution (D), using a parameter-estimation method (E). This D/E-procedure is then used to make statistical inferences about the estimated design flood values.

4.7.1 CHOICE OF STATISTICAL DISTRIBUTION (D)

In choosing a statistical distribution (D) for fitting AMF or POT flood data sets, some of the points that the hydrologist needs to consider are:

(1) *Number of distribution parameters.* This factor is important because the "descriptive" and "predictive" abilities of a distribution (Cunnane, 1987) largely depend on the number of parameters entering into the distribution's p.d.f.. Three-parameter distributions are considered by many hydrologists to be sufficiently flexible to fit a wide variety of AMF observed data sets. The Pearson Type 3 (PIII), log-Pearson Type 3 (LPIII), three-parameter lognormal (LN3) and the general extreme value (GEV) distributions have attained some popularity in fitting AMF series. However, with the short AMF samples that are typically available, two-parameter distributions such as the extreme value type 1 (EV1) (also known as the Gumbel (GU) distribution), two-parameter lognormal (LN), two-parameter PIII (gamma), may in many cases be preferred for prediction, because their lower number of parameters tends to reduce the variance of flood-quantile estimators in the extrapolation range.

Table 4.1 presents the p.d.f. and some characteristics of some commonly used distributions in AMF flood frequency modelling. For further details about these and other statistical distributions commonly used for fitting AMF data sets, the following general references can be consulted: W.M.O. (1989), Kite (1988), Bobée and Ashkar (1991), Maidment (1993), Yevjevich (1972), Haan (1977), NERC (1975), ACH (1989), Sevruk and Geiger (1981), Sokolov et al. (1976), Ashkar et al. (1995).

(2) *"Desirable" properties of a flood distribution.* A general feature considered to be desirable when picking a POT exceedance distribution, but which is generally not considered as such when choosing an AMF distribution, is the monotonic decrease of the distribution p.d.f. (see Figures 4.3 and 4.4). Some common distributions that have this characteristic are the exponential and the Pareto (PA) (sometimes referred to as the "generalized Pareto"), but also the Weibull (WE) and gamma distributions can share this property for certain choices of their shape parameter. However, even in modelling POT exceedance distributions, it may be necessary to consider modal distributions, especially when the distribution is to be used for descriptive purposes.

Table 4.2 presents the p.d.f. and some characteristics of some commonly used distributions in POT flood frequency modelling. For more detail about these and other statistical distributions commonly used for fitting POT exceedance magnitudes, the following general references can be consulted: Miquel (1984), W.M.O. (1989), NERC (1975), ACH (1989), Maidment (1993), Sevruk and Geiger (1981), Rosbjerg (1993), Rasmussen et al. (1994).

(3) *Functional relationships between distribution candidates.* The hydrologist who needs to explore in detail what distributions are best fit for representing hy-

TABLE 4.1
The p.d.f. and some characteristics of distributions commonly used in flood frequency analysis in relation with the AMF model (note that since flood flows can take only positive values, only the case $\alpha > 0$ is physically plausible when fitting a GG3 or PIII distribution)

Distribution	Probability density function $f(x)$, or cumulative distribution function $F(x)$	Range of variate (x) and value of parameters
GG3(s, α, λ) Generalized Gamma 3-parameter	$f(x) = \dfrac{\|\alpha s\| e^{-(\alpha x)^S}(\alpha x)^{S\lambda-1}}{\Gamma(\lambda)}$	if $\alpha > 0, x \geq 0$ if $\alpha < 0, x \leq 0$ $\lambda > 0, s \neq 0$
PIII(α, λ, m) Pearson Type III	$f(x) = \dfrac{\|\alpha\| e^{-[\alpha(x-m)]}[\alpha(x-m)]^{\lambda-1}}{\Gamma(\lambda)}$	if $\alpha > 0, x \geq m$ if $\alpha < 0, x \leq m$ $\lambda > 0,$ $-\infty < m < +\infty$
LPIII(α, λ, m) Log-Pearson Type III	$f(x) = \dfrac{\|\alpha\| e^{-\alpha(\ln x - m)}[\alpha(\ln x - m)]^{\lambda-1}}{x\Gamma(\lambda)}$	if $\alpha > 0, x \geq e^m$ if $\alpha < 0, 0 \leq x \leq e^m$, $\lambda > 0$
GEV(α, λ, m) Generalized extreme value	$f(x) = \dfrac{1}{\alpha}[1 - \lambda(x - \dfrac{m}{\alpha})^{(1-\lambda)/\lambda}]$ $\times \exp\{-[1 - \lambda(\dfrac{x-m}{\alpha})]^{1/\lambda}\}$ $F(x) = \exp\{-[1 - \lambda(\dfrac{x-m}{\alpha})]^{1/\lambda}\}$	if $\lambda < 0, x \geq m + \alpha/\lambda$ if $\lambda > 0, x \leq m + \alpha/\lambda$ $\alpha > 0$
GU(ξ, s) Gumbel	$f(x) = s \exp\{-e^{-s(\xi+x)} - s(x + \xi)\}$ $F(x) = \exp\{-e^{-s(\xi+x)}\}$	$-\infty < x < +\infty$, $-\infty < \xi < +\infty$, $s > 0$
LGU(ξ, s) Log-Gumbel	$f(x) = s e^{-s\xi} x^{-(s+1)} \exp\{-x^{-s} e^{-s\xi}\}$ $F(x) = \exp\{-e^{-s(\ln x + \xi)}\}$	$x \geq 0$, $-\infty < \xi < +\infty$, $s > 0$
$N(\mu, \sigma)$ Normal	$f(x) = \dfrac{1}{\sigma\sqrt{2\pi}} \exp\{-\dfrac{(x-\mu)^2}{2\sigma^2}\}$	$-\infty < x < +\infty$ $-\infty < \mu < +\infty$ $\sigma > 0$
LN(μ, σ) Log-normal 2-parameter	$f(x) = \dfrac{1}{\sigma x \sqrt{2\pi}} \exp\{-\dfrac{(\ln x - \mu)^2}{2\sigma^2}\}$	$x > 0$, $-\infty < \mu < +\infty$, $\sigma > 0$

Extreme Floods

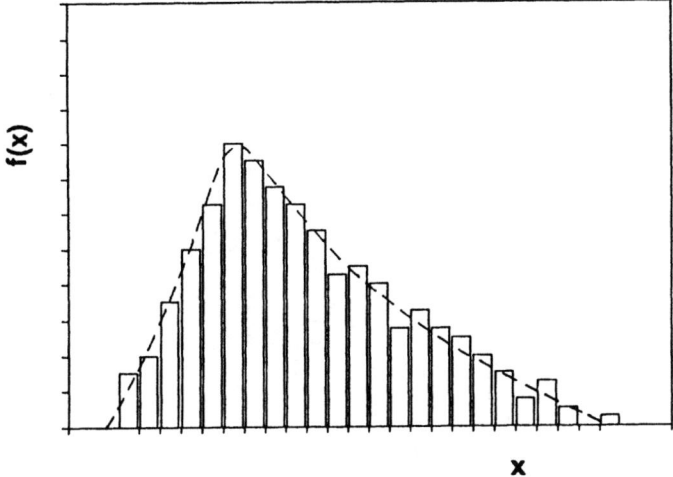

Fig. 4.3. Distribution pdf for AMF series.

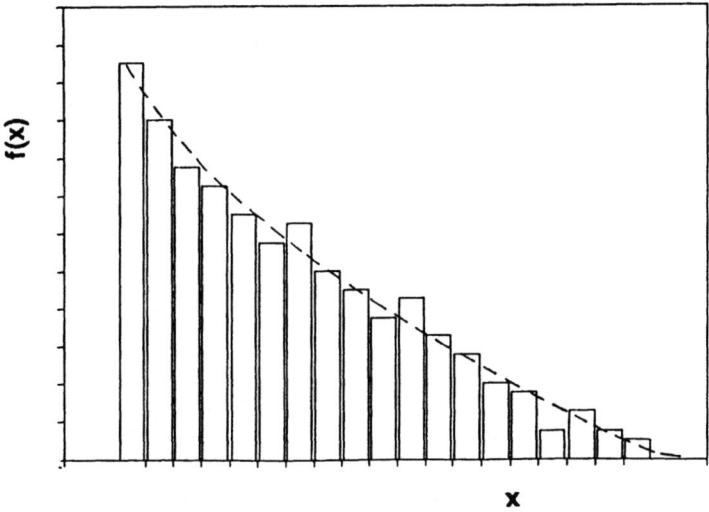

Fig. 4.4. Distribution pdf for POT exceedances.

drologic extremes such as flood data, needs to have a good knowledge of the mathematical relations that exist between distribution candidates. For example, it is important to know how a logarithmic transformation produces a "light-tailed" distribution from one that is "thick-tailed". Bobée and Ashkar (1991) and Ashkar et al. (1995) highlight some of the functional relationships that exist between an important group of statistical distributions commonly used in hydrological frequency analysis (one-, two-, three-, and fourparameter distributions are considered). These

TABLE 4.2
The p.d.f. and some characteristics of distributions commonly used in flood frequency analysis in relation with the POT model

Distribution	Probability density function $f(x)$, or cumulative distribution function $F(x)$	Range of variate (x) and value of parameters
PA(α, m) Pareto ("Type I")	Form A: $f(x) = |\alpha| e^{\alpha m} x^{-(\alpha+1)}$ if $\alpha > 0$, $F(x) = 1 - e^{\alpha m} x^{-\alpha}$ if $\alpha < 0$, $F(x) = e^{\alpha m} x^{-\alpha}$	if $\alpha > 0, x \geq e^m$ if $\alpha < 0, 0 < x \leq e^m$ $-\infty < m < +\infty$
	Form B: $f_x(x) = \beta^{-1}(1 - \frac{kx}{\beta})^{(1/k)-1}$ $F_x(x) = 1 - (1 - \frac{kx}{\beta})^{1/k}; k \neq 0$	if $k < 0, 0 \leq x < \infty$ if $k > 0, 0 \leq x \leq \beta/k$ $\beta > 0$
EX2(α, m) Exponential 2-parameter	$f(x) = |\alpha| e^{-\alpha(x-m)}$ if $\alpha > 0$: $F(x) = 1 - e^{-\alpha(x-m)}$ if $\alpha < 0$: $F(x) = e^{-\alpha(x-m)}$	if $\alpha > 0, x \geq m$ if $\alpha < 0, x \leq m$ $-\infty < m < +\infty$
WE(s, α) Weibull 2-parameter	$f(x) = \alpha|s| e^{-(\alpha x)^s} (\alpha x)^{s-1}$ if $s > 0$: $F(x) = 1 - e^{-(\alpha x)^s}$ if $s < 0$: $F(x) = e^{-(\alpha x)^s}$	$x \geq 0$, $s \neq 0, \alpha > 0$

two references also provide a helpful diagram that visually displays the functional relationships between these commonly used distributions.

(4) *Shape flexibility of distributions.* Moment ratio diagrams (MRD's) have been used to show the shape flexibility of various distribution types commonly employed in hydrological frequency analysis (GEV, PIII, Weibull, Pareto, ...). An MRD may generally be defined as a graph depicting the relationship between two dimensionless moments of a distribution. It is often helpful and convenient to present the graphs of more than one distribution on the same MRD. When the coefficient of kurtosis (C_k) is plotted against the coefficient of skewness (C_s), a $C_s - C_k$ MRD is obtained, whereas a plot of the coefficient of skewness versus the coefficient of variation (C_v), produces a $C_s - C_v$ MRD. Recently, "L-moment ratio diagrams" have been introduced, and have become quite widely used in hydrology (Hosking, 1990). For more detail about moment ratio diagrams see, e.g., W.M.O. (1989), or Bobée et al., (1993b).

Another question with important practical applications in flood frequency analysis is how to classify the tail behaviour of statistical distributions in order to better assess the adequacy of the various tail shapes for representing extreme empirical flood data. Ashkar et al. (1995) shed some light on this problem and classify, in a tabular form, some of the important distributions used in hydrology according to their right-tail characteristics (asymptotic properties).

(5) *Criteria for selecting between distribution candidates.* One area of continual discussion and research in hydrology is how to arrive at schemes and criteria to

compare distributions and distribution fitting methods for the purpose of estimating hydrological extremes such as floods, that are as broadly agreed upon among hydrologists, as possible. To select between distribution candidates for the estimation of design flood values, clear criteria are still needed, especially for choosing between distribution candidates that have different degrees of flexibility (different "descriptive" and "predictive" abilities).

Examples of criteria that have been used for comparing distributions and D/E procedures are statistical criteria such as minimum mean square error and minimum bias of quantile estimates; however, it would be beneficial to consider a multiple-criteria decision approach that allows for other criteria to be incorporated. Such criteria include computational ones (e.g., computer program availability, ease of use), or a user-related criterion such as acceptability or personal preference and past experience (Duckstein et al., 1991; Ashkar et al., 1995).

(6) *Guidelines by government agencies.* Some countries have introduced guidelines specifying particular types of statistical distributions to be used by their government agencies and private enterprises for floodflow frequency analysis. These guidelines have most frequently been concerned with the fitting of AMF rather than POT series. The LPIII distribution has been proposed in the USA (Benson, 1968) and Australia (I.E.A., 1977), and the three-parameter generalized gamma (GG3) distribution has been recommended in the republics of the exUSSR (Kritsky and Menkel, 1969) while in the UK, the use of the GEV distribution has been proposed (NERC, 1975). These guidelines are helpful in achieving a certain uniformity in the management of water resources, but need to be applied with special care.

4.7.2   CHOICE OF PARAMETER-ESTIMATION METHOD (E)

The maximum likelihood (ML) method has been favoured by many statisticians for fitting a wide variety of statistical distributions. However, this method has often failed to give the desired accuracy in estimating extremes from short hydrological data sets. It is known that for many statistical distributions, the ML method yields asymptotically optimal parameter estimators (unbiased, minimum variance), but in small samples it has not been found to particularly outperform other commonly used methods. Two important points need to be made about the ML method: (1) Its optimal properties are based on the assumption that the flood sample on which it is applied actually comes from the distribution that is assumed in the procedure (i.e., the "hypothesized" distribution, D), but if this hypothesized distribution is not the "true" one, the optimality of the ML method is not guaranteed; (2) If the domain of variation of the random variable X under consideration, with p.d.f. $f(x; a, b, \ldots)$, depends on one or more of the distribution's parameters, $a, b, \ldots$, then the optimality of the ML method is again not assured.

Therefore, hydrologists have sought other methods for fitting distributions to flood data, among which the method of moments (MM) has often been favoured for its simplicity. In the case of three-parameter distributions, the classical use of this method is based on the sample's mean, variance, and coefficient of skewness. However, recent tendencies have been to avoid the use of sample moments of order 3 or more in the estimation (particularly, the sample coefficient of skewness which is function of moments of orders 3 and 2), in favour of lower-order moments which are less variable. Among the lower-order moments that have been proposed are the first- and second-order moments in "log space" (moments of logarithms of the observed flood values), as in the "method of mixed moments" (Rao, 1980; Phien and Hira, 1983). On the other hand, the geometric and harmonic means in "real space" have also been proposed, as in the "generalized method of moments" (Ashkar et al., 1988). Greenwood et al. (1979) have suggested the use of "probability weighted moments" (PWM), which are linear combinations of order statistics (Landwehr et al., 1979; Hosking et al., 1985, Hosking and Wallis, 1987). The use of "L-moments", which are derived from PWM's, but which have a clearer statistical interpretation, has been proposed by Hosking (1990).

However, these are but few of the methods that have been suggested for fitting probability distributions to flood data. Among the other methods employed, are ones based on least squares, maximum entropy, "incomplete means", and others (see, for example, Jain and Singh, 1987; Arora and Singh, 1989; Phien, 1987; Raynal and Salas, 1986).

Two important points are in order about these different parameter estimation methods: (1) they can differ substantially in the weights they give to the different values in the sample, and this can lead to significant differences in flood quantile estimators in the upper tail of the distribution (this is especially true when exceptional values are present in the data); and (2) the optimal choice of distribution fitting method is closely associated with the type of distribution p.d.f. that is being considered (LN, gamma, Weibull, etc.).

4.7.3 SOURCES OF UNCERTAINTY IN ESTIMATING DESIGN FLOODS

The shortness of the commonly available AMF and POT flood records and the diversity in their characteristics can pose serious difficulties in selecting an "appropriate" D/E procedure for analyzing them. Typical record lengths of river discharge measurements around the world rarely exceed fifty years. Uncertainties in the estimation of design flood values come from many sources such as modelling errors, sampling errors, and other factors such as measurement, recording, and processing errors and inconsistencies. The combined uncertainty from all these sources can become especially important when flood quantiles beyond the range covered by observed flood values need to be estimated ("extrapolation" into the range of "rare" or "very rare" flood events). Experience has shown that when interest lies in the tail area of a distribution (extrapolation area), frequency estimates can differ substantially under alternate models that may fit the data well in the area of relatively ordinary events ("interpolation" area). In extrapolations, modelling error occupies the major part of the global error, but in interpolation, the global uncertainty is mainly dominated by sampling errors. Due to these compounded errors that enter into extrapolation from short hydrological data sets, Canadian practice in the area of flood frequency analysis, for example, limits specification of return periods that can be estimated from currently available streamflow records, to periods of 200 years or less, "except in special circumstances" (ACH, 1989, p. 41).

4.7.4 STATISTICAL TOOLS FOR SELECTING D/E PROCEDURES

As mentioned above, selecting between two D/E procedures for design flood estimation can pose special difficulties when the two candidate distributions differ in their number of parameters (differences in flexibility). Three complementary tools have been commonly employed in hydrology for D/E-procedure choice (Bobée and Ashkar, 1991): (1) visual techniques (e.g., plotting the flood sample values along with the fitted probability distribution on appropriate probability paper and assessing the goodness of fit), (2) use of moment-ratio diagrams (MRD's), which have been mentioned above, and (3) use of statistical tests of goodness of fit (e.g., Kolmogorov–Smirnov, Chi-Square, Anderson–Darling, Cramer–Von Mises or some other test). However, these are not powerful tools for the choice of a D/E procedure when small samples are involved. It has been argued that goodness-of-fit tests tend to put more emphasis on goodness of fit than on uncertainty due to any additional parameters that might have been entered into the model. In this way, they tend to be biased towards the most flexible distribution, or the one that has the largest number of parameters (Ashkar et al., 1994; Chow and Watt, 1992). It should be remembered that for a fixed number of observations, the more parameters in a model, the less efficient is the estimation of the parameters. The principle of *parsimony* in model building states that a model should be chosen that adequately fits the historical data without using any unnecessary parameters.

4.8 Regional Flood Frequency Estimation

The previous sections were concerned with single-station statistical flood frequency analysis procedures (i.e., "local" or "at-site procedures"). However, it is only when the at-site record is exceptionally long, and the return period of the flood to be estimated is not very large, that such procedures can be sufficient by themselves for the estimation of design flood values. In many situations, adequate station data for single-site analysis do not exist near the location of interest or within the drainage basin, making it essential to look for additional information allowing to improve the at-site estimation of flood quantiles. Regional flood frequency estimation procedures have been effectively used (1) to improve the estimation quality of flood quantiles at gauged sites with data that is inadequate for single-site analysis (this is done by combining the station estimates with estimates based on regional frequency analysis of data from other sites); and (2) to predict such quantiles for ungauged sites. By incorporating regional information, these procedures have been shown to be efficient in reducing model and sampling uncertainties inherent in at-site flood frequency estimations. It is noted, however, that POT models have until now been very rarely used in a regional context in hydrology, so there is a strong need for regional procedures based on these important types of models to be developed and applied.

Generally, the regional flood frequency estimation methods that are presently used are based on the idea that some sort of "hydrologically homogeneous" region exists, in which AMF populations at different sites within the region can be considered to be "similar" with respect to some specific statistical characteristics that are not dependent on catchment size. These procedures generally involve some kind of "pooling" of data from sites within the homogeneous region, in order to improve the estimation of flood quantiles at the specific site of interest. The various regional flood frequency estimation procedures differ in the way they transfer information to the specific site, from surrounding stations. In applying many of these methods, it has often been assumed that a regional parent distribution function is applicable, and a parameter estimation technique is selected. In these applications, one or two parameters of the flood population at the required site are often obtained from the at-site record, and the remainder of the required information is obtained by some regional averaging procedure (e.g., regional averaging of dimensionless at-site quantile estimates or dimensionless moments such as C_v and/or C_s).

The *index flood* technique (NERC, 1975; ACH, 1989; Maidment, 1993; Wallis and Wood, 1985; Landwehr et al., 1987; Potter and Lettenmaier, 1990, among others) is the regionalization technique with the longest history in hydrology. This technique assumes that the distributions of floods at all sites within a homogeneous region are the same, except for a scale or index-flood parameter. The mean μ_x of the at-site recorded flood values is the index-flood parameter most commonly used. To estimate the T-year flood x_T, the ratio x_T/μ_x is estimated from regional information, and multiplied by μ_x.

The *regression* regionalization procedure is another technique widely used. It relates hydrologic statistics (e.g., flood means, standard deviations, quantiles) to physiographic basin characteristics and meteorologic variables affecting the flood process. The regression technique is especially useful at ungauged sites. Ordinary least squares regression has been widely applied, but generalized least squares (GLS) regression has also been recommended. Examples of GLS applications in flood frequency analysis are given by Stedinger and Tasker (1985, 1986), and Tasker and Stedinger (1989).

The *record augmentation* procedure (Hirsch, 1979, 1982; Tasker, 1983; Vogel and Stedinger, 1985; Vogel and Kroll, 1991) is another technique that can be employed when floods at a short-record site are highly cross-correlated with floods at a nearby site with a longer record. The general idea behind this technique is to improve the estimates of at-site lower-order moments such as the mean and the variance by transferring information from the site with longer record.

Among the important problems in the area of regional flood frequency estimation on which much research has concentrated is how to identify groups of stations that can be considered sufficiently "homogeneous" for the efficient transfer of spacial data. A number of statistical "homogenization methods" have been proposed, such as principal component analysis, cluster analysis, canonical correlations analysis, and others (Acreman and Sinclair, 1984; Wiltshire, 1986a,b,c; Burn, 1990; Cavadias, 1990; Bobée et al., 1994). Two closely related problems to the development of homogenization methods are: (1) how to identify the characteristics of the stations that are most relevant for the purposes of at-site estimation, and (2) how to compare the degrees of homogeneity between two groups of stations that have been identified as homogeneous by two different homogenization methods.

To identify stations that have similar flood characteristics, it has been suggested to construct "measures of similarity" between stations based on statistical and physical flood-related "attributes" (Burn, 1990). Statistical attributes related to flood measurements, such as the coefficient of variation or skewness of the AMF values, have been found useful, as well as physical and geographical attributes of the basins such as rainfall pattern, drainage area, percentage of forests, percentage of lakes, basin slope, soil type, and others. Similarity in the values of statistical attributes is considered important because it indicates a similarity in the form of the parent distribution function at different stations in a homogeneous region. On the other hand, similarity in geographical attributes implies similarity in flood-generating mechanisms such as snowmelt or rainfall, and similarity in physical attributes is considered to indicate similarity in the basins' "responses" to such mechanisms. To avoid introducing biases due to scaling differences for different attributes, standardization is usually performed on these attributes. One way to do this is to divide the row data by the standard deviation of attribute values from stations in the proximity.

In summary, a regional flood frequency estimation procedure is composed of: (a) a "homogenization method" that is coupled with (b) a way of transferring information to a specific site from stations within the homogeneous region. Comparing two different procedures means deciding which of them uses the available data more efficiently.

4.9 Conclusions

The attention that has been given over the past 40 to 50 years to the problem of flood frequency analysis and the comparisons that have been made of the various AMF and POT methodologies have resulted in some general recommendations concerning "appropriate" D/E procedures to be used. However, these recommendations still contain points that need further statistical and/or physical justification, or that require further discussion. Bobée et al., (1993a), and Ashkar et al., (1992, 1994) suggest that there is a certain degree of confusion among practitioners as to which flood frequency estimation procedures are best to use. This confusion is partially caused by some misleading terminology that has sometimes been employed in the literature. According to Bobée et al., (1993a), a comprehensive and systematic approach to the comparison of flood-frequency-analysis procedures is needed.

It was stressed in the present chapter that treatment of hydrologic data may be done by a deterministic approach, by a statistical approach, or by both, and according to Yevjevich (1972, p. 3) "the use of one approach independently of the other may lead to an analysis of data without a sound theoretical background or to a theoretical treatment without the proper verification of hypotheses and of discovered regularities". The problem of frequency analysis of hydrological extremes, and particularly that of flood frequency analysis for the purpose of design flood estimation, is a challenging area in hydrology in which all possible physical and statistical evidence should be gathered and analyzed if the data are to be used in the most efficient manner.

It is strongly recommended that all additional information capable of improving the estimation of design flood values from a single at-site systematic hydrometric record be incorporated into the flood frequency estimation process. In particular, historical flood information, when available at or near the site, should definitely be considered for inclusion into the estimation. Much research remains to be done on regional flood frequency estimation methods which allow for the estimation of flood risk at a location with limited data, by using information from basins with similar characteristics or "responses". Research should be pursued for improved methods on how to include information from surrounding stations in the at-site estimation and on techniques for forming regions using similarities in statistical, physical, and geographic flood-related measures from different sites or basins.

Acknowledgements

The author would like to thank Dr. Taha B.M.J. Ouarda for his kind help and suggestions concerning the preparation of this chapter. Financial support provided by the Natural Science and Engineering Research Council of Canada (NSERC) is gratefully acknowledged.

References

Acreman, M.C. and Sinclair, C.D. (1984). 'Classification of drainage basins according to their physical characteristics; an application for flood frequency analysis in Scotland'. J. Hydrol., **84**: 365-380.

Arora, K. and V.P. Singh (1989). 'A comparative evaluation of the estimators of the log Pearson type 3 distribution'. J. Hydrol., **105**, 19-37.

Ashkar, F. (1995). 'On the statistical frequency analysis of hydrological extremes'. In Singh/Kumar (eds.), Proceedings of the International Conference on Hydrology and Water Resources, New Delhi, India, Dec. 20-22, 1993, Kluwer Academic Publishers, Vol. 1.

Ashkar, F., Bobée, B., Leroux, D. and Morisette, D. (1988). 'The generalized method of moments applied to the generalized gamma distribution'. Stochastic Hydrol. Hydraul., **2**, 161-174.

Ashkar, F., Bobée, B., and Bernier, J. (1992). 'Separation of skewness: reality or regional artifact?'. J. Hydraul. Eng., ASCE, **118**(3), 460-475.

Ashkar, F., Bobée, B. and T.B.M.J. Ouarda (1995). 'Functional relationships and asymptotic properties of distributions of interest in hydrologic frequency analysis'. In Singh/Kumar (eds.), Proceedings of the International Conference on Hydrology and Water Resources, New Delhi, India, Dec. 20-22, 1993, Kluwer Academic Publishers, Vol. 1.

Ashkar, F., Bobée, B., Rasmussen, P. and D. Rosbjerg (1994). 'A perspective on the annual maximum flood approach to flood frequency analysis'. In K.W. Hipel (ed.), Stochastic and Statistical Methods in Hydrology and Environmental Engineering, Kluwer Academic Publishers, 3-14.

ACH (Associate Committee on Hydrology) (1989). Hydrology of floods in Canada: A guide to planning and design, Chapter 5: Statistical Frequency Analysis of Hydrologic Data (Ottawa: National Research Council Canada).

Benson, M.A. (1968). 'Uniform Flood-Frequency Estimating Methods for Federal Agencies'. Water Resour. Res., **4**(5), 891-908.

Bernier, J. (1967). 'Sur la théorie du renouvellement et son application en hydrologie'. Electricité de France, Hyd. **67**, 10.

Bobée, B. and Ashkar, F. (1991). The Gamma Family and Derived Distributions Applied in Hydrology. Water Resources Publications, Littleton, Colo., 217 p.

Bobée, B., Cavadias, G., Ashkar, F., Bernier, J., and Rasmussen, P. (1993a). 'Towards a systematic approach to the comparison of distributions used in flood frequency analysis'. J. Hydrol., 142, 121-136.

Bobée, B., Ashkar, F. and L. Perreault (1993b). 'Two kinds of moment ratio diagrams and their applications in hydrology'. Stochastic Hydrol. Hydraul., 7, 41-65.

Bobée, B., Adamowski, K., Ashkar, F., Nguyen, V.T.V. Rousselle, J. and P. Rasmussen (1994). 'Comparison of regional flood frequency procedures for canadian rivers: an overview'. In K.W. Hipel (ed.), Stochastic and Statistical Methods in Hydrology and Environmental Engineering, Kluwer Academic Publishers, 209-215.

Burn, D.H. (1990). 'Evaluation of regional flood frequency analysis with a region of influence approach'. Water Resour. Res., 26(10), 2257-2265.

Cavadias, G. S. (1990). 'The canonical correlation approach to regional flood estimation'. Proceedings of the Ljubljana symposium. Regionalization in Hydrology. April. IAHS Publ. No. 191, 171-178.

Chow, K.C.A. and Watt, W.E. (1992). 'Use of Akaike information criterion for selection of flood frequency distribution'. Canadian Journal of Civil Engineering, 19, 616-626.

Condie, R. and Lee, K. (1982). 'Flood frequency analysis with historic information'. J. Hydrol., 58(1/2), 47-61.

Cruise, J.F. and Arora, K. (1990). 'A hydroclimatic application strategy for the Poisson partial duration model'. Water Resour. Bull., 26(3), 431-442.

Cunnane, C. (1987). 'Review of Statistical Models for Flood Frequency Estimation'. In V.P. Singh (ed.), Hydrologic Frequency Modelling (Dordrecht, Holland: Reidel), 49-95.

Duckstein, L., Bobée, B. and Ashkar, F. (1991). 'A multiple criteria decision modelling approach to selection of estimation techniques for fitting extreme floods'. Stochastic Hydrol. Hydraul., 5(3), 227-238.

Greenwood, J.A., Landwehr, J.M., Matalas, N.C. and Wallis, J.R. (1979). 'Probability weighted moments: Definition and relation to parameters of several distributions expressable in inverse form'. Water Resour. Res., 15(5), 1049-1054.

Grubbs, F. E. and Beck, G. (1972). 'Extension of sample sizes and percentage points for significance tests of outlying observations'. Technometrics, 14, 847-854.

Haan, C.T. (1977). Statistical Methods in Hydrology. Iowa State University Press, Ames, Iowa.

Hirsch, R.M. (1979). 'An evaluation of some record reconstruction techniques'. Water Resour. Res., 15(6), 1781-1790.

Hirsch, R.M. (1982). 'A comparison of four record extension techniques'. Water Resour. Res., 18(4), 1081-1088.

Hirsch, R.M. and J.R. Stedinger (1987). 'Plotting positions for historical floods and their precision'. Water Resour. Res., 23(4), 715-727.

Hosking, J.R.M. (1990). 'L-moments: Analysis and estimation of distributions using linear combination of order statistics'. J. Roy. Stat. Soc., Series B, 52(1), 105-124.

Hosking, J.R.M. and J.R. Wallis (1986). 'The value of historical data in flood frequency analysis'. Water Resour. Res., 22(11), 1606-1612.

Hosking, J.R.M. and Wallis, J.R. (1987). 'Parameter and quantile estimation for the generalized Pareto distribution'. Technometrics, 29(3), 339-349.

Hosking, J.R.M., Wallis, J.R. and Wood, E.F. (1985). 'Estimation of the generalized extreme value distribution by the method of probability weighted moments'. Technometrics, 27(3), 251-261.

Institution of Engineers (1977). Australian Rainfall and Runoff: flood analysis and design, I.E.A., 149 pp.

Jain, D. and V.P. Singh (1987). 'Estimating parameters of EV1 distribution for flood frequency analysis'. Water Resour. Bull., 23(1), 59-145.
Kite, G. W. (1988). Frequency and Risk Analysis in Hydrology. Water Resources Publications, Littleton, Colorado.
Kritsky, S.N. and M.F. Menkel (1969). 'On principles of estimation methods of maximum discharge'. Floods and their computation. I.A.H.S., No 84.
Landwehr, J.M., Matalas, N.C. and Wallis, J.R. (1979). 'Probability weighted moments compared with some traditional techniques in estimating Gumbel parameters and quantiles'. Water Resour. Res., 15(5), 1055-1064.
Landwehr, J.M., Tasker, G.D., and Jarrett, R.D. (1987). Discussion of 'Relative accuracy of log Pearson 3 procedures', by J.R. Wallis and E.F. Wood. J. Hydr. Eng., ASCE, 111(7), 1206-1210.
Maidment, D.R., editor in chief (1993). Handbook of Hydrology. McGrawHill, New York.
Mann, H.B. and Whitney, D.R. (1947). 'On the test of whether one of two random variables is stochastically larger than the other'. Ann. Math. Statist., 18, 50-60.
Miquel, J. (1984). Guide Pratique d'Estimation des Probabilités de Crues. Editions Eyrolles, Paris, 160 pp.
NERC (Natural Environment Research Council, LONDON) (1975). *Flood Studies Report, Vols. 1-5*, 1100 p.
Noether, G.E. (1976). Introduction to Statistics: A Nonparametric Approach. 2nd edition. Houghton Mifflin, Boston.
North, M. (1980). 'Time-dependent stochastic model of floods'. J. Hyd. Div., ASCE, 106, 649-665.
Phien, H.N. (1987). 'A review of methods of parameters estimation for the extreme value type 1 distribution'. J. Hydrol., 90, 251-268.
Phien, H.N. and Hira, M.A. (1983). 'Log-Pearson type 3 distribution: parameter estimation'. J. Hydrol., 64, 25-37.
Potter, K.W. and D.P. Lettenmaier (1990). 'A comparison of regional flood frequency estimation methods using a resampling method'. *Water Resear. Res.*, 26(3), 415-424.
Rao, D.V. (1980). 'Log Pearson type 3 distribution: method of mixed moments'. J. Hydraul. Div., ASCE, 106(6), 999-1019.
Rasmussen, P.F. and Rosbjerg, D. (1991). 'Prediction uncertainty in seasonal partial duration series'. Water Resour. Res., 27, 2875-2883.
Rasmussen, P., Ashkar, F., Rosjberg, D. and B. Bobée (1994). 'The POT method for flood estimation: A review' In K.W. Hipel (ed.), Stochastic and Statistical Methods in Hydrology and Environmental Engineering, Kluwer Academic Publishers, 15-26.
Raynal, J.A. and Salas, J.D. (1986). 'Estimation procedures for the type 1 extreme value distribution'. J. Hydrol., 87, 315-336.
Rosbjerg, D. (1993). Partial duration series in water resources. Institute of Hydrodynamics and Hydraulic Engineering, Technical University of Denmark. Doctor Technices thesis.
Sevruk, B. and Geiger, H. (1981). 'Selection of distribution types for extremes in precipitation'. W.M.O., Operational Hydrology Report No. 15, Geneva, Switzerland, 65 p.
Shane, R.M. and Lynn, W.R. (1964). 'Mathematical model for flood risk evaluation', J. Hydraul. Div., ASCE, 90, 1-20.
Sokolov, A.A., Rantz, S.E. and Roche, M. (1976). 'Floodflow computation: methods compiled from world experience'. Studies and Reports in Hydrology, No. 22, The Unesco Press, Paris, France.
Stedinger, J. and T.A. Cohn (1986). 'flood frequency analysis with historical and paleoflood information'. water resour. res., 22(5), 785-793.

Stedinger, J.R. and G.D. Tasker (1985). 'Regional hydrologic regression. 1. Ordinary, weighted, and generalized least squares compared'. Water Resour. Res., 21(9), 1421-1432.

Stedinger, J.R. and Tasker, G.D. (1986). 'Correction to "Regional hydrologic regression", 1. Ordinary, weighted, and generalized least squares compared'. Water Resour. Res., 22(5), 844.

Stedinger, J.R. and Tasker, G.D. (1986). 'Regional hydrologic analysis, 2. Model error estimates, estimation of sigma, and logPearson 3 distributions'. Water Resour. Res., 22(10), 1487-1499.

Tasker, G.D. (1983). 'Effective record length for the T-year event'. J. Hydrol., 64, 39-47.

Tasker, G.D., and Stedinger, J.R. (1989). 'An operational GLS model for hydrologic regression'. J. Hydrol., 111, 361-375.

Tasker, G.D. and Thomas W.O. (1978). 'Flood-frequency analysis with prerecord information'. J. Hydraul. Div., ASCE, 104(HY2), 249-259.

Todorovic, P. (1970). 'On some problems involving random number of random variables'. Annals of Mathematical Statistics, 41(3), 1059-1063.

Todorovic, P. and Rousselle, J. (1971). 'Some problems of flood analysis'. Water Resour. Res., 7, 1144-1150.

Todorovic, P. and Zelenhasic, E. (1970). 'A stochastic model for flood analysis'. Water Resour. Res., 6, 1641-1648.

U.S. Interagency advisory committee on water data, (1982). *Guidelines for determining flood flow frequency, Bull. 17B*, U.S. Department of the Interior, U.S. Geological Survey, Office of Water Data Coordination, Reston, Va.

U.S. Water resources council (USWRC) (1967). A Uniform Technique for determining flood flow frequencies, Bull. No. 15 (Washington, DC).

Vogel, R.M., and Kroll, C.N. (1991). 'The value of streamflow augmentation procedures in low flow and flood-flow frequency analysis'. J. Hydrol., 125, 259-276.

Vogel, R.M., and Stedinger, J.R. (1986). 'Minimum variance streamflow record augmentation procedures'. Water Resour. Res., 21(5), 715-723.

W.M.O. (World meteorological organization) (1989). Statistical distributions for flood frequency analysis. Operational hydrology, Report N° 33. WMO Publ. N° 718, WMO, Geneva, Switzerland.

Wald, A. and Wolfowitz, J. (1943). 'An exact test for randomness in the non-parametric case based on serial correlation'. Ann. Math. Stat., 14, 378-388.

Wallis, J.R. and Wood, E.F. (1985). 'Relative accuracy of Log Pearson 3 procedures'. J. Hydr. Eng., ASCE, 111(7), 1043-1057 [with discussion and closure, J. Hydr. Eng., 113(7), 1205-1214, 1987].

Waylen, P. and Woo, M.K. (1982). 'Prediction of annual floods generated by mixed processes'. Water Resour. Res., 18, 1283-1286.

Wiltshire, S.E. (1986a). 'Identification of homogeneous regions for flood frequency analysis'. J. Hydrol., 84: 287-302.

Wiltshire, S.E. (1986b). 'Regional flood frequency analysis 1: homogeneity statistics'. Hydrological Sciences Journal, 31(3): 321-333.

Wiltshire, S.E. (1986c). 'Regional flood frequency analysis 2: multivariate classification of drainage basins in Britain'. *Hydrological Sciences Journal*, 31(3): 335-346.

Yevjevich, V. (1972). Probability and statistics in hydrology. Water Resources Publications. Fort Collins, Colorado, 302 p.

CHAPTER 5

Dam-Breach Floods

D.L. Fread

5.1 Introduction

Dams provide society with essential benefits such as water supply, flood control, recreation, hydropower, and irrigation. However, catastrophic flooding occurs when a dam fails and the impounded water escapes through the breach to cause death and destruction of people and their developments existing in the downstream valley. Usually, the magnitude of the flow greatly exceeds all previous floods and the response time available for warning the populace is much shorter than for precipitation-runoff floods. According to reports by the International Commission on Large Dams (ICOLD, 1973) and the United States Committee on Large Dams in cooperation with the American Society of Civil Engineers (ASCE/USCOLD, 1975), about 38 percent of all dam failures are caused by overtopping of the dam due to inadequate spillway capacity and by spillways being washed out during large inflows to the reservoir from heavy precipitation runoff. About 33 percent of dam failures are caused by seepage or piping through the dam or along internal conduits, while about 23 percent of the failures are associated with foundation problems, and the remaining failures are due to slope embankment slides, damage or liquefaction of earthen dams from earthquakes, and overtopping of the dam by landslide-generated waves within the reservoir. Middlebrooks (1952) describes earthen dam failures that occurred within the United States prior to 1951. Johnson and Illes (1976) summarize 300 dam failures throughout the world.

The potential for catastrophic flooding due to a dam failure (breach) was brought to the attention of politicians, emergency action personnel, engineers, and portions of the general populace within the United States during the 1970's by several catastrophic floods due to dam failures, i.e., the Buffalo Creek coal-waste dam in 1972, the Teton Dam in 1976, the Laurel Run Dam in 1977, and the Kelly Barnes Dam in 1977.

The Buffalo Creek coal-waste dam collapsed (Davies et al., 1975) on the Middle Fork, a tributary of Buffalo Creek in southwestern West Virginia near Saunders. Most of the dam was eroded away very rapidly on February 26, 1972, due to overtopping waters; the breached dam released about 500 acre-ft of impounded waters into Buffalo Creek valley, causing the most catastrophic flood in the state's history with the loss of 118 lives, 500 homes, and property damage exceeding $50 million.

The Teton Dam near Sugar City, Idaho, a 300 ft high earthen dam with 250,000 acre-ft of stored water, failed (Ray et al., 1976) on June 5, 1976, due to internal piping, killing 11 people, making 25,000 homeless, and inflicting about $400 million in damages to the downstream Teton-Snake River Valley.

The 45-ft high earthen embankment Laurel Run Dam near Johnstown, Pennsylvania, was overtopped and breached (Chen and Armbruster, 1980) July 20, 1977, releasing 450 acre-ft of stored water. This resulted in the death of 40 people and heavy property damages.

The Kelly Barnes Dam near Toccoa, Georgia, an earthen embankment dam reconstructed several times, finally reaching a height of about 35 ft, with 600 acre-ft of storage, was overtopped and subsequently breached (Federal Investigative Board, 1977) November 6, 1977. This resulted in the death of 39 people who resided about 0.75 mile downstream of the dam.

Within the United States, as well as in many nations throughout the world, there are many dams that are 30 or more years old, and many of the older dams are a matter of serious concern because of increased hazard potential due to downstream development and increased risk of failure due to structural deterioration and/or inadequate spillway capacity. A report by the U.S. Army (1981) gives an inventory of the approximately 70,000 dams within the United States with heights greater than 25 ft or storage volumes in excess of 50 acre-ft. The report also classifies some 20,000 of these as being "so located that failure of the dam could result in loss of human life and appreciable property damage ..."

In addition to the man-made dams described above, natural formed dams can also produce dam-breach floods. Occasionally dams are formed naturally when a landslide blocks a river that traverses through rugged terrain. Eventually the landslide-formed dam is overtopped by the blocked and ponded river flow, and a breach is eroded through the naturally formed dam creating a dam-breach flood.

A distinguishing feature of dam-breach or dam-break floods is the great magnitude of the peak discharge when compared to any precipitation runoff-generated floods that could occur in the same valley. The dam-break flood is usually many times greater (an order of magnitude or more) than the runoff flood of record. Another distinguishing characteristic of dam-break floods is the extremely short time from beginning of rise until the occurrence of the peak and very short total duration time of the flood. The time to peak, in almost all instances, is synonymous with the interval of time required for the breach (failure) to develop completely once it starts to form. This time of failure is of the order of minutes for most dams, although some very large dams may have a time of failure of an hour or greater. This characteristic, along with the great magnitude of the peak discharge, causes the dam-breach flood wave to have acceleration components of a far greater significance that those associated with a precipitation runoff-generated flood and helps to produce significant wave peak attenuation.

5.2 Breach Description

The breach is the opening formed in the dam as it fails. The actual failure mechanics are understood only partially for earthen dams and less for concrete dams. Prior to about 1970, efforts to predict downstream flooding due to dam failures usually assumed that the dam failed completely and instantaneously, e.g., Ritter (1892), Schocklitsch (1917), Ré (1946), Dressler (1954), Stoker (1957), Su and Barnes (1969), and Sakkas and Strelkoff (1973). Others, such as the Army Corps of Engineers (1960) recognized the need to assume a partial rather than complete breach; however, it was still assumed the breach occurred instantaneously. The assumptions of instantaneous and complete breaches were used for reasons of convenience when applying certain mathematical techniques for analyzing dam-breach flood waves. The assumptions are somewhat appropriate for concrete arch dams, but are not appropriate for earthen dams and concrete gravity dams. For these dams, as well as well concrete arch dams, the breach should be considered (1) to develop over a finite interval of time (τ) and (2) to encompass only a portion of the dam except for concrete arch dams (Fread and Harbaugh, 1973; Fread, 1977).

Partial dam breaches with $\tau > 0$ result in considerably smaller dam-breach floods than instantaneous ($\tau = 0$) and complete breaches. It is readily apparent that a smaller breach will allow less peak outflow than a larger breach; however, it is not quite as apparent that a larger failure time results in less peak outflow. As the dam breach forms, the outflow through the breach reduces the reservoir storage contained by the dam, resulting in a reduction of the reservoir water level. The rate of flow through the breach is proportional to the height (head) of the water above the breach bottom (as in weir-type flow). Therefore, as the breach forms, the water level reduces; and when the breach is fully formed, the resulting head of water is less than that if the breach formed instantaneously or even at a faster rate. The smaller head of water available to produce flow through the breach when it completely forms (both in the vertical and horizontal directions) results in a smaller peak outflow and a smaller dam-breach flood. The extent of flood peak reduction due to a larger failure time is directly proportional to the magnitude of the final breach width and inversely proportional to the magnitude of the reservoir storage volume.

5.2.1 Mathematical Description of Breach

The breach may be described mathematically using the following parameters: the time of failure (τ), the terminal bottom width parameter (b), another parameter (z) which provides for breach shapes of rectangular, triangular, or trapezoidal. The parametric approach is convenient in predicting dam-breach floods for reasons of simplicity, generality, wide applicability, and the uncertainty in the actual failure mechanism. The parametric approach to the breach description follows that used by Fread and Harbaugh (1973) and later by Fread (1977, 1985, 1988).

The shape parameter (z) is the side slope of the breach, i.e., 1 vertical:z horizontal. The value for z varies from 0 to about unity. Its value depends on the

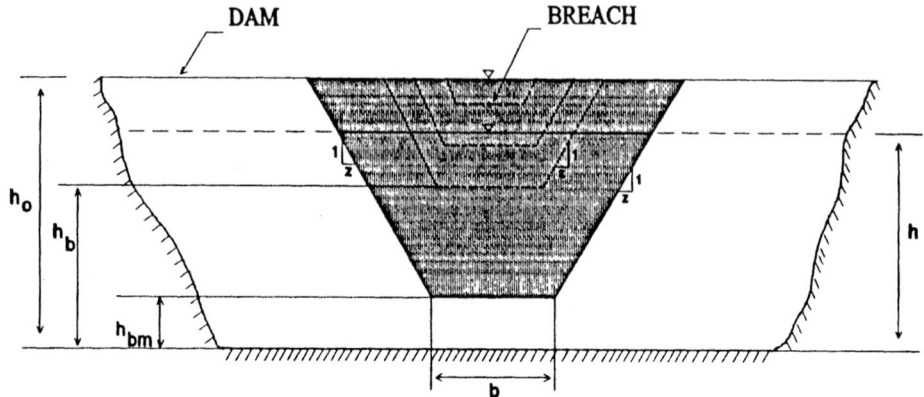

Fig. 5.1. Front view of dam showing formation of breach.

angle of repose of the compacted and wetted materials composing the dam and through which the breach develops. Rectangular, triangular, or trapezoidal shapes may be specified by using various combinations of values for z and b, e.g., $z = 0$ and $b > 0$ produces a rectangular-shaped breach, and $z > 0$ and $b = 0$ yields a triangular-shaped breach. The terminal bottom width (b) is related to the average width of the breach ($\bar{b}$) by the following:

$$b = \bar{b} - zH_d, \tag{5.1}$$

in which H_d is the height of the dam. In the parametric description of the breach, the breach bottom width starts at a point at the crest of the dam (see Fig. 5.1) and enlarges at a linear or nonlinear rate over the failure time (τ) until the terminal bottom width (b) is attained and the breach bottom has eroded to the minimum elevation, h_{bm}. The instantaneous bottom elevation of the breach is described as a function of time (t_b) according to the following:

$$h_b = h_d - (h_d - h_{bm})(t_b/\tau)^\rho, \qquad 0 \leq t_b \leq \tau, \tag{5.2}$$

in which h_d is the elevation of the top of the dam, h_{bm} is the final elevation of the breach bottom which is usually, but not necessarily, the bottom of the reservoir or outlet channel bottom, t_b is the time since beginning of breach formation, and ρ is the parameter specifying the degree of nonlinearity, e.g., $\rho=1$ is a linear formation rate, while $\rho=2$ is a nonlinear quadratic rate; the range for ρ is $1 \leq \rho \leq 4$, with the linear rate usually assumed. The instantaneous bottom width (b_i) of the breach is given by the following:

$$b_i = b(t_b/\tau)^\rho, \qquad 0 \leq t_b \leq \tau \tag{5.3}$$

When simulating a dam failure, the actual breach formation can commence when the reservoir water surface elevation (h) exceeds a specified value, h_f. This feature permits the simulation of an overtopping of a dam in which the breach does not form until a sufficient amount of water has passed over the crest of the dam to have eroded away the downstream face of the dam. The breach can also commence when a specified start-of-failure (t_f) time is reached in the simulation. A piping failure may also be simulated by specifying the initial centerline elevation of the pipe, using Eqs. (5.2)–(5.3), letting the top of the pipe form at the same rate as the bottom of the pipe, and letting $z = 0$. It is possible to also limit the breach formation to only the spillway section of the dam.

5.2.2 CONCRETE DAMS

Concrete gravity dams tend to have a partial breach as one or more monolith sections formed during the construction of the dam are forced apart and overturned by the escaping water. The time (τ) for breach formation is in the range of a few minutes depending on the number of monoliths that fail in succession. It is difficult to predict the number of monoliths which may be displaced or fail; however, by using a dam-breach flood prediction model such as described later in Section 5.3.1.5, and making several separate applications of the model wherein the breach width parameter (b) representing the combined lengths of assumed failed monoliths is varied in each application, the resulting reservoir water surface elevations can be used to indicate the extent of reduction of the loading pressures on the dam. Since the loading diminishes as b is increased, a limiting safe loading condition which would not cause further failure may be estimated. Unlike the concrete gravity dams, concrete arch dams tend to fail completely and are assumed to require only a few minutes for the breach formation. The shape of the breach is usually approximated as rectangular for either gravity or arch concrete dams; this is accomplished by using a value of zero for the shape parameter (z).

5.2.3 EARTHEN DAMS

Earthen dams which exceedingly outnumber all other types of dams do not tend to completely fail, nor do they fail instantaneously. The fully formed breach in earthen dams tends to have an average width ($\bar{b}$) in the range ($H_d \leq \bar{b} \leq 5H_d$) where H_d is the height of the dam. The middle portion of this range for $\bar{b}$ is supported by the summary report of Johnson and Illes (1976) and the upper range by the report of Singh and Snorrason (1982). Breach widths for earthen dams are therefore usually much less than the total length of the dam as measured across the valley. Also, the breach requires a finite interval of time (τ) for its formation through erosion of the dam materials by the escaping water. Total time of failure (for overtopping) may be in the range of a few minutes to usually less than an hour, depending on the height of the dam, the type of materials used in construction, the extent of compaction of the materials, and the magnitude and duration of the overtopping flow of the escaping water. The time of failure (τ) as used herein is the duration of time between the first breaching of the upstream face of the dam until the breach is fully formed. For overtopping failures, the beginning of breach formation at the upstream face of the dam occurs after the downstream face of the dam has eroded away and the resulting crevasse has progressed back across the width of the dam crest to reach the upstream face.

Piping failures occur when initial breach formation takes place at some point below the top of the dam due to erosion of an internal channel through the dam by the escaping water. Times of failure are usually considerably longer for piping than overtopping failures. As the erosion proceeds, a larger and larger opening is formed; this is eventually hastened by caving-in of the top portion of the dam.

Poorly constructed coal-waste (mine tailings) dams which impound water tend to fail more rapidly than well-designed dams and have average breach widths in the upper range of those for the earthen dams.

5.2.3.1 Statistically-Based Breach Predictors

Some statistically derived predictors for $\bar{b}$ and τ have been presented in the literature, i.e., MacDonald and Langridge-Monopolis (1984) and Froelich (1987, 1995) using data of the properties of 63 breaches of dams ranging in height from 15 to 285 ft, with 6 of the dams greater than 100 ft, obtained the following predictive equations:

$$\bar{b} = 9.5 k_o (V_r H_d)^{0.25}, \tag{5.4}$$

$$\tau = 0.3 V_r^{0.53} / H_d^{0.9}, \tag{5.5}$$

in which $\bar{b}$ is average breach width (ft), τ is time of failure (hrs), $k_o = 0.7$ for piping and 1.0 for overtopping, V_r is volume (acre-ft) and H_d is the height (ft) of water over the breach bottom (H_d is usually about the height of the dam). Standard error of estimate for $\bar{b}$ was ±82 ft which is an average error of ±56 percent of $\bar{b}$, and the standard error of estimate for τ was ±0.9 hrs which is an average error of ±74 per cent of τ.

5.2.3.2 Physically-Based Breach Erosion Models

Another means of determining the breach properties is the use of physically-based breach erosion models. Cristofano (1965) modeled the partial, time-dependent breach formation in earthen dams; however, this procedure required critical assumptions and specification of unknown critical parameter values. Also, Harris and Wagner (1967) used a sediment transport relation to determine the time for breach formation, but this procedure required specification of breach size and shape in addition to two critical parameters for the sediment transport relation. Then, Ponce and Tsivoglou (1981) presented a rather computationally complex breach erosion model which coupled the Meyer-Peter and Muller sediment transport equation to the one-dimensional differential equations of unsteady flow (Saint-Venant equations) and sediment conservation. They compared the model's predictions with observations of a breached landslide-formed dam on the Mantaro River in Peru. The results were substantially affected by the judicious selection of the breach channel hydraulic friction factor (Manning n), an empirical breach width-flow relation parameter, and an empirical coefficient in the sediment transport equation.

Another physically-based breach erosion model (BREACH) for earthen dams was developed (Fread, 1984, 1987) which substantially differed from the previously mentioned models. It predicted the breach characteristics (size, shape, time of formation) and the discharge hydrograph emanating from a breached earthen dam which was man-made or naturally formed by a landslide. The model was developed by coupling the conservation of mass of the reservoir inflow, spillway outflow, and breach outflow with the sediment transport capacity of the unsteady uniform flow along an erosion-formed breach channel. The bottom slope of the breach channel was assumed to be the downstream face of the dam. The growth of the breach channel was dependent on the dam's material properties (D_{50} size, unit weight, internal friction angle, cohesive strength). The model considered the possible existence of the following complexities: (1) core material properties which differ from those of the outer portions of the dam; (2) formation of an eroded ditch along the downstream face of the dam prior to the actual breach formation by the overtopping water; (3) the downstream face of the dam could have a grass cover or be composed of a material such as rip-rap or cobble stones of larger grain size than the major portion of the dam; (4) enlargement of the breach through the mechanism of one or more sudden structural collapses of the breaching portion of the dam due to the hydrostatic pressure force exceeding the resisting shear and cohesive forces; (5) enlargement of the breach width by collapse of the breach sides according to slope stability theory; and (6) the capability for initiation of the breach via piping with subsequent progression to a free-surface breach flow. The outflow hydrograph was obtained through a computationally efficient time-stepping iterative solution. This breach erosion model was *not* subject to numerical stability/convergence difficulties experienced by the more complex model of Ponce and Tsivoglou. The model's predictions were favorably compared with observations of a piping failure of the large man-made Teton Dam in Idaho, the piping failure of the small man-made Lawn Lake Dam in Colorado, and an overtopping activated breach of a large landslide-formed dam in Peru. Model sensitivity to numerical parameters was minimal; however, it was somewhat sensitive to the internal friction angle of the dam's material and the extent of grass cover when simulating man-made dams; and it was sensitive to the cohesive strength of the material composing landslide-formed dams. A reasonable variation of cohesion and internal friction angle parameters produced less than ±20 percent variation in the breach properties.

Other physically-based breach erosion models include the following: (1) the BEED model (Singh and Quiroga, 1988) which is similar to the BREACH model except it considers the effect of saturated soil in the collapse of the breach sides and it routes the breach outflow hydrograph through the downstream valley using a simple diffusion routing technique (Muskingum-Cunge) which neglects backwater effects and can produce significant errors in routing a dam-breach hydrograph when the channel/valley slope is less than 0.003; (2) a numerical model (Macchione and Sirangelo, 1988) based on the coupling of the one-dimensional unsteady flow (Saint-Venant) equations with the continuity equation for sediment transport and the Meyer-Peter and Muller sediment transport equation; (3) a numerical model (Bechteler and Broich, 1993) based on the coupling of the two-dimensional unsteady flow equations with the sediment continuity equation and the MeyerPeter and Muller sediment transport equation; and (4) a series of analytical models (Singh and Quiroga, 1988) requiring calibration of critical parameters. A more detailed description of the BREACH model (chosen as a practical representative of physically-based breach models) follows.

5.2.3.3 BREACH Model

The BREACH model utilizes the principles of soil mechanics, hydraulics, and sediment transport to simulate the erosion and bank collapse processes which form the breach. Reservoir inflow, storage, and spillway characteristics, along with the geometrical and material properties of the dam (D_{50} size, cohesion, internal friction angle, porosity, and unit weight) are utilized to predict the outflow hydrograph. The essential model components are described as follows.

Reservoir level computation. Conservation of mass is used to compute the reservoir water surface elevation (h) due to the influence of a specified reservoir inflow hydrograph (Q_i), spillway overflow (Q_{sp}) as determined from a spillway rating table, broadcrested weir flow (Q_o) over the crest of the dam, broad-crested weir flow (Q_b) through the breach, and the reservoir storage characteristics described by a surface area (S_a)-elevation table. Letting Δh represent the change in reservoir level during a small time interval (Δt), the conservation of mass requires the following relationship:

$$\Delta h = \frac{0.0826 \Delta t}{S_a}(\bar{Q}_i - \bar{Q}_b - \bar{Q}_{sp} - \bar{Q}_o), \tag{5.6}$$

in which the units of Δh, Δt, Q and S_a are ft, sec, ft^3/sec, and acre-ft, respectively, and the bar ($\bar{}$) denotes the average value during the Δt time interval. Thus, the reservoir elevation (h) at time (t) can easily be obtained since $h = h' + \Delta h$, in which h' is the reservoir elevation at time ($t - \Delta t$). If the breach is formed by overtopping, the breach outflow is simulated using a broad-crested weir flow equation, i.e.,

$$Q_b = 3 A_b (h - h_b)^{0.5}. \tag{5.7}$$

If the breach is formed by piping, a short-tube orifice flow equation is used to simulate the breach outflow, i.e.,

$$Q_b = A_b [2g(h - h_p)/(1 + fL/d_m)]^{0.5}, \tag{5.8}$$

in which A_b is the area (ft^2) of flow over the weir or orifice area, h_b is the elevation of the bottom of the breach at the upstream face of the dam, h_p is the specified center-line elevation of the pipe, f is the Darcy friction factor which is dependent on the D_{50} grain size, L is the length of the pipe, and d_m is the diameter or width of the pipe.

Breach width. Initially the breach is considered rectangular with the width (B_o) based on the assumption of optimal channel hydraulic efficiency, $B_o = B_r y_c$, in which y_c is the critical depth of flow at the entrance to the breach; i.e., $y_c =$

$2/3(h - h_b)$. The factor B_r is 2 for overtopping and 1 for piping. The initial rectangular-shaped breach can change to a trapezoidal shape when the sides of the breach collapse due to the breach depth exceeding the limits of a freestanding cut in soil of specified properties of cohesion (C), internal friction angle (ϕ), unit weight (γ), and existing angle (θ') that the breach cut makes with the horizontal. The collapse occurs when the effective breach depth (d'') exceeds the critical depth (d_c), i.e.,

$$d_c = 4C \cos \phi \sin \theta' / [\gamma - \gamma \cos(\theta' - \phi)]. \tag{5.9}$$

The effective breach depth (d'') is determined by reducing the actual breach depth (d) by $y_c/3$ to account for the supporting influence of the water flowing through the breach. The θ' angle reduces to a new angle (θ'') upon collapse which is simply $\theta'' = (\theta' + \phi)/2$.

Breach erosion. Erosion is assumed to occur equally along the bottom and sides of the breach except when the sides of the breach collapse. Then, the breach bottom is assumed not to continue to erode downward until the volume of collapsed material along the length of the breach is removed at the rate of sediment transport occurring along the breach at the instant before collapse. After this characteristically short pause, the breach bottom and sides continue to erode. Material above the wetted portion of the eroding breach sides is assumed to simultaneously collapse as the sides erode. Once the breach has eroded to the specified bottom of the dam, erosion continues to occur only along the sides of the breach. The rate at which the breach is eroded depends on the capacity of the flowing water to transport the eroded material. The Meyer-Peter and Muller sediment transport relation, as modified by Smart (1984) for steep channels, is used, i.e.,

$$Q_s = 3.64(D_{90}/D_{30})^{0.2} \frac{D^{2/3}}{n} PS^{1.1}(DS - 0.0054 D_{50} \tau_c), \tag{5.10}$$

in which Q_s is the sediment transport rate, D_{90}, D_{30}, and D_{50}, are the grain sizes in (mm) at which 90 percent, 30 percent, and 50 percent respectively of the total weight is finer, D is the hydraulic depth of flow computed from Manning's equation for flow along the breach channel at any instant of time, S is the breach bottom slope which is assumed to always be parallel to the downstream face of the dam, τ_c is Shield's critical sheer stress that must be exceeded before erosion occurs, and n is the Manning friction (roughness) coefficient which can be computed from the Strickler equation (Chow, 1959) for sand-bed channels or for gravel-bed channels (Jarrett, 1984) or simply estimated. The Δd incremental thickness eroded from the breach bottom and each side during a very short interval of time (Δt) is given by:

$$\Delta d' = Q_s \Delta t / [PL(1-p)] \tag{5.11}$$

in which P is the total perimeter of the breach, L is the length of the breach through the dam, and p is the porosity of the breach material.

Computational algorithm. The sequence of computations in the model are iterative since the flow into the breach is dependent on the bottom elevation of the breach and its width while the breach dimensions are dependent on the erosion depth (Δd) which is dependent on the sediment transport capacity of the breach flow; and the sediment transport capacity is dependent on the breach size and flow. A simple iterative algorithm is used to account for the mutual dependence of the flow, erosion, and breach properties. An estimated incremental erosion depth (Δd) is used at each time step to start the iterative solution. This estimated value can be extrapolated from previously computed values. Convergence occurs when $\Delta d'$, computed from Eq. (5.11), differs from the estimated Δd by less than an acceptable specified tolerance. Typical applications of the BREACH model require less than a minute on microcomputers. The computations show very little sensitivity to a reasonable variation in the specified time step size. The model is numerically robust, i.e., it has not shown any numerical instability or convergence problems.

Applications. BREACH was applied to the piping initiated failure of the earthfill Teton Dam which breached in June 1976, releasing an estimated peak discharge (Q_p) of 2.2 million cfs having a range of 1.6 to 2.6 million cfs. The simulated breach hydrograph is shown in Fig. 5.2. The computed final top breach width (W) of 645 ft compared well with the observed value of 650 ft. The computed slide slope of the breach was 1:1.06 compared to 1:1.00. The computed time (T_p) to peak flow was 2.2 hr. Additional information on this and another successful application of BREACH to the overtopping failure of a naturally formed landslide dam on the Mantaro river in Peru, which breached in June 1974, can be found elsewhere (Fread, 1984). The model has also been satisfactorily verified with the piping-initiated failure of the 28 ft high Lawn Lake Dam in Colorado (Jarrett and Costa, 1982).

5.2.4 Assessment of Breach Parameters

A method for quickly checking the overall reasonableness of the selected breach parameters ($\bar{b}$ and τ) uses the following equations:

$$Q_p^* = 370(V_r H_d)^{0.5}, \tag{5.12}$$

$$Q_p = 3.1\bar{b}\left(\frac{C}{\tau + C/\sqrt{H_d}}\right)^3, \tag{5.13}$$

in which Q_p^* and Q_p are the expected peak discharge (cfs) through the breach, V_r and H_d are the reservoir volume (acre-ft) and height (ft) of dam, respectively, and $C = 23.4 A_s/\bar{b}$ in which A_s is the surface area (acres) of the reservoir at the top of the dam. Eq. (5.12) was developed by Hagen (1982) from historical data of 14 dam failures; it provides a maximum envelope of all 14 of the observed discharges. It over-estimates the peak discharges for each of some 21 observed dam failures (including the previously mentioned 14 failures) by an average of 130 percent. Eq. (5.13) was developed by Fread (1981) and is used in the NWS Simplified Dam Break Model, SMPDBK (Wetmore and Fread, 1984). Eq. (5.13) yields peak discharges within a few percent of those produced by a more exact numerical method based on reservoir level-pool routing described later in Section 5.3.1.5.

Dam-Breach Floods

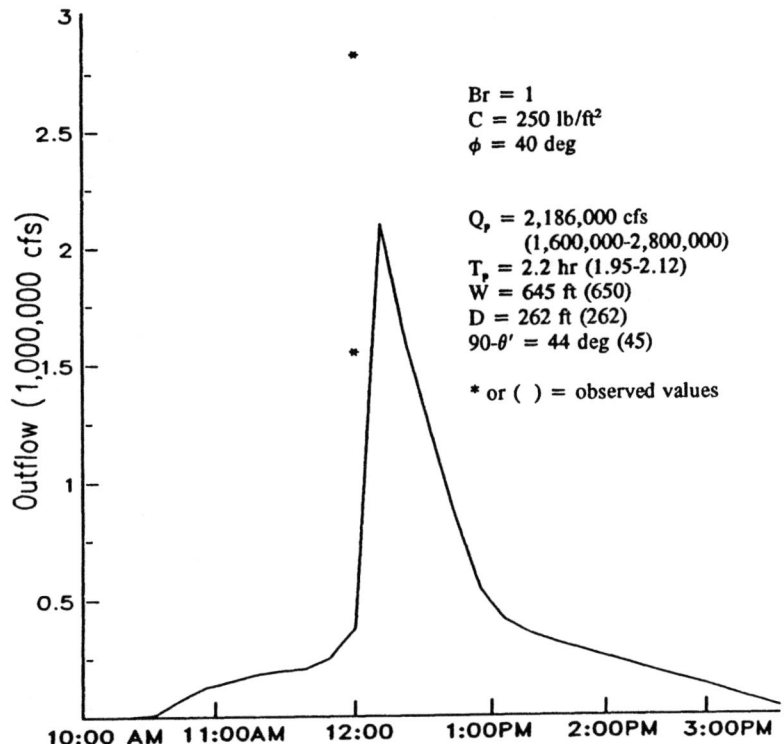

Fig. 5.2. Teton Dam: Predicted and observed breach outflow hydrograph and breach properties.

After selecting $\bar{b}$ and τ, Eq. (5.13) is used to compute Q_p which then can be compared with Q_p^* from Eq. (5.12). If $Q_p > Q_p^*$, then probably either $\bar{b}$ is too large and/or τ is too small.

Selection of breach parameters introduces a varying degree of uncertainty in the downstream flooding predictions produced by a dam-breach flood model; however, errors in the breach description and thence in the resulting peak outflow rate are damped-out as the flood wave advances downstream, i.e., variations in Q_p due to different breach parameters are reduced as the flood peak advances downstream. The extent of damping is related to the size of the downstream floodplain, i.e., the wider the floodplain, the greater the extent of damping. Sensitivity tests on the breach parameters are best determined using a dam-breach flood model and then comparing the variation in simulated flood peaks at critical downstream locations. In this way, the real uncertainty in the breach parameter selections will be determined.

For conservative predictions which err by creating too large of flood waves, values for b and z should produce an average breach width $\bar{b}$ in the uppermost range of probable values for a certain type of dam. The time of failure (τ) should be selected in the lower range of probable values to produce a maximum outflow.

Also, Eq. (5.13) can be used conveniently to test the sensitivity of $\bar{b}$ and τ for a specific reservoir having properties of V_r, H_d, and A_s. For example, using Eq. (5.13) for a moderately large reservoir (V_r = 250,000 acre-ft, H_d = 250 ft, A_s = 2,000 acres) it can be shown that Q_p varies in proportion to the variation in $\bar{b}$; however, Q_p only varies by less than 1/5 of the variation in τ. Although for a fairly small reservoir (V_r = 500 acre-ft, H_d = 40 ft, A_s = 10 acres), it can be shown, using Eq. (5.13), that Q_p varies less than 20 percent for a variation in $\bar{b}$ of 50 percent; however, Q_p varies about 40 percent for a variation in τ of 50 percent. Thus, it may be generalized, that for large reservoirs Q_p is quite sensitive to $\bar{b}$ and rather insensitive to τ, while for very small reservoirs Q_p is relatively insensitive to $\bar{b}$ and quite sensitive to τ.

5.3 Dam-Breach Flood Routing

5.3.1 Dynamic Routing

Flood waves produced by the breaching (failure) of a dam are known as dam-breach flood waves. They are much larger in peak magnitude, considerably more sharp-peaked, and generally of much shorter duration with acceleration components of a far greater significance than flood waves produced by precipitation runoff. The prediction of the extent and time of occurrence of flooding in the downstream valley is known as flood routing. The dam-breach wave is modified (attenuated, lagged, and distorted) as it flows (is routed) through the downstream valley due to the effects of valley storage, frictional resistance to flow, floodwave acceleration components, flow losses, and downstream channel constrictions and/or flow control structures. Modifications to the dam-break flood wave are manifested as attenuation (reduction) of the flood peak magnitude, spreading-out or dispersion of the temporal varying flood-wave volume, and changes in the celerity (propagation speed) or travel time of the flood wave. If the downstream valley contains significant storage volume such as a wide floodplain, the flood wave can be extensively attenuated and its propagation speed greatly reduced. Even when the downstream valley approaches that of a relatively narrow uniform rectangular-shaped section, there is appreciable attenuation of the flood peak and reduction in the wave celerity as the wave progresses through the valley.

There are two basic types of flood routing methods, hydrologic and hydraulic routing. (See Fread (1985, 1992) for a more complete description of the two types of routing methods.) The hydrologic methods usually provide a more approximate analysis of the progression of a flood wave through a river reach than do the hydraulic methods. The hydrologic methods are used for reasons of convenience and economy. They are most appropriate, as far as accuracy is concerned, when the flood wave is not rapidly varying, i.e., the flood-wave acceleration effects are negligible compared to the effects of gravity and channel friction. Also, they are best used when the flood wave is very similar in shape and magnitude to previous flood waves for which stage and discharge observations are available for calibrating the hydrologic routing parameters (coefficients), and when unsteady backwater effects are negligible.

In routing dam-break flood waves, a particular hydraulic routing method known as dynamic routing is most appropriate because of its ability to provide more accuracy in simulating the dam-break flood wave than that provided by the hydrologic methods, as well as, other less complex hydraulic methods such as the kinematic wave and the diffusion wave methods (Fread, 1985, 1992). Of the many available hydrologic and hydraulic routing techniques, only dynamic routing accounts for the acceleration effects associated with the dam-break wave and the influence of downstream unsteady backwater effects produced by channel constrictions, dams, bridge-road embankments, and tributary inflows. Also, dynamic routing can be used economically, i.e., the computational time can be made rather insignificant if advantages of certain "implicit" numerical solution techniques are utilized.

Dynamic routing is based on the complete one-dimensional equations of unsteady flow which are used to route the dam-break flood hydrograph through the downstream valley. The complete one-dimensional equations are an expanded version of the original equations developed by Barré de Saint-Venant (1871). The only coefficient that must be extrapolated beyond the range of past experience is the coefficient of flow resistance. Guidance for the selection and sensitivity of this parameter is discussed later in Sections 5.4.3 and 5.6.3, respectively.

5.3.1.1 Saint-Venant Equations

A modified and expanded form (Fread, 1988, 1992) of the original one-dimensional Saint-Venant equations (Saint-Venant, 1871; Henderson, 1966; Chow et al., 1988) consist of a conservation of mass equation, i.e.,

$$\frac{\partial Q}{\partial x} + \frac{\partial s_c(A + A_o)}{\partial t} - q = 0 \tag{5.14}$$

and the momentum equation, i.e.,

$$\partial(s_m Q)/\partial t + \partial(\beta Q^2/A)/\partial x + gA(\partial h/\partial x + S_f + S_{ec} + S_i) + L + W_f B = 0, \tag{5.15}$$

where h is the water-surface elevation, A is the active cross-sectional area of flow, A_o is the inactive (off-channel storage) cross-sectional area which may be preferred omitted when used to represent heavily wooded floodplains, and its effect represented by a higher frictional resistance for that portion of the cross section, s_c and s_m are area-weighted and conveyance-weighted sinuosity factors, respectively (Delong, 1989) which correct for the departure of a sinuous in-bank channel from the x-axis of the floodplain, x is the longitudinal mean flow-path distance measured along the center of the watercourse (channel and floodplain), t is time, q is the lateral inflow or outflow per lineal distance along the watercourse (inflow is positive and outflow is negative), β is the momentum coefficient for nonuniform velocity distribution within the cross section, g is the gravity acceleration constant, S_f is the boundary friction slope, S_{ec} is the expansion/contraction (large eddy loss) slope, and S_i is the viscous dissipation slope.

Friction slope. The boundary friction slope (S_f) is evaluated from Manning's equation for uniform, steady flow, i.e.,

$$S_f = n^2|Q|Q/(\mu^2 A^2 R^{4/3}) = |Q|Q/K^2, \tag{5.16}$$

in which n is the Manning coefficient of frictional resistance, R is the hydraulic radius, μ is a units conversion factor (1.49 for US units and 1.0 for SI), and K is the channel conveyance factor. The absolute value of Q is used to correctly account for the possible occurrence of reverse (negative) flows. The conveyance formulation is preferred (for numerical and accuracy considerations) for composite channels having wide, flat overbanks or floodplains in which K represents the sum of the conveyance of the channel (which is corrected for sinuosity effects by dividing by s_m), and the conveyances of left and right floodplain areas.

When the conveyance factor (K) is used to evaluate S_f, the river channel/valley cross-sectional properties are designated as left floodplain, channel, and right floodplain rather than as a composite channel/valley section. Special orientation for designating left or right is not required as long as consistency is maintained. The conveyance factor is evaluated as follows:

$$K_\ell = \frac{\mu}{n_\ell} A_\ell R_\ell^{2/3}, \tag{5.17}$$

$$K_c = \frac{\mu A_c R_c^{2/3}}{n_c s_m^{1/2}}, \tag{5.18}$$

$$K_r = \frac{\mu}{n_r} A_r R_r^{2/3}, \tag{5.19}$$

$$K = K_\ell + K_c + K_r, \tag{5.20}$$

in which the subscripts ℓ, c, and r designate left floodplain, channel, and right floodplain, respectively.

Sinuosity Factors. The area-weighted and conveyance-weighted sinuosity factors (s_c and s_m, respectively) in Eqs. (5.14), (5.15) and (5.18) represent the ratio(s) of the flow-path distance along a meandering channel to the mean flow-path distance along the floodplain. They vary with depth of flow according to the following relations:

$$s_{cJ} = \frac{\sum_{k=2}^{k=J} \Delta A_{\ell k} + \Delta A_{ck} s_k + \Delta A_{rk}}{A_{\ell J} + A_{cJ} + A_{rJ}}, \tag{5.21}$$

$$s_{mJ} = \frac{\sum_{k=2}^{k=J} \Delta K_{\ell k} + \Delta K_{ck} s_k + \Delta K_{rk}}{K_{\ell J} + K_{cJ} + K_{rJ}}, \tag{5.22}$$

in which $\Delta A = A_{J+1} - A_J$, and s_k represents the sinuosity factor for a differential portion of the flow between the Jth depth and the $J+1$th depth, and K is the conveyance factor.

Expansion/contraction effects. The term (S_{ec}) is computed as follows:

$$S_{ec} = k_{ec}\Delta(Q/A)^2/(2g\Delta x), \qquad (5.23)$$

in which k_{ec} is the expansion/contraction coefficient (negative for expansion, positive for contraction) which varies from $-1.0/0.4$ for an abrupt change in section geometry to $-0.3/0.1$ for a very gradual, curvilinear transition between cross sections. The Δ represents the difference in the term $(Q/A)^2$ at two adjacent cross sections separated by a distance Δx. If the flow direction changes from downstream to upstream, k_{ec} can be automatically changed (Fread, 1988).

Since dam-break floods usually have much greater velocities, it is important, especially for nonuniform channels (Rajar, 1978) to include in the Saint-Venant momentum Eq. (5.15) the expansion/contraction losses via the S_{ec} term defined by Eq. (5.23). The ratio of expansion/contraction losses (form losses) to the friction losses can be in the range of $0.01 < S_{ec}/S_f < 1.0$. The larger ratios occur for very irregular channels with relatively small n values.

Momentum correction coefficient. The momentum correction coefficient (β) for nonuniform velocity distribution is:

$$\beta = \frac{K_\ell^2/A_\ell + K_c^2/A_c + K_r^2/A_r}{(K_\ell + K_c + K_r)^2/(A_\ell + A_c + A_r)}, \qquad (5.24)$$

in which K is conveyance, A is wetted area, and the subscripts ℓ, c, and r denote left floodplain, channel, and right floodplain, respectively. When floodplain properties are not separately specified and the total cross section is treated as a composite section, β can be approximated as $1.0 \leq \beta \leq 1.06$ in lieu of Eq. (5.24).

Lateral flow momentum. The term (L) in Eq. (5.15) is the momentum effect of lateral flows, and has the following form: (a) lateral inflow, $L = -qv_x$, where v_x is the velocity of lateral inflow in the x-direction of the main channel flow; (b) seepage lateral outflow, $L = -0.5qQ/A$; and (c) bulk lateral outflow, $L = -qQ/A$ (Strelkoff, 1969).

Mud or debris flows. The term (S_i) is included in the momentum equation (5.15) in addition to S_f to account for viscous dissipation effects of non-Newtonian flows such as mud or debris flows. Mine tailings dams, where the viscous contents retained by the dam have non-Newtonian properties, are dam-breach flood applications requiring the use of S_i in Eq. (5.15). This effect becomes significant only when the solids concentration of the flow is in the range of about 40 to 50 percent by volume. For concentrations of solids greater than about 50 percent, the flow behaves more as a landslide and is not governed by the Saint-Venant equations. S_i is evaluated for any non-Newtonian flow as follows:

$$S_i = \frac{\kappa}{\gamma}\left[\frac{(b+2)Q}{AD^{b+1}} + \frac{(b+2)(\tau_0/\kappa)^b}{2D^b}\right]^{1/b}, \qquad (5.25)$$

in which γ is the fluid's unit weight, τ_0 is the fluid's yield strength, D is the hydraulic depth (A/B), $b = 1/m$ where m is the exponent of the power function that fits the fluid's stress (τ_s)-strain (dv/dy) properties, and κ is the apparent viscosity or scale factor of the power function, i.e., $\tau_s = \tau_o + \kappa(dv/dy)^m$. The viscous properties, τ_o and κ, can be estimated from the solids concentration ratio of the mud flow (O'Brien and Julien, 1984).

Wind effects. The last term (W_fB) in Eq. (5.15) represents the resistance effect of wind on the water surface (Fread, 1985, 1992); B is the wetted topwidth of the active flow portion of the cross section; and $W_f = V_r|V_r|c_w$, where the wind velocity relative to the water is $V_r = V_w \cos w + V$, V_w is the velocity of the wind (+) if opposing the flow velocity and (−) if aiding the flow, w is the acute angle the wind direction makes with the x-axis, V is the velocity of the unsteady flow, and c_w is a wind friction coefficient. This modeling capability can be used to simulate the effect of potential dam overtopping due to wind set-up within a reservoir by applying the Saint-Venant equations to the unsteady flow in a reservoir.

5.3.1.2 Implicit Four-Point, Finite-Difference Approximations

The extended Saint-Venant Eqs. (5.14) and (5.15) constitute a system of partial differential equations with two independent variables, x and t, and two dependent variables, h and Q; the remaining terms are either functions of x, t, h, and/or Q, or they are constants. The partial differential equations can be solved numerically by approximating each with a finite-difference algebraic equation; then the system of algebraic equations are solved in conformance with prescribed initial and boundary conditions.

Of various implicit, finite-difference solution schemes that have been developed, a four-point scheme first used by Preissmann (1961) and later a weighted version by many others (Fread, 1977, 1985, 1988; Cunge et al., 1980) is most advantageous. It is readily used with unequal distance steps, its stability-convergence properties are conveniently modified, and boundary conditions are easily applied.

Space-time plane. In the weighted four-point implicit scheme, the continuous x-t region in which solutions of h and Q are sought is represented by a rectangular grid of discrete points as shown in Fig. 5.3. The x-t plane (solution domain) is a convenient method for expressing relationships among the variables. The grid points are determined by the intersection of lines drawn parallel to the x- and t-axes. Those parallel to the t-axis represent locations of cross sections; they have a spacing of Δx, which need not be the same between each pair of cross sections. Those parallel to the x-axis represent time lines; they have a spacing of Δt, which also need not be the same between successive time lines. Each point in the rectangular network can be identified by a subscript (i) which designates the x-position or cross section and a superscript (j) which designates the particular time line.

Numerical approximations. The time derivatives are approximated by a forward-difference quotient at point M' (Fig. 5.3) centered between the i and $i+1$ lines along the x-axis, i.e.,

$$\partial \phi / \partial t \simeq (\phi_i^{j+1} + \phi_{i+1}^{j+1} - \phi_i^j - \phi_{i+1}^j)/(2\Delta t_j) \tag{5.26}$$

where ϕ represents any dependent variable or functional quantity (Q, s_c, s_m, A, A_o, q, h). Spatial derivatives are approximated at point M' by a forward-difference

Dam-Breach Floods

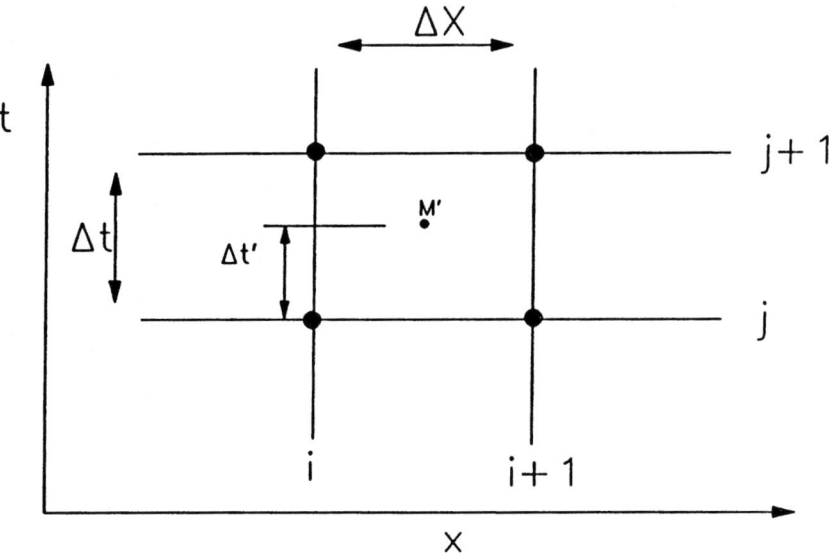

Fig. 5.3. The x-t solution domain for the weighted four-point implicit scheme.

quotient located between two adjacent time lines according to weighting factors of θ (the ratio $\Delta t'/\Delta t$ shown in Fig. 5.3) and $1 - \theta$, i.e.,

$$\partial \phi / \partial x \simeq \theta(\phi_{i+1}^{j+1} - \phi_i^{j+1})/\Delta x_i + (1 - \theta)(\phi_{i+1}^{j} - \phi_i^{j})/\Delta x_i. \tag{5.27}$$

Non-derivative terms are approximated with weighting factors at the same time level (point M') where the spatial derivatives are evaluated, i.e.,

$$\phi \simeq \theta(\phi_i^{j+1} + \phi_{i+1}^{j+1})/2 + (1 - \theta)(\phi_i^{j} + \phi_{i+1}^{j})/2. \tag{5.28}$$

Numerical stability. The weighted four-point implicit scheme is unconditionally, linearly stable for $\theta \geq 0.5$ (Fread, 1974); however, the sizes of the Δt and Δx computational steps are limited by the accuracy of the assumed linear variations of functions between the grid points in the x-t solution domain. Values of θ greater than 0.5 dampen parasitic oscillations which have wave lengths of about $2\Delta x$ that can grow enough to invalidate or destroy the solution. The θ weighting factor causes some loss of accuracy as it departs from 0.5, a box scheme, and approaches 1.0, a fully implicit scheme. This effect becomes more pronounced as the magnitude of the ratio $(T_r/\Delta t)$ decreases where T_r is the time of rise of the hydrograph (time interval from beginning of rise to peak of the hydrograph). Usually, a θ weighting factor of 0.60 is used to minimize the loss of accuracy while avoiding the possibility of weak (pseudo) instability for θ values of 0.5 when frictional effects are minimal.

Selection of Δt and Δx computational parameters. The computational time step (Δt) can be either specified or automatically determined to best suit the most rapidly rising hydrograph occurring within the system of rivers containing one or more breaching dams. The time step is selected according to the following:

$$\Delta t = T_r/M, \tag{5.29}$$

where T_r is the minimum time of rise of any hydrograph that has been specified at upstream boundaries or in the process of being generated at a breaching dam. M is user specified according to the following guidance (Fread, 1993):

$$M \simeq 2.67 \left[1 + \mu' n^{0.9}/(q^{0.1} S_o^{0.45})\right], \tag{5.30}$$

in which $\mu' = 3.97$ (3.13 SI units), n is the Manning friction coefficient, q is the peak flow per unit channel width, and S_o is the channel bottom slope. M usually varies within the range, $6 \leq M \leq 40$, with M often assumed to be approximately 20.

The Δx computational distance step can be specified or automatically determined according to the smaller of two criteria (Fread, 1993). The first criterion is:

$$\Delta x \leq cT_r/20, \tag{5.31}$$

in which c is the bulk wave celerity (the celerity or velocity associated with an essential characteristic of the unsteady flow such as the peak of the hydrograph). In most applications, the wave velocity is well approximated as a kinematic wave, and c is estimated as $3/2V$ (V is the flow velocity) or c can be obtained by dividing the distance between two points along the channel by the difference in the times of occurrence of the peak of an observed flow hydrograph at each point. Since c can vary along the channel, Δx may not be constant along the channel.

The second criterion for selecting Δx is the restriction imposed by rapidly varying cross-sectional changes along the x-axis of the watercourse. Such expansion/contraction is limited to the following inequality (Samuels, 1985):

$$0.635 < A_{i+1}/A_i < 1.576. \tag{5.32}$$

This condition results in the following approximation for the maximum computational distance step:

$$\Delta x \leq L'/N, \tag{5.33}$$

where:

$$N = 1 + 2|A_i - A_{i+1}|/\hat{A}, \tag{5.34}$$

in which L' is the distance between two adjacent cross sections differing from one another by approximately 50 percent or greater, A is the active cross-sectional area, i and $i+1$ are index counters, $\hat{A} = A_{i+1}$ if $A_i > A_{i+1}$ (contracting reach) or $\hat{A} = A_i$ if $A_i < A_{i+1}$ (expanding reach), and N is rounded to the nearest integer value.

Significant changes in the bottom slope of the watercourse also require small distance steps in the vicinity of the change. This is required particularly when the flow changes from subcritical to supercritical or conversely along the watercourse. Such changes can require computational distance steps in the range of 50 to 200 ft.

Automatic interpolation. It is essential for a dam-breach flood routing model to automatically provide linearly interpolated cross sections at a user specified spatial resolution in order to increase the spatial frequency at which solutions to the Saint-Venant equations are obtained. This is often required for purposes of attaining numerical accuracy stability when (a) routing very sharp-peaked hydrographs such as those generated by breached dams, (b) when adjacent cross sections either expand or contract by more than about 50 percent, and (c) where mixed flow changes from subcritical to supercritical or vice versa.

Algebraic routing equations. Using the finite-difference operators of Eqs. (5.26) to (5.28) to replace the derivatives and other variables in Eqs. (5.14) and (5.15), the following weighted four-point, implicit finite-difference algebraic equations are obtained:

$$\theta \left[\frac{Q_{i+1}^{j+1} - Q_i^{j+1}}{\Delta x_i} \right] - \theta q_i^{j+1} + (1-\theta) \left[\frac{Q_{i+1}^j - Q_i^j}{\Delta x_i} \right] - (1-\theta)q_i^j$$

$$+ \left[\frac{s_{c_i}^{j+1}(A+A_o)_i^{j+1} + s_{c_i}^{j+1}(A+A_o)_{i+1}^{j+1} - s_{c_i}^j(A+A_o)_i^j - s_{c_i}^j(A+A_o)_{i+1}^j}{2\Delta t_j} \right]$$

$$= 0, \tag{5.35}$$

$$\left[\frac{(s_{m_i}Q_i)^{j+1} + (s_{m_i}Q_{i+1})^{j+1} - (s_{m_i}Q_i)^j - (s_{m_i}Q_{i+1})^j}{2\Delta t_j} \right]$$

$$+\theta \left[\frac{(\beta Q^2/A)_{i+1}^{j+1} - (\beta Q^2/A)_i^{j+1}}{\Delta x_i} \right.$$

$$\left. + g\bar{A}_i^{j+1} \left(\frac{h_{i+1}^{j+1} - h_i^{j+1}}{\Delta x_i} + \bar{S}_{f_i}^{j+1} + S_{ec_i}^{j+1} + S_{i_i}^{j+1} \right) + L_i^{j+1} + (W_f\bar{B})_i^{j+1} \right]$$

$$+(1-\theta) \left[\frac{(\beta Q^2/A)_{i+1}^j - (\beta Q^2/A)_i^j}{\Delta x_i} + g\bar{A}_i^j \left(\frac{h_{i+1}^j - h_i^j}{\Delta x_i} + \bar{S}_{f_i}^j + S_{ec_i}^j + S_{i_i}^j \right) \right.$$

$$\left. + L_i^j + (W_f\bar{B})_i^j \right] = 0, \tag{5.36}$$

where:

$$\bar{A}_i = (A_i + A_{i+1})/2, \tag{5.37}$$
$$\bar{S}_{f_i} = n^2 \bar{Q}_i |\bar{Q}_i|/(\mu^2 \bar{A}_i^2 \bar{R}_i^{4/3}) = \bar{Q}_i |\bar{Q}_i|/\bar{K}_i^2, \tag{5.38}$$
$$\bar{Q}_i = (Q_i + Q_{i+1})/2, \tag{5.39}$$
$$\bar{R}_i \simeq \bar{A}_i/\bar{B}_i, \tag{5.40}$$
$$\bar{B}_i = (B_i + B_{i+1})/2, \tag{5.41}$$
$$\bar{K}_i = (K_i + K_{i+1})/2, \tag{5.42}$$

The terms L and $W_f B$ are defined in Eq. (5.15); terms associated with the jth time line are known from initial conditions or previous time-step computations; and μ in Eq. (5.38) is defined in Eq. (5.16). The Δx distance between cross sections is measured along the mean flow path of the (channel/valley) watercourse.

5.3.1.3 Solution Procedure

The flow equations are expressed in finite-difference form for all Δx_i reaches between the first and last (Nth) cross section ($i = 1, 2, \ldots, N$) along the channel/floodplain and then solved simultaneously for the unknowns (Q and h) at each cross section. In essence, the solution technique determines the unknown quantities (Q and h at all specified cross sections along the watercourse) at various times into the future; the solution is advanced from one time to a future time over a finite time interval (time step) of magnitude Δt. Thus, applying Eqs. (5.35) and (5.36) recursively to each of the $(N - 1)$ rectangular grids in Fig. 5.3 between the upstream and downstream boundaries, a total of $(2N - 2)$ equations with $2N$ unknowns are formulated. Then, prescribed boundary conditions for subcritical flow (Froude number less than unity, i.e., $\text{Fr} = Q/(A\sqrt{gD}) < 1$), one at the upstream boundary and one at the downstream boundary, provide the two additional and necessary equations required for the system to be determinate. Since disturbances can propagate only in the downstream direction in supercritical flow ($\text{Fr} > 1$), two upstream boundary conditions and no downstream boundary condition are required for the system to be determinate. The boundary conditions are described later. Due to the nonlinearity of Eqs. (5.35) and (5.36) with respect to Q and h, an iterative, highly efficient quadratic solution technique such as the Newton–Raphson method is frequently used. Other solution techniques linearize Eqs. (5.35) and (5.36) via a Taylor series expansion or other means. Convergence of the iterative technique is attained when the difference between successive solutions for each unknown is less than a relatively small prescribed tolerance. Convergence for each unknown at all cross sections is usually attained within about one to five iterations. A more complete description of the solution method may be found elsewhere (Fread, 1985).

The solution of $2N \times 2N$ simultaneous equations requires an efficient technique for the implicit method to be feasible. One such procedure requiring 38N computational operations (+, –, *, /) is a compact, penta-diagonal Gaussian elimination method (Fread, 1971, 1985) which makes use of the banded structure of the coefficient matrix of the system of equations. This is essentially the same as the double sweep elimination method (Liggett and Cunge, 1975; Cunge et al., 1980).

When flow is supercritical, the solution technique previously described can be somewhat simplified. Two boundary conditions are required at the upstream boundary and none at the downstream boundary since flow disturbances cannot propagate upstream in supercritical flow. The unknown h and Q at the most upstream cross section are determined from the two boundary equations. Then, cascading from upstream to downstream, Eqs. (5.35) and (5.36) are solved for the two unknowns (h_{i+1} and Q_{i+1}) at each cross section by using Newton–Raphson iteration applied recursively to the two nonlinear equations, Eq. (5.35) and Eq. (5.36).

5.3.1.4 Initial Conditions

Values of water-surface elevation (h) and discharge (Q) for each cross section must be specified initially at time $t = 0$ to obtain solutions to the Saint-Venant equations. Initial conditions may be obtained from any of the following: (a) observations at gaging stations and interpolated values between gaging stations for intermediate cross sections in large rivers; (b) computed values from a previous unsteady flow solution (used in real-time flood forecasting); and (c) computed values from a steady-flow backwater solution. The backwater method is most commonly used, in which the steady discharge at each cross section is determined by:

$$Q_{i+1} = Q_i + q_i \Delta x_i, \quad i = 1, 2, 3, \ldots, N-1, \tag{5.43}$$

in which Q_1 is the assumed steady flow at the upstream boundary at time $t = 0$, and q_i is the known average lateral inflow or outflow along each Δx reach at $t = 0$. The water-surface elevations (h_i) are computed according to the following steady-flow simplification of the momentum equation, Eq. (5.15):

$$(Q^2/A)_{i+1} - (Q^2/A)_i + g\bar{A}_i(h_{i+1} - h_i + \Delta x_i \bar{S}_{f_i}) = 0, \tag{5.44}$$

in which $\bar{A}$ and $\bar{S}_{f_i}$ are defined by Eqs. (5.37) and (5.38), respectively. The computations proceed in the upstream direction ($i = N-1, \ldots, 3, 2, 1$) for subcritical flow (they must proceed in the downstream direction for supercritical flow). The starting watersurface elevation (h_N) can be specified or obtained from the appropriate downstream boundary condition for the discharge (Q_N) obtained via Eq. (5.43). The Newton-Raphson iterative solution method for a single equation and/or a simple, less efficient, but more stable bi-section iterative technique can be applied to Eq. (5.44) to obtain h_i. The initial water surface profile can also be obtained from steady-flow backwater models such as HEC-2 (Hydrologic Engineering Center, 1982). Due to friction, small errors in the initial conditions will dampen-out after several computational time steps during the solution of the Saint-Venant equations.

5.3.1.5 Upstream Boundary

Values for the unknowns at external boundaries (the upstream and downstream extremities of the routing reach) of the channel/floodplain, must be specified in order to obtain solutions to the Saint-Venant equations. In fact, in most unsteady flow applications, the unsteady disturbance is introduced at one or both of the external boundaries.

Discharge hydrograph. A specified discharge time series (hydrograph) of inflow to the upstream reservoir is used as the upstream boundary condition. The hydrograph should not be affected by downstream flow conditions. This hydrograph may be obtained from the following: (1) historical observations, (2) assumed design hydrograph, or (3) a runoff hydrograph from specified rainfall-runoff model using calibrated or estimated model parameters. The upstream boundary is expressed mathematically as follows:

$$Q_1^{j+1} - Q(t) = 0, \tag{5.45}$$

in which $Q(t)$ is the specified discharge time series and the subscript indicates the discharge at the first cross section, i.e., the upstream boundary. Eq. (5.45) is used for the upstream boundary if dynamic routing (based on the discretized Saint-Venant equations) commences at this location. However, if the most upstream cross section represents the inlet to an upstream reservoir, a simple routing procedure (reservoir level-pool routing) can be used rather than the considerably more complex dynamic routing if (1) the reservoir is not excessively long and (2) the inflow hydrograph $Q(t)$ is not rapidly changing with time. Level-pool routing errors (E_q), with a magnitude of less than about 5 percent, can usually be tolerated.

Level-pool routing. In level-pool routing, the reservoir is assumed always to have a horizontal (level) water surface throughout its entire length; hence, level-pool. The water-surface elevation (h) changes with time (t), and the outflow from the reservoir is assumed to be a function of $h(t)$. This is the case for reservoirs with uncontrolled overflow spillways such as the ogee-crested, broad-crested weir, and morning-glory types. Gate controlled spillways can be included in level-pool routing if the gate setting (height of the gate bottom above the gate sill) is a predetermined function of time, since the outflow is a function of h and the extent of gate opening. Reservoirs, wherein the dam fails and produces a breach outflow hydrograph, can also be included in the level-pool routing approach.

The upstream boundary condition for this situation is represented by the following expression:

$$Q_1^{j+1} - Q(t) + 43560 \bar{S}_a \Delta h_1^{j+1}/\Delta t^j = 0, \qquad (5.46)$$

where:

$$\bar{S}_a = (S_a^j + S_a^{j+1})/2, \qquad (5.47)$$
$$\Delta h^{j+1} = h_1^{j+1} - h_1^j, \qquad (5.48)$$

In this approach, the first cross section is located immediately upstream of the dam, and the second cross section is located immediately downstream of the dam in the tailwater area. Two internal boundary equations (described later) are used to govern the flow through the dam, between the first and second cross sections.

Accuracy of level-pool routing. The accuracy of level-pool routing relative to the more accurate distributed dynamic routing model based on the Saint-Venant equations is shown in Fig. 5.4. The error (in percent) associated with level-pool routing is expressed as a normalized error for the rising limb of the outflow hydrograph. The peak outflow is used as the normalizing parameter. The normalized error (E_q) is:

$$E_q = \frac{100}{Q_{D_P}} \sqrt{\frac{\sum_{i=1}^{N'}(Q_{L_i} - Q_{D_i})^2}{N'}}, \qquad (5.49)$$

in which Q_{L_i} is the level-pool routed flow; Q_{D_i} is the dynamic routed flow peak, and N' is the number of computed discharges comprising the rising limb of the routed hydrograph. Since level-pool routing is based on the assumption of a horizontal water surface along the length of the reservoir at all times, the error

Dam-Breach Floods

(E_q) associated with level-pool routing increases as (a) reservoir mean depth (D_r) decreases, (b) reservoir length (L_r) increases, (c) time of rise (T_r) of inflow hydrograph decreases, and (d) inflow hydrograph volume decreases. These effects can be represented by three dimensionless parameters, σ_ℓ, σ_t, σ_v; where $\sigma_\ell = D_r/L_r$, $\sigma_t = L_r/[3600T_r(gD_r)^{1/2}]$ in which g is the gravity acceleration constant and T_r is the time (hrs) from beginning of rise until the peak of the inflow hydrograph, and σ_v = hydrograph volume/reservoir volume. As shown in Fig. 5.4, E_q increases as σ_t increases and as σ_ℓ and σ_v decrease; also the influence of σ_v increases as σ_ℓ decreases. Level-pool routing is not recommended when the inflow hydrograph is one generated from an upstream dam failure.

5.3.1.6 Downstream Boundary

For subcritical flow, a specified discharge or water-surface elevation time series, or a tabular relation between discharge and water-surface elevation (single-valued rating curve) can be used as the downstream boundary condition.

Loop rating. Another downstream boundary condition can be a computed loop-rating curve based on the Manning equation, i.e.,

$$Q_N^{j+1} - \mu/n A_N^{j+1} (R_N^{j+1})^{2/3} (S_{f_N}^j)^{1/2} = 0. \tag{5.50}$$

The loop is produced by using the friction slope (S_f) rather than the channel bottom slope (S_o) in the Manning equation. The friction slope exceeds the bottom slope during the rising limb of the hydrograph while the reverse is true for the recession limb. The friction slope (S_f) is approximated by using Eq. (5.15) where L and W_f are assumed to be zero while s_m and β are assumed to be unity (Fread, 1985, 1988, 1992), i.e.,

$$\begin{aligned}S_{f_N}^j \simeq &-(Q_N^j - Q_N^{j-1})/(gA_N^j \Delta t^j) - [(Q^2/A)_N^j \\ &-(Q^2/A)_{N-1}^j]/(gA_N^j \Delta x_{N-1}) \\ &-(h_N^j - h_{N-1}^j)/\Delta x_{N-1}.\end{aligned} \tag{5.51}$$

The loop-rating boundary equation allows the unsteady wave to pass the downstream boundary with minimal disturbance by the boundary itself, which is desirable when the routing is terminated at an arbitrary location along the channel/floodplain and not at a location of actual flow control such as a dam or waterfall, or where the flow is affected by downstream backwater conditions produced by tidal action, reservoirs, or tributary inflow.

Critical flow. The downstream boundary condition can also be a critical flow section such as the entrance to a waterfall or a steep channel reach, i.e.,

$$Q_N^{j+1} - \sqrt{g/B_N^{j+1}} (A_N^{j+1})^{3/2} = 0. \tag{5.52}$$

Critical flow occurs when the bottom slope (S_o) equals or exceeds the critical slope (S_c) which can be easily computed as follows:

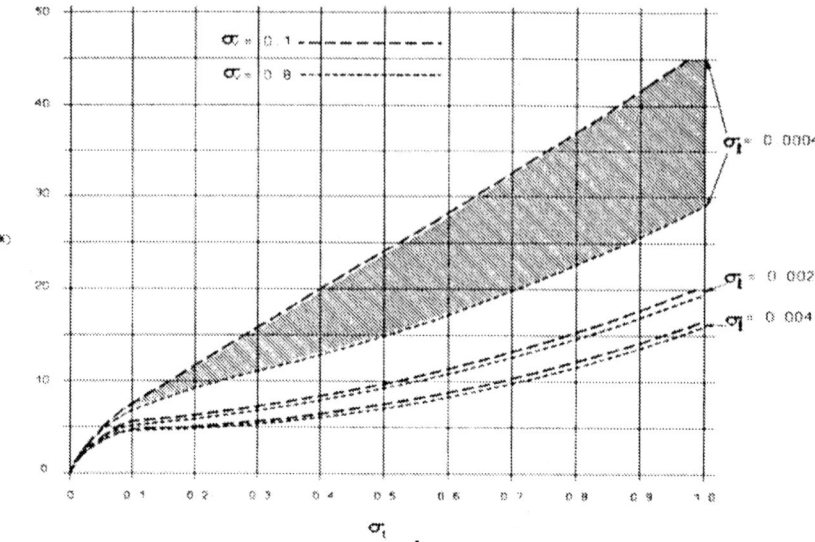

Fig. 5.4. Level-pool routing compared to dynamic routing showing the normalized error (E_q) of the outflow hydrograph as a function of dimensionless parameters σ_t, σ_ℓ, σ_v.

$$S_c = \hat{\mu} n^2 / D^{1/3}, \tag{5.53}$$

where $\hat{\mu} = 14.6$ for US units and $\hat{\mu} = 9.8$ for SI units.

Rating curve. When the downstream boundary is a stage/discharge relation (rating curve), the flow at the boundary should not be otherwise affected by flow conditions further downstream. Although there are often some minor effects due to the presence of cross-sectional irregularities downstream of the chosen boundary location, these usually can be neglected unless the irregularity is so pronounced as to cause significant backwater or drawdown effects. Reservoirs, major tributaries, or tidal effects located below the downstream boundary which cause backwater effects at the boundary should be avoided. When either of these situations are unavoidable, the routing reach should be extended downstream to the dam in the case of the reservoir or to a location downstream of where the major tributary enters. Sometimes the routing reach may be shortened by moving the downstream boundary to a location further upstream where backwater effects are negligible.

5.3.1.7 Internal Boundaries

Often along the channel/floodplain, there are locations such as a dam, bridge, or waterfall (short rapids) where the flow is rapidly varied in space rather than gradually varied. At such locations (internal boundaries), the Saint-Venant equations are not applicable since gradually varied flow is a necessary condition for their derivation. Empirical water elevation-discharge relations such as weir-flow are utilized for simulating rapidly varying flow. At internal boundaries, cross sections are specified for the upstream and downstream extremities of the section where rapidly varying flow occurs. The Δx reach containing an internal bound-

Dam-Breach Floods

ary requires two internal boundary equations; since, as with any other Δx reach, two equations equivalent to the Saint-Venant equations are required. One of the required internal boundary equations represents conservation of mass with negligible time-dependent storage, i.e.,

$$Q_i^{j+1} - Q_{i+1}^{j+1} = 0. \tag{5.54}$$

Dam. The second equation is usually an empirical rapidly varied flow relation. If the internal boundary represents a dam, the following equation can be used:

$$Q_i^{j+1} - (Q_s + Q_b)^{j+1} = 0, \tag{5.55}$$

in which Q_s and Q_b are the spillway and dam-breach flow, respectively. In this way, the flows Q_i and Q_{i+1} and the elevations h_i and h_{i+1} are in balance with the other flows and elevations occurring simultaneously throughout the entire flow system which may consist of additional downstream dams which are treated as additional internal boundary conditions via Eqs. (5.54) and (5.55). In fact, this approach can be used to simulate the progression of a dam-break flood through an unlimited number of reservoirs located sequentially along the valley. The downstream dams may also breach if they are sufficiently overtopped. The spillway flow (Q_s) is computed from the following expression:

$$Q_s = c_s L_s (h_i - h_s)^{1.5} + c_g A_g (h_i - h_g)^{0.5} + c_d L_d (h_i - h_d)^{1.5} + Q_t, \tag{5.56}$$

in which c_s is the uncontrolled spillway discharge coefficient, h_s is the uncontrolled spillway crest, c_g is the gated spillway discharge coefficient, h_g is the center-line elevation of the gated spillway, c_d is the discharge coefficient for flow over the crest of the dam, L_s is the spillway length, and Q_t is a constant outflow term which is head independent or it may be a specified discharge time series. The uncontrolled spillway flow or the gated spillway flow can also be represented as a table of head-discharge values. The gate flow may also be specified as a function of time via a known time series for $A_g(t)$. The breach outflow (Q_b) is computed as broad-crested weir flow (Fread, 1977, 1985, 1988, 1992; Fread and Lewis, 1988), i.e.,

$$Q_b = c_v k_s [3.1 b_i (h_i - h_b)^{1.5} + 2.45 z (h_i - h_b)^{2.5}], \tag{5.57}$$

in which c_v is a small correction for velocity of approach, b_i is the instantaneous breach bottom width, h_i is the elevation of the water surface just upstream of the structure, h_b is the elevation of the breach bottom as described by Eq. (5.2) in which h_b is assumed to be a linear function of time (t_b) from beginning of the breach formation time (τ), z is the side slope of the breach, and k_s is the submergence correction factor due to the downstream tailwater elevation (h_t), i.e.,

$$k_s = 1.0, \qquad h^* \leq 0.67, \tag{5.58}$$

$$k_s = 1.0 - 22.3(h^* - 0.67)^3, \qquad h^* > 0.67, \tag{5.59}$$

where:

$$h^* = (h_t - h_b)/(h_i - h_b). \tag{5.60}$$

If the breach is formed by piping, Eq. (5.57) is replaced by an orifice equation:

$$Q_b = 4.8 A_p (h_i - h_p)^{1/2}, \tag{5.61}$$

where:

$$A_p = [b_i + z(h_p - h_b)](h_p - h_b), \tag{5.62}$$

in which h_p is the specified center-line elevation of the pipe. Each of the terms in Eq. (5.56) may be modified by a submergence correction factor similar to k_s which can be computed by Eq. (5.59) in which h_b is replaced by h_s, h_g, and h_d, respectively.

Bridge. If the internal boundary represents highway/railway bridges together with their earthen embankments which cross the floodplain, Eqs. (5.54) and (5.55) can still be used although Q_s in Eq. (5.55) is computed by the following contracted bridge flow expression:

$$Q_s = C_b \sqrt{g}\, A_{i+1}(h_i - h_{i+1})^{0.5} + C_d k_s (h_i - h_c)^{1.5}, \tag{5.63}$$

in which C_b is a coefficient of bridge flow, C_d is the coefficient of flow over the crest of the road embankment, h_c is the crest elevation of the embankment, and k_s is similar to Eqs. (5.58)–(5.60) except h_b is replaced by h_c. A breach of the embankment is treated the same as with dams.

5.3.1.8 Levee Overtopping/Floodplain Interactions

Flows which overtop levees located along either or both sides of a main-stem river and/or its principal tributaries can be treated as lateral flow (q) in Eqs. (5.14)–(5.15) where the lateral flow diverted over the levee is computed as broad-crested weir flow. This overtopping flow is corrected for submergence effects if the floodplain water-surface elevation sufficiently exceeds the levee-crest elevation. After the flood peak passes, the overtopping flow may reverse its direction when the floodplain water-surface elevation exceeds the river water-surface elevation, thus allowing flow to return to the river. The overtopping broad-crested weir flow is computed according to the following:

$$q = -c_\ell k_s (h - h_c)^{3/2}, \tag{5.64}$$

where k_s, the submergence correction factor, is computed as in Eqs. (5.58)–(5.60) except $h^* = (h_{fp} - h_c)/(h - h_c)$, in which c_ℓ is the weir discharge coefficient, h_c is the levee-crest elevation, h is the watersurface elevation of the river, and h_{fp} is the water-surface elevation of the floodplain. Flow in the floodplain can affect overtopping flows via the submergence correction factor. Flow may also pass from the waterway to the floodplain through a time-dependent crevasse (breach) in the

Dam-Breach Floods

levee via a breach-flow equation similar to Eq. (5.57). The floodplain, which is separated from the principal routing channel (river) by the levee, may be treated as: (a) a deadstorage area (A_o) in the Saint-Venant equations; (b) a tributary which receives its inflow as lateral flows (the flows from the river which overtop the levee-crest) which are simultaneously dynamically routed along the floodplain; and (c) the flows and water-surface elevations can be computed by using a level-pool routing method particularly if the floodplain is divided into compartments by levees (dikes) or elevated roadways located somewhat perpendicular to the river levee(s).

5.3.1.9 Supercritical/Subcritical Mixed Flow

Flow can change with either time or distance along the routing reach from supercritical to subcritical while passing through critical flow, or conversely. This "mixed flow" requires special treatment to prevent numerical instabilities in the solution of the Saint-Venant equations. This difficulty can be addressed by using a concept based on avoiding the use of the Saint-Venant equations at the point where mixed flow occurs. An enhanced mixed flow algorithm automatically subdivides the total routing reach into sub-reaches wherein only subcritical or supercritical flows occur (Fread, 1983, 1985, 1988). The transition locations where flow changes from subcritical to supercritical or vice versa are treated as boundary conditions thus avoiding the application of the Saint-Venant equations to the transition flow and subsequent numerical solution difficulties. The mixed-flow algorithm has two components, one for obtaining the initial condition of discharge and water elevation at $t = 0$ and another which functions during the unsteady flow solution. The Froude number (Fr) is used to determine the supercritical reaches, for which Fr > 1. At each time step, the solution commences with the most upstream sub-reach, and proceeds sub-reach by subreach in the downstream direction. Hydraulic jumps are allowed to move upstream (downstream) at the end of a time step according to the relative values of supercritical (subcritical) sequent depth and the adjacent downstream subcritical (upstream supercritical) depth.

An alternative for treating mixed flows (Fread et al. 1996) is to provide a "local partial inertia" filter $(1 - Fr^m)$ which multiplies the first two (inertia) terms in the momentum Eq. (5.15). Fr is the Froude number of the flow in any ith Δx-reach and the exponent (m) varies from 1 to 10 with 5 visually preferred. The filter takes on a value of zero when Fr > 1. The local partial inertia filter avoids numerical difficulties associated with mixed flows while introducing negligible errors, less than about 2%, for almost all flow conditions.

5.3.1.10 Flow Through a River System

A river system consisting of a main-stem river and one or more principal tributaries is efficiently solved using an iterative relaxation method (Fread, 1973, 1985) in which the flow at the confluence of the main-stem and tributary is treated as the lateral inflow/outflow (q) in Eqs. (5.14)–(5.15). If the river consists of bifurcations such as islands and/or complex dendritic systems with tributaries connected to tributaries, etc., a network solution technique is used (Fread, 1985), wherein three internal boundary equations conserve mass and momentum at the confluence. This system of algebraic equations uses a special sparse matrix Gaussian elimination technique for an efficient solution (Fread, 1983).

5.4 Dam-Breach Flood Routing Data

5.4.1 Cross-Sectional Properties

Much of the uniqueness of a specific dam-breach flood routing application is due to the properties of the cross sections located at selected points along the downstream channel/valley shown in Fig. 5.5. The computation of discharge (Q) and water surface elevation (h) by solving the Saint-Venant equations occurs at the locations where cross sections are selected.

5.4.1.1 Active Sections

That portion of the channel cross section in which flow occurs is called active. Cross sections may be of regular or irregular geometrical shape. As indicated in Fig. 5.5, each cross section can be described by tabular values of channel topwidth and water-surface elevation which constitute a piece-wise linear relationship. Generally about 4 to 12 sets of topwidths and associated elevations provide a sufficiently accurate description of the cross section. Area-elevation tables can be generated initially from the specified topwidth-elevation data. Areas or widths associated with a particular water-surface elevation are linearly interpolated from the tabular values. Cross sections at gaging station locations are generally used as computational points, as well as those locations along the river where significant cross-sectional or flowresistance changes occur or at locations where major tributaries enter. The spacing of cross sections can range from a few hundred feet to a few miles apart. Typically, cross sections are spaced farther apart for large rivers than for small streams, since the degree of variation in the cross-sectional characteristics is greater for the small streams. It is essential that the selected cross sections, with the assumption of linear variation between adjacent sections, represent the volume available to contain the flow along the watercourse.

5.4.1.2 Inactive (Dead) Sections

There can be inactive portions of a cross section where the flow velocity in the x-direction is negligible relative to the velocity in the active portion. The inactive portion is also called off-channel (dead) storage; it is represented by the term (A_o) in Eq. (5.14). Off-channel storage areas can be used to effectively account for adjacent embayments, ravines, or tributaries (see Fig. 5.5) which connect at some elevation with the flow channel but do not convey flow in the x-direction; they serve only to store some of the passing flow. Sometimes, off-channel storage can be used to simulate a heavily wooded floodplain which primarily stores some of the flood waters while conveying a very minimal portion of the flow. Dead storage cross-sectional properties can be described by width (dead storage) vs. elevation tables.

5.4.2 SINUOSITY FACTORS

A meandering or sinuous channel provides a longer flow path than that provided by the floodplain. This effect is simulated via the sinuosity factors (s_c and s_m) in Eqs. (5.14)–(5.15). The sinuosity factor is specified for each reach between two adjacent specified cross sections. The sinuosity factor, which is always ≥ 1.0, is the ratio of the flow-path distance along the meandering channel to the mean flow-path distance along the floodplain. For those elevations used to describe the topwidth at bankfull elevation and below, the sinuosity factor is as previously defined; however, at elevations above bankfull, the sinuosity factor for each layer of flow between specified elevations is decreased such that for those flow layers, say 5 to 10 feet above bankfull, the sinuosity factor is reduced to unity. This indicates that the floodplain flow has fully captured the upper layers of flow directly above the channel.

The sinuosity factor, as used in the finite-difference SaintVenant Eqs. (5.35)–(5.36), is depth-weighted according to Eqs. (5.21)–(5.22). The depth-weighting results in a sinuosity factor which only approaches unity, even for the upper elevations associated with large floodplain flows. This occurs since the total flow is still comprised of the relatively small flow within bankfull which follows the meandering channel, as well as the larger portion of the total flow which follows the floodplain flow path.

5.4.3 MANNING n FRICTION COEFFICIENTS

The resistance to flow in the channel/valley may be parameterized by the Manning n or some other friction (roughness) coefficient which represents the effect of roughness elements of the channel bank and bed particles as well as form losses attributed to dynamic alluvial bed-forms and vegetation of various types (grass, shrubs, field crops, brush, and trees) located along the banks and overbanks (floodplain). Also river bend losses are often included as components of the Manning n. The Manning n is defined for each channel reach and specified as a tabular piece-wise linear function of stage or discharge, with linear interpolation used to obtain values intermediate to the tabulated values.

5.4.3.1 Estimation

The Manning n varies with the magnitude of flow. As the flow increases and more portions of the bank and overbank become inundated, the vegetation located at these elevations causes an increase in the resistance to flow. Also, the Manning n may be larger for small floodplain depths than for larger depths due to flattening of the brush, thick weeds, or tall grass as the flow depths and velocities increase. This effect may be reversed in the case of wooded overbanks where, at the greater depths, the flow impinges against the leaved branches rather than only against the tree trunks, thus increasing the Manning n. The Manning n may also decrease with increasing discharge when the increase in the overbank flow area is relatively small compared to the increase of flow area within the banks, as the case of wide rivers with levees situated closely along the natural river banks, or when floods

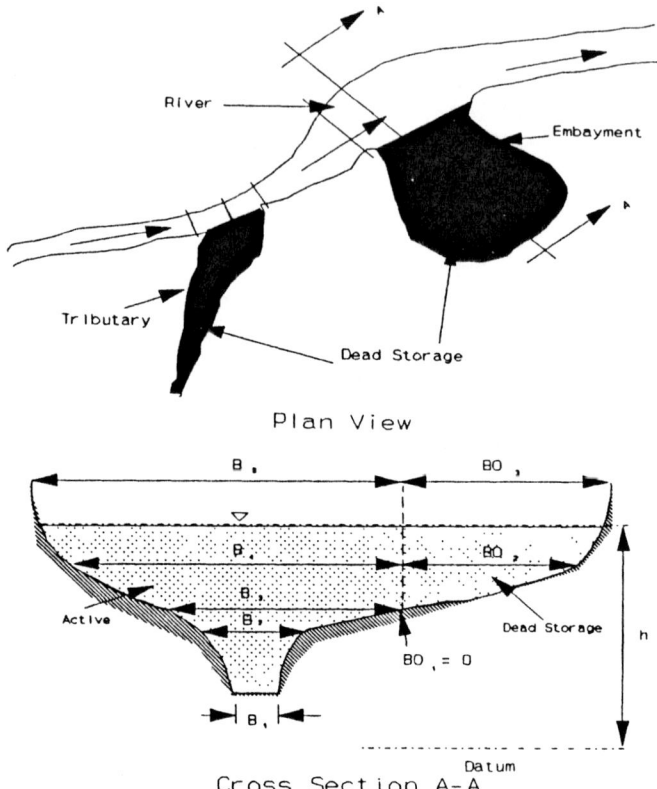

Fig. 5.5. Plan view of a river with active and dead storage areas, and a cross-section view.

remain confined within the channel banks. Seasonal influences (leaves and weeds occur in summer but not in winter) may also affect the selection of the Manning n.

Basic references for selecting the Manning n may be found in Chow (1959), Barnes (1967), and Chow et al. (1988); Arcement and Schneider (1984) can be used for wooded and urbanized floodplains; and Jarrett (1984) proposed the following predictor for the Manning n for in-bank flows of relatively steep ($0.002 \leq S_o \leq 0.040$) streams with gravel/cobble/boulder beds, i.e.,

$$n = 0.39 S_o^{0.38}/R^{0.16}, \tag{5.65}$$

in which S_o is the bottom slope (ft/ft) and R is the hydraulic radius (ft). Manning n values for flows less than bankfull are approximately 0.015–0.035 for large rivers (Mississippi, Ohio, Missouri, Illinois), 0.03–0.04 for moderate sized rivers and streams, 0.04–0.07 for mountain streams, and 0.04–0.25 for overbank flows (Fread, 1989a).

Unfortunately, the flow observations used in developing the Manning n predictive methodologies have been confined to floods originating from rainfall/snowmelt-runoff. The much greater magnitude of a dam-break flood pro-

Dam-Breach Floods

duces greater velocities and results in the inundation of portions of the floodplain never before inundated.

Also, the dam-break flood is much more capable than the lesser runoff-generated flood of creating and transporting large amounts of debris, e.g., uprooted trees, demolished houses, vehicles, etc. The higher velocities of the dam-breach flood will cause additional energy losses due to temporary flow obstructions formed by transported debris which impinge against some more permanent feature along the river such as a bridge or other man-made structure. Therefore, the Manning n values often need to be increased in order to account for the additional energy losses associated with the dam-break flows such as those due to the temporary debris dams which form and then disintegrate when ponded water depths become too great. The extent of the debris effects, of course, is dependent on the availability and amount of debris which can be transported and the existence of man-made or natural constrictions where the debris may impinge behind and form temporary obstructions to the flow.

5.4.3.2 Calibration

The Manning n for the range of flows associated with previously observed floods may be selected via a trial-and-error calibration methodology. With observed stages and flows, preferably continuous hydrographs from a previous large flood, an unsteady flow routing model can be used to determine the Manning n values as follows: (1) use the observed flow hydrograph as the upstream boundary condition and select an appropriate downstream boundary (an observed stage hydrograph at the downstream boundary could be used if available); (2) estimate the Manning n values throughout the routing reach; (3) obtain computed h and Q from the solution of the Saint-Venant equations; (4) compare the computed elevations with the observed elevations at the upstream boundary and elsewhere; (5) if the computed elevations are lower than the observed, increase the estimated Manning n values; or if the computed elevations are higher than the observed, decrease the estimated Manning n values; (6) repeat steps (3)–(5) until the computed and observed elevations are approximately the same. The final Manning n values are sufficient for the range of flows used in the calibration; however, the Manning n values for those flow elevations exceeding the observed must be estimated as previously discussed. The calibrated Manning n values, however, provide an initial estimate from which the unknown Manning n values may be extrapolated or ultimately approximated.

5.4.4 LEVEE PROPERTIES

The levee properties required for modeling their effects are the elevation (h_c) of the top of the levee and an estimated discharge coefficient (c_ℓ) which has the range, $2.6 < c_\ell < 3.1$. These properties need to be specified for each reach between selected cross sections.

5.4.5 LATERAL FLOWS

Specified unsteady flows associated with tributaries that are not dynamically routed can be added to the unsteady flow along the routing reach. This is accomplished via the term q in Eqs. (5.14) and (5.15). The total tributary flow which is a known function of time, i.e., $Q_i(t)$ which is a specified time series, is distributed along a single Δx_i reach, i.e., $q_i(t) = Q_i(t)/\Delta x_i$. Backwater effects of the routed flow on the tributary flow are ignored, and the lateral flow is usually assumed to enter perpendicular to the routed flow. Known outflows can be simulated by using a negative sign with the specified $Q_i(t)$. Numerical difficulties in solving the Saint-Venant equations sometimes arise when the ratio of lateral inflow to channel flow, q_i/Q_i, is too large; this can be overcome by increasing Δx_i for this reach.

5.5 Teton Dam-Breach Flood Case Study

A case study using the DAMBRK Model (Fread, 1977, 1988, 1989; Chow et al., 1988) for the Teton dam-breach flood is presented. The DAMBRK Model is based on the dam breach and flood routing equations presented in Sections 5.2 and 5.3. It has been favorably reviewed (Land, 1980; Wurbs, 1986) and has received wide applications in the United States and in many countries throughout the world. DAMBRK model results are compared with observed downstream peak stages, discharges, and travel times.

The Teton Dam, a 300 ft high earthen dam with a 3,000 ft long crest and 250,000 acre-ft of stored water, failed on June 5, 1976, killing 11 people, making 25,000 homeless, and inflicting about $400 million in damages to the downstream Teton-Snake River Valley. Data from a U.S. Geological Survey Report by Ray, et al. (1977) provided observations on the approximate development of the breach, a description of the reservoir storage, downstream cross sections and estimates of Manning n approximately every 5 miles, indirect peak discharge measurements at two sites and rating curves at two other sites, flood-peak travel times, and floodpeak elevations at frequent locations along the downstream channel/valley. The inundated area was as much as 9 miles in width about 16 miles downstream of the dam.

The following breach parameters were used to reconstitute the downstream flooding due to the failure of the Teton Dam: $\tau = 1.4$ hrs, $b = 81$ ft, $z = 1.04$, $h_{bm} = 0.0$, $h_d = H_d = 261.5$ ft. They were obtained from the BREACH model (Fread, 1984, 1989b). The time of failure (τ) was obtained by using the average of two values, i.e.,

$$\tau = 0.5(\tau_1 + \tau_2), \tag{5.66}$$

where τ is computed by rearranging Eq. (5.13) with Q_p, $\bar{b}$, H_d computed by the BREACH model, i.e.,

$$\tau_1 = C[(3.1\bar{b}/Q_p)^{1/3} - 1/H_d^{0.5}r], \tag{5.67}$$

in which $Q_p = 2{,}200{,}000$ cfs, $\bar{b} = 353$ ft, and $C = 2.34 S_a/\bar{b}$, in which $S_a = 1936$ acres. The term (τ_2) is derived by equating the integrated area of the computed

Dam-Breach Floods

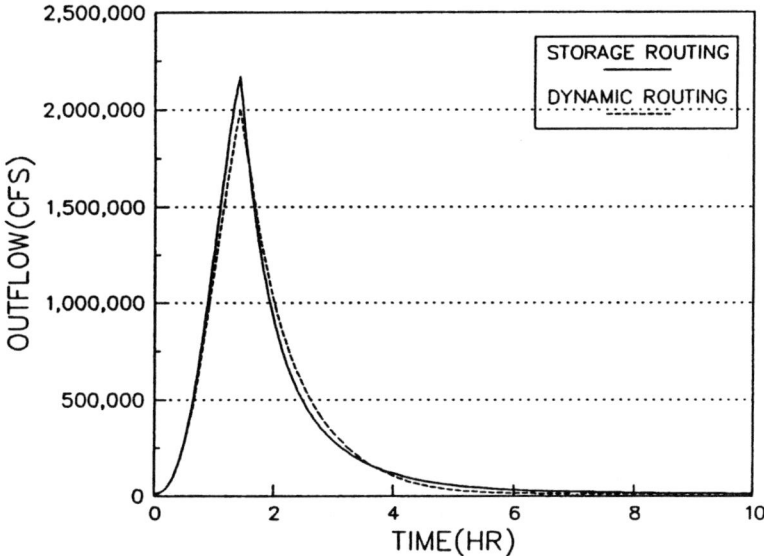

Fig. 5.6. Outflow hydrograph from the Teton Dam failure.

outflow hydrograph $Q(t)$ (from beginning of breach to T_p, the time when the peak outflow occurs) to a triangle with Q_p as the peak and τ_2 as the base, i.e.,

$$\tau_2 = \frac{2}{Q_p} \int_0^{T_p} Q(t)\, dt. \tag{5.68}$$

Cross-sectional properties were used at 12 locations along the 60-mile reach of the Teton-Snake River Valley below the dam. Five topwidths were used to describe each cross section. The downstream valley consisted of a narrow canyon (approx. 1,000 ft wide) for the first 5 miles and thereafter a wide valley which was inundated to a maximum width of about 9 miles. Manning n values ranging from 0.038 to 0.047 were provided from field estimates by the Geological Survey. Computational distance steps (Δx) between cross sections were assigned values that gradually increased from 0.5 miles near the dam, to a value of 1.4 miles near the downstream boundary at the Shelly gaging station (valley mile 59.5 downstream from the dam). The reservoir surface area-elevation values were obtained from U.S. Geological Survey topographic maps. The downstream boundary was assumed to be channel flow control as represented by a loop-rating curve given by Eq. (5.50).

The computed outflow hydrograph using reservoir level-pool routing is shown as the solid line in Fig. 5.6 with a peak value of 2,172,000 cfs, a time to peak of 2.15 hrs, and a total duration of significant outflow of about 6 hrs. This peak discharge is about 30 times greater than the flood of record approximately 45 miles downstream at Idaho Falls on the Snake River. The temporal variation of the computed time-integrated outflow volume compared within 3 percent of the observed. Also, in Fig. 5.6, a comparison is presented of Teton reservoir outflow hydrographs computed via reservoir dynamic routing as shown by the dashed line.

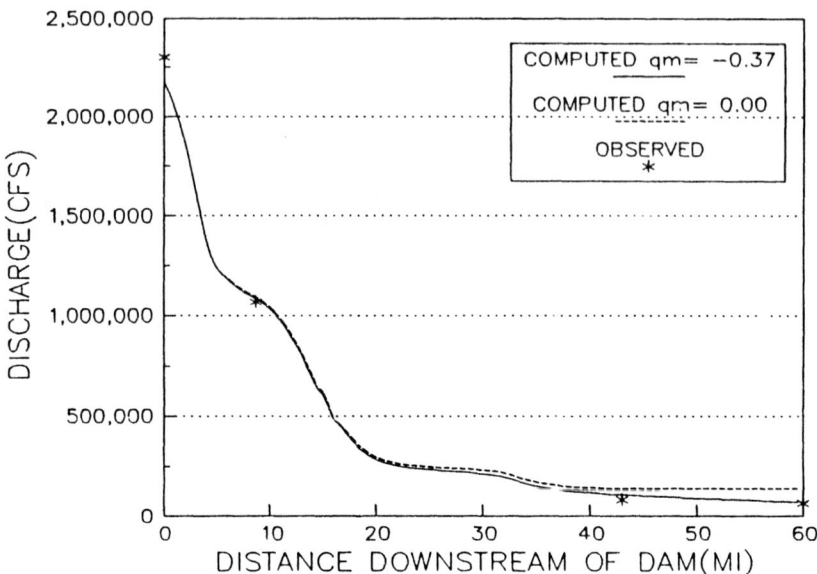

Fig. 5.7. Profile of peak discharge from the Teton Dam failure.

Since the breach of the Teton Dam formed gradually over approximately a two hour interval, a steep negative wave did not develop in the reservoir. Also, the inflow to the reservoir was insignificant. For these reasons, the reservoir surface remained essentially level during the reservoir drawdown and the dynamic routing yielded almost the same outflow hydrograph as the level-pool routing technique.

The computed peak discharge values along the 60-mile downstream valley are shown in Fig. 5.7 along with four observed values (two by indirect measurement; two by rating curves) at downstream miles 2.0, 8.5, 43.0, and 59.5. The average absolute difference between the computed and observed values is 5.2 per cent. Most apparent is the extreme attenuation of the peak discharge as the flood wave propagates through the valley. Two computed curves are shown in Fig. 5.7; one in which no flow losses were assumed, and a second in which the flow losses amounting to about 30 percent of the reservoir outflow volume, which were due to infiltration and detention storage behind irrigation levees distributed throughout large portions of the inundated floodplain, were accounted for in the routing via the q term in Eqs. (5.14)–(5.15).

The a priori selections of the breach parameters (τ and b) cause the greatest uncertainty in forecasting dam-break flood waves. The sensitivity of downstream peak discharges to reasonable variations in τ and b are shown in Fig. 5.8. Although there are large differences in the discharges (+75 to –42 percent) near the dam, these rapidly diminish in the downstream direction. After 8.5 miles the variation is about ±17 percent, and after 22 miles the variation has further diminished to about ±6 percent. The tendency for extreme peak attenuation and rapid damping of differences in the peak discharge is accentuated in the case of Teton Dam due to the presence of the very wide downstream valley. Had the narrow canyon extended

Dam-Breach Floods

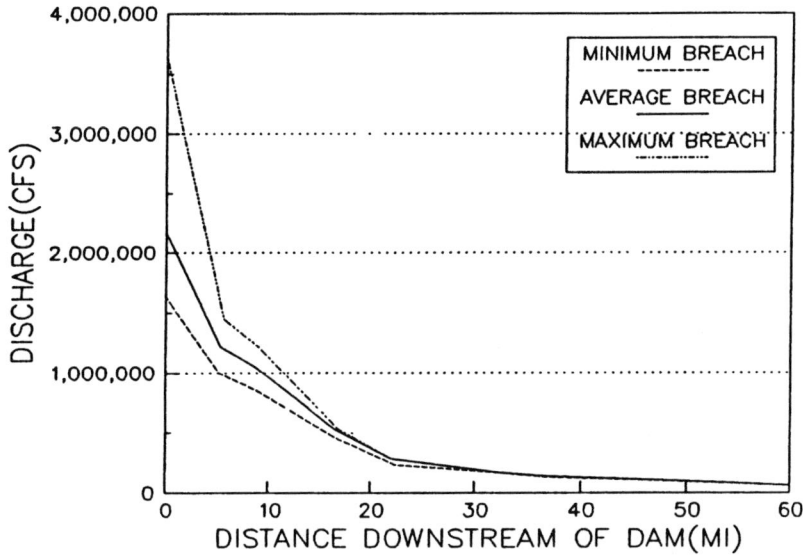

Fig. 5.8. Profile of peak discharge from the Teton Dam failure showing the sensitivity of various breach parameters.

all along the 60-mile reach to Shelly, the peak discharge would not have attenuated as much and the differences in peak discharges due to variations in τ and $\bar{b}$ would be more persistent. In this instance, the peak discharge would have attenuated to about 750,000 rather than 67,000 as shown in Fig. 5.8, and the differences in peak discharges at mile 59.5 would have been about ±17 percent as opposed to ±5 percent as shown in Fig. 5.8.

Computed peak elevations compared favorably with observed values, as shown in Fig. 5.9. The average absolute error was 1.9 ft, while the average arithmetic error was only +0.8 ft.

The computed flood-peak travel times and three observed values are shown in Fig. 5.10. The differences between the computed and observed travel times at mile 59.5 are about 5 percent for the case of using the estimated Manning n values and about 13 percent if the Manning n values are arbitrarily increased by 20 percent.

As stated previously in Section 5.4.3.1, the Manning n must be estimated, especially for the flows above the flood of record. The sensitivity of the computed water elevations and discharges of the Teton flood due to a substantial change (20 percent) in the Manning n was found to be as follows: (1) 0.3 ft in computed peak water surface elevations or about 1 percent of the maximum flow depths, (2) 13 percent deviation in the computed peak discharges, (3) 0.5 percent change in the total attenuation of peak discharge incurred in the reach from the Teton Dam to the Shelly gaging station, and (4) 13 percent change in the flood-peak travel time at Shelly. These results indicate that Manning n has little effect on peak elevations or depths; however, the travel time is affected by more than one-half of the percentage change in the Manning n values.

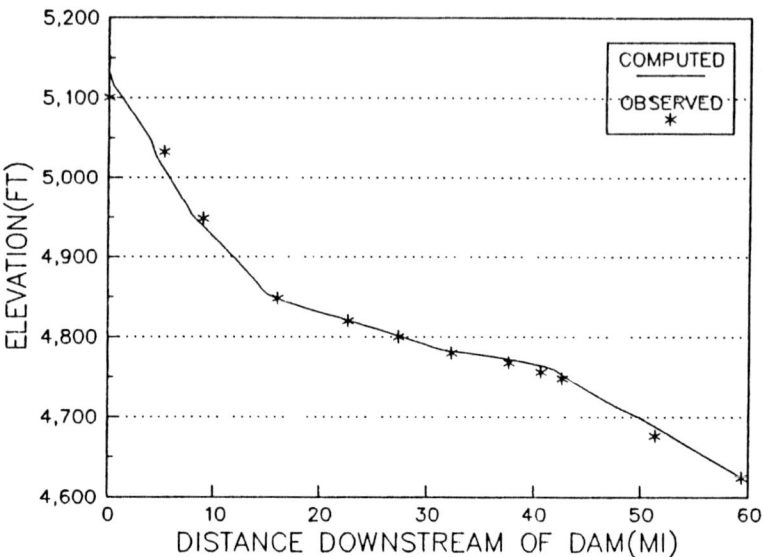

Fig. 5.9. Profile of peak flood elevation from the Teton Dam failure.

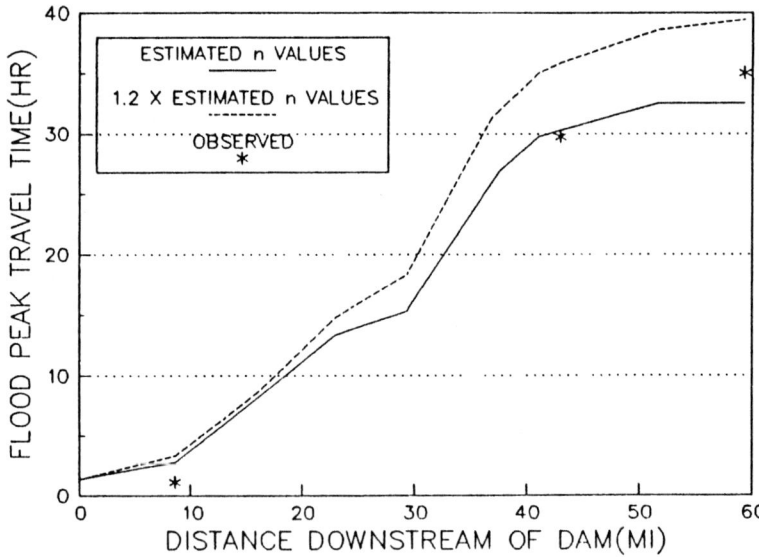

Fig. 5.10. Travel time of the flood peak from the Teton Dam failure.

A typical simulation of the Teton flood as described above involved 73 Δx reaches, 55 hrs of prototype time, and an initial time step (Δt) of 0.07 hrs which automatically increased gradually to 0.58 hrs. The simulation required only 30 seconds on the latest PC micro-computer.

5.6 Uncertainties of Dam-Breach Flood Modeling

Dam-breach flood modeling is subject to uncertainties due to the governing equations and the lack of exact specification of some of the model's parameters.

5.6.1 TWO-DIMENSIONAL EFFECTS

When the governing equations for routing hydrographs (unsteady flows) are the one-dimensional Saint-Venant equations, there are some instances where the flow is more nearly two-dimensional than one-dimensional, i.e., the velocity of flow and water surface elevations vary not only in the x-direction along the river/valley but also in the transverse direction perpendicular to the x-direction. Neglecting the two-dimensional nature of the flow can be important when the flow first expands onto an extremely wide and flat floodplain after having passed through an upstream reach which severely constricts the flow. In many cases where the wide floodplain is bounded by rising topography, the significance of neglecting the transverse velocities and water surface variations is confined to a transition reach in which the flow changes from one-dimensional to two-dimensional and back to one-dimensional along the x-direction. In this case, the use of radially defined cross sections along with judicious off-channel storage widths can minimize the two-dimensional effect neglected within the transition reach. The radial cross sections appear in plan-view as concentric circles of increasing diameter in the downstream direction which is considered appropriate for radial flow expanding onto a flat plane. The cross sections become perpendicular to the x-direction for the reach downstream of the transition reach. Where the very wide, flat floodplain appears unbounded, the radial representation of the cross sections is at best only an approximation which varies from reality the farther from the constricted section and the greater the variability of the floodplain topography and friction.

5.6.2 CROSS-SECTIONAL DEGRADATION

The high velocity flows associated with dam-break floods can cause significant scour (degradation) of alluvial channels. This enlargement in channel cross-sectional area is neglected since the equations for sediment transport, sediment continuity, dynamic bed-form friction, and channel bed armoring are not included among the governing equations. The significance of the neglected alluvial channel degradation is directly proportional to the channel/floodplain conveyance ratio, since the characteristics of most floodplains along with their much lower flow velocities cause much less degradation within the floodplain. As this ratio increases, the degradation could cause a significant lowering of the water surface elevations until the flows are well within the recession limb of the dam-break hydrograph; however, in many instances this ratio is fairly small and remains such until the dambreak flood peak has attenuated significantly at locations far downstream of the dam, and where this occurs the maximum flow velocities also have attenuated. However, narrow channels with minimal floodplains are subject to overestimation of water elevations due to significant channel degradation. The alluvial fill (aggradation) occurring in the floodplain or in the channel during receding flows are considered to have relatively small effects on the flood conditions.

5.6.3 MANNING n

The uncertainty associated with the selection of the Manning n can be quite significant for dam-break floods due to: (1) the great magnitude of the flood produces flow in portions of floodplains which were never before inundated; this necessitates the selection of the n value without the benefit of previous evaluations of n from measured elevation/discharges or the use of calibration techniques for determining the n values; (2) the effects of transported debris can alter the Manning n. Although the uncertainty of the Manning n may be large, this effect is considerably damped or reduced during the computation of the water surface elevations. Based on the Manning equation, the relationship between the error or uncertainty in the Manning n and the resulting flow depth is as follows:

$$d_e/d = (n_e/n)^{b'}, \qquad (5.69)$$

where:

$$b' = 3/(3m + 5), \qquad (5.70)$$

in which d_e is the flow depth associated with an erroneous n_e value, d is the flow depth associated with the correct n value, and m is a cross section shape factor, i.e., $m = 0$ for rectangular sections, $m = 0.5$ for parabolic, $m = 1$ for triangular, and $1 < m < 3$ for channels with floodplains (the wider and more flat the floodplain, the greater the m value). Since for channels with wide floodplains ($m \simeq 2$), the exponent b' as defined by Eq. (5.70) is equal to 0.27; and from an inspection of Eq. (5.69) it is evident that the difference between d_e and d is substantially damped relative to the difference between n_e and n. In fact, if $n_e/n = 1.5$, then $d_e/d = 1.12$, which illustrates the degree of damping. Thus for rivers with wide floodplains the uncertainty in the Manning n results in considerably less uncertainty in the flow depths.

The propagation speed (c) of the floodwave is related to the uncertainty in the Manning n according to the following:

$$c_e/c = (n_e/n)^{0.67b'-1}, \qquad (5.71)$$

in which c_e is the propagation speed associated with an erroneous n_e value. If $n_e/n = 1.5$, then $c_e/c = 0.72$, which indicates less damping than that associated with Eq. (5.69). Thus errors in the Manning n affect the rate of propagation more than the flow depth, but in each instance the error is not proportional to the n_e error, but rather the error is damped.

When the range of probable Manning n values is fairly large, a sensitivity test should be made using a dam-breach flood routing model to simulate the flow, first with the lower estimated n values and then with the higher estimated n values. The resulting high water profiles computed along the river/valley for each simulation represent an envelope of possible flood peak elevations within the range of uncertainty associated with the estimated n values.

5.6.4 DEBRIS EFFECTS

Dam-break floods create a large amount of transported debris; this may accumulate at constricted cross sections such as bridge openings where it acts as a temporary dam and partially or completely restricts the flow. The maximum magnitude of this effect, i.e., the upper envelope of the flood peak elevation profile, can be approximated by using a dam-breach flood routing model to simulate the blocked constriction as a downstream dam having an estimated elevation-discharge relation approximating the gradual flow stoppage. Also, the downstream dam can be used to simulate the later rapid increase due to the release of the ponded waters when the debris dam is allowed to breach.

5.6.5 BREACH PROPERTIES

The uncertainty associated with the breach parameters, especially $\bar{b}$ and τ, also cause uncertainty in the flood peak elevation profile and arrival times. The best approach is to perform a sensitivity test using minimum, average, and maximum values for $\bar{b}$ and τ. The maximum flood is produced by selecting the maximum probable $\bar{b}$ and minimum probable τ, whereas the minimum flood is produced by using the minimum probable $\bar{b}$ and maximum probable τ values. The differences in flood peak properties (flow, elevation, time of arrival) at any section downstream of the dam due to variations in the breach parameters reduces in magnitude or is damped as the dam-break flood propagates through the downstream river/valley.

5.6.6 FLOW LOSSES

There is uncertainty associated with volume losses incurred by the flood as it propagates downstream and inundates large floodplains where infiltration and detention storage losses may occur. Such losses are difficult to predict. The conservative approach is to neglect such losses, unless very good reasons justify their consideration, e.g., observed losses associated with several previous large floods in the same floodplain.

References

Arcement, G.J., Jr. and Schneider, V.R. (1984). Guide for Selecting Manning's Roughness Coefficients for Natural Channels and Flood Plains, Report No. RHWA-TS-84-204, U.S. Geological Survey for Federal Highway Administration, National Tech. Information Service, PB84-242585, 61 pp.
ASCE/USCOLD (1975). Lessons from Dam Incidents, USA, American Society of Civil Engineers, New York.
Barnes, H.H., Jr. (1967). Roughness Characteristics of Natural Channels, Geological Survey Water-Supply Paper 1849, U.S. Government Printing Office, Washington, DC, 213 pp.
Bechteler, W. and Broich, K. (1993). 'Computational Analysis of the Dam-Erosion Problem,' Advances in Hydro-Science and Engineering, Vol. 1, Wang, S.S.Y. (editor), Ctr. for Computational Hydroscience and Engineering, Univ. of Mississippi, pp. 723-728.

Chen, C.L. and Armbruster, J.T. (1980). 'Dam-Break Wave Model: Formulation and Verification,' J. Hydraul. Div., ASCE, Vol. 106, No. HY5, May, pp. 746-767.
Chow, V.T. (1959). Open-Channel Hydraulics, McGraw-Hill, New York.
Chow, V.T., Maidment, D.R., and Mays, L.W. (1988). Applied Hydrology, McGraw-Hill, New York.
Cristofano, E.A. (1965). Method of Computing Rate for Failure of Earth Fill Dams, Bureau of Reclamation, Denver, CO, April.
Cunge, J.A., Holly, F.M., Jr., and Verway, A. (1980). Practical Aspects of Computational River Hydraulics, Pitman, Boston, MA.
Davies, W.E., Bailey, J.F., and Kelly, D.B. (1972). 'West Virginia's Buffalo Creek Flood: A Study of the Hydrology and Engineering Geology,' Geological Survey Circular 667, U.S. Geological Survey, 32 pp.
DeLong, L.L. (1989). 'Mass conservation: 1-D Open Channel Flow Equations,' J. Hydraul. Div., Vol. 115, No. HY2, pp. 263-268.
Dressler, R.F. (1954). 'Comparison of Theories and Experiments for the Hydraulic Dam-Break Wave,' Internat. Assoc. Sci. Pubs., 3, No. 38, pp. 319-328.
Federal Investigative Board (1977). Report of Failure of Kelly Barnes Dam, Toccoa, Georgia, U.S. Army Corps of Engineers, Atlanta, GA.
Fread, D.L. (1971). 'Discussion of Implicit Flood Routing in Natural Channels,' by M. Amein and C. S. Fang, J. Hydraul. Div., ASCE, Vol. 97, No. HY7, pp. 1156-1159.
Fread, D.L. (1973). 'Technique for Implicit Dynamic Routing in Rivers with Tributaries,' Water Resources Research, Vol. 9, No. 4, pp. 918-926.
Fread, D.L. (1974). Numerical Properties of Implicit Four-Point Finite Difference Equations of Unsteady Flow, HRL-45, NOAA Tech. Memo NWS HYDRO-18, Hydrologic Research Laboratory, National Weather Service, Silver Spring, MD.
Fread, D.L. (1977). 'The Development and Testing of a Dam-Break Flood Forecasting Model,' Proc. of Dam-Break Flood Modeling Workshop, U.S. Water Resources Council, Washington, DC, pp. 164-197.
Fread, D.L. (1981). 'Some Limitations of Contemporary Dam-Break Flood Routing Models,' Preprint 81-525: Annual Meeting of American Society of Civil Engineers, Oct. 17, 1982, St. Louis, MO, Oct. 27, 15 pp.
Fread, D.L. (1983). 'Computational Extensions to Implicit Routing Models,' Proceedings of the Conference on Frontiers in Hydraulic Engineering, ASCE, MIT, Cambridge, MA, pp. 343-348.
Fread, D.L. (1984). 'A Breach Erosion Model for Earthen Dams,' Proceedings of Specialty Conference on Delineation of Landslides, Flash Flood, and Debris Flow Hazards in Utah, Utah State Univ., Logan, UT, June 15, 30 pp.
Fread, D.L. (1985). 'Channel Routing,' Hydrological Forecasting, (Eds: M.G. Anderson and T.P. Burt), John Wiley and Sons, New York, Chapter 14, pp. 437-503.
Fread, D.L. (1987). BREACH: An Erosion Model for Earthen Dam Failures, Hydrologic Research Laboratory, NOAA, NWS, U.S. Dept. of Commerce, Silver Spring, MD, June, 34 pp.
Fread, D.L. (1988). The NWS DAMBRK Model: Theoretical Background/User Documentation, HRL-256, Hydrologic Research Laboratory, National Weather Service, Silver Spring, MD, 315 pp.
Fread, D.L. (1989a). 'Flood Routing and the Manning n,' Proc. of the International Conference for Centennial of Manning's Formula and Kuichling's Rational Formula, (Ed: B.C. Yen), Charlottesville, VA, pp. 699-708.
Fread, D.L. (1989b). 'National Weather Service Models to Forecast Dam-Breach Floods,' Hydrology of Disasters (Eds: O. Starosolszky and O.M. Melder), Proc. of the World Meteorological Organization Tech. Conf., Nov. 1988, Geneva, Switzerland, pp. 192-211.

Fread, D.L. (1992). 'Flow Routing,' Handbook of Hydrology (Ed. D. Maidment), McGraw-Hill, New York, Chapter 10, pp. 10.1-10.36.
Fread, D.L. (1993). 'Selection of Δx and Δt Computational Steps for Four-Point Implicit Non-linear Dynamic Routing Models,' Proceedings, National Hydraulic Engineering Conference, ASCE, San Francisco.
Fread, D.L. and Harbaugh, T.E. (1971). 'Open Channel Profiles by Newton's Iteration Technique,' J. Hydrol., Vol. 13, pp. 70-80.
Fread, D.L. and Lewis, J.M. (1988). 'FLDWAV: A Generalized Flood Routing Model,' Proc. of National Conf. on Hydraulic Engr., ASCE, Colorado Springs, CO, pp. 668-673.
Fread, D.L., Jin, M., Lewis, J.M. (1996).
Froehlich, D.C. (1987). 'Embankment-Dam Breach Parameters,' Proc. of the 1987 National Conf. on Hydraulic Engr., ASCE, New York, August, pp. 570-575.
Froehlich, D.C. (1995).
Hagen, V.K. (1982). 'Re-evaluation of Design Floods and Dam Safety,' Paper Presented at Fourteenth ICOLD Congress, Rio de Janeiro.
Harris, G.W. and Wagner, D.A. (1967). Outflow from Breached Dams, Univ. of Utah.
Henderson, F.M. (1966). Open Channel Flow, Macmillan Co., New York, pp. 285-287.
Hydrologic Engineering Center (1982). HEC-2 Water Surface Profiles Users Manual, U.S. Army Corps of Engineers, Davis, CA.
ICOLD (1973). Lessons from Dam Incidents, Abridged Edition, USCOLD, Boston, MA.
Jarrett, R.D. (1984). 'Hydraulics of High-Gradient Streams,' J. Hydraul. Div., ASCE, Vol. 110, No. HY11, Nov., pp. 1519-1539.
Jarrett, R.D. and Costa, J.E. (1982). Hydrology, Geomorphology, and Dam-Break Modeling of the July 15, 1982, Lawn Lake Dam and Cascade Lake Dam Failures, Larimer County, Co., U.S. Geological Survey, Open File Report 84-62, 109 pp.
Johnson, F.A. and Illes, P. (1976). 'A Classification of Dam Failures,' Water Power and Dam Construction, Dec., pp. 43-45.
Land, L.F. (1980). 'Evaluation of Selected Dam-Break Flood-Wave Models by Using Field Data,' U.S. Geological Survey, Water Resources Investigations 80-44, NSTL Station, MS, 54 pp.
Liggett, J.A. and Cunge, J.A. (1975). 'Numerical Methods of Solution of the Unsteady Flow Equations,' Unsteady Flow in Open Channels, Vol. I, (Eds: K. Mahmood and V. Yevjevich), Vol. I, Chapt. 4, Water Resource Pub., Fort Collins, CO, pp. 89-182.
Macchione, F. and Sirangelo, B. (1988). 'Study of Earth Dam Erosion due to Overtopping,' Hydrology of Disasters, Proc. of Tech. Conf. in Geneva, November 1988, Starosolszky, O. and Melder, O.M. (editors), James and James, London, pp. 212-219.
MacDonald, T.C. and Langridge-Monopolis, J. (1984). 'Breaching Characteristics of Dam Failures,' J. Hydraul. Div., ASCE, Vol. 110, No. HY5, May, pp. 567-586.
Middlebrooks, T.A. (1952). 'Earth-Dam Practice in the United States,' Centennial Transactions, ASCE, Paper No. 2620, pp. 697-722.
O'Brien, J.S. and Julien, P. (1984). 'Physical Properties and Mechanics of Hyperconcentrated Sediment Flows,' Delineation of Landslide, Flash Flood, and Debris Flow Hazards in Utah, Utah State Univ., Utah Water Research Laboratory, Logan, UT, (Ed: D.S. Bowles), General Series UWRL/G-85/03, pp. 260-279.
Ponce, V.M. and Tsivoglou, A.J. (1981). 'Modeling of Gradual Dam Breaches,' J. Hydraul Div., ASCE, Vol. 107, No. HY6, pp. 829-838.
Preissmann, A. (1961). 'Propagation of Translatory Waves in Channels and Rivers,' in Proc., First Congress of French Assoc. for Computation, Grenoble, France, pp. 433-442.
Rajar, R. (1978). 'Mathematical Simulation of Dam-Break Flow,' J. Hydraul. Div., ASCE, Vol. 104, No. HY7, pp. 1011-1026.

Ray, H.A., Kjelstrom, L.C., Crosthwaite, E.G., and Low, W.H. (1976). 'The Flood in Southeastern Idaho from the Teton Dam Failure of June 5, 1976,' Unpublished open file report, U.S. Geological Survey, Boise, ID.

Ré, R. (1946). 'A Study of Sudden Water Release from a Body of Water to Canal by the Graphical Method,' La Houille Blanche (France), No. 3, pp. 181-187.

Ritter, A. (1892). 'The Propagation of Water Waves,' Ver. Deutsch Ingenieure Zeitschr. (Berlin), 36, Pt. 2, No. 33, pp. 947-954.

Sakkas, J.G. and Strelkoff, T. (1973). 'Dam-Break Flood in a Prismatic Dry Channel,' J. Hydraul. Div., ASCE, Vol. 99, No. HY12, Dec. pp. 2195-2216.

Saint-Venant, Barré de (1871). 'Theory of Unsteady Water Flow, with Application to River Floods and to Propagation of Tides in River Channels,' Computes rendus, Vol. 73, Acad. Sci., Paris, France, pp. 148-154, 237-240. (Translated into English by U.S. Corps of Engrs., No. 49-g, Waterways Experiment Station, Vicksburg, MS, 1949.)

Samuels, P.G. (1985). Models of Open Channel Flow Using Preissmann's Scheme, Cambridge Univ., Cambridge, England, pp. 91-102.

Schocklitsch, A. (1917). 'On Waves Created by Dam Breaches,' Adak, Wiss. (Vienna) Proc., 126, Pt. 2A, pp. 1489-1514.

Singh, K.P. and Snorrason A. (1982). 'Sensitivity of Outflow Peaks and Flood Stages to the Selection of Dam Breach Parameters and Simulation Models,' University of Illinois State Water Survey Division, Surface Water Section, Champaign, IL, June, 179 pp.

Singh, V.P. and Quiroga, C.A. (1988). 'Dimensionless Analytical Solutions for Dam Breach Erosion,' J. of Hydraul. Res., Vol. 26, No. 2, pp. 179-197.

Singh, V.P., Scarlatos, P.D., Collins, J.G. and Jourdan, M.R. (1988). 'Breach Erosion of Earthfill Dams (BEED) Model,' Natural Hazards, Vol. 1, pp. 161-180.

Smart, G.M. (1984). 'Sediment Transport Formula for Steep Channels,' J. Hydraul. Div., ASCE, Vol. 110, No. HY3, pp. 267-276.

Stoker, J.M. (1957). Water Waves, Interscience, New York, pp. 452-455.

Strelkoff, T. (1969). 'The One-dimensional Equations of OpenChannel Flow,' J. Hydraul. Div., ASCE, Vol. 95, No. HY3, pp. 861874.

Su, S.T. and Barnes, A.H. (1970). 'Geometric and Frictional Effects on Sudden Releases,' J. Hydraul. Div., ASCE, Vol. 96, No. HY11, Nov., pp. 2185-2200.

U.S. Army Corps of Engineers (1961). 'Floods Resulting from Suddenly Breached Dams – Conditions of High Resistance, Hydraulic Model Investigation,' Misc. Paper 2-374, Report 2, WES, Nov., 121 pp.

U.S. Army Corps of Engineers (1975). National Program of Inspection of Dams, Bul. I-4, Dept. of the Army, Office of Chief of Engineers, Washington, DC.

Wetmore, J.N. and Fread, D.L. (1984). 'The NWS Simplified Dam Break Flood Forecasting Model for Desk-Top and Hand-Held Microcomputers,' Printed and Distributed by the Federal Emergency Management Agency (FEMA), 1984, 122 pp.

Wurbs, R.A. (1986). Comparative Evaluation of Dam-Breach Flood Wave Models, Texas A&M Univ., College Station, TX, May, pp. 13-20.

CHAPTER 6

Extreme Droughts

M.L. Kavvas and M.L. Anderson

6.1 Introduction

Extreme drought is a natural disaster as recently illustrated in the United States by economic and environmental damage (Mayer et al., 1988, and Kennedy et al., 1991), and in parts of Africa by severe famine (Glantz, 1987). Such extreme droughts can influence transportation, water-resource systems, and agriculture as well as intensify anthropogenic impacts on ecosystems potentially causing desertification. In order to illustrate the effects of droughts, Table 6.1 shows a sample of drought impacts on the United States in 1988.

Attempts to define droughts have not led to a unique and universally accepted definition of the term. Scientists see droughts from the perspective of their particular field of study and attempt to quantify observations made from their particular field which leads to confusion. This confusion can be associated with the intrinsic nature of droughts, which exist only because the effects they produce exist — effects that are more readily observable than the physical characteristics of the phenomenon itself, which involve planetary scales and long observational times. In other words, droughts are to be considered as relative phenomena to which only a conventional meaning of the term can be associated. Therefore, one must be aware of the limitations associated with any definition of a drought.

Description of droughts has long occupied researchers in several fields. Hydrologists, atmospheric scientists, climatologists, statisticians and others have examined various aspects of droughts in attempts to define characteristics or patterns of a drought and its effects. In hydrologic practice, empirical or conceptual relations were sought in order to allow the fitting of observations. Various coefficients were defined in order to quantify ambiguous terms such as 'drought severity', 'tendency to return to normal climatic characteristics', etc. A pioneering attempt to apply such procedures was given by Palmer (1965) who, being a meteorologist, saw drought as a purely meteorological phenomenon. Matalas (1963) investigated four probability distributions in order to test the appropriateness of their application

TABLE 6.1
1988 Drought Impacts on the United States (from Mayer et al., 1988)

Agriculture	$13 billion in direct agricultural losses to GNP
Forestry	Estimated 73,000 fires burned 5 million acres (2 times the acreage as of 1987 for the same number of fires)
Wildlife	75% reduction in the number of young waterfowl born in the central United States Decline in economic and sport fish populations in central and western United States
Transportation	Limits on barge traffic on Mississippi, Missouri, and lower Ohio river systems
Power Generation	Corps of Engineers hydropower generation down approximately 23%

to low-stream flow analysis. Dracup et al. (1980) presented a statistical analysis of droughts, utilizing drought magnitude, duration, and intensity as variables for the description of droughts. Sen (1980) introduced a mathematical approach to calculating lengths of droughts utilizing the statistical theory of extremes and Markov processes. Wijayaratne and Golub (1991) utilized synthetically generated data along with historical data in fitting a probability distribution to drought events. Paulson et al. (1985) used thirteen watershed and climatic characteristics in a regional frequency analysis of droughts.

Because of the complexity of the atmospheric-hydrologic-oceanic system with its nonlinear feedback processes, a unique probability distribution which could describe all drought phenomena does not seem possible. Nathan and McMahon (1990) point out the difficulties in obtaining useful data sets and in determining appropriate parameters for fitting a single probability distribution in low-flow frequency analysis. Instead, it may be more appropriate to study the physical processes pertaining to drought phenomena.

Droughts, as physical occurrences, obey the general rules that apply to the evolution of the hydrologic, atmospheric and oceanic systems. Studying droughts from such a rigorous point of view is yet to be accomplished, possibly due to the huge complexity of the processes that are involved.

In fact, only recently, with the development of global climate models, has the scientific community been able to recognize the influence of hydrology on atmospheric processes (see for example Manabe and Wetherald, 1975). The parameterization of the atmospheric boundary layer, i.e. the lowest kilometer or so of air in contact with the ground surface, plays an important role because by means of this layer, the atmosphere communicates with the oceanic and land systems (Mintz,

1984). Not being able to consider the atmospheric system as an independent component of the earth system adds much complexity to climate studies and shows that phenomena with a typical time-scale of the order of years can not be considered purely meteorological. It can be concluded then, that droughts too, being long time-scale phenomena, can not be interpreted correctly unless the influence of the hydrologic and oceanic systems are taken into account. Exchanges of water vapor (evapotranspiration) and of heat will have to be correctly modeled if they are to be used to study the evolution and duration of droughts. This can be done only by considering hydrology as a fundamental component of the climate system.

This chapter focuses on the current state of knowledge pertaining to the physical processes present in extreme droughts. The different physical systems involved in a drought are identified first, followed by an overview of the evolution of an extreme drought. Modeling of droughts is then discussed where various types of models are briefly reviewed, and parameterizations which are used in a simplified climate model, are presented. The chapter concludes with a discussion on extreme droughts.

6.2 Physical Systems Involved in a Drought

As stated earlier, a drought is a complex physical phenomenon involving many nonlinear feedback mechanisms in the interaction of the atmospheric, hydrologic and oceanic systems. It involves a range of spatial scales from the local scale where drought impacts are felt, to the synoptic and global scales where atmospheric and oceanic processes which evolve at these scales, influence the time scale of a drought. An overview of the atmospheric, oceanic and hydrologic systems is given in this section including the factors relating to their interaction.

6.2.1 ATMOSPHERIC SYSTEM

The atmospheric system extends from the surface of the earth upward to a height of about 100 km. This 100 km range is divided into regions based upon physical /chemical composition, mass, and energetic characteristics. In the ascending order from the earth's surface to the top of the atmosphere these regions are called the troposphere, the stratosphere, the mesosphere, and the thermosphere. These regions are separated conceptually by interfaces, called the tropopause (between troposphere and stratosphere), the stratopause (between stratosphere and mesosphere), and the mesopause (between mesosphere and thermosphere).

Of the 100 km of atmosphere, 99% of the mass is found in the bottom 30 km where the troposphere and the lower stratosphere are found. The troposphere is the first 10 km of atmosphere above the earth's surface and is characterized by a negative temperature lapse rate, the predominant concentration of atmospheric water vapor, and instabilities in the mean flow at earth's midlatitudes, called weather systems. The dynamics of the atmosphere with respect to weather systems occur on relatively short time scales, that is, on the order of days to weeks. A set of equations, called the primitive equations (see Pedlosky, 1987), have been used to describe the dynamics of the lower atmosphere. These continuity, momentum, and thermodynamic energy equations contain information on the atmospheric processes which occur at a multitude of space and time scales. In order to filter out the processes which are not important at large space and time scales, a systematic simplification of the primitive equations, known as quasigeostrophic theory, has been developed.

Quasigeostrophic theory consists of a systematic simplification of the governing equations (also known as the primitive equations) of conservation of mass (continuity), momentum and thermodynamic energy. The simplifications are decided by means of scale analysis. Only the leading terms in the equations are kept, whereas other terms that are estimated to be smaller than 1% (or sometimes even 10%) of the leading terms are systematically removed. Such an operation leads, naturally, to a loss of precision and puts a scale constraint on the problem to be studied. However, the gain in physical insight on atmospheric processes compensates for this loss. In other words, with this approximation it becomes possible to make qualitative analyses, as the one that is presented here, which is quite useful for a proper understanding of the drought phenomena and the mechanisms which support them.

One of these simplified equations is the omega equation (Holton, 1979) which can be written as

$$\underbrace{\left(\nabla^2 + \frac{f_0^2}{\sigma}\frac{\partial^2}{\partial p^2}\right)\omega}_{A} = \underbrace{\frac{f_0}{\sigma}\frac{\partial}{\partial p}\left[\mathbf{V_g} \cdot \nabla\left(\frac{\nabla^2\phi}{f_0} + f\right)\right]}_{B} + \underbrace{\frac{1}{\sigma}\nabla^2\left[\mathbf{V_g} \cdot \nabla\left(-\frac{\partial\phi}{\partial p}\right)\right]}_{C},$$

(6.1)

where f_0 is the average Coriolis parameter, considered constant; σ is the stability parameter, considered constant; p is pressure, ω is the 'pressure velocity' dp/dt, closely related to the vertical component of velocity w with the sign changed ($\omega \simeq -w\rho g$); $\mathbf{V_g}$ is the geostrophic component of velocity, i.e. the velocity that results from the balance of pressure gradient and Coriolis forces alone; ϕ is the geopotential, defined by $d\phi = g\,dz$ and representing the work needed to raise a unit parcel of air from a reference datum to a height z; f is the Coriolis parameter which varies with latitude, namely $f = 2\Omega \sin\psi$, Ω being the terrestrial angular velocity and ψ the latitude.

The omega equation is useful in the sense that it can be utilized to give a qualitative interpretation of droughts when one recognizes the physical significance of the three terms indicated as A, B, and C in the above equation (1). Such a qualitative interpretation was given by Bravar and Kavvas (1991a) by first idealizing the 'typical' behavior of the atmosphere as a sinusoidal wave with high and low pressure centers. With this idealization, it becomes possible to make a rough estimate of the behavior of the terms in the omega equation. The first term, A, is proportional to $-\omega$ because the quadratic operator $[\nabla^2 + (f_0^2/\sigma)(\partial^2/\partial p^2)]$ is applied to ω. The second derivative of a sinusoidal wave is proportional to itself, with the sign changed. As A is proportional to the pressure velocity with the sign changed, A is proportional to the vertical velocity of air, w. We can say something

Extreme Droughts

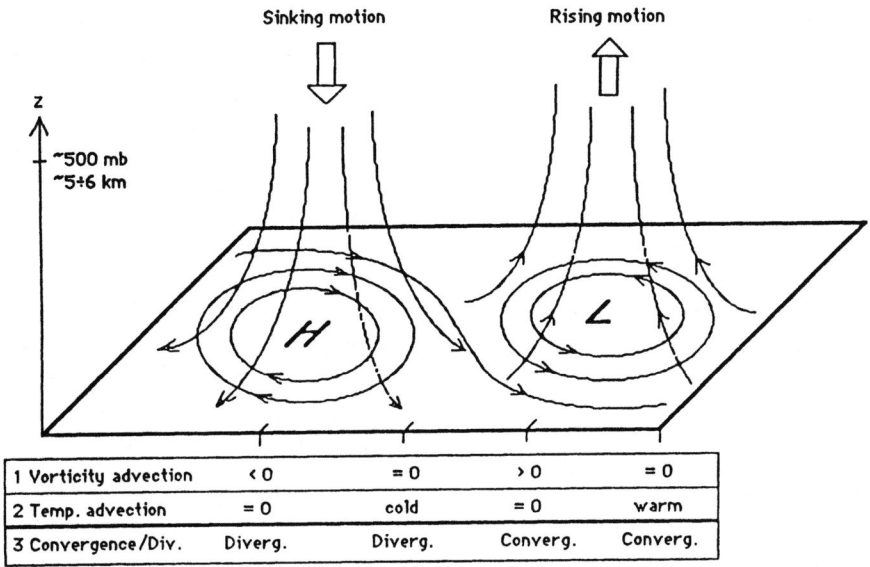

Fig. 6.1. Qualitative deduction of air movement characteristics under the cyclones (low pressure centers) and anticyclones (high pressure centers), obtained from the omega equation (from Bravar and Kavvas, 1991a).

more about A by exploiting the continuity equation. The continuity equation can be written in the form:

$$\frac{\partial w}{\partial z} = -\left(\frac{\partial u}{\partial x} + \frac{\partial v}{\partial y}\right), \qquad (6.2)$$

where u, v, w are the components of velocity along the x, y, z axes, the last one being the vertical axis. Note that the form of the continuity equation is exact even for a compressible fluid, if one takes pressure as the vertical coordinate and therefore substitutes ω in place of w.

The quantity in parentheses in the continuity equation (6.2) is the horizontal divergence, or convergence if the negative sign is included. Since $w = 0$ at the ground surface, if $w > 0$ at some height (e.g., at half the depth of the troposphere, that is, 5 km or so) then also $\partial w/\partial z > 0$ on the average. Therefore, from equation (6.2) it can be deduced that the upward motion of air, $w > 0$, is associated with an area of horizontal convergence at the ground, whereas the downward motion, $w < 0$, corresponds to an area of divergence (see Figure 6.1).

We conclude that term A is proportional not only to vertical velocity, but also to horizontal convergence at the surface. Therefore, A represents the information we seek. In fact, from the sign of w one can have a rough idea of the distribution of precipitation. When w is positive, air is moving upward, encounters smaller and smaller pressures and undergoes an (almost) adiabatic expansion that cools it down, possibly to the point where condensation can occur, and thereby, rain can occur too. When w is negative, the air will undergo an adiabatic warming which makes condensation conditions impossible. Furthermore, in the first case we have horizontal convergence at the ground which collects and concentrates the moisture released at the surface, whereas in the latter case the divergence will dissipate the moisture input from evapotranspiration.

The term B is proportional to the rate of increase with height $(-\partial/\partial p)$ of absolute vorticity advection $(-\mathbf{V}_g \cdot \nabla[\ ])$. Actually, the quantity $[(\nabla^2 \phi/f_0) + f]$ can be shown to be the sum of relative vorticity (strictly related to the square gradient of geopotential) and vorticity of the frame of reference (the Coriolis parameter f is the component of the terrestrial angular velocity along the local vertical axis). Typically, winds increase in intensity with height, so that B can be said to be proportional simply to vorticity advection. In our idealization of weather patterns, positive vorticity is advected following air parcels that move from the high pressure center toward the lower pressures, and vice versa. Obviously, in between, we expect the vorticity advection to be negligible (see Fig. 6.1).

Term C still needs to be analyzed. As the squared gradient ∇^2 is present, assuming a sinusoidal behavior of the weather disturbance, C is proportional to the argument of the operator ∇^2, but with the sign changed. This means that C is proportional to the advection of the quantity $(-\partial \phi/\partial p)$ which can be shown to be roughly proportional to temperature. Therefore, C is proportional to warm air advection and will be positive when air moves northward, as on the right of the low pressure center; it will be negative when air is moved southward, as on the right of a high pressure center (we assume that east is displayed on the right in Fig. 6.1).

At this point it is possible to rewrite the omega equation in qualitative terms by exploiting what has been discussed so far:

$$\text{ground} \begin{bmatrix} \text{convergence} \\ \text{divergence} \end{bmatrix} \alpha[\pm] \text{ vorticity advection} + \begin{bmatrix} \text{warm} \\ \text{cold} \end{bmatrix} \text{ air advection.}$$

The omega equation in this form enables us to deduce the main characteristics of air movement associated with 'usual' cyclones and anticyclones, as is shown in Fig. 6.1. For example, it is by means of this qualitative equation that one can explain the presence of clouds and precipitation not directly associated with fronts but instead with the whole area of low pressure. The processes of precipitation and evaporation are the means through which the oceanic, atmospheric and hydrologic systems interact.

Oceanic System

The component of the earth system where water and energy are stored and transported within the oceans is the oceanic system. The oceans have circulations which are much slower than the atmosphere. Because of this fact, changes in the ocean occur over longer time scales than changes in the atmosphere. However, because extreme droughts occur over time periods on the order of years, and because the ocean and atmosphere form a strongly coupled system (through heat and momentum exchange between the two systems, evaporation transport from the ocean, and precipitation from the atmosphere), the oceans should be considered as a part of the physical system pertaining to a drought.

Changes in the ocean on the same time scale as droughts occur in the surface layer of the ocean which can be defined as the layer of the ocean from the surface to the thermocline, or about the first 100 m of depth (Peixoto and Oort, 1992). Within this layer, the variable with the largest influence is sea surface temperature (SST). Regional SST fluctuations can occur due to fluctuations in radiation, advection of surface waters from other regions, and upwelling of deeper, colder waters (Philander, 1982). Such fluctuations impact evaporation and precipitation patterns as well as energy transport throughout the ocean basins. This is significant since the ocean surface covers two-thirds of the earth's surface and is a significant portion of the lower boundary of the atmosphere.

Hydrologic System

The hydrologic system can be considered as the system wherein water is stored and transported on land sections of earth. Storage of water in this system can be in the form of snowpack, groundwater, rivers, and surface impoundments such as reservoirs and lakes. Anthropogenic factors have a large impact on this system. The hydrologic system receives input from the atmospheric system in the form of precipitation, and returns the moisture through evapotranspiration which is dependent in part on atmospheric processes. It also discharges water into oceans by means of rivers, and receives humidity from oceans through the atmospheric system in the form of precipitation. In some sense, the hydrologic system responds to atmospheric forcings impacting the ecosystems and anthropogenic systems. In turn, the hydrologic system can have significant influence on the atmospheric system.

Above, it was shown that a low pressure center provokes air convergence at the ground, convergence being in the horizontal sense. The water vapor that leaves the ocean system, or the land/vegetation system, is therefore accumulated and forced to move upward. It is already discussed that condensation is associated with the upward movement of air. Therefore, it is expected that water subtracted from the surface will return to it, although not necessarily at the same geographic position. For mid-latitude regions being considered, prevalent winds are westerlies (Wallace and Hobbs, 1977), so that it is possible to conclude that the passage of a low pressure center induces a displacement of oceanic/hydrologic water towards the east. Such a displacement though, will mean a net gain of moisture for regions which are downwind from oceans, such as, for example, California (as Fig. 6.2 suggests). It is also noted that the net gain of moisture is in terms of different time-scales and intensities: low intensity but steady action of evapotranspiration, high intensity but discontinuous rain events. What are called as 'normal conditions' are therefore the result of the balance of these two forcing factors; the moisture, released by land and vegetation in months, is returned by precipitation in a matter of days. When rainy days are missing, moisture depletion will occur from the upper soil layers and a reduction of relative humidity will occur eventually in the atmosphere, establishing the drought conditions.

Quite opposite is the situation associated with a high pressure system. It provokes horizontal divergence at the ground surface. Therefore, water vapor entering

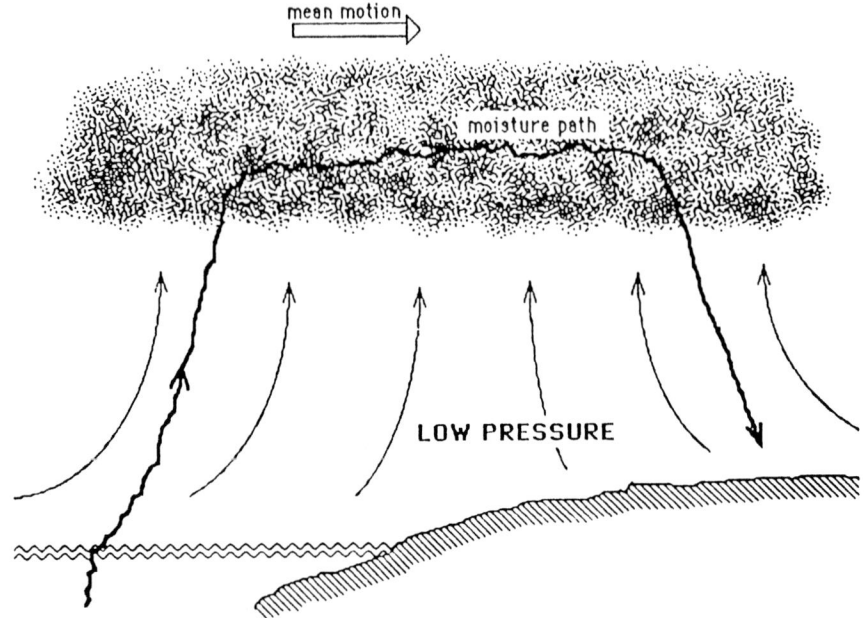

Fig. 6.2. Moisture displacement due to a low pressure, converging area (from Bravar and Kavvas, 1991a).

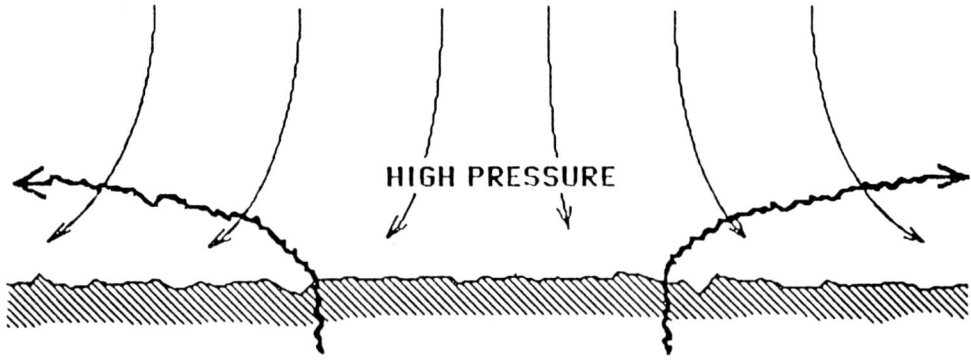

Fig. 6.3. Moisture loss due to a high pressure, diverging area (from Bravar and Kavvas, 1991a).

the atmosphere through evapotranspiration processes is dispersed and eventually will leave the geographic area carried by winds. As was seen in the previous section, the downward motion associated with high pressure centers prohibits formation of rain. Therefore, the moisture is carried away by air currents (see Figure 6.3).

6.3 The Evolution of Extreme Drought

Because of the complex interactions involved with drought phenomena, a complete explanation of the physics of the evolution of a drought is extremely difficult and is yet to be accomplished. At this stage the evolution of a drought can be described in terms of three distinct phases: initiation, growth, and recovery. The characteristics of each phase can be defined with respect to four variables: 500 mb atmospheric temperature, atmospheric moisture content, surface temperature, and water storage over the land area. In this section, an overview is given of the characteristics that mark each phase of a drought's evolution through space and time.

6.3.1 INITIATION

If it is assumed that an atmospheric process is the initiation mechanism of droughts, such a process must persist longer than passing weather systems. Anomalous patterns called blocking patterns do exist in the atmosphere, and persist longer than passing weather systems.

The effect of a blocking pattern on the local atmospheric conditions is the increase in atmospheric temperature and decrease in atmospheric moisture content. Processes that affect atmospheric temperature are radiation, latent heat fluxes due to evapotranspiration and condensation, and surface temperature.

Because the initiation mechanisms of a drought involve atmospheric processes that affect precipitation patterns, a drought can be initiated when the normal precipitation patterns for a region are interrupted. In order for this to occur, some anomalous atmospheric condition which prevents moisture-laden weather systems from moving over a region, must exist for an extended period of time . One such condition is called a blocking event (see Palmen and Newton, 1969).

An atmospheric blocking event can be thought of as having the following characteristics (from Rex, 1950):

1. Split jet stream extending for at least 45 degrees of longitude.

2. Appreciable mass flow in the meridional direction instead of the zonal direction.

3. Persistence of pattern for at least ten days.

Figure 6.4 shows the patterns of streamfunction lines associated with some blocking events. Notice the position of the high pressure system and the flow around it. Weather systems will follow the streamfunction lines around the high pressure region, associated with the blocking event, preventing precipitation from occurring in the region below the high pressure. Atmospheric and surface temperatures will be higher in the region under the anomalous high pressure, and precipitation will be nonexistant during the duration of the blocking event (Rex, 1950).

The alternation of high and low pressure centers will balance out over periods with a time-scale on the order of 1 year, as long as the frequency of alternation is high enough. It is appropriate to ask what this frequency depends upon, and the

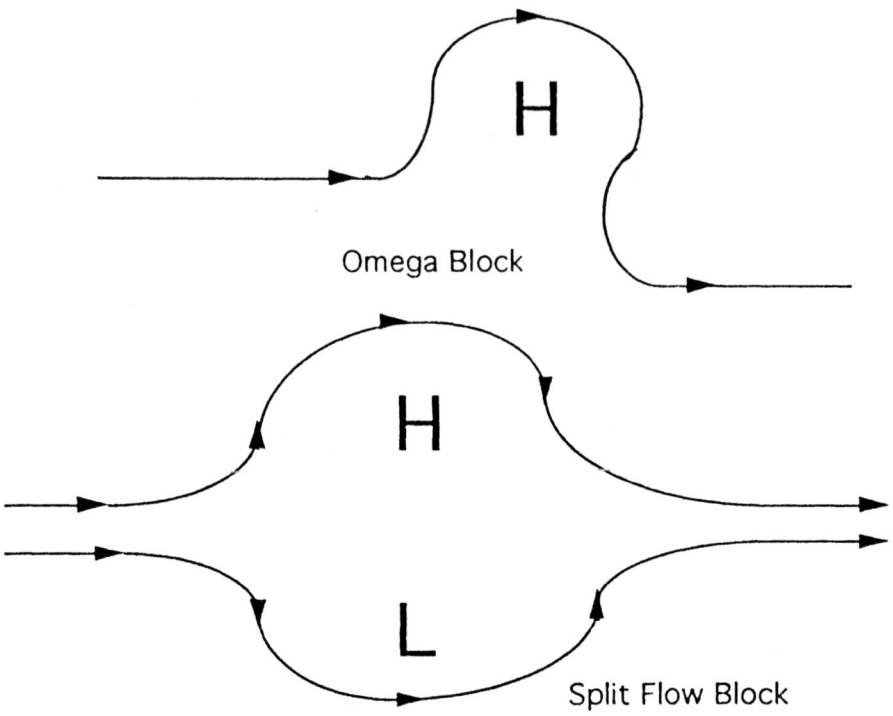

Fig. 6.4. Schematic description of atmospheric blocking patterns (from Rex, 1950).

answer lies in the study of atmospheric synoptic scale wave motions, known as Rossby waves (see, e.g. Holton, 1979). Without considering the presence of frontal systems (baroclinic waves) and filtering out gravity waves, it can be shown that vorticity of the atmosphere is conserved. Assuming further that the main contribution to vorticity comes from variations of longitudinal velocity v, atmospheric motions satisfy the following equation:

$$\frac{\partial^2 v}{\partial t \partial x} + u\frac{\partial^2 v}{\partial x^2} + \beta v = 0, \tag{6.3}$$

where $\beta = (\partial f / \partial y)$ is considered constant, and x and y are the coordinates of longitude and latitude.

The solution to the linearized version of equation (6.3) is called a Rossby wave and is characterized by a wave speed (celerity) c given by:

$$c = u - \frac{L^2 \beta}{4\pi^2}, \tag{6.4}$$

where L is the wavelength. Rossby waves correctly predict, for usual wavelengths $L = 3000$–4000 km, a celerity of the order of 10–15 m/s, which is the observed translatory speed of atmospheric disturbances. Furthermore, for a wavelength of the order of 10000 km at mid-latitudes, Rossby waves become stationary ($c = 0$).

Thus, when a wave forms at the planetary scale, it will not move for a long time. Such waves are actually observed from time to time. They extend from low latitudes to polar regions and, because of their shape and behavior, are called 'omega blocks'. The origin of such large-scale phenomena is yet unexplained, although it has been recognized that they are linked to the sea surface temperature anomalies at the Pacific Ocean. Such omega blocks form long anticyclonic events which can trigger droughts. This mechanism will be explored in the Modeling of Droughts section below.

Long lasting weather patterns do not really balance out in terms of soil surface moisture content as their shorter duration counterparts do. A long-lasting high pressure causes a steady depletion of moisture from the ground, even though the intensity of moisture loss will decrease in time as surface water becomes less available for evapotranspiration (surface and stomatal resistances play an important role in conserving moisture). On the other hand, a long-lasting low pressure system will produce most of its precipitation in the first stage, after which the atmosphere will have to wait in order to accumulate enough humidity for condensation conditions to occur again.

The presence of high pressure centers for long times, therefore, provides a primer for drought occurence, but they can not explain why dry periods survive much longer than the anticyclone from which they originate. Indeed, the reason lies in the non-linearity of the phenomenon.

For a blocking pattern to initiate a drought, it must be stable enough to exist long enough for the feedback mechanisms to take place. This constitutes the growth phase of a drought. Charney and DeVore (1979) indicate the possibility of multiple equilibrium states in the atmosphere, one of which could correspond to anomalous flow patterns called blocking events. If blocking is indeed an equilibrium state of the atmosphere, then it is possible for the blocking event to persist long enough to initiate a drought.

A blocking event can initiate a drought because of the physical processes that occur during its existence. Associated with a blocking event is a high pressure system that will sit over a region for the duration of the blocking event. Such high pressure systems mean that temperatures will be anomalously high accompanied by divergent sinking motions. These conditions, accompanied by the lack of precipitation, are responsible for initiating the feedback processes contributing to the growth and persistence of a drought.

It may be observed that the primer of a drought need not be a persistent high pressure system. Consider what happens if forest vegetation is removed over large areas. Rain water that was initially stored by the plant system will be transported away by surface flow and groundwater flow. Less water vapor will be available at ground level, and weather disturbances that should release precipitation will miss the primer for rain production. The net result will be a reduction of soil moisture content rendering, in turn, a reduction in relative humidity resulting from a decrease in evapotranspiration. It is quite imaginable that, if the area of removed vegetation is large enough, the process becomes irreversible, and the phenomenon of desertification will take over.

Charney (1975) and Charney et al. (1975) discussed another mechanism by which desertification is enhanced, namely the increase in albedo owing to reduction of plant cover. An increase in surface albedo provokes a decrease in the incoming solar radiation. Lessening of net incoming radiation decreases evapotranspiration rates (and, thereby, the latent heat and moisture fluxes) into the atmosphere. In addition, air tends to be cooled radiatively and, to maintain thermal equilibrium, must sink, so that rainfall is suppressed. Therefore, the analogy between the physics involved in droughts and in desertification processes is limited to the primer mechanism: droughts are reversible phenomena, whereas desertification is not.

6.3.2 Growth

Consider the 'typical' mid-latitude storm that has the possibility, under normal conditions, to return part of the moisture to the land/vegetation system that lost moisture owing to a long-lasting high pressure center. The fuel of such storm disturbance is the humidity contained in the air. In fact, by condensation of the water vapor, large amounts of energy are furnished to the atmosphere which utilizes it to avoid dissipation of the disturbance. The energy is released in the form of latent heat of condensation. However, the process of condensation depends non-linearly on the water vapor content in the atmosphere. It is necessary that humidity be present in sufficient quantity, in order that saturation conditions be reached as a result of lifting currents at some height called the condensation level. If the presence of water vapor is limited, it is quite possible that saturation conditions will not be reached, and no rain will be produced in such a case.

After a dry period (that is, after a long-lasting high pressure center) the ability of the surface to release moisture by evapotranspiration is limited. The piezometric levels in aquifers are lower, which makes evaporative fluxes smaller. In addition, the vegetation conserves its water. The passage of a low pressure system with the associated convergence zone, although it induces lifting currents that cool the air, may not have a water content high enough to result in a condensation level sufficiently low. As evapotranspiration moisture is missing, the disturbance will not be able to unload the water it is transporting because the relative humidity level necessary for condensation has not been reached. Note that the moisture released at the ground is immediately available to the atmospheric system because of the presence of the atmospheric boundary layer. This layer is able to transport the water vapor vertically across its whole depth (1–2 km) in tens of minutes because of the turbulence that characterizes it.

The lack of precipitation into a region for an extended period of time deprives that region of an influx of moisture. Evaporation and streamflow out of that region drain existing moisture from the region. As a result the local system becomes dryer which can be associated with drought intensification. If an otherwise normal storm system passes over such a region, there may not be enough local moisture to enable precipitation to occur. As a result, the system dries further. Even though the initiation mechanism may no longer be present, the drought can still intensify.

The growth of a drought in space can be thought of from the standpoint that the precipitation patterns of one region affect a nearby region because of interactions

Extreme Droughts

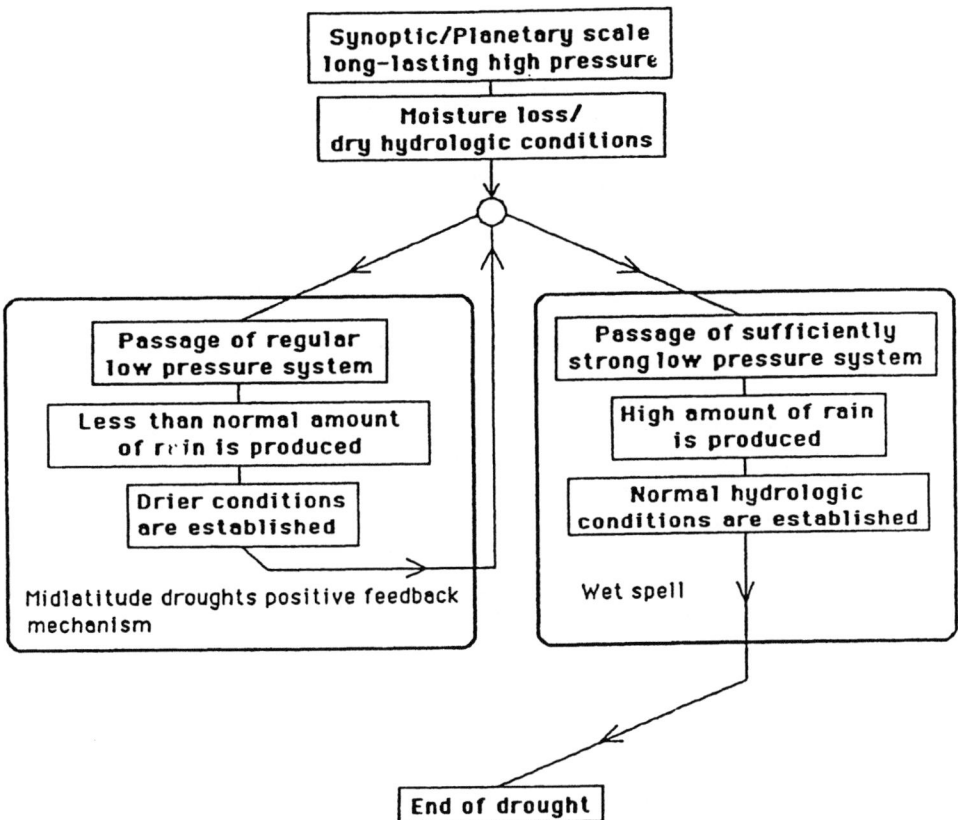

Fig. 6.5. Primer and self-supporting mechanism of mid-latitude droughts (from Bravar and Kavvas, 1991a).

with the atmosphere. If a region is experiencing extreme drought conditions, then the dry atmospheric conditions can be advected to neighboring regions. As a result, evapotranspiration may increase which contributes to the drying of the neighboring region. If conditions become sufficiently dry, normal weather systems may not provide enough moisture for precipitation to occur. If precipitation does occur it may be less than normal. As a result, drought conditions begin to form in the neighboring region. If such conditions persist, feedback mechanisms can start for the neighboring region as well, leading to a drought in that region. It may be noted that many factors other than atmospheric advection influence precipitation and, therefore, influence the physical processes involved in the spatial growth of droughts. The growth of droughts in time and space ends only when atmospheric conditions allow a larger than normal influx of moisture to the affected region.

6.3.3 RECOVERY

Based on this reasoning, we can deduce that a disturbance which under normal conditions would help replenish the land reservoirs with water, will produce little if any precipitation during an extreme drought period and, therefore, will not improve the situation. Clearly, what has been just described is a self-supporting, positive feedback mechanism: the less water is present, the less probable rain becomes, and therefore the water content diminishes even more. Only disturbances that carry sufficient moisture from outside the dry area will be able to produce rain and help end the drought period. However, not all of the disturbances have such characteristics. Furthermore, the longer the dry period, the less moisture will be available at the soil surface, and thus the more intense the disturbance will have to be to restore normal conditions (see Fig. 6.5).

Recovery from a drought, in the sense that all depleted moisture is returned to a region, can occur only if wetter-than-normal storm systems pass over the drought-stricken region. In this case, the storm system provides all of the necessary moisture for precipitation to occur. Feedback mechanisms are possible during the recovery phase of a drought in the sense that the timing of the passing of successive storms can influence how much precipitation occurs. If a series of wet storms pass over a region, the moisture from the previous storm can influence the amount of precipitation during the next storm. If several powerful storms occur in succession, even flooding can become a possibility, depending on the state of the region's water resources at the time of the heavy precipitation. In this respect, the management of water resources within a drought-stricken region can influence how quickly the region recovers from a drought.

6.3.4 EXAMPLE

An example from Bravar and Kavvas (1991a) is given here in order to compare the effects of the passage of two disturbances over the west coast of USA to show that, although two atmospheric systems look much alike, their effects, in terms of rain production, are very different. From the start it may be stated that it is extremely difficult to find two meteorological situations that are exactly the same. However, in the example reported in Figure 6 the disturbances show some comparable features. First of all, both of them refer to the first week of May, although one is in 1986 while the other is in 1988. Both of them originate in the Pacific Ocean and develop in the same number of days, with frontal areas (shaded in Fig. 6.6) moving practically at the same speed and direction. They show quite comparable 500 mb maps (reported together with their daily weather maps in Fig. 6.6), which means that in both cases the mid-tropospheric wind fields are comparable. Hence, one expects that both disturbances will produce comparable effects in terms of rain. However, this is not the case.

The first system refers to 1986, which is to be considered 'normal' in terms of precipitation events over California; the second is dated 1988, which was a very dry year in California, with significantly less than normal amounts of rain produced during the winter of 1988. It is important to note that while the two disturbances have similar initial conditions, they have different boundary conditions, the latter

Extreme Droughts

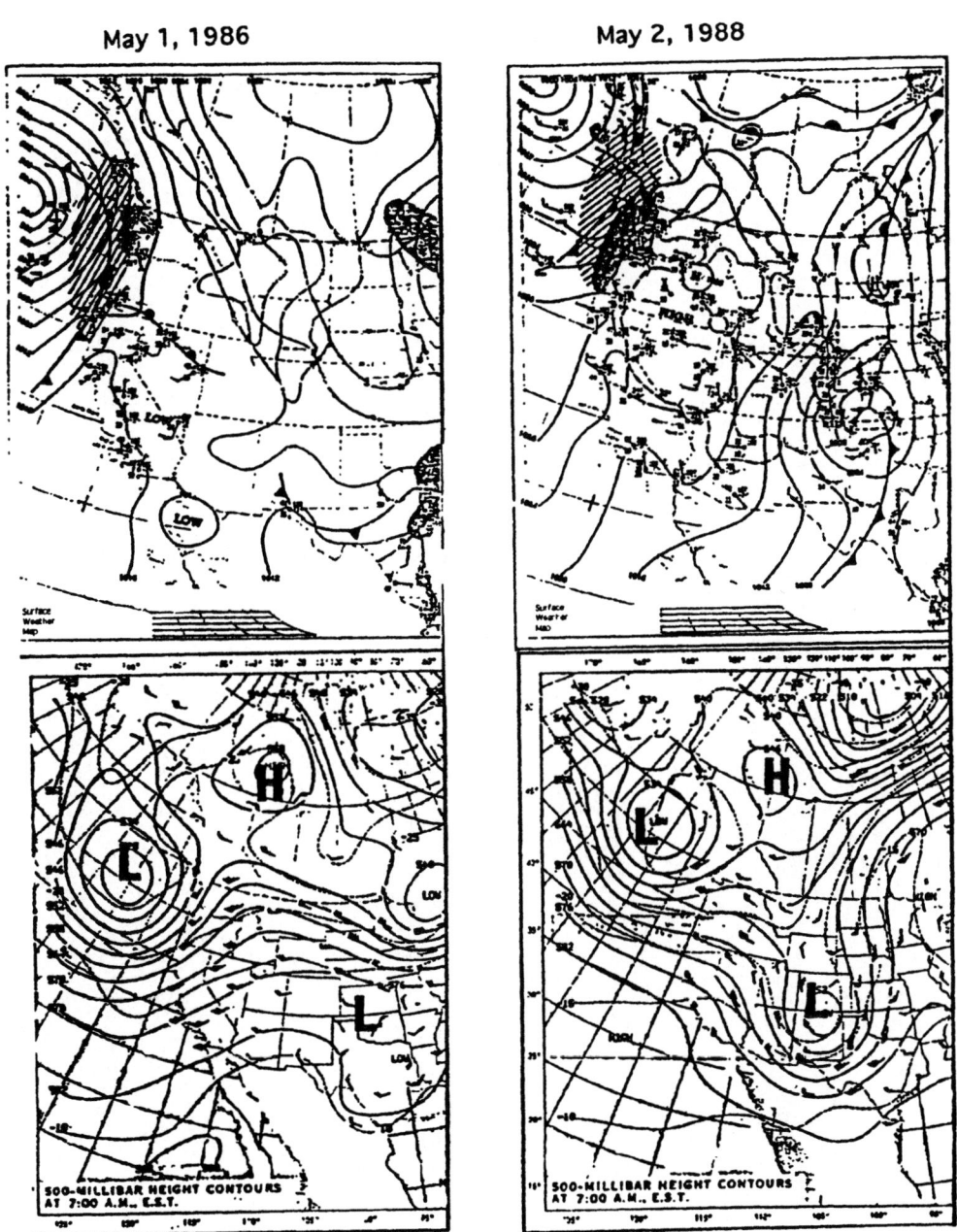

Fig. 6.6. Synoptic weather maps of two similar low pressure systems over California during 1–4 May 1986 and 2–5 May 1988 (from Bravar and Kavvas, 1991a).

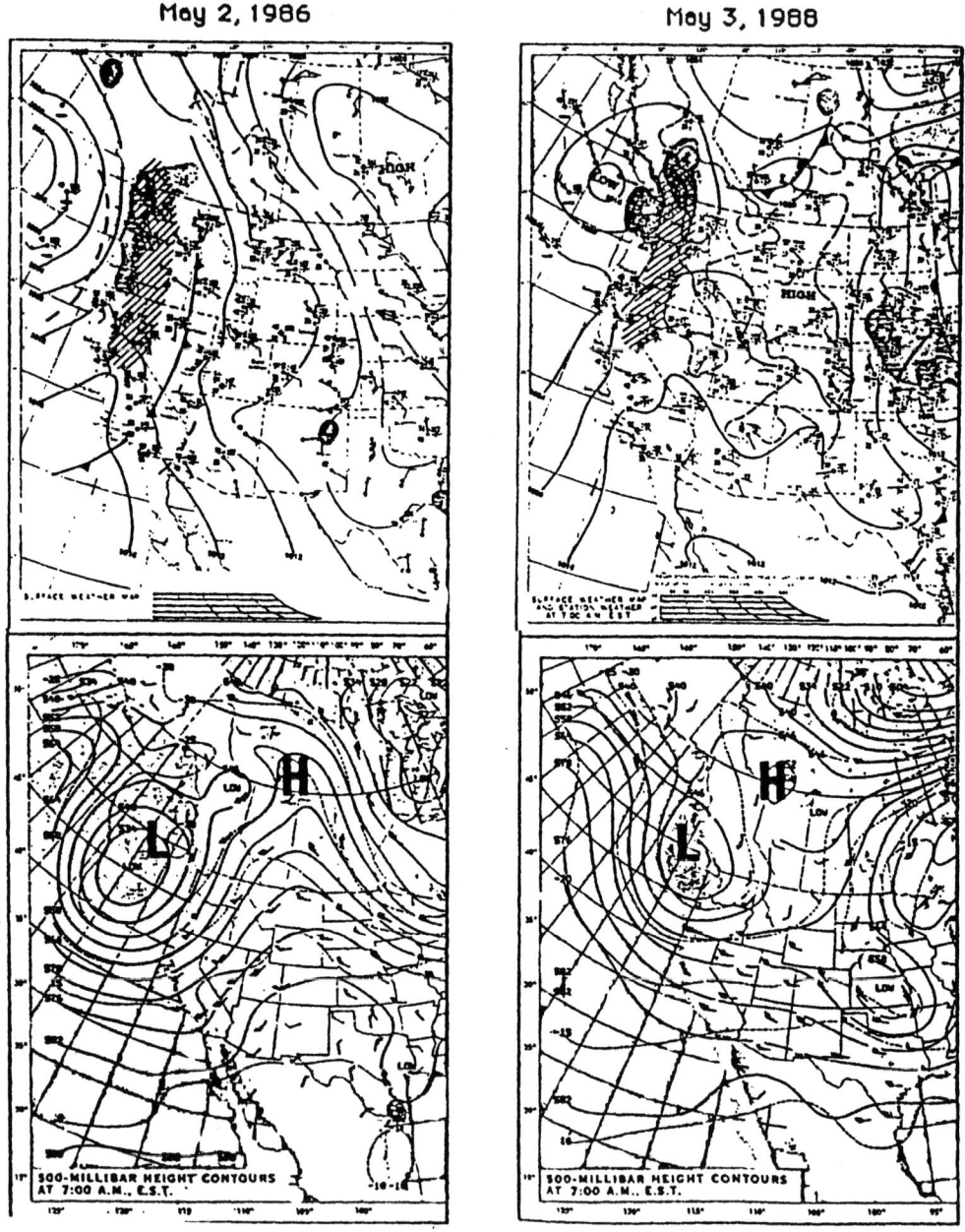

Fig. 6.6. *Continued.*

Extreme Droughts

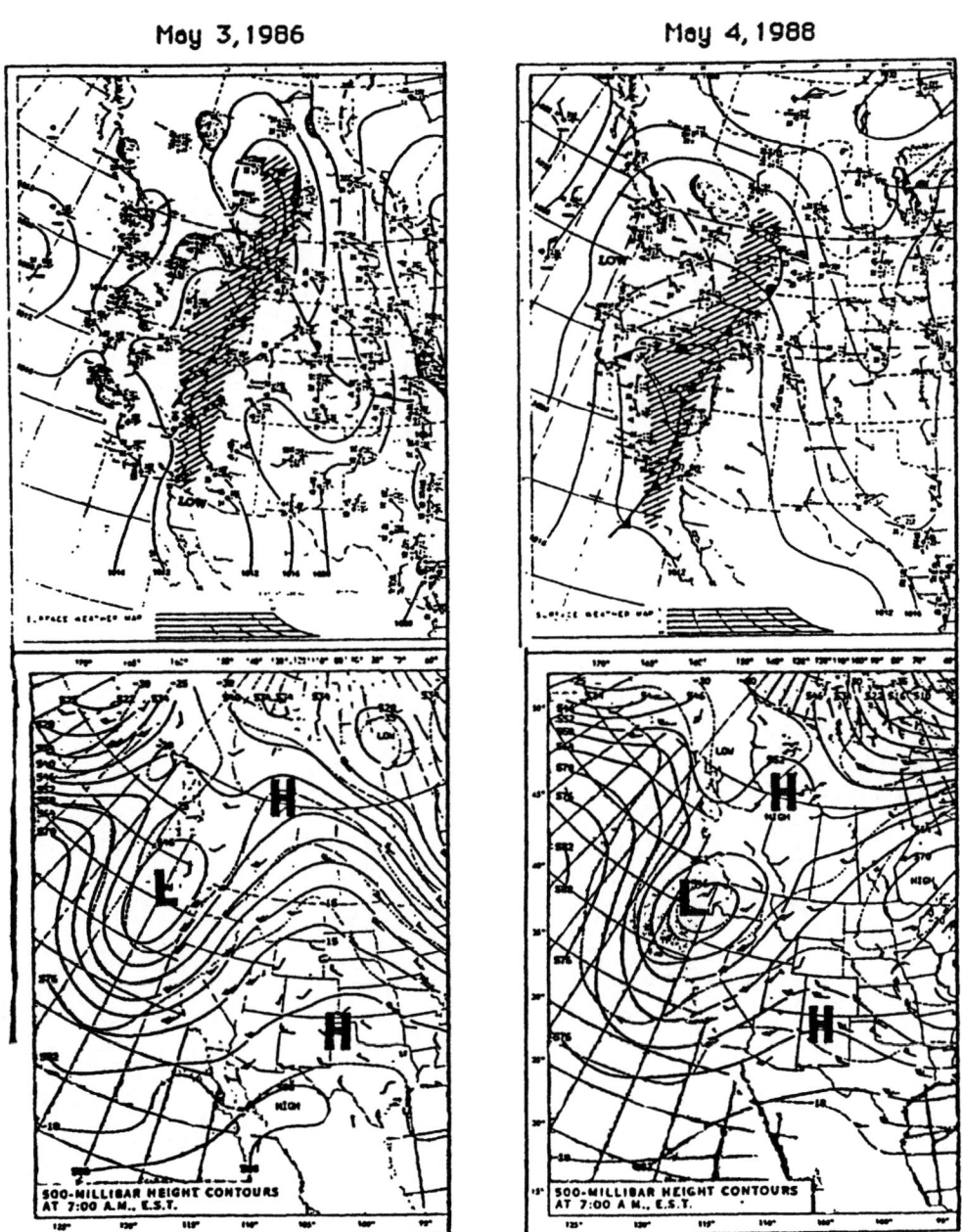

Fig. 6.6. *Continued.*

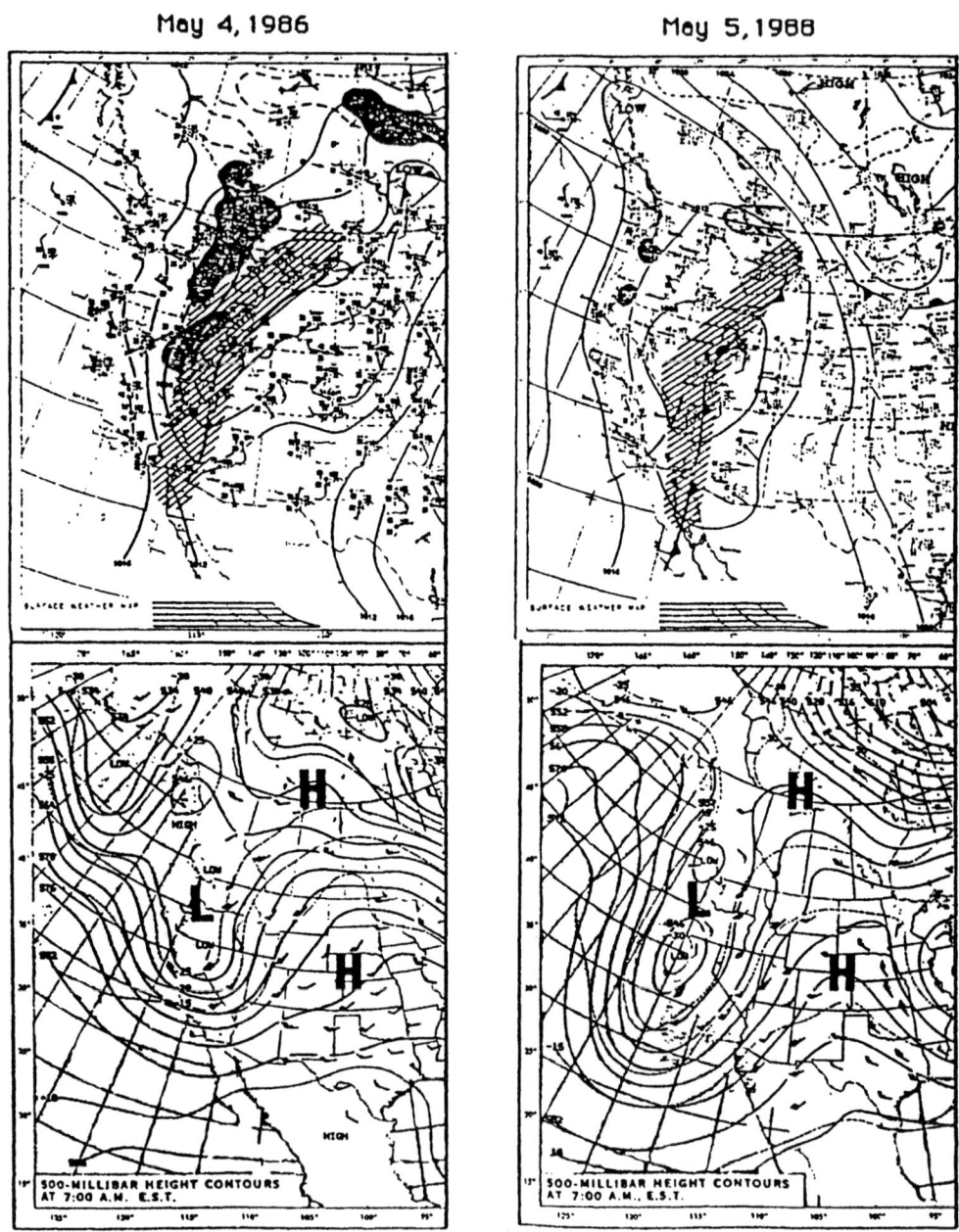

Fig. 6.6. *Continued.*

Extreme Droughts 145

being given by the different land and vegetation moisture fluxes. Therefore, the differences in the hydrologic boundary conditions correlate well with those in the amount of overall rain produced.

However, it must be noted that there is also a difference in the initial conditions between the two disturbances as they display different dewpoint temperatures. A higher dewpoint temperature will render a higher water vapor content (moisture) within the air mass. How much of this difference is due to previous planetary-scale circulations, as opposed to previous differences in boundary conditions is difficult to say. It may be noted that the dry conditions existed for months before the 1988 disturbance had developed. Nevertheless, the differences between actual temperatures and dewpoint temperatures are roughly the same for the two disturbances, which mean that both systems have comparable potential for rain production. In fact, the height of the condensation level in a rain cloud is strictly related to this difference in temperature. The larger this difference is, the higher an air parcel must be lifted (and therefore cooled) to obtain condensation conditions.

The comparison of the effects of the two low pressure systems shows a significant difference. The 1986 disturbance produces higher amounts of rain, both near the coast and inland (Table 6.2). (In Table 6.2, Eureka is on the northern coast and San Francisco is on the central coast of California, Red Bluff is located in the northern sector and Fresno is located in southern sector of the Central Valley in the inland region of California.) One possible explanation is to relate this phenomenon to the differences in soil moisture content. Although along the coast one might attribute the difference in rain entirely to the previous history of air currents, the significant differences in the amount of rain produced inland may be explained by the changes in soil and vegetation water content which is significantly smaller in 1988. This reduction in soil water content, in turn, causes a reduction in atmospheric relative humidity, as indicated by lower dewpoint temperatures in 1988, due to a decrease in evapotranspiration rates. As the relative humidity level which is necessary for condensation, is not reached, the rain does not occur during the considered May period in 1988, although weather conditions very similar to 1986 exist.

It may be observed that a lower temperature accompanies the 1988 disturbance, while evapotranspiration is an increasing function of temperature (Bravar and Kavvas, 1991a). Lower temperature renders lower evapotranspiration rates. However, vapor flux is necessary for condensation and rain production. Thus, the drought periods are not necessarily associated with temperatures higher than normal.

In Table 6.2, the framed data refer to inland meteorological stations whereas other data refer to stations along the coast. For almost all stations the amount of precipitation is higher for the 1986 disturbance.

6.4 Modeling an Extreme Drought

One way of gaining insight into the physical processes involved in the evolution of an extreme drought is through the use of computer models. This section examines some of the ongoing work in the area of drought modeling. A review of two types

TABLE 6.2
Temperatures, dew point temperatures and rainfall over some California meteostations during 1–4 May 1986 and 2–5 May 1988 (from Bravar and Kavvas, 1991b)

	EUREKA		S.FRANC.		RED-BL.		FRESNO		Description
Yr	86	88	86	88	86	88	86	88	
T	57	45	49	49	54	47	58	29	System is approaching
Td	42	38	40	44	37	38	41	20	California
R	-	-	-	-	-	-	-	-	
T	53	52	57	50	57	52	53	48	System is entering
Td	49	43	49	47	49	35	46	42	California
R	.30	.12	-	-	T	-	-	-	
T	46	47	55	49	53	51	58	49	System is passing
Td	46	40	52	41	34	27	49	38	(note drop in Td)
R	.86	T	.03	-	.11	-	-	-	
T	51	49	52	49	42	47	49	45	System is leaving
Td	45	41	49	42	42	41	41	37	
R	1.65	.03	.11	-	-	.03	.16	-	

of models, utilized in drought research, is given, followed by a presentation of the parameterizations used in a simplified model. A discussion of the results of a drought study by means of a simplified model is also given.

6.4.1 GENERAL CIRCULATION MODELS AND ENERGY BALANCE MODELS

General Circulation Models (GCMs) can be used to study the atmospheric circulation patterns that may lead to a drought over a region. Such models utilize the primitive equations of atmospheric dynamics (Pedlosky, 1988) in order to model the spatial and temporal change of the atmosphere in three dimensions over the entire earth. GCMs are the most comprehensive models for modeling atmospheric processes and climate. Mo et al. (1991) use a GCM in a study of the 1988 United States drought and Oglesby (1991) uses a GCM to study soil moisture and climatic

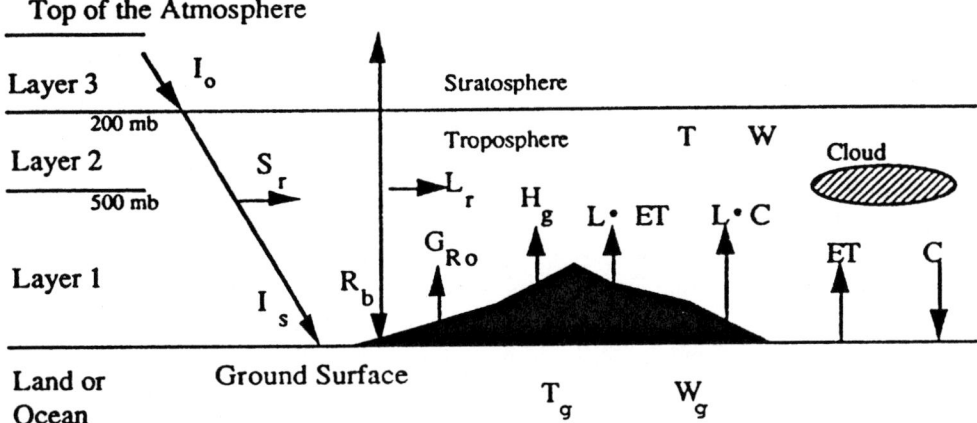

Fig. 6.7. Schematic description of the climate system in the energy balance climate model of Nakatsugawa and Kavvas (1992).

variability in association with North American droughts. Other studies relating to drought research using GCMs are given in Hunt (1991).

Simplified climate models utilize averaging processes in one or more dimensions in order to shorten computation time or to focus on certain specific processes (Gates, 1988). Among the simplified climate models is the energy balance model (EBM) which was developed first by Budyko (1958) and Sellers (1969). Energy balance models have been used in climate studies by North and Cahalan (1981), Adem (1991), Jentsch (1991), Kim and North (1991, 1992), Bravar and Kavvas(1991b), and Nakatsugawa and Kavvas (1992). Of these EBM studies, only the one by Bravar and Kavvas (1991b) pertains directly to drought research.

Because of the extent by which the GCMs model the system, computations require extensive amounts of CPU time, making the choice of a simplified climate model more appealing. This is because a sufficient number of realizations can be run by a simplified model to obtain realistic statistical properties for the different parameters or processes. A simplified model also allows the researcher to investigate interactions between processes with less effort and more clarity. North (1988) reviews energy balance climate models and their use in climate studies, pointing out that the use of such simplified climate models can be "elegant, comprehensive, and economical."

6.4.2 Energy Balance Model Parameterization

The following presentation pertains to the one-dimensional simplified climate model which was originally developed by Bravar and Kavvas (1991b) for the study of droughts, and later modified by Nakatsugawa and Kavvas (1992). The process parameterizations presented include solar radiation, ocean heat storage, ocean evaporation, land evapotranspiration, precipitation, land water storage, atmospheric water content including clouds, and the thermal processes and structure of the atmosphere. A schematic of these processes is shown in Figure 6.7. It is noted here that only four variables are considered by this model for analysis and modeling of droughts. These four variables are atmospheric temperature, atmospheric water content, surface temperature, and hydrologic system water storage. Their parameterizations are given below along with the parameterizations for the other related processes.

The one-dimensional climate model of Nakatsugawa and Kavvas (1992) is made up of two vertical layers for the troposphere, and one layer for the stratosphere. The first tropospheric layer extends from the earth surface to the 500 mb pressure layer and the other from 500 mb to the 200 mb pressure layer (see Fig. 6.7). The stratosphere is assumed to be in a thermal equilibrium state.

The temperature lapse rate is considered to be constant in the troposphere with a value of $\gamma = 0.0042$ K/m. This allows temperature to vary linearly with elevation. Topographic effects are modeled by relating temperature, pressure and elevation using the hydrostatic approximation and gas equations, as follows

$$z(p) = \frac{T_0}{\gamma}\left[1 - \left(\frac{p}{p_0}\right)^{R_d\gamma/g}\right], \qquad (6.5)$$

$$p(z) = P_0\left[\frac{T_0 - \gamma z}{T_0}\right]^{g/(R_d\gamma)}, \qquad (6.6)$$

$$T(z) = T_0 - \gamma z, \qquad (6.7)$$

where $z(p)$ is elevation corresponding to pressure p, p_0 is standard pressure (1000 mb), $T(z)$ is temperature at elevation z, T_0 is the temperature at 1000 mb level, R_d is gas constant for dry air (287 J/kg/K), g is gravitational acceleration, γ is the lapse rate, and $p(z)$ is pressure at elevation z.

Atmospheric temperature T is determined using the first law of thermodynamics which is integrated over the depth of the entire troposphere in order to express the heat transport in the west-east direction around the globe. It is expressed as:

$$C_p M \left(\frac{\partial T}{\partial t} + V_T \cdot \nabla T\right) + \omega Z_0 = L_r + S_r + H_g + LC, \qquad (6.8)$$

where T and ω are, respectively, depthwise averaged temperature and pressure velocity in the troposphere, C_p is the specific heat at constant pressure, and M is the mass of a vertical air column between ground surface and top of troposphere over a 1 m^2 surface area. The right-hand-side of (6.8) expresses the external (diabatic) heating terms. The diabatic heating terms consist of long wave radiation absorbed in the troposphere, L_r, short wave radiation absorbed in the troposphere, S_r, sensible heat flux, H_g, and latent heat releases from condensation, $L \cdot C$ (please refer to Fig. 6.7 for a schematic description of these terms). We shall discuss the parameterization of these terms below.

Besides being transported zonally around the globe, heat is also transported vertically in the atmosphere. In order to model this transport, it is necessary to parameterize the land surface thermal balance, the ocean surface thermal balance, sensible heat flux, latent heat flux (evapotranspiration) both from ocean surfaces and land surfaces, and the various components of radiation.

Extreme Droughts

Assuming a parabolic temperature profile in a soil layer of 0.5 m thickness, the land surface thermal balance may be expressed as

$$\frac{1}{3}z_g C_g \frac{dT_g}{dt} = R_n - H_g - L \cdot \text{ET}, \qquad (6.9)$$

where T_g is soil surface temperature, z_g is depth of soil layer being considered, C_g is soil thermal capacity, R_n is net radiation, L is latent heat of evaporation, ET is the evapotranspiration flux, and other terms are as defined before. Similarly, assuming a parabolic temperature profile in a top ocean layer of 70 m thickness, the ocean surface thermal balance may be expressed as

$$\frac{dT_0}{dt} = \frac{3}{z_0 c_0}[R_n - H_g - L \cdot \text{ET}], \qquad (6.10)$$

where T_0 is ocean surface temperature, z_0 is thickness of ocean layer being considered, c_0 is the ocean thermal capacity and other terms are as defined above.

Among the diabatic heating terms in the heat transport equation, following Nakatsugawa and Kavvas (1992), the shortwave radiation absorbed in the troposphere, S_r, may be parameterized as a function of atmospheric temperature T, depthwise integrated atmospheric water content W and incoming insolation I_0. Longwave radiation absorbed in the troposphere, L_r, may be parameterized as a function of T, T_g or T_0, W, and atmosperic carbon dioxide concentration (see Fig. 6.7).

Net radiation, R_n, in ocean and land surface thermal balances may be parameterized as follows (Bravar and Kavvas, 1991),

$$R_n = I_s + R_b + G_{R_0}, \qquad (6.11)$$

where I_s is incoming solar radiation absorbed by the Earth surface, R_b is longwave radiation emitted by the atmosphere to the surface and G_{R_0} is longwave radiation emitted by the earth surface (see Fig. 6.7). Following Nakatsugawa and Kavvas (1992), I_s may be parameterized as a function of T, W and I_0; R_b may be parameterized as a function of T, T_g or T_0, and W; and G_{R_0} may be parameterized as a function of T_g or T_0.

The sensible heat flux, H_g, which appears in the atmospheric heat transport equation, and in land surface and ocean surface thermal balance equations, may be parameterized as follows (Washington and Williamson, 1977),

$$H_g = c_p C_d \rho_b |V_b|(T_g - T_b), \qquad (6.12)$$

where c_p is specific heat at constant pressure, C_d is drag coefficient, ρ_b is air density at the bottom of atmosphere, $|V_b|$ is the surface wind speed, and the temperature at the bottom of troposphere, T_b, is calculated from

$$T_b = T + 0.5\gamma(z_t - z_b). \qquad (6.13)$$

In (6.13) γ is the lapse rate in the troposphere, z_t is the elevation at the top of the troposphere and z_b is the elevation of the ground surface.

The evapotranspiration (latent heat flux) which appears in the heat transport equation and in the land and ocean thermal balances, may be parameterized as follows for the land surfaces (Budyko, 1958; Churchill et al., 1982),

$$\text{ET} = \frac{\text{GW} - \text{GW}_{\min}}{\text{GW}_{\max} - \text{GW}_{\min}} C_{\text{dw}} \rho_b |V_b|(w_{\text{gs}} - w_b), \tag{6.14}$$

where $\text{GW}_{\max}$ represents the maximum storage of water for which the evapotranspiration achieves its potential value, $\text{GW}_{\min}$ is the value of GW at which the evapotranspiration is assumed to become negligible, C_{dw} is drag coefficient, w_{gs} is the saturation mixing ratio of water vapor at the ground temperature, and w_b is the mixing ratio of water vapor in the surface boundary layer. As seen from (6.14), evapotranspiration, ET, is based upon potential evaporation multiplied by an evaporation coefficient in order to account for the moisture supply restriction over land surfaces. For evaporation from ocean surfaces one can utilize a similar parameterization as follows (Churchill et al., 1982),

$$\text{ET}_p = C_{\text{dw}} \rho_b |V_b|(w_{\text{ws}} - w_b), \tag{6.15}$$

where w_{ws} is the saturation mixing ratio at the water surface at temperature T_0.

Atmospheric water content, W, is seen above to play an important role in various parameterizations besides being the state variable in atmospheric water balances. It is defined as the total mass of water vapor in a vertical column of air with unit base area. It is determined through conservation of atmospheric moisture, written as (Rasmussen, 1977),

$$\frac{\partial W}{\partial t} = V_w \cdot \nabla W = \text{ET} - C, \tag{6.16}$$

where $V_w \cdot \nabla W$ is the advection term for water vapor, ET is evapotranspiration rate from land or ocean surfaces (described above) and C is precipitation rate (kg/m²-sec). In the simplified model of Nakatsugawa and Kavvas (1992) precipitation, C, only occurs when supersaturated conditions exist in the atmosphere. It is formulated by the Clausius–Clapeyron equation

$$C = \frac{w - w_s}{1 + L^2 w_s/(C_p R_v T^2)} \frac{m}{\Delta t}, \tag{6.17}$$

where w is average mixing ratio of water vapor in the troposphere, w_s is average saturation mixing ratio in the troposphere, m is fractional weight of water vapor in an atmospheric column at which precipitation starts, L is latent heat of condensation, R_v is gas constant for moist air and other terms are as defined before. Precipitation is modeled to occur when $w > w_s$.

Hydrologic system water storage, GW, is determined by the water balance equation (Thorntwaite and Matter, 1955),

$$\frac{d\text{GW}}{dt} = -k\,\text{GW} - \text{ET} + C, \tag{6.18}$$

TABLE 6.3
Comparison of simplified climate model-simulated climatic averages with actual data taken from Sellers (1969) and Budyko (1982) (from Bravar and Kavvas, 1991b)

Quantity	Climate Model	Actual Data (30°–50°N)
Precipitable water (kg m^{-2})	16	17
Air temperature at surface (°C)	13.2	13
Storage of water (GW) (kg m^{-2})	1050	–
Soil temperature (°C)	13.8	–
Cloudiness over land	0.49	0.50
Cloudiness over oceans	0.37	–
Evaporative heat flux (LET) over land (W m^{-2})	35	32
Evaporative heat flux (LET) over ocean (W m^{-2})	103	93
Overall evapotranspiration (mm/year)	876	822
Overall precipitation (mm/year)	918	890
Net radiation over land (W m^{-2})	60	70
Net radiation over ocean (W m^{-2})	93	90
Sensible heat flux (H_g) over land (W m^{-2})	25	19
Sensible heat flux (H_g) over ocean (W m^{-2})	54	39

where the time rate of change of hydrologic system water storage, dGW/dt, is computed as a balance of outflow corresponding to water movement out of the land hydrologic system to the oceans, k GW, and to evapotranspiration, ET, and inflow corresponding to precipitation, C. In this equation, evapotranspiration and precipitation are parameterized values discussed above, and the coefficient k is the water storage recession coefficient.

6.4.3 Simulation of an Extreme Drought by an Energy Balance Climate Model

Application of the climate model, developed by Bravar and Kavvas (1991b), to the simulation of an extreme drought is presented in detail in their paper. A review of their study is presented here. The goal of the study of Bravar and Kavvas (1991b) was to reproduce reasonably well the characteristics of an extreme continental drought by means of a simplified climate model. In order to accomplish this task, the following procedure was followed.

First, the model was calibrated for V_T, V_w, $|V_b|$, C_d and m through a 300 year simulation of monthly climatic averages at 1000 km grid scale. The climate model required slight calibration in order to attain average values that would match actual averages of the main variables. Table 6.3 shows that the model is capable of reproducing correctly the order of magnitude of all fundamental variables which describe climate.

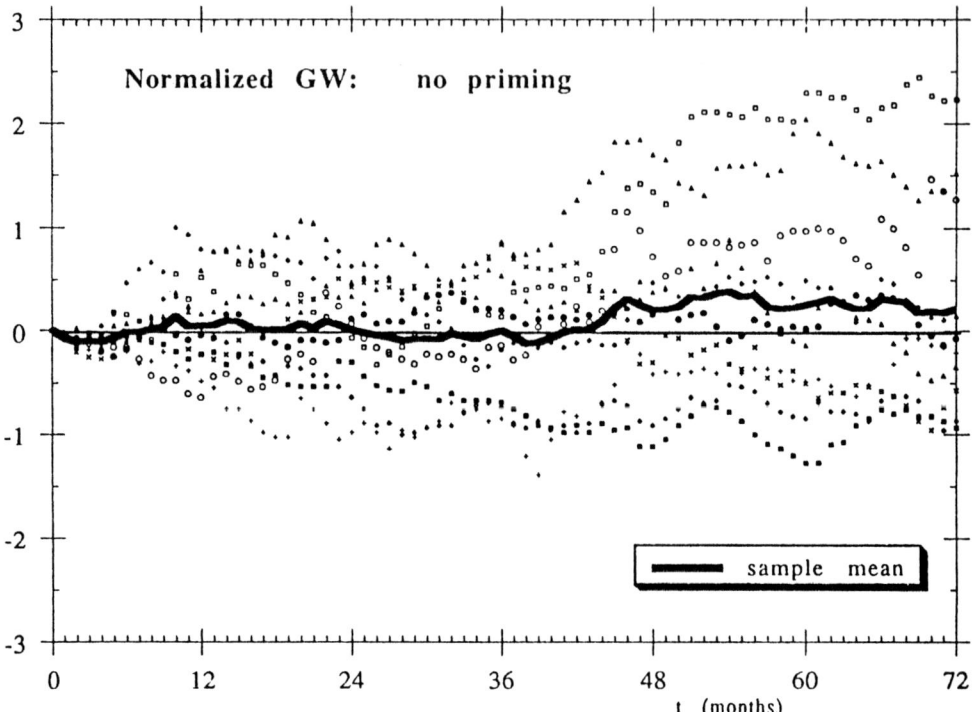

Fig. 6.8. Ten different realizations and their sample mean for hydrologic system water storage GW which were simulated by the simplified climate model. All of the simulations started with the average climatic conditions on December 31. (from Bravar and Kavvas, 1991b)

The simulation runs all started on December 31 from initial conditions corresponding to the average climate conditions at all geographical locations, except over western USA grid where a stationary temperature wave of amplitude T_p (deg C) is hypothesized to be present for a duration of d_p (months). For this stationary wave the longest temperature wave (taken at a wavelength of 10000 km by Bravar and Kavvas, 1991b) must be used because stationary disturbances are observed only for very long wavelengths (omega blocks).

Even though the initial conditions for the atmospheric temperature T and water vapor content W are fixed, one is left with complete freedom about the initial position of T and W waves, which are initially of zero amplitude. Therefore, the same run can be repeated starting with exactly the same inital conditions but with null waves placed at different positions. The outputs of the climate model look similar for the first few months but they differ dramatically afterwards. In reality, by placing a T or a W wave at some region, it is implicitly assumed that there is the potential for the growth of one of these waves in that region; whether the wave will develop or not depends on the state of the earth system.

Extreme Droughts

First, it is verified that the initial conditions used for all the climate model simulations, do not induce by themselves an average trend in atmospheric and/or hydrologic conditions. To test this issue the climate model was run for 10 different inital positions of the T and W waves which were taken at their climatic average values. Figure 6.8 shows the results of each run realization together with their sample mean (ensemble average) trajectory in terms of water storage GW. It is seen that no particular trend develops, meaning that unless a stationary temperature wave is imposed on the initial conditions for some period of time, drought conditions do not realize. However, it may be noted that this is valid only in terms of the average trend, since single outputs do display quite dry conditions established after some time. The values displayed in this and other similar figures refer to monthly averaged quantities for a grid size of 1000 km located over the western USA. These values are in standardized form. Specifically, the difference between simulated value and climatological mean is divided by the climatological standard deviation in order to obtain the displayed value.

Next, simulations were run by inducing a long temperature wave (wavelength of 10000 km) which had an amplitude increase of T_p over the climate average and which persisted for d_p months as the initial condition over western USA. After d_p months the temperature wave was brought back to its climatic average state and the simulation of the climate system was continued for six years. For each type of simulation, multiple runs were made (by changing slightly the initial location of the T and W waves within the western USA grid) and average trends of the multiple runs were used for analysis. This simulation exercise was performed in order to study the effect of a short duration (one to three months) Rossby-type stationary temperature wave anomaly over western USA on an extreme drought over western USA. Specifically, the question being addressed was whether a stationary temperature wave anomaly of a few months duration can cause an extreme drought of several years over western USA. In Figure 6.9 the simulation results of Bravar and Kavvas (1991b), corresponding to a temperature amplitude increase of $T_p = 1°C$ for a duration of $d_p = 1$ month as the initial condition over western USA, are displayed. With respect to the mean trends, it may be noted that a one-month temperature anomaly of only 1°C caused a severe decrease in hydrologic system water storage (GW) and a severe reduction in rainfall (a reduction of –1.4 in standardized rainfall means practically no rainfall) during a period of nine months (starting on December 31) over western USA. This is due to the nonlinear feedbacks among the hydrologic and atmospheric components of the climate system, as may be seen when the four parts of Fig.6.9 are analyzed simultaneously. An increase in atmospheric temperature (due to induced temperature anomaly) during the first month also increases the saturation mixing ratio in the atmosphere. Hence, although there may be an appreciable amount of water vapor in the troposphere, as seen from Fig. 6.9b precipitation decreases significantly since supersaturation can not be reached due to high saturation mixing ratio in the troposphere. From Fig. 6.9c it is seen that initially the soil temperature increases significantly when the temperature wave is increased by an amplitude of 1°C during the first month of simulation. Due to increase in atmospheric temperature, cloudiness increases which, in turn, decreases the net radiation on the earth surface. Decrease in net

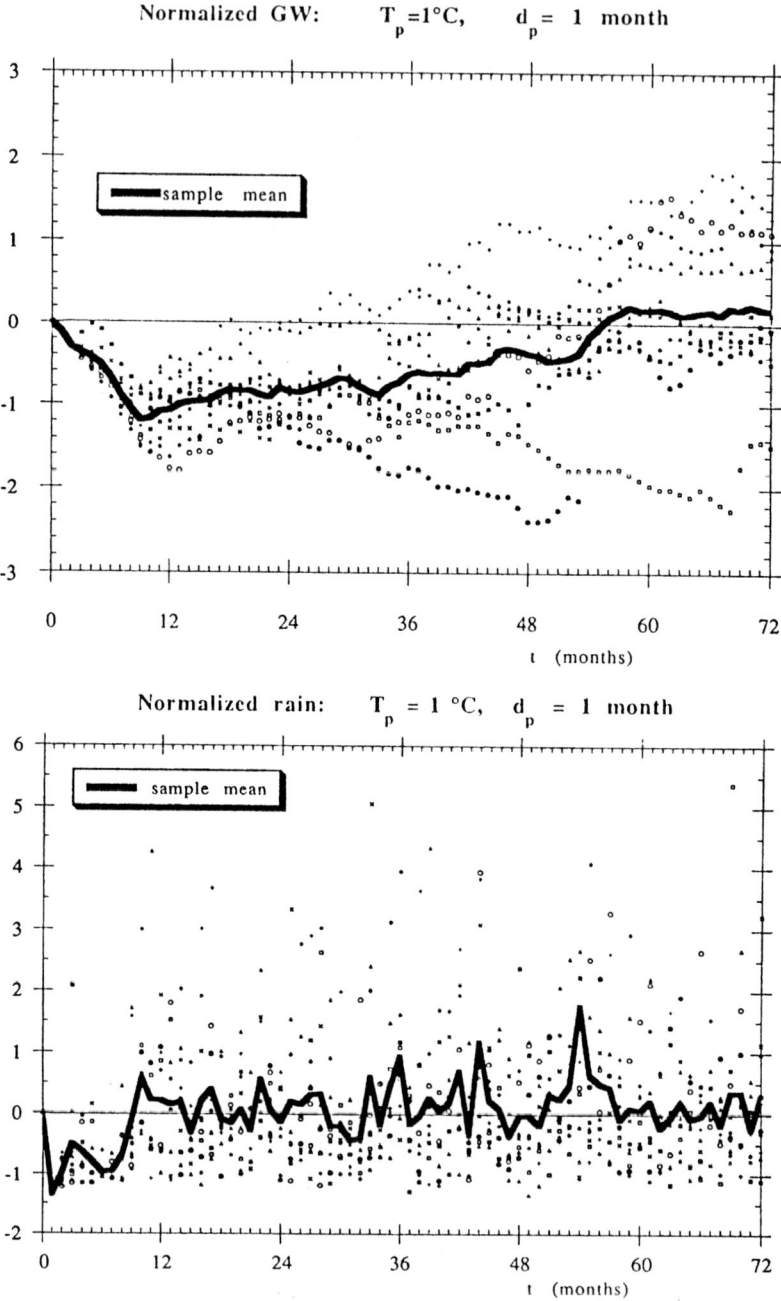

Fig. 6.9. Standardized realizations and their sample mean for the various components of the climate system which were simulated by the simplified climate model under an initial heat wave with an amplitude 1°C above the climatic average and which lasted for 1 month. (a, *top*) Hydrologic system water storage, GW, (b, *bottom*) precipitation, (c) soil temperature, (d) evapotranspitation. (From Bravar and Kavvas, 1991b)

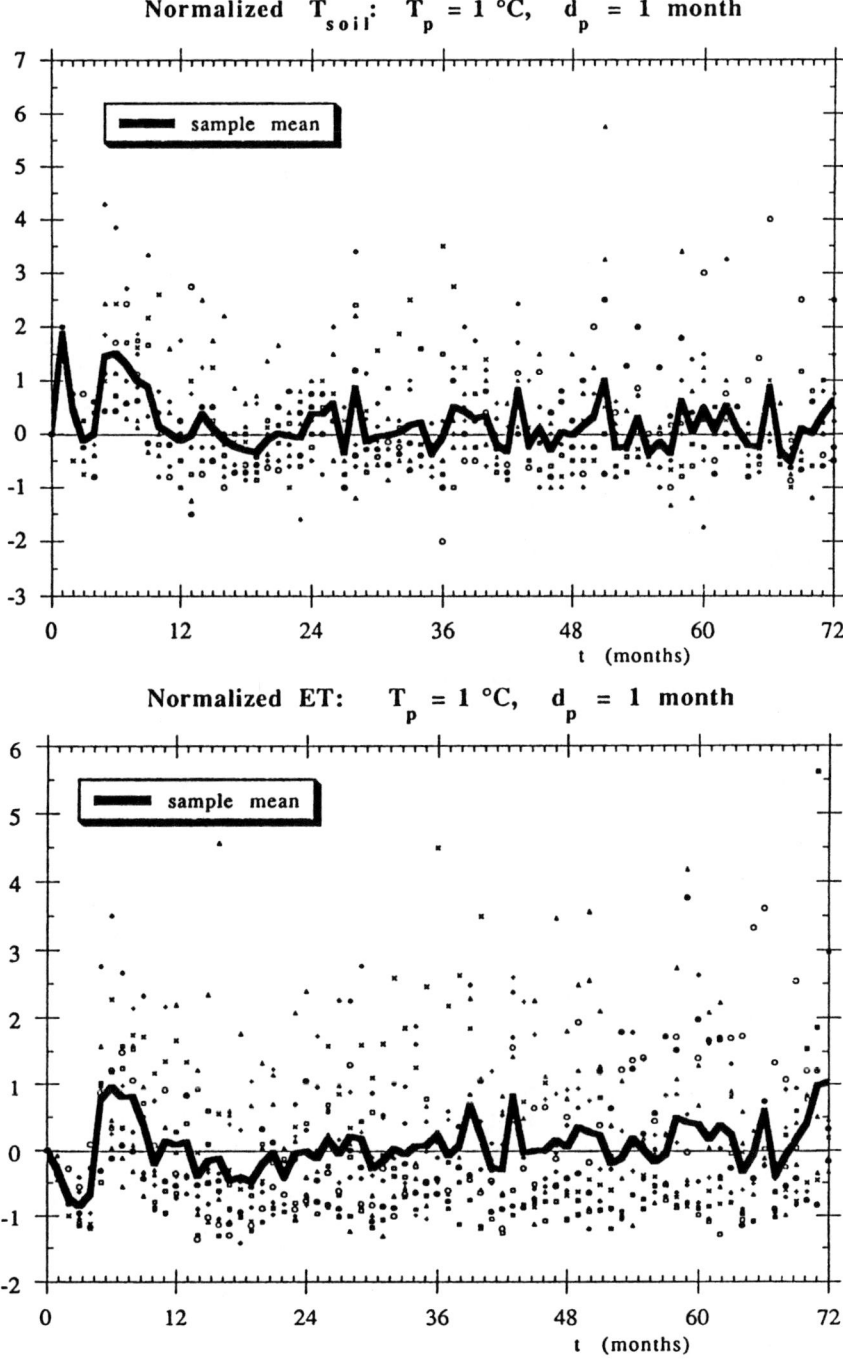

Fig. 6.9. *Continued*; (c, *top*) and (d,*bottom*)

radiation, in turn, decreases the evapotranspiration, as seen during the first few months of Fig. 6.9d. It also decreases soil temperature, as seen during the second and third months of Fig. 6.9c. Increased cloudiness also increases solar radiation which is absorbed by troposphere, thereby, heating the air further. This increase in air temperature, in turn, increases incoming longwave radiation towards the earth surface and decreases the sensible heat flux from the earth to the atmosphere. These changes render increases in soil temperatures and evapotranspiration rates, as seen during the fourth, fifth and sixth months of Figures 6.9c and 6.9d. Also, the increase in air temperature helps keep the saturation mixing ratio at a high level so that it is difficult to attain supersaturation conditions necessary for rainfall. This explains the continued shortage in rainfall, as seen during the second through ninth months of Fig. 6.9b. Eventually, as the evapotranspiration rates increase, the local moisture supply to the atmosphere increases and supersaturation conditions are finally attained for rainfall. One may continue in this fashion, explaining the feedbacks among the various components of the hydrologic-atmospheric system.

After soil temperatures, evapotranspiration and precipitation are back to normal in about ten months (as seen in Figures 6.9b–d), the hydrologic system is left with a depletion of water storage (as seen in Fig. 6.9). A reduced hydrologic water storage induces evapotranspiration fluxes which are less than their climatic average. Therefore, even with average rainfall rates the hydrologic water storage recovers slowly towards the climatic average conditions. However, as seen from Fig. 6.9a, this recovery from the time of largest water deficit (nine months after the start of simulation) takes about four years. Therefore, a temperature anomaly of one month duration caused a drought period of 4.5 years duration which is 54 times longer than the duration of the heat wave. Of course, this conclusion is based upon the ensemble average of the simulated realizations and not on individual realizations which, as seen from Fig. 6.9, show a chaotic behavior. This chaotic behavior is another manifestation of the fundamentally nonlinear nature of the feedbacks among the hydrologic and atmospheric components of the earth's climate. These nonlinear positive feedbacks are the main reason why short period anomalies can produce climate changes at the meso-climatic time-scale, as demonstrated in the study of Bravar and Kavvas (1991b). It may be noted that this study was confined to only the atmospheric and hydrologic components of the earth system. However, the oceans have a profound influence on the climate. Therefore, future studies on droughts need to consider the nonlinear dynamic interaction among all the three components of the earth system (oceanic, atmospheric and hydrologic) in order to have more realistic descriptions of the extreme drought phenomena.

6.5 Discussion and Conclusions

Extreme droughts are complex, nonlinear physical processes which have the potential for great destruction. Such destruction of crops and ecosystems does not occur instantaneously but evolves over a time frame on the order of years. The physical processes involved are based on the atmospheric, oceanic, and hydrologic systems occurring on spatial scales ranging from the local scale where drought effects are felt, to the global scale where oceanic and atmospheric processes which govern the initiation and evolution of droughts, evolve.

The current technology of hydrology in dealing with droughts is based on the development of probability distributions of droughts in order to quantify the drought risk for a drought plan. In the current technology such probability distributions describe droughts either in terms of rainfall characteristics or in terms of streamflow characteristics. They are constructed by means of a statistical analysis-stochastic modeling approach which has traditionally utilized rainfall and streamflow time-series records over the geographical region of interest. More recently, such records were extended to longer periods in terms of tree-ring data in the hope that such long historical records would contain all the possible drought scenarios. Although such probability distributions proved to be very valuable as tools for planning of droughts before their occurrence, they did not prove to be particularly useful in the real-time management of resources in drought-stricken areas during the actual drought period. For the real-time management of resources during the occurence of droughts it is necessary to predict the onset, evolution and termination of drought conditions. Prediction of the onset of a drought would provide an early warning to the society at risk in order to enable this society to take measures for mitigating the future impacts of the upcoming drought. Prediction of the evolution and of the termination of a drought would enable optimal real-time operation of the financial, agricultural and water resources for minimizing the impact of the drought. Since the aforementioned statistical approaches are not derived physically from the interacting dynamics of the atmosphere- ocean- land system, they can not yield the conditional probabilities which are necessary for the prediction of the onset of droughts when an individual drought-causing mechanism (such as an anomaly in sea surface temperatures (SST) in the tropical regions of Pacific) or a combination of drought-causing mechanisms (such as the simultaneous occurence of SST anomalies and land surface anomalies) are observed to occur prior to a drought. Similarly, the aforementioned statistical approaches can not provide conditional probabilistic descriptions of the time-space evolution of various facets of a drought over a geographical region as the evolution of the atmosphere- ocean-land system over the globe is observed during the ongoing drought.

Presently, a methodology which is based on the physics of the atmosphere-ocean-land system, and which can be utilized for the development of conditional probabilities that are necessary for the prediction of the onset, evolution and termination of droughts, is nonexistent. Such a methodology can be based on climate models which have interacting oceanic-atmospheric-hydrologic components. General circulation models (GCMs) are the most comprehensive climate models. However, due to their complexity, it would be difficult to determine what interactions are responsible for the positive feedback mechanisms. They are also costly to make enough runs in order to allow reasonable statistical analysis. Because of these limitations with CGMs, simplified climate models have been developed where one or two dimensions of the GCMs are averaged. Such simplified models can be used to study droughts by focusing in on the processes involved in droughts. An example of such an approach is the study by Bravar and Kavvas (1991b) who utilized a one-dimensional simplified climate model to study the physical mechanisms of extreme droughts. However, their model can only be considered as a preliminary step in the direction of the physically-based modeling of droughts in order to obtain the aforementioned probability distributions which are necessary for the planning and real-time management of extreme droughts. Fundamental contributions to the understanding of the physical mechanisms involved in extreme droughts, and to their realistic modeling in order to help solve the related problems is yet to be realized.

Acknowledgements

This work was supported by NSF Grant No. CMS-9318339. This support is gratefully acknowledged.

References

Adem, J. (1991), "Review of the development and application of the Adem thermodynamic climate model, " Climate Dynamics, Vol. 5, Springer-Verlag, pp. 145-160.
Bravar, L., and M. L. Kavvas (1991 a), "On the physics of droughts. I. A conceptual framework, " Journal of Hydrology, Vol. 129, 281-297.
Bravar, L., and M. L. Kavvas (1991 b), "On the physics of droughts. II. Analysis and simulation of the interaction of atmospheric and hydrologic processes during droughts, " Journal of Hydrology, Vol. 129, 299-330.
Budyko, M. I. (1958), Heat Balance at the Earth's Surface, US Weather Bureau, Washington, DC.
Budyko, M. I. (1982), The Earth's Climate: Past and Future, Academic Press.
Charney, J. G., 1947, "The dynamics of long waves in a baroclinic westerly current, " Journal of Meteorlogy, Vol. 4, pp 135-163.
Charney, J. G., 1975, ?Dynamics of deserty and droughts in the Sahel, ? Quarterly Journal of the Royal Meteorlogical Society, Vol 101, pp193-202.
Charney, J. G., P. H. Stone, and W. J. Quirz, 1975, ?Drought in the Sahara: A biogeophysical feedback mechanism, ? Science, Vol. 1987, 434-435.
Charney, J. and J. G. DeVore (1979), "Multiple flow equilibria in the atmosphere and blocking, " Journal of Atmospheric Sciences, V36, No. 7, American Meteorological Society, pp. 1205-1216.
Churchill, J. N., Ellyett, C. D. and Holmes, J. W. (1982), "A study of regional evapotranspiration using remotely sensed and meterorological data-contrasts between forest and grassland", Aust. Water Res. Counc., Tech. Pap. No. 70, Australian Govn. Pub. Service.
Dole, R. (1989), "Life cycles of persistent anomalies. Part I: Evolution of 500 mb height fields, " Monthly Weather Review, Vol. 117, American Meteorological Society, pp. 177-211.
Dracup, J. A. et al. (1980), "On the statistical characteristics of drought events, " Water Resources Research, Vol. 16, No. 2, American Geophysical Union, pp. 289-296.
Gates, W. L. (1988), "Climate and the climate system, " Physically-Based Modeling and Simulation of Climate and Climatic Change, Part 1, edited by M. E. Schlesinger, NATO ASI Series, Series C: Mathematical and Physical Sciences - Vol. 243, pp. 3-21.
Glantz, M. H. (1987), "Drought and economic development in sub-saharan Africa, " Planning for Drought, Donald A. White, Willian E. Easterling, and Deborah A. Wood, editors, Westview Press, Boulder, CO., pp. 297-316.

Holton, J. R. (1979), An Introduction to Dynamic Meteorology, Second Edition, International Geophysics Series, Volume 23, Academic Press.
Hunt, B. G. (1991), "The simulation and prediction of drought," Vegetatio, Vol. 91, Kluwer Academic Publishers, pp. 89-103.
Jentsch, V. (1991), "An energy balance climate model with hydrological cycle 1. model description and sensitivity to internal parameters," Journal of Geophysical Research, Vol. 96, No. D9, 17169-17179.
Kennedy, D. N., S. Gutterfield, D. F. Priest, J. Minton, M. D. Roos and L. K. Gage (1991), California's Continuing Drought 1987-1991: Summary of Impacts and Conditions as of December 1, 1991, California Department of Water Resources Report.
Kim, K., and G. R. North (1991), "Surface temperature fluctuations in a stochastic climate mode," Journal of Geophysical Research, Vol. 96, No. D10, American Geophysical Union, pp. 18573-18580.
Kim, K. and G. R. North (1992), "Seasonal cycle and second-moment statistics of a simple coupled climate system," Journal of Geophysical Research, Vol. 97, No. D18, American Geophysical Union, pp. 20437-20448.
Lenjenas, H., and R. Madden (1992), "Traveling Planetary-Scale Waves and Blocking," Monthly Weather Review, Vol. 120, American Meteorological Society, pp. 2821-2830.
Manabe, S. and Wetherald, R. T. (1975), "The effects of doubling the CO_2 concentration on the climate of a general circulation model", J. Atmos. Sci., 32, 3-15.
Matalas, N. (1963), "Probability distribution of low flows," Professional Paper 434-A, United States Geological Survey, Reston, VA.
Mayer, L., K. Collins, R. Meekhof, R. Milton, N. Strommer, J. Vertrees, P. Longsworth, J. Cook, M. Snow, R. Silberman, R. Walsh, E. Dickey, Z. Montzai, D. DiBuono, G. Campbell, H. Tohler, D. Sumner, A. Nravcak, N. Conklin, R. Brooks, and A. Judd (1988), " The drought of 1988", final report of the President's interagency drought policy committee, Washington, D. C.
McClurg, S. (1992), "California's Chinook salmon: upstream battle to restore the resource, " Western Water, Water Education Foundation, November/December issue.
McNab, A. L. (1989), "Climate and Drought", EOS, 70(40), 873-888, 1989.
Mintz, Y., 1984, "The sensitiity of numerically simulated climates to land-surface boundary conditions," In: J. T. Houghton (Editor), The Global Climate, Cambridge University Press, Cambridge, pp. 79-105.
Mo, K., J. R. Zimmerman, E. Kalnay and M. Kanamitsu (1991), "A GCM study of the 1988 United States drought," Monthly Weather Review, Vol. 119, American Meteorological Society, pp. 1512-1232.
Nakatsugawa, M. and M. L. Kavvas (1992), "Simplified climate model with combined atmospheric-hydrologic processes," Proceedings of the Workshop on the Effects of Global Climate Change on Hydrology and Water Resources at the Catchment Scale, February 3-6, pp. 99-108.
Nathan, R. J. and T. A. McMahon (1990), "Practical aspects of low-flow frequency analysis, " Water Resources Research, Vol. 26, No. 9, pp. 2135-2141.
North, G. R. (1988), "Lessons from energy balance models," Physically-Based Modeling and Simulation of Climate and Climatic Change, Part 2, NATO ASI Series C: Mathematical and Physical Sciences - Vol. 243, pp. 627-651.
North, G. R. and R. F. Cahalan (1981), "Predictability in a solvable stochastic climate model," Journal of the Atmospheric Sciences, Vol. 38, pp. 504-513.
Oglesby, R. J. (1991), "Springtime soil moisture, natural climatic variability, and North American drought as simulated by the NCAR Community Climate Model 1," Journal of Climate, Vol. 4, pp. 890-897.

Palmen, E. and C. W. Newton (1969), Atmospheric Circulation Systems, Academic Press, Inc., New York.
Palmer, W. C. (1965) Meteorological Drought. Weather Bureau Res. Pap., No. 45, Washington, DC, 58 pp.
Paulson, E. G., J. Sadeghipour and J. Dracup (1985), "Regional frequency analysis of multiyear droughts using watershed and climatic information, " Journal of Hydrology, Vol. 77, 55-76.
Pedlosky, J. (1987), Geophysical Fluid Dynamics, second edition, Springer-Verlag.
Peixoto, J. P. and A. H. Oort (1992), Physics of Climate, American Institute of Physics, New York.
Philander, S. G. (1990), El Nino, La Nina, and the Southern Oscillation, Academic Press, Inc., New York.
Rasmussen, E. M. (1977), "Hydrological application of atmospheric vapor-flux analyses", Operational Hydrology Rept. No. 11, WMO, Geneva, 50 pp.
Rex, D. F. (1950), "Blocking action in the middle troposphere and its effect upon regional climate", Tellus, 2(3), 195-211.
Sellers, W. D. (1969), "A global climatic model based on the energy balance of the earth-atmosphere system", J. Appl. Meteor., 8, 392-400.
Sen, Z. (1980), "Statistical analysis of hydrologic critical droughts, " ASCE Journal of the Hydraulics Division, Vol. 106(HY1), pp. 99-115.
Thorntwaite, C. W. and J. R. Matter (1955), "The water balance", Pub. in Climatology, 8(1), Lab. of Climatology, Centerton, NJ, 86pp.
Wallace, J. M. and P. V. Hobbs (1977), Atmospheric Science, An Introductory Survey, Academic Press, Inc., New York.
Wijayaratne, L. H. and E. Golub (1991), "Multiyear drought simulation, " Water Resources Bulletin, Vol. 27, No. 3, American Water Resources Association, pp. 387-395.
Yevjevich, V., W. A. Hall and J. D. Salas (1978), Drought Research Needs, Proceedings of the Conference on Drought Research Needs, held at Colorado State University, Fort Collins, Colorado, December 12-15, 1977, Water Resources Publications.

CHAPTER 7

Mud and Debris Flows

Peggy A. Johnson and Richard H. McCuen

ABSTRACT: Debris flows are a natural hazard that occur in many parts of the world. They have been responsible for deaths as well as destruction of roadways, bridges, and homes. Debris flows are typically associated with steep, mountainous areas, depositing large amounts of sediment on the alluvial fan. The initiation of a debris flow is a very complex process; a number of factors, including antecedent moisture, precipitation, vegetative cover, topography, geology, and soil type, all interact to form the debris flow. Debris flows can move at very high rates of speed and typically carry large boulders and other debris, such as trees and cars. A variety of models have been proposed to predict the magnitude and frequency of debris flows. These range from statistical models to theoretical models and are typically applicable only for the region where they are developed. Mitigation methods include warning systems, passive methods, such as zoning, and active methods, such as deflection structures and debris basins.

7.1 Introduction

Debris flows are a natural hazard that occur in many parts of the world, particularly in steep mountainous regions. Steep slopes and torrential rainfalls combine in a complex process to cause slope failures that produce a debris-laden mud that flows downslope at avalanche speeds. Debris flows have been responsible for deaths and destruction of property in such places as California, Utah, British Columbia, parts of South America, China, and Japan.

There are many examples of the destructive powers of debris flows. For example, a series of debris flows totalling approximately 35- to 55,000 m^3 flowed along a stretch of highway north of Los Angeles, California (Cronin et al., 1990). The debris flow, containing boulders and small trees, moved fully loaded tractor-trailer rigs and swept one car more than 2 km down the highway. The occupant of the car was killed. Johnson (1984) lists many examples of debris flow destruction, including one event in Virginia caused by Hurricane Camille in 1969 that killed about 150 people and caused tens of millions of dollars in damage. In 1982, 25 people were killed and $66 million in damage resulted from debris flows in the San Francisco Bay region (Keefer et al. 1987). As another example, the Harrison Canyon debris basin in southern California failed in the winter of 1980. As designed, the capacity of the basin was exceeded by at least 34,000 cubic meters. During a series of debris flow events, the wall of the basin was overtopped, followed by failure. Although there were no deaths, more than 30 homes were damaged (Slosson et al., 1991). Other reports of debris basin failures in the western U.S. are reported by Rice and Foggin (1971), Costa and Jarrett (1981), and Keefer et al. (1987). In British Columbia, Alberta, and the Yukon, more than 17 deaths and $100 million of damage have been attributed to debris flows (Van Dine, 1985).

Pierson and Scott (1985), Fairchild (1987), and Takahashi (1987) discuss relatively large debris flows that occur on the slopes of volcanoes. In these cases the debris flow is called a lahar. Pierson (1985) studied lahars triggered by the May, 1980, eruption of Mount St. Helens in Washington. Peak lahar discharge was reported at 250,000 m^3/s. During the monsoon season following the 1992 Mt. Pinatubo eruption in the Philippines, a series of lahars flowed down the mountainsides carrying ash deposited by the eruption. Numerous bridges and homes downstream were destroyed or damaged.

Debris flows come in a variety of sizes and travel at a wide range of speed. Debris flows can travel at avalanche speeds of more than 12 m/s, but have also been observed as slow moving, sluggish flow. In a single year, as much as 60,000 m^3 have flowed into a single debris basin in the Los Angeles area.

Debris flows are also disruptive to riparian and stream environments. A stream in the Cascade Mountains, Oregon, was catastrophically affected by a debris flow in, 1986 (Lamberti, 1991). The debris flow altered channel geomorphology and destroyed riparian vegetation for 500 m. The result, according to Lamberti, was a reach with short, disordered channel units, low hydraulic retention, and an open canopy.

In this chapter, the physical process of debris flows, prediction of magnitude and frequency, and mitigation methods are described.

7.2 Physical Processes

Debris flows are defined as a mixture of granular solids, water, and air that flow over low to moderate slopes (Johnson, 1970). The flow is typically 70 to 90 percent solids by weight and carries large amounts of gravel, cobbles, and boulders with only minor amounts of clay and other cohesive materials. The clay and much of the silt in a debris flow are an intrinsic part of the interstitial fluid (Pierson and Scott, 1985). The movement and appearance of a debris flow has often been compared to that of wet cement.

7.2.1 FACTORS AFFECTING DEBRIS FLOW INITIATION

The primary factors that effect the potential for debris flow include the geology and topography of the watershed, soil type, climate, runoff, antecedent moisture conditions, and ground cover. Each of these will be discussed below.

Geology and Topography

Debris flows occur in many parts of the world. The geology of each of these areas is varied. However, there are a number of factors that each site has in common. Debris flows tend to occur in areas that are geologically active, characterized by steep, unstable slopes. Campbell (1975) found that debris flows in southern California are typically initiated on slopes of 26 to 45 percent. Similarly, Clark (1987) found that debris flows in the Appalachians of the eastern United States occur on slopes between 18 and 44 degrees. In British Columbia, debris flows occur on glacially oversteepened slopes (Church and Miles, 1987). Wieczorek et al. (1988) used a digital elevation model of San Mateo County, California, to determine the effect of slope steepness on debris flow distribution. They found that most debris flows in the area were initiated on slopes of approximately 25° to 29°. Although debris flows may occur on higher slopes, they are not as frequent, possibly because of the inability of debris to form and remain on steeper slopes. They also concluded that the slope at which debris flows are generated is somewhat dependent on the geologic unit.

Debris flows require a minimum thickness of colluvium (loose, incoherent deposits at the foot of a steep slope) for initiation. Reneau et al. (1984) found that debris flows frequently occur in colluvium-filled bedrock hollows and that groundwater was a factor in increasing the pore-water pressure that increased the potential for a debris flow. Ala and Mathewson (1990) also found groundwater moving through fractured bedrock to be a factor in initiating debris flows. Flows mobilized along the Wasatch Front in Utah were initiated in about 2 m thickness of coarse sand colluvium overlying bedrock (Santi, 1989). Additional colluvium was mobilized as the flow moved downstream through a channel within the canyon. In studies conducted after debris flow occurrences, Santi concluded that a minimum of 0.4 m of colluvium must be affected by saturation conditions before a debris flow will occur.

Debris flows occur in colluvium produced from a variety of bedrock. In San Mateo County, California, Wieczorek et al. (1988) analyzed more than 4,000 debris flows occurring over 53 geologic units. The geologic units were then ranked as high, medium, and low incidence. Ten units were ranked as high incidence, accounting for more than 50 percent of the debris flows and less than 20 percent of the area. These units are made up of a variety of rock types, including volcanic rocks, mudstone, granite, and sandstone. Wieczorek et al. (1988) concluded that bedrock geology strongly influences debris flow distribution. Other research, however, has concluded that there is no correlation between bedrock geology and debris flow occurrence (for example, see Govi and Sorzana (1980)). It may be assumed then that, while particular bedrock may be susceptible to debris flows in one region, the correlation cannot be extended to other regions.

Soil Type

Debris flows occur in many different soil types. Ellen and Fleming (1987) conducted a study of 50 soil samples from debris flows that occurred in the San Francisco Bay region in California in 1982. They found that a common characteristic was a low clay content. For all 50 samples, the clay content ranged from 8 to about 30 percent. The fast-moving debris flows all had clay contents less than 25 percent. The low clay content is required for mobilization. In British Columbia, Church and Miles (1987) also found that the soils at debris flow sites contained low clay content.

Studies conducted following the devastating debris flows in Virginia caused by Hurricane Camile showed that soils in unstable basins were loose, silty sands with a very low plasticity (Auer and Shakoor 1989). The soil contained about 20 percent gravel, 55 percent sand, and 25 percent fines (silt and clay). In stable basins, the soil consisted of slightly plastic silts that contained 37 percent sand, of which 20 percent was fine sand. They also found that unstable basins had significantly thinner soils than stable basins.

Climate, Antecedent Moisture Conditions and Runoff

The climate of the regions in which debris flows occur is as varied as the geology. Although many of the debris flows reported in the literature occur in the desert southwest of the United States, debris flows also occur in more humid climates, such as the eastern United States. Debris flows also occur where there is significant and rapid snowmelt, such as the Cascades of Oregon. The important indicators of debris flow potential are the antecedent moisture conditions and runoff.

In addition to the duration and intensity of a storm that ultimately produces a debris flow, antecedent rainfall is an important meteorological characteristic (Campbell, 1975; Johnson, 1984; Wieczorek, 1987; Keefer et al., 1988). The antecedent rainfall saturates the soil, increases the pore pressure, and reduces the frictional resistance at the failure plane, with the debris-generating rainfall triggering the failure. The significant period of antecedent rainfall may vary from days to months, depending on local conditions. Caine (1980) collected worldwide storm data of duration and rainfall intensities that triggered debris flows. He developed an expression relating storm intensity to duration that yields threshold values for debris flow initiation:

$$I_r = 14.82 D^{-0.39}, \tag{7.1}$$

where I_r is rainfall intensity (millimeters per hour) and D is duration (hours). Campbell (1975) found that debris flows in the Los Angeles area usually occurred when the seasonal antecedent rainfall exceeded 25 cm followed by a storm with an intensity of at least 6 mm per hour. Wiezorcek (1987) conducted a study of the rainfall intensity-duration relationship that triggered debris flows in the Santa Cruz Mountains, California. He found that no flows were triggered before 28 cm of rainfall had accumulated each season, suggesting that prestorm soil-moisture conditions are important. Blodgett (1989) reported numerous debris flows in the San Francisco Bay area following greater than normal antecedent rainfall together with a prolonged heavy rain (40 cm). For this same area, Mark and Newman (1988) analyzed data from the 1982 storm and found that debris flows followed antecedent rainfall for the three-month period preceding the storm of 25 to 40 cm and storm precipitation of more than 25 cm.

In some areas, snowmelt can be an important factor in producing antecedent moisture. The snowmelt can be responsible for building up pore-water pressures to the point where a debris flow may be initiated. Pierson (1985) found that the main source of water for lahar initiation at Mount St. Helens was from eroded snow and ice incorporated into the flow from turbulent mixing. In the Wasatch Front in Utah, Santi (1989) found that several debris flows had been initiated by snow melt.

Johnson and Sitar (1990) conducted a field study of hydrologic conditions preceding debris flow initiation. A field site was instrumented and monitored for two (wet) winter seasons. They found that the hydrologic response was highly dependent on antecedent moisture conditions. They concluded that traditional models of hillslope hydrology do not fully account for the positive pore-pressure pulses that preceded debris flow initiation.

Ground cover

Another factor that is an important determinant of debris movement is the land cover of the watershed, particularly when fire destroys the vegetative cover. Erosion is greatly accelerated during the rainy season following a fire (Wells et al., 1987). Postfire debris flows can occur during relatively small storms and require less antecedent rainfall than on a stable watershed. The heat generated by the fire sears the surface, which then limits infiltration. Thus, the loss of the naturally rough cover and the decreased potential for infiltration cause significant increases in the volume and velocities of the runoff, which initiate rill erosion (Johnson, 1970; Wells et al., 1987). The water that does infiltrate cannot transpire because of the lack of a vegetal cover and, thus, remains in the soil profile, which increases the stress. The lack of a viable plant root system in the denuded watershed compounds the problem since the resistance to tangential shear is reduced. These changes to the soil profile can increase the rate of debris movement from the surface, as well as increasing the tangential stresses along the shear plane. McNabb et al. (1989) discussed the effects of watershed burn on infiltration, water repellency, and soil moisture. They indicated that three factors are related to water repellency: soil texture, intensity of the burn, and the species of the fuel consumed. To induce intense water repellency, susceptible soils must be heated to temperatures between 350° and 400°F. It then takes about five months for the infiltration rate to recover to pre-burn rates. Helvey (1980) evaluated the effects of fire on runoff and sediment production in North-Central Washington. He pointed out that the inability to measure sediment lost during debris flows causes underestimations of the effect of burn on sediment production. For three watersheds used in his analysis, he found that sediment production was increased more than eight times during the first five years following fires. Sidle et al. (1985) reports a three-fold increase in erosion rates for the southern California area. Shuirman et al. (1985) provide a brief description of the effects of fire on the watershed and indicate that case histories in southern California show that erosion may increase by a factor approaching 50 in the first year following watershed burn. Robison (1990) found that fire damage increased debris flow potential by about 700 to 800 percent.

It is often difficult to accurately assess the effect of vegetation type on debris flows. Vegetation is also effected by bedrock type, slope, and rainfall; however, it is clear that removal of vegetation increases the probablity of debris flow occurrences. In a study conducted on debris flows in the San Francisco Bay region, Wieczorek et al. (1988) found that clearing of vegetation by logging made that area more susceptible to debris flows.

Debris flows typically destroy most, if not all, vegetation in its path. Recovery of the vegetation is important to restabilization of the area and to recovery of the local biota. Recovery of vegetation following a debris flow depends on surface deposit characteristics and intensity of scour (Gecy and Wilson, 1990). In a study of three debris flows in the western Oregon Cascades, Gecy and Wilson (1990) found that recovery was most rapid where scour intensity was lightest and on fine-grained debris deposits. In another study in the same area, Lamberti (1991) found that the biota of a stream disrupted by debris flow had mostly recovered within one year.

7.2.2 Laboratory and Mathematical Studies

A debris flow typically exhibits non-Newtonian behavior. Debris flows have been modeled by various researchers using Bingham plastic, dilatant, or viscoplastic models (e.g., Takahashi, 1978; Takahashi et al., 1987; Johnson, 1984; Chen, 1988). A Newtonian fluid, such as water, will flow under any applied shear stress and the viscosity of the fluid will remain constant at a given temperature. A Bingham fluid behaves like a Newtonian fluid except that a certain applied stress is required to initiate flow. Pseudoplastic and dilatant fluids may also require an initial applied stress, but the viscosity changes with the changing shear rate. A pseudoplastic fluid is shear-thinning, and a dilatant fluid is shear-thickening (Pierson and Costa, 1987).

A number of laboratory studies have been conducted to determine various characteristics associated with debris flows. Davies (1990) conducted a set of experiments in a moveable-bed flume and concluded that the behavior of the debris-flow wave is predominantly controlled by the larger grains within the flow. He also found that the front of the debris-flow pulse is highly erosive, while the tail is nonerosive or depositional, and the maximum depth of the flow is a function of the velocity gradient within the flow. Major and Pierson (1992) conducted experiments to determine the rheological behavior of debris flows. Debris flow material collected from the Mount St. Helens area was used for the experiments. The influence of sediment concentration and sand proportion on the rheological behavior of the fine-grained debris was then tested using a viscometer. They found that at shear rates greater than about 5 $\sec^{-1}$, the flow behavior was a function of clay, silt, and water content and essentially can be modeled as a Bingham fluid. At shear rates less than about 5 $\sec^{-1}$, the flow behavior is determined by the sediment concentration and the proportion of sand. They concluded that debris flows cannot be characterized by a single rheological model.

Takahashi et al. (1992) developed a computer simulation model of debris flow, which he then verified using experimental data. The model determines the discharge hydrograph and solids concentration. Debris flow is generated at the head of a channel. As it is routed downstream, the larger particles in the debris flow move upward and forward. Routing of the flow is based on continuity, erosional processes, and erosional speed. Takahashi used experimental data to verify the model; however, there was considerable scatter in the data so only trends could actually be verified. The movement of large particles upward and forward was also studied by Zhaoyin and Xinyu (1990). They conducted a flume experiment with flows consisting of various concentrations of clay and gravels (4 to 25 mm). The mud was allowed to flow over the gravel, incorporating gravel. They observed that the front of the flow was always steep and high, and that the larger particles moved forward toward the snout of the flow.

7.3 Methods of Prediction: When, Where, and How Much

Prediction of debris flows has proven to be a difficult task. Although all debris flows share some characteristics in common, soils that are stable in one region may not be stable in another. Methods of predicting where debris flows might occur depends on a ground reconaissance of the characteristics previously described. Prediction of magnitude and frequency is important for establishing policies for land management and for mitigation design. Predicting when debris flows will occur is extremely important in residential areas so that such areas can be safely evacuated and affected bridges or roadways closed.

7.3.1 MAGNITUDE

Methods of estimating the magnitude of debris flows have been developed by researchers in the U.S., Japan, Canada, and Austria. Debris volume is a function of a combination of parameters, typically including one or more of the following: drainage area, channel length, channel slope, channel width, vegetation (or lack of, as a function of burn, logging, etc.), relief ratio, and hypsometric ratio. Although it is generally acknowledged that soil type is also an important parameter in estimating magnitude, most methods do not explicitly account for soil type; therefore, those models are valid only within the area where they are developed. Van Dine (1985) provides a table showing the parameters included in the estimation of debris flow magnitude for 11 methods used in Japan, Canada, and Austria.

As a first approximation, the magnitude of a debris flow can be determined simply as a function of drainage area (Hungr et al., 1984). However, because variation in the debris flow volume due to other factors is very significant, methods based only on the drainage area may only provide planning estimates. Van Dine (1985) reported of a study conducted by Thurber Consultants (1983) where the volume of debris flows (m^3) in 24 creeks was plotted against drainage area (km^2), with the following best-fit line: $Y = 10000A$, where Y is debris flow yield or volume and A is drainage area. However, a similar analysis of 68 data points from 29 watersheds in southern California performed by the first author of this paper showed no significant correlation between debris flow volume and drainage area. This is due to the tremendous variation in magnitude within a single watershed from one event to the next. A relationship between Y and A can only be considered as a planning estimate, as opposed to a design estimate, because it ignores the frequency that is associated with the volume, as well as other factors that have a physical relationship with the volume.

Typically, there are inadequate debris flow magnitude data to statistically treat the problem of prediction in a manner similar to that of precipitation or streamflow discharge. In addition, the uniqueness of each site and the complexity of the processes of debris flow initiation and movement are reflected in the large variation observed in event-to-event flow volumes on a single watershed. Thus, estimation of a design debris flow volume requires an initial estimate of the volume, possibly predicted from an equation described below, followed by modification of that estimate by other limiting or contributing factors for the particular watershed of interest. These factors are best determined by a field investigation of the watershed.

Hungr et al. (1984) developed a method of estimating debris flow volume, which was later modified to include other factors. The magnitude of the flow is given as

$$M = LBe, \tag{7.2}$$

where M = magnitude (m^3), L = channel length with uniform erodibility (m), B = channel width (m), and e = mean erosion depth (m). Equation (7.2) can then be modified by dividing the stream into n sections

$$M = \sum_{i=1}^{n} A_i^{1/2} L_i e_i, \tag{7.3}$$

where L_i = sector length (m), e_i = channel erodibility coefficient (m^3/(m·km^2)), and A_i = area (km^2). Hungr et al. (1984) provided a table of estimated erodibility coefficients as a function of channel gradient, bed material, side slopes, and stability condition prior to a debris flow event. Hungr et al. indicated that the erodibility coefficients are applicable only on a narrow regional basis and incorporate climatic, geological, and biological factors. The method developed by Hungr et al. has been modified to include inspection of the debris flow path and noting width, slope, particle size of the creek bed, height, angle, and degree of stability of the banks (Hungr et al., 1987). The design magnitude is then empirically estimated from these parameters.

Using data from 29 watersheds in the Los Angeles area, Johnson et al. (1991a) developed equations for predicting the seasonal volume of debris flows for watersheds with drainage areas up to 8 km^2. The watersheds are located in the Santa Monica, San Gabriel, Verdugo, and Santa Susana Mountains. Elevations range from 150 to 900 m in the Santa Monica's to 1500 to 2700 m in the San Gabriel's. Fifteen variables, including precipitation, antecedent rainfall, drainage area, hypsometric index, relief ratio, SCS curve number, drainage density, stream length, and years since the watershed was last burned, were compiled and statistically analyzed for each of these basins. Rainfall is restricted almost entirely to the winter months, November through March. The relief ratio is defined as the maximum basin relief (m) divided by the longest horizontal distance of the basin measured parallel to the major stream (km). The relief ratio indicates the overall steepness of the basin (Ritter, 1978). The hypsometric index is defined as the relative height at which a watershed may be divided into two equal ground surface areas by a given contour. A subset of the predictor variables was then selected based on a combination of principal components analysis, an assessment of the correlation matrix, and an understanding of the relative importance of the variables in the physical processes. Using a power model and numerical optimization, the debris yield (cubic yards per square mile per year) was regressed on the relief ratio (ft/mi), hypsometric index, and the number of years since the most recent burn in which at least 40 percent of the vegetated area of the watershed burned. The result was a set of equations for seasonal debris flow volumes for various return periods from 2 to 100 years. A debris basin designed for a single "maximum" event, rather than a seasonal volume will be more likely to be overtopped since there is a likelihood that smaller, more frequent flows will either precede or follow the design flow. In addition to the seasonal-volume models, Johnson et al. (1991b) provided a method for estimating single-event debris yields.

Tatum (1963) provided a method to estimate the capacity of a debris basin for one major storm that would be exceeded in magnitude only on rare occasions. Tatum assumed an ultimate debris production of 1.9 million cubic yards per square mile; this value is multiplied by empirically derived correction factors for the slope, drainage density, hypsometric-analysis index, and the 3-hour rainfall for the debris producing storm. Using this adjusted value, a graphical adjustment for drainage area is made, which yields another estimate in millions of cubic yards per square mile. A burn adjustment is then made using a linear equation that is a function of the area of the watershed burned, the number of years since the burn occurred, and the area not burned. This method, developed over 30 years ago, tends to over-predict debris flow volumes and does not assign a frequency to the computed magnitude.

The methods of magnitude estimation are often criticized for not including enough information regarding the soil type in the watershed. However, if the equations are used within the area for which they are developed, factors such as soil type and geology become imbedded in the coefficients of the method.

Methods have been developed in various regions to incorporate geologic factors into the prediction. For example, Nelson and Rasely (1990) used the Pacific Southwest InterAgency Committee (PSIAC) Sediment Yield Rating Model to estimate potential debris flow volumes for an area near Salt Lake City, Utah. In this model, factors are assigned according to geology, soils, climate, runoff, topography, ground cover, land use, upland erosion, and channel erosion. The factors are then summed and converted to sediment yield using a relationship developed by Renard (1972).

7.3.2 Frequency of Occurrence

Whether developing a policy concerning debris flows or computing a design volume for a debris mitigation scheme, an estimate of the frequency of occurrence of the design volume is necessary. The lack of debris flow data in many parts of the world where debris flows occur prohibits the statistical treatment of frequency in those areas. In some areas, such as debris-prone areas of southern California, frequency is not a design factor.

Debris flow frequencies do not generally coincide with precipitation patterns (Hungr et al., 1984; Van Dine, 1985). Since historical data are usually insufficient to establish frequency curves, Hungr et al. suggested a qualitative assessment of the frequency of occurrence based on geologic evidence and streambed types, such as bedrock or unconsolidated deposits. Van Dine (1985) also used geologic and surficial evidence to determine frequency. He described the probability of occurrence qualitatively as very high, high, moderate, low, or no risk depending on creek characteristics, history of debris flows, creek gradient, size of deposition fan, debris source, logging, and bank stability. Van Dine provides a table describing drainage areas in each of the five categories of risk. For example, if a creek is classified in the *very high probability* category, then it is likely that debris flows of less than the design magnitude will occur frequently and the design flow should be assumed to occur within the short term. A *moderate probability* would indicate that the design debris flow would likely occur during the life of a structure, such as a house or bridge. Assuming that the design life of such a structure is typically 75 years, then a moderate probability would correspond to a 75-year return interval for the design debris flow.

The volume of debris for a particular site may vary by orders of magnitude from one event to the next. If adequate historical data exist, return periods can be assigned to the various magnitudes. Johnson et al. (1991a) analyzed 19 years of debris flow data for watersheds in southern California. A log-normal population was assumed to underlie the physical processes. The flows ranged in magnitude from about 2,300 to 58,500 m^3. The means and standard deviations were used with the log-normal distribution to compute the debris flow volumes in cubic yards per square mile per year for exceedance frequencies of 2, 5, 10, 25, 50, and 100 years. For a given set of conditions, the magnitude-frequency equations could be used to compute a debris-flow frequency curve, from which debris volumes could be estimated for any other exceedence probability.

Godfrey (1985) attempted to use a magnitude-frequency approach to characterize landslides across the Wasatch Plateau. He concluded that the frequency of the total annual precipitation for that region can be used to indicate the recurrence interval of landslide events; however, because of the scarcity of data, a relationship between magnitude and frequency could not be developed. Godfrey provided some indications of the risk involved.

Hollingsworth and Kovacs (1981) developed a method that does not specify either a magnitude or frequency but does provide a qualitative indication of the uncertainty. Their system leads to a numerical value that is transformed into an ordinal-scale measurement that indicates the potential for slope failure, with the outcomes of nil, low, high, and extreme. This may be an adequate measure of frequency because the method is primarily intended for the protection of individual structures. In contrast, the methods of Johnson et al. and Tatum are intended for the design of watershed-scale protection structures.

7.4 Debris Flow Mitigation

There are a number of ways to guard against debris flows. Although none of the methods can halt erosion of the mountains, they can be used to reduce the number of fatalities and minimize destruction caused by debris flows. In residential areas, warning systems can be established so that an area can be evacuated before an event occurs. Passive mitigation measures, such as zoning and land use management, can be used to protect against debris flows. Residential areas and highway crossings can be protected by debris basins, deflection structures, and check dams. Each of these methods are described in this section.

7.4.1 Warning Systems

Warning systems are used to warn residents of the potential for debris flow or of an actual debris flow occurrence. Pre-event warning systems are typically based on rainfall in a debris-prone area (Hungr et al. 1987). In the San Francisco Bay region, a warning system, consisting of a network of 45 telemetering rain gages, was developed following the tragic 1982 storms in which 25 people were killed mainly by debris flows (Keefer et al., 1987). The rain gage data are transmitted to the U.S. Geological Survey in Menlo Park where the incoming data is continuously monitored during a storm. The rainfall data are then compared to a set of rainfall intensity and duration curves for debris flow initiation, similar to Eq. (7.1). Based on this information plus the forecast for the next six-hour period, a judgment is made by U.S. Geological Survey geologists whether or not to issue a warning. The warning is transmitted to local radio and television stations in special weather statements from the National Weather Service, in which residents are advised to watch for increasing rainfall and failing slopes.

Rainfall alone is often not adequate to reasonably determine when a debris flow may occur, as previously described. Hungr et al. (1987) described a sophisticated weather observation system that was placed in five locations in a debris-prone area north of Vancouver. Although the system has not been adequately tested at this time, the purpose is to provide decision-makers with a detailed weather picture so that a decision can be made as to whether or not a debris flow warning should be issued.

Warning systems can also be installed for the purpose of warning residents and motorists of an actual event. Debris flows can travel at very high velocities; therefore, an event warning system provides only a very limited amount of time to react. The usefulness of such a device for evacuation is questionable. Event-based warning systems are probably most useful for bridges and other roadway crossings (Hungr et al., 1987).

7.4.2 Passive Mitigation Measures

Passive mitigation measures are those that do not include construction. These methods include zoning and hazard mapping, land management, and vegetative cover. Proper zoning may be implemented so that the building of homes and roads in the likely paths of debris flows is avoided. Zoning, of course, is the most effective method of avoiding debris flow damage, although not always practical. If zoning laws have not been established prior to the initiation of building in an area, it is very difficult to implement zoning restrictions where homes have already been built.

Land management has been shown to have an influence on the potential for debris flows. For example, cattle and sheep grazing and land clearing and burning was a factor in triggering debris flows in the Wasatch Front in Utah (Mathewson and Keaton, 1990). Logging operations can also be a factor in debris flow initiation. In both cases, land is cleared of vegetation, allowing for increased sediment movement. To avoid debris flows caused by poor land management practice, guidelines must be set to minimize debris flows. For example, in British Columbia, the Ministry of Forests has set guidelines for logging and forest road construction (Hungr et al., 1987). The guidelines include buffer strips, plan and shape of clear cuts, replanting, and culverts and troughs. Maintenance is also an important aspect of the guidelines.

7.4.3 Active Mitigation Measures

There are many different mitigation measures that have been tried in various parts of the world. The success of any of these structures is dependent on the intended use (e.g., to redirect flow or to halt flow) and the estimate of debris flow magnitude for the design. Channel improvements have proved ineffective because channels quickly become blocked causing the debris to take different flow paths (Waldron, 1967). Hollingsworth and Kovacs (1981) discussed the following control methods: retaining walls, deflection walls, stem walls, debris basins, and debris fences. Retaining, deflection, and stem walls are used for the protection of individual buildings such as private homes. Debris fences are used to retard the flow rate of the debris flow and break up the flowing mass. The improper placement of these fences usually results in failure. Johnson and McCuen (1989) developed a design method for slit dams to retard the flow rate of debris flows and trap the larger boulders. The use of check dams constructed in series have been used in Japan and Europe to retard the flow of large debris flows and lahars. The construction of these dams can be difficult along steep channels and may take an extended period of time to complete, sometimes more than a decade (Hungr et al., 1987). The expense of constructing check dams is often prohibitively high. Waldron (1967) describes the failure of six small check dams designed to retard the debris flow rate; debris flows destroyed the dams when they were nearly completed.

In the San Francisco Bay area, extensive grading is a common mitigation measure (Montgomery et al., 1989). The grading may involve removal of an entire colluvial deposit. This is an expensive effort, aesthetically unpleasing, and may actually accelerate erosion if not properly graded.

Debris basins are larger structures often constructed at the base of narrow canyons. The failure of a debris basin is usually due to an insufficient capacity such that the basin cannot contain the incoming debris flow. Debris basins are effective when they are designed properly and maintained regularly; however,

Mud and Debris Flows

proper design requires an estimate of the magnitude, frequency, and impact force of a design debris flow. In addition, the design must permit passage of water discharge and should provide an emergency spillway to prevent failure of the structure. A debris basin must be maintained; the basin cannot function as designed if it is not maintained. Material deposited in the basin from previous events must be removed on a regular basis so that there is adequate volume remaining in the basin for another debris flow of the design magnitude or less. In designing a debris basin for debris flow mitigation, an estimate of the magnitude of the expected debris flows, the frequency of the design or lesser flows, and the impact force of the flow on the basin wall should be determined. The magnitude, or volume, of the design debris flow (or flows if the basin is to contain multiple events) is needed to properly size the basin. The number of events that may occur during a season should also be considered so that the basin can be designed to contain multiple events that might occur before the deposited material can be removed. The impact force of the debris flow on the basin wall is needed to determine the required strength of the wall. Johnson and McCuen (1993) provide a summary of various debris basin designs.

There are several major problems associated with active mitigation. First, the structures tend to be aesthetically unpleasing, detracting from the beauty of the mountains. Second, the structures are expensive. The burden of the cost is typically shared by all the taxpayers of a particular area, rather than those who have decided to live in such unstable areas. Third, the structures must be emptied to prepare for the next event. This maintenance is not only very expensive, but a problem arises in determining where to deposit the sediment removed from the basin. Fourth, the structures may give a false sense of security to local residents and actually increase the risk of hazard by holding back smaller events and then failing for larger events during which all the sediment is released.

7.5 Statistical Modeling of Debris Flows

Debris-flow disasters are preventable. Engineering solutions require estimates of debris volumes, which can be provided by models. Models that have been fitted to and tested with measured volumes of debris are more likely to provide accurate estimates of future debris volumes than uncalibrated models. Where such data have not been collected, data collection programs can be designed to collect the necessary data.

There are two general approaches to modeling debris flows: theoretical models and statistical analysis of measured data. Theoretical, or mathematical, modeling was discussed briefly in Section 7.2.2. Further separation of the statistical approach can be made based on the complexity of the analysis, which is partially reflected in the type of output provided. The previously cited works of Hungr et al. (1984) and Tatum (1963) provide models that yield volumes of debris for set conditions. Unfortunately, the methods do not provide a measure of the frequency of occurrence, thus limiting their use in engineering design, landuse planning, and economic decision-making.

Where sufficient measured data are available, prediction methods that provide estimates of debris-flow volumes for selected frequencies can be developed. These methods have the advantage of providing the means of developing a frequency curve for debris volumes. While the general methodology used for peak-discharge regionalization can be applied in the development of regional debris volume-frequency equations, there are some important differences in the regionalization of peak-discharge and debris-volume data.

7.5.1 THE REGIONALIZATION PROCESS

Regional models are useful because they provide the hydrologist with a model that can be used for design at locations where measurements of the hydrologic variable are not available at the time when a design is needed. The following steps provide a brief general outline of the regionalization process:

1. Define the criterion variable and all potential predictor variables.

2. Identify all watersheds in a homogeneous region where sufficient data are available.

3. From the data, obtain values of the criterion variable for the selected interval (one year for annual maximum series).

4. Perform a frequency analysis of the annual debris volumes for each watershed.

5. From each frequency curve, obtain values of the criterion variable for selected frequencies.

6. Determine values of potential predictor variables for each watershed.

7. Select a model structure to use for the prediction equations; the power model is commonly used.

8. Form a data matrix, with the n rows representing the n watersheds and the $(p + 1)$ columns representing the p predictor variables and the criterion variable.

9. Fit prediction equations for each frequency selected in step 5; use stepwise modeling to eliminate unimportant predictor variables.

10. Smooth the fitted coefficients to ensure that the equations will provide rational frequency curves.

The ten-step regionalization process can be used to develop a model that provides magnitude-frequency estimates for use at locations where actual measurements are not available.

7.5.2 Regionalization of Debris Volumes

While the above ten-step procedure is applicable to debris-volume estimation, several points should be emphasized because of distinct characteristics of debris data. Debris measurements are rarely made either after every storm or on a regular, annual basis. Measurements are usually made only when debris basins are cleaned out, so the debris volumes are the result of several storms over a nonconstant duration. In periods where rainfall is low, the interval between basin cleanings can be several years. Thus, it is difficult to obtain individual event volumes, and so models are usually developed for annual or seasonal volumes for which estimates are easier to compute. Annual rates are obtained by dividing the total volume by the period between measurements. It is preferable to make annual measurements especially where the computed annual rates are the result of one or two major storm events rather than a more uniform rate of debris generation over the period between measurements. The intent is to have a debris flow series in which the individual values are independent because frequency analysis assumes independent measurements. When a debris basin is cleaned out more than once a water year, only one value should be used per year if an annual-series analysis is the intent. While the clean-out period is often greater than one year, the measured debris volumes can be expressed as an annual rate, i.e., cubic meters per year; this would use the ratio of the debris volume to the duration (years) between cleanings. It is possible that single-event volumes could be computed from measured data if extensive rainfall data are available and a reliable relationship between debris volume and rainfall is available; this is rarely possible.

A second important element of the regionalization process is the selection of a probability function to use in representing the frequency curve. This aspect has not been widely studied. Johnson et al. (1991a,b) found that data from watersheds in southern California fit a log-normal distribution. However, other distributions such as in log extreme value or log Pearson type III could be used. The distribution can influence the magnitudes computed for the more extreme frequencies. Thus, where 50-yr and 100-yr debris volumes are needed, careful consideration should be given in the selection of a distribution to use in the frequency analyses.

The selection of predictor variables is an important step in the regionalization process. Typically data will be collected so there is at least one predictor variable for each of the factors listed previously. Slope is the most commonly used predictor variable to represent topography. Variables based on the hyposometric curve are often selected; however, these are more difficult to quantify. The clay or sand content could be used to represent soil type. The 3-day or 7-day antecedent rainfall depth can be used to represent the antecedent moisture. The volume (depth) of rainfall can be used as a predictor. However, for multiple-event debris volumes, where there is some variation in rainfall intensity-duration-frequency characteristics within the region, a design-storm depth or intensity can be used. The time between fires or clearing of the ground cover can be used as a measure of ground cover. The variables should be selected in a way that assures that values for the predictors can be readily estimated at watersheds where design estimates will be needed.

In addition to selecting the predictor variables to include in the analysis, it is also necessary and important to delete those that are unimportant. If a linear or power model structure is to be used, a stepwise regression program can be used to select variables (McCuen, 1993). It is important to use the same predictor variables for each of the regression equations for the individual exceedence frequencies. Otherwise, the estimates may be inconsistent.

The results of the regionalization will be a set of prediction equations for different exceedence probabilities (e.g., 0.5, 0.2, 0.1, 0.04, 0.02, and 0.01). The coefficients of each of the equations are values of random variables. The coefficients for any one predictor variable should show either nearly-equal values or a systematic trend across the equations for different exceedence frequencies. If this does not exist, then the coefficients will need to be changed so that they are equal or show a trend. Otherwise, the prediction equations may yield a higher debris-flow estimate for a more frequent event. Therefore, coefficient smoothing may be necessary. The accuracy of the prediction equations, as measured by the standard error of estimate or the correlation coefficient, should be computed after coefficient smoothing.

As a predictor variable, rainfall deserves special mention because of its obvious relationship with flow volumes. If the model is intended to predict volumes for real-time forecasts, then actual rainfall depths would be needed along with the actual debris volumes. For real-time forecast models, variables based on the rainfall intensity-duration-frequency (IDF) curve would be inappropriate because real-time models are based on actual measured rainfalls during the progress of a storm. When modeling debris volumes for design estimation, a rainfall IDF characteristic may be an appropriate predictor variable because an exceedence probability can then be associated with the computed debris volume. However, unless there is significant variation in its value over the region being studied, it may not show up as a significant predictor in the regression analysis. In such cases, it should be dropped from the equations. This does not imply that rainfall is not an important factor in the physical process of debris generation. Actually, rainfall is one of the more important factors. It fails to show up as a statistically significant predictor only because of the small variation in rainfall characteristics within the region. The hydrologic effects of rainfall are still reflected in the equation. Specifically, the regression coefficients reflect the effect of the rainfall characteristics of the region. The debris volumes reflect the rainfall, which in turn is reflected in the coefficients, especially the intercept coefficient.

7.5.3 DATA REQUIREMENTS FOR REGIONALIZATION

Systematic collection of debris-flow volumes are rarely available in regions where the potential exists for debris disasters. If a regional government is interested in developing models for use in making debris-volume estimates, a systematic data-collection program should be developed. A regional program has the advantage of providing information from many sites rather than one long record at a single site. This allows for assessment of causative factors and for providing a regional

Mud and Debris Flows

model in the shortest possible time. Regionalization of debris volumes requires a systematic data-collection program.

Data required to develop a regional model of debris volumes using the procedure outlined above includes measured rates of debris production and values of predictor variables such as those discussed in Section 7.2. Existing debris basins should be monitored on a regular basis. For the larger basins, annual measurements of debris volumes can be made. In years during which very large volumes are deposited, the basin may need to be cleared of debris more than once. All measurements should be recorded. For the smaller basins, maintenance after each debris-generating event may be necessary. If the capacity of the debris basin is exceeded, estimates of the overflow should be made, if possible. In periods or years in which no debris is generated, the period should still be denoted as a zero-volume period. Data collected on an annual or multi-event basis can be used to develop seasonal prediction models. Single-event models can be developed when data are collected from individual storm events. There is a trade-off between single-event and seasonal models. Large basins require less maintenance but have larger construction costs. Small basins are cheaper to construct, but they require more maintenance and they offer less protection. A better data base will result from collecting data from each debris-generating event even from the larger debris basins.

In addition to debris volumes, rainfall records should be maintained. When possible, hourly data should be collected; however, daily rainfall depths can provide useful measures of antecedent rainfall. Local intensity-duration-frequency data should also be developed.

Changes in vegetation can have a significant effect on debris volumes. As indicated above, studies have shown that debris volumes are higher following events when vegetation has been destroyed by fire. For larger watersheds, a fire event may only destroy vegetation on part of the watershed. Records of the fraction of the watershed that suffered burn should be maintained and used in the modeling effort. Records on vegetation growth following fire destruction can also be used in developing regional models.

For each watershed, watershed and soils data should be collected. Watershed area and slope are the primary variables. Hypsometric curves should be developed. When the primary site or sites from which debris was generated can be identified, specific data about these sites should be collected, including topographic, vegetation, and soils data.

7.6 Conclusions

Debris flows are naturally-occurring, highly destructive hazards that have been responsible for loss of life and destruction of property, bridges, and roadways. Although debris flows are a naturally-occurring erosional process, this process can be accelerated by human activity through poor or improper land management. Prediction of debris flow magnitude and frequency is required for warning systems, zoning, and other mitigation measures, but prediction is not yet at a reliable state. The debris flow process is extremely complex; models developed for one site or region are not extendable to other sites or regions. Theoretical models have also been limited in their usefulness because of detailed input requirements and necessary assumptions and simplifications.

Mitigation of debris flows can be passive or active. Debris basins and other structures have saved many roads, bridges, and homes from destruction, but there have also been plenty of failures. Certainly, the most effective mitigation measure is to avoid building in debris-prone areas. There are two main problems with this simple solution. First, many debris-prone areas are already developed with homes, bridges, and roadways. Second, debris-prone areas and frequency of occurrence of debris flows in those areas must be identified for zoning purposes. This is not a simple task since debris flows are a function of many factors.

There is still considerable research to be conducted and data to be collected on debris flows to improve our ability to predict magnitudes and frequencies for the purpose of mitigation and zoning.

References

Ala, S., and Mathewson, C.C., 1990. Structural control of ground-water induced debris flows. Proceedings of the Hydraulics/Hydrology of Arid Lands Symposium, ASCE, San Diego, California, 590-595.

Auer, K., and Shakoor, A., 1989. Geotechnical characterization of drainage basin stability with respect to debris avalanches in central Virginia. Bulletin of the Association of Engineering Geologists, 26(3), 387-395.

Blodgett, J.C., 1989. Flood of January 1982 in the San Francisco Bay Area, California. USGS Water Resources Investigations Report 88-4236.

Caine, N. 1980. The rainfall intensity-duration control of shallow landslides and debris flows. Geografiska Annaler, 62A(1-2), 23-27.

Campbell, R.H. 1975. Soil slips, debris flows, and rainstorms in the Santa Monica Mountains and vicinity, Southern California. USGS Professional Paper 851, 51 p.

Chen, C.L., 1988. Generalized viscoplastic modeling of debris flow. Journal of Hydraulic Engineering, ASCE, 114(HY3), 237-258.

Clark, G.M., 1987. Debris slide and debris flow historical events in the Appalachians south of the glacial border. in Debris Flows/Avalanches: Process, Recognition, and Mitigation, J.E. Costa, and G.F. Wiezorek (Eds.), Geological Society of America Reviews in Engineering Geology, VII, 125-138.

Church, M., and Miles, M.J., 1987. Meteorological antecedents to debris flow in southwestern British Columbia; some case studies. in Debris Flows/Avalanches: Process, Recognition, and Mitigation, J.E. Costa, and G.F. Wiezorek (Eds.), Geological Society of America Reviews in Engineering Geology, VII, 63-80.

Clark, G.M., 1987. Debris slide and debris flow historical events in the Appalachians south of the glacial border. in Debris Flows/Avalanches: Process, Recognition, and Mitigation, J.E. Costa, and G.F. Wiezorek (Eds.), Geological Society of America Reviews in Engineering Geology, VII, 125-138.

Costa, J.E. and Jarrett, R.D. 1981. Debris flows in small mountain stream channels of Colorado and their hydrologic implications. Bull. Assoc. Engineering Geologists. XVIII (3), 309-322.

Cronin, V.S., Slosson, J.E., Slosson, T.L., and Shuirman, G., 1990. Deadly debris flows on I-5 near Grapevine, CA. Proceedings of the Hydraulics/Hydrology of Arid Lands Symposium, ASCE, San Diego, California, 78-83.

Davies, T.R.H., 1990. Debris-flow surges - experimental simulation. Journal of Hydrology, 29(1), 18-46.
Ellen, S.D., and Fleming R.W., 1987. Mobilization of debris flows from soil slops, San Francisco Bay region, California. in Debris Flows/Avalanches: Process, Recognition, and Mitigation, J.E. Costa, and G.F. Wiezorek (Eds.), Geological Society of America Reviews in Engineering Geology, VII, 31-40.
Fairchild, L.H. 1987. The importance of lahar initiation processes. in Debris Flows/Avalanches: Process, Recognition, and Mitigation, J.E. Costa, and G.F. Wieczorek (Eds.), Geological Society of America Reviews in Engineering Geology, VII, 51-62.
Friday, J., 1983. Debris flow hazard assessment for the Oregon Caves National Monument. USGS Water Resources Investigations Report 83-4100.
Gecy, J.L., and Wilson, M.V., 1990. Initial establishment of riparian vegetation on after disturbances by debris flows in Oregon. American Midland Naturalist, 123(2), 282-291.
Godfrey, A.E., 1985. Utah's Landslides of 1983 as a Magnitude-Frequency Event with a Finite Return Probability. pp 67-85 of Delineation of Landslides, Flash Flood, and Debris Flow Hazards in Utah, D.S. Bowles(ed.), Utah State University, Logan.
Govi, M., and Sorzana, P.F., 1980. Landslide susceptibility as a function of critical rainfall amount in Piedmnot Basin (North-Western Italy). Studia Geomorphologica CarpathoBalcanica, 14, 43-61.
Helvey, J.D., 1980. Effects of a North Central Washington Wildfire on runoff and sediment production. Water Resources Bulletin, 16(4), 627-634.
Hollingsworth, R. and Kovacs, G.S., 1981. Soil slumps and debris flows: prediction and protection. Bull. Assoc. Engineering Geologists, XVIII (1), 17-28.
Hungr, O., Morgan, G.C., and Kellerhals, R., 1984. Quantitative analysis of debris torrent hazards for design of remedial measures. Can. Geotech. J., 21, 663-777.
Hungr, O., Morgan, G.C., VanDine, D.F., and Lister, D.R., 1987. Debris flow defenses in British Columbia. in Debris Flows/Avalanches: Process, Recognition, and Mitigation, J.E. Costa and G.F. Wieczorek (Eds.), Geological Society of America Reviews in Engineering Geology, VII, 201-222.
Johnson, A.M., 1970. The ability of debris, heavily freighted with coarse clastic materials, to flow on gentle slopes. Sedimentology, 23, 213-234.
Johnson, A.M., 1984. Debris flow. in Slope Instability. D. Brunsden and D.B. Prior (Eds.), Wiley and Sons, New York, Chapter 8, 257-361.
Johnson, P.A., and McCuen, R.H., 1989. Slit dam design for debris flow mitigation. Journal of Hydraulic Engineering, ASCE, 115(9), 1293-1296.
Johnson, P.A., McCuen, R.H., and Hromadka, T.V., 1991a. Magnitude and frequency of debris flows. J. Hyd., 123, 69-82.
Johnson, P.A., McCuen, R.H., and Hromadka, T.V., 1991b. Debris basin policy and design. J. Hyd., 123, 83-95.
Johnson, P.A., and McCuen, R.H., 1993. Effect of debris flows on debris basin design. Critical Reviews in Environmental Control, 22(1/2), 137-149.
Johnson, K.A., and Sitar, N., 1990. Hydrologic conditions leading to debris-flow initiation. Canadian Geotechnical Journal, 27(6), 789-801.
Keefer, D.K., et al., 1987. Real-time landslide warning during heavy rainfall. Science, 238, 921-925.
Lamberti, G.A., Gregory, S.V., Ashkenas, L.R., and Wildman, R.C., 1991. Stream ecosystem recovery following a catastrophic debris flow. Canadian Journal of Fisheries and Aquatic Sciences, 48(2), 196-208.
Major, J.J., and Pierson, T.C., 1992. Debris flow rheology: experimental analysis of fine-grained slurries. Water Resources Research, 28(3), 841-857.

Mark, R.K., and Newman, E.B., 1988. Rainfall totals before and during the storm: distribution and correlation with damaging landslides. U.S. Geological Survey Professional Paper 1434, 17-26.

Mathewson, C.C., and Keaton, J.R., 1990. Multiple phenomena of debris-flow processes: a challenge for hazard assessments. Proceedings of the Hydraulics/Hydrology of Arid Lands Symposium, ASCE, San Diego, California, 549-553.

McCuen, R.H., 1993. Microcomputer Applications in Statistical Hydrology. Prentice-Hall, Inc., Englewood Cliffs, New Jersey.

McNabb, D.H., Gaweda, F., and Froehlich, H.A., 1989. Infiltration, Water Repellency, and Soil Moisture Content after Broadcast Burning a Forest Site in Southwest Oregon, J. Soil and Water Conservation, 44(1), 87-90.

Montgomery, D.R., Booth, T., and Wright, R.H., 1989. Preventative debris flow mitigation. Proceedings of the 16th Annual Conference, Water Resources Planning and Management, ASCE, Sacramento, California, 267-270.

Nelson, C.V., and Rasely, R.C., 1990. Debris flow potential and sediment yield analysis following wild fire events in mountainous terrain. Proceedings of the Hydraulics/Hydrology of Arid Lands Symposium, ASCE, San Diego, California, 54-59.

Pierson, T.C., 1985. Initiation and flow behavior of the 1980 Pine Creek and Muddy River Lahars, Mount St. Helen, Washington. Geological Society of America Bulletin, 96(8), 1056-1069.

Pierson, T.C., and Costa, J.E., 1987. A rheologic classification of subaerial sediment-water flows. in Debris Flows/Avalanches: Process, Recognition, and Mitigation, J.E. Costa, and G.F. Wiezorek (Eds.), Geological Society of America Reviews in Engineering Geology, VII, 1-12.

Pierson, T.C., and Scott, K.M., 1985. Downstream dilution of a lahar: transition from debris flow to hyperconcentrated streamflow. Water Resources Research, 21(10), 1511-1524.

Renard, K.G., 1972. Sediment problems in the arid and semiarid Southwest. Soil Conservation Society of America 27th Annual Meegin Proceedings, 225,232.

Reneau, S.L., Dietrich, W.E., Wilson, C.J., and Rogers, J.D., 1984. Colluvial deposits and associated landslides in the northern San Francisco Bay area, California, USA. Proceedings, IV International SYmposium on Landslides, Toronto, Canadian Geotechnical Society, 425-430.

Rice, R.M., and Foggin III, G.T., 1971. Effect of high intensity storms on soil slippage on mountainous watersheds in southern California. Water Resources Research, 7(6), 14851496.

Ritter, D.F., 1978. Process Geomorphology. W.C. Brown, Dubuque, IA, p.196.

Robison, R.M., 1990. Potential sediment yield from a burned drainage - An example from the Wasatch Front, Utah. Proceedings of the Hydraulics/Hydrology of Arid Lands Symposium, ASCE, San Diego, California, 60-65.

Santi, P.M., 1989. The kinematics of debris flow transport down a canyon. Bulletin of the Association of Engineering Geologists, 26(1), 5-9.

Shuirman, G., J.E. Slossen, and D. Yoakum, 1985. Relationship of Fire/Flood to Debris Flows, pp. 178-194 in Delineation of Landslides, Flash Flood, and Debris Flow Hazardous in Utah, D.S. Bowles (ed.), Utah State University, Logan.

Sidle, R.C., Pearce, A.J., and O'Loughlin, C.L., 1985. Hillslope stability and land use. Water Resourc. Monogr. Ser. No. 11, Am. Geophys. Union, Washington, DC, 140 pp.

Slosson, J.E., Havens, G.W., Shuirman, G., 1991. Harrison Canyon debris flows of 1980. Env. Geol. and Wat. Sci., 18, 27-38.

Takahashi, T., 1978. Mechanical characteristics of debris flow. J. of the Hydraulics Division, ASCE, HY8, 1153-1169.

Takahashi, T., et al., 1992. Routing debris flows with particle segregation. J. of Hyd. Eng., ASCE, 118(11), 1490-1507.

Takahashi, T., Nakagawa, H. and Kuang, S., 1987. Estimation of debris flow hydrograph on varied slope bed. Erosion and Sedimentation in the Pacific Rim, R.L. Beschta, T. Blinn, G.E. Grant, F.J. Swanson, and G.G. Ice (Eds.), Publication No. 165, International Association of Hydrological Sciences, IAHS Press, Wallingford, UK, OX10 8BB, 167-177.

Tatum, F.E., 1963. A new method of estimating debris-storage requirements for debris basins. Second National Conference on Sedimentation of the Subcommittee on Sedimentation, ICWR, Jackson, Mississippi.

Thurber Consultants LTD., 1983. Debris torrents and flooding hazards, Highway 99, Howe Sound. Report to the Ministry of Transportation and Highways, British Columbia. Unpublished, 25 p.

Van Dine, D.F., 1985. Debris flows and debris torrents in the southern Canadian Cordillera. Can. Geotech. Journal, 22(1), 44-68.

Waldron, H.W., 1967. Debris flow and erosion control problems caused by the ash eruptions of Irazu Volcano, Costa Rica. U.S. Geological Survey Bulletin, 1241-I, 1-37.

Wells, W.G., P.M. Wohlgemuth, and A.G. Campbell, 1987. Postfire Sediment Movement by Debris Flows in the Santa Ynez Mountains, California, pp. 275-276 in Erosion and Sedimentation in the Pacific Rim, (R.L. Beschta et al., eds.), IAHS Publ. No. 165, Wallingford, UK.

Wieczorek, G.F., 1987. Effect of rainfall intensity and duration on debris flows in central Santa Cruz Mountains, California. in Debris Flows/Avalanches: Process, Recognition, and Mitigation, J.E. Costa and G.F. Wieczorek (Eds.), Geological Society of America Reviews in Engineering Geology, VII, 93-104.

Wieczorek, G.F., Harp, E.L., and Mark, R.K., 1988. Debris flows and other landslides in San Mateo, Santa Cruz, Contra Costa, Alameda, Napa, Solano, Sonoma, Lake, and Yolo counties, and factors influencing debris-flow distribution. U.S. Geological Survey Professional Paper 1437, 133-162.

CHAPTER 8

Landslides

T.P. Gostelow

ABSTRACT. This chapter reviews the chief factors which are responsible for the distribution of landslides which have been triggered by hydrological mechanisms. Emphasis is placed on rainfall and subaerial slides and it is concluded that apart from climate, susceptibility to failure depends on topography, geology, land-use and the initial groundwater conditions. A large number of hydrogeological models and empirical triggering factors have been proposed for hazard prediction, but it is suggested that geological heterogeneity reduces their usefulness in most cases. Spatial information can however be usefully stored in modern Geographical Information Systems, (GIS) and the application of simple rules can assist in the identification of areas where landslides may have disastrous socio-economic consequences.

8.1 Introduction

Landslides can be defined, following Skempton and Hutchinson (1969), as downslope gravitational movements of soil or rock which occur primarily as a result of discrete shear failure. They are triggered by a variety of mechanisms, but amongst these, there can be little doubt that hydrological factors, especially rainfall, have caused, and contributed to some of the world's worst landslide disasters, resulting in loss of life and property, (Housner 1989).

This chapter reviews some of the factors which have been responsible for the distribution of hydrologically induced slides, and puts forward suggestions for identifying and mapping the geological, topographic and climatic conditions responsible for them.

8.2 Hydrological Triggering Mechanisms

Hydrological triggering mechanisms operate in two fundamental ways:-

1. Reducing the effective shear strength, s, of rocks and soils through weathering or by raising pore, or discontinuity water pressures under conditions of constant total stress. There are five processes,

(i) Rainfall during either infiltration, groundwater recharge, or discharge.

(ii) Fluctuations in river, lake or reservoir level causing groundwater waves in adjacent slopes.

(iii) Snow/glacier/ice melt during infiltration, or following groundwater recharge, or discharge.

(iv) Externally in submarine slopes by wave induced stresses during storms, either in oceans or lakes.

(v) Weathering; internal seepage erosion.

2. Increasing gravitational shear stresses in slopes,

(i) Externally, by rivers, either by downcutting or lateral removal (e.g. during flooding).;

(iii) Externally by surface runoff, (overland flow) gulleying and erosion.

In this review emphasis is placed on rainfall and subaerial landslides. Only brief mention is made of weathering, erosional effects, external water levels and submarine slope failures.

Hydrological triggering mechanisms, reduce the limit equilibrium factor of safety of slopes (F) to 1.0, where F is given by,

$$F = \frac{\text{Average Shear Strength (s)}}{\text{Average Shear Stress}}. \tag{8.1}$$

The average effective *shear strength*, s, of saturated slope forming materials can be expressed in general by,

$$s = c' + (\sigma - u)\tan\phi' \tag{8.2}$$

where c' is the material cohesion, σ is the total normal stress on a potential or actual shear plane, u is the pore water pressure, and $\tan\phi'$ is the effective friction angle. A number of alternative shear strength failure laws are available which take into account non-linearity, and these can be selected, if required, on the basis of the engineering character of the slope forming materials.

If they consist of soils and are unsaturated, then following Fredlund (1987), the strength s, can be expressed as,

$$s = c' + (\sigma - u_a)\tan\phi' + (u_a - u_w)\tan\phi_b, \tag{8.3}$$

where u_a is the pore air pressure, and u_w is the pore water pressure. The friction angle, ϕ_b is equal to the slope of a plot of matric suction $(u_a - u_w)$ versus shear strength when $(\sigma - u_a)$ is held constant.

The in-situ *shear stresses* in slopes depend primarily on the geometry and unit weight of the materials, and increase, with increasing slope height and steepness.

Landslides

The cohesion and friction angle are primarily functions of lithostratigraphy, while the effective stresses, $(\sigma - u)$ are controlled by the distribution and depths of the unsaturated and saturated *shear strength* zones, and regional groundwater flow patterns.

A loss of shear strength with strain reflects a high material brittleness, and this may encourage large and rapid post-failure landslide displacements. The strength loss can occur in either the c', ϕ', $(\sigma - u_a)$, $(u_a - u_w)$ or $(\sigma - u)$ terms of the equations, although the latter is perhaps more typical of hydrological landslides which become disasters.

8.3 Rainfall and Landslide Disasters

Most disastrous first-time slides have occurred in urban or industrial areas, usually on cut and fill or landscaped slopes, where design has failed to take into account local hydrological and hydro-geological conditions. An extreme, individual case was the Aberfan disaster in South Wales, UK which occurred on a glaciated valley slope in an area of high mean annual rainfall. Loosely placed coal waste was sited over an area of groundwater discharge, (Bishop, 1973) and although the rainfall for the preceeding two days had only totalled 70 mm, the average monthly rainfalls for the year, (1966) were greatly in excess of the mean annual totals. These figures were sufficient to provide groundwater which contributed to subsequent liquefaction of the coal waste, and a slide movement of 600 m down a 12.5° slope, killing 140 people.

Major individual natural slides in more remote mountainous terrain have sometimes caused dams which have had disastrous secondary effects. For example in 1933, the Deixi landslide dam in central China failed, and the subsequent flooding resulted in the deaths of 2400 people, (Costa and Schuster, 1988). However, these individual landslide dam disasters are perhaps less common than those associated with multiple failure. The slides are often smaller, but because they involve a larger cumulative area, greater economic losses occur. A storm in 1889 in the Totsu river area of Japan caused extensive flooding and loss of life from the failure of 53 landslide dams, (Costa and Schuster, 1988). They occurred where susceptible ground conditions had also been affected by anthropogenic development.

The rainfalls which trigger landslide disasters may be associated with frontal, convective, orographic, or cyclonic, (convergent) weather sequences. Generalisation is not possible, but the most damaging high intensity rains tend to be found wherever orographic influences accentuate convectional fronts or cyclones. An example of the latter has been described from central Virginia, USA by Williams and Guy (1973) and Gryta and Bartholomew, (1989) where, in 1969, more than 150 people died following an 8 hour rainfall of 710 mm which fell from the remnants of hurricane Camille. The majority of the 1107 slides occurred below a single high intensity rain cell of about 30 km^2 which passed over geologically susceptible slopes of 35°+. The area was subjected again to a low pressure cyclonic storm between 3–5 November 1985, with a maximum rainfall of 325 mm, (Jacobson et al., 1993). On this occasion central Virginia was less affected by mass movement, but the storm triggered over 3000 slides in eastern Virginia, with numbers 200%

higher on cleared rather than forested land. The total social cost amounted to 70 lives lost and $1.3 billion in damage to homes. Jacobson et al. (1993) noted that rainfall intensities were moderate, reaching 38 mm/hour, but averaging 8 mm/hour over 8 hours. They suggested that landslide thresholds depended on geology, but a 48 hour rainfall of 200 mm could be used in the future as a regional predictive trigger. In 1977, Pennsylvania, also in the Appalachians, was subjected to 300 mm of rain in 9 hours. This also triggered multiple debris slides which contributed to the $300 million of storm damage which made 50000 people homeless (Pomeroy, 1980).

Similar meteorological patterns occur in the mountainous ranges of western North America and recurring disastrous events are well known. For example, their socio-economic impacts in the Los Angeles area have been summarised by Cooke, (1984), and in Vancouver by Eisbacher and Clague, (1981). One of the worst storms of recent years in January 1982, occurred in and around Los Angeles, (Ellen and Wieczorek, (1988). An estimated 18000 slides were triggered after 440 mm of rain fell in 32 hours, which caused 25 deaths and $70 million of damage. In 1993, winter rains again caused landslide damage to homes, streets and utilities in Los Angeles, estimated at a further $60 million, (Barrows et al., 1993); between 1862 and 1982 there are records of 60 other comparable events in this area (Brown, 1988).

Accounts of multiple cyclonic landslide disasters are available in most mountainous urban areas. Woo (1992) has described 30 over the last 20 years in Korea. For example in Seoul, a 2 day rainfall total of 352 mm in July 1987 with intensities of up to 52 mm/hour caused 30 landslides, the loss of 36 lives and 40 ha of developed land. In 1987 alone he estimated that total damage from storms in Korea was worth c. $ 1.5 billion. In May 1982 650 mm of rain fell in 4 days with intensities approaching 110 mm/hour in Hong Kong, (Malone, 1988). The event triggered 1500 landslides, resulted in 22 deaths and left some 8000 people homeless. There have been 13 similar major cyclonic storms between 1963 and 1983 (Brand, 1984).

In common with flood forecasting, there is thus a need for further work on the distribution and frequency of landslide events with high intensity rainfalls, so that warning systems can be established and vulnerable areas protected. However, while extreme rainfalls and floods are closely related, landslides are controlled by a number of interactive temporal variables, making empirical relationships difficult to establish. Moreover, it is the location and potential post-failure displacements of slides which are just as important as forecasts of return periods, and this will depend on local factors such as slope angle and geology. Nevertheless, despite these apparent difficulties, more reviews of historical data are required for future risk analysis, disaster prevention, planning, and design. Following Wilson (1990), the approach should include,

1. Identification of the events or sequence of events which may lead to slope failure.

2. Identification of specific features of a landslide, or susceptible area which might initiate a failure.

3. Analysis of the liklihood of combinations of 1 and 2.

4. Assessment of the economic, social and environmental costs of each combination in 3.

Whilst there are established techniques for mapping rainfall extremes for different durations, there is less knowledge of the impact of the rainfall from those durations on the stability of slopes, and the features which make them susceptible to failure from those events. The following sections thus concentrate on ground conditions rather than hydrometeorology, or aspects of rainfall frequency analysis.

8.4 Regional Groundwater Flow

8.4.1 GENERAL

The 'steady state' elevation of the saturated zone or water table depends on topography, geology and climate. Toth (1963) has identified three types of steady state groundwater flow system which are driven by differences in topography, i.e.

1. A local system with a recharge area at a topographic high and a discharge area at an adjacent topographic low.
2. An intermediate system with one or more topographic highs between the recharge and discharge areas.
3. A regional system where the recharge area occupies the drainage basin's water divide, and the discharge area lies at the bottom of the basin.

Freeze and Witherspoon (1967), and Hodge and Freeze, (1977) have shown how geology affects equipotentials and flow patterns in the three systems, while Toth and Millar (1983), and Neuzil (1986) have discussed the possibilties of non-steady state regional flow caused by tectonics and erosion. Their results suggested that the interactions of flow patterns and pressure systems caused by meteoric gravitational water, and non-steady state compaction or swelling are not well defined, and in some neo-tectonic settings it is conceivable that slow tectonic deformation leading to surface mass movement might occur as a result of extreme hydrological events. Despite this possibility, most large pre-existing landslide complexes seem to be associated with 'steady state' groundwater discharge areas, usually within local topographic groundwater systems, although some may have more complex controls. Erosion and/or rainfall input to the different parts of these three flow systems triggers instability, and hence as an initial condition, they must always be taken into account in assessing mass movement susceptibility. Three time-dependent situations regarding rainfall can be recognised,

(i) Infiltration into an unsaturated zone.
(ii) Recharge to a water table.
(iii) Groundwater discharge.

8.4.2 GROUNDWATER IN MOUNTAINOUS SETTINGS

Hydrological landslide disasters frequently occur in mountainous settings; in areas with high relative relief, narrow valleys, steep slopes, high elevation heads and hydraulic gradients. Pre-existing landslide debris is frequently involved, which has developed perhaps during previous colder climates, (Schroder, 1971). Forster and Smith (1988) have reviewed a number of factors controlling water table levels in these areas using 2-dimensional finite element models of homogeneous slopes. They demonstrated that rock hydraulic conductivities have the greatest impact, while variations in infiltration rate, presence of glaciers, slope profiles, and basal heat flux were of lesser importance.

A variety of mass movement types are triggered, which depend on topography and the distribution and elevation of materials with a high hydraulic conductivity. However, the shape and size of landslides are usually constrained by geological heterogeneity and internal kinematics, (Skempton and Hutchinson, 1969). In this review two broad groups will be considered; deep-seated, stratigraphically-controlled slides and translational mass movements.

8.5 First-time and reactivated landslides

An important distinction must be made between first-time failure, and reactivations of old slide complexes. The latter are often large, deep-seated, and are susceptible to comparatively small, individual movements after some hydrological triggering threshold has been exceeded. Their position, size and shape usually reflect a geohydrological pre-disposition to failure, such as an eroding valley slope, fault, or a combination of aquifers and materials of low shear strength. In contrast, first-time slides are shallow, translational, and take place as a direct result of erosion, rainfall infiltration and through-flow in weathered soils of low plasticity. Post-failure displacements are rapid and large, and this is perhaps a typical feature of disastrous landslides reported by the media, especially from urban areas such as Hong Kong and Los Angeles where they are an annual economic problem.

Reactivated shallow slides are not unknown in higher plasticity soils with pre-existing shear surfaces and low frictional strengths. For example they are comparatively common on slope angles above 10° in the clay formations of southern England, UK which have been covered in layers of relict periglacial slope deposits.

8.6 First-time Translational Slides

8.6.1 EXAMPLES

Gostelow (1991a) has summarised some of the worldwide examples of translational debris slides which have taken place following extreme rainfall events. As discussed below, anthropogenic factors are important, but many seem to be the result of a strength reduction following direct recharge into superficial weathered soils. However, the following examples show that actual failure locations on slope faces has often been hydrogeologically controlled, i.e. by perched water tables in bedrock aquifers, (Hack and Goodlett, 1960) or by the growth of saturated zones following groundwater throughflow in topographic depressions, (Tsukamoto et al., 1982).

Hong Kong. In Hong Kong, the damaging rainfall-induced landslides are mostly translational with large post-failure displacements. In a typical example, between the 17th and 18th June, 1972, 650 mm of rain fell with intensities approaching 100 mm/hour. There were 285 slides in weathered granites and volcanics, and one of these demolished a building and killed 67 people. Nevertheless despite the high rainfall totals, hydrological modelling by Leach and Herbert (1982) suggested that average rainfall recharge was too low to sustain the measured piezometric levels. It was suggested additional sources which contributed to the failure came from both groundwater, and leakages from services, (water mains, sewers and storm drains).

India. The Darjeeling Hills, which form part of the foreland graben of the Sikkim Himalaya in India, reach 2500 m-3000 m in height, and receive some of the highest rainfall intensities in the world. They consist of deeply weathered, Precambrian and Palaeozoic phylites and gneisses, which outcrop on 20°-40°, deforested valley slopes, up to 700 m high. Because of the long history of subsequent tea plantations in the area, a close network of climatic recording stations has been set up, with records of both mass movement events and rainfall going back 100 years. Average rainfall is 3,000 mm, 95% of which falls in the 6 months of the summer monsoon. Between the 2nd and 5th October, 1968, Starkel (1972) recorded that up to 1000 mm of rain fell with intensities reaching 40-50 mm/hour. Towards the end of the storm widespread landslide activity (mostly first time, Starkel, 1972, noted that old landslide scars tended to remain stable), occurred in the area. Descriptions at the time, discussed by Starkel (1972) refer to a 'sudden' release of groundwater from slopes ('gushing out under pressure'), with translational slips taking place in small slope hollows, at the sites of springs, subsurface erosion, and piping. Some 25% of tea bushes in the area were lost during this single storm.

United Kingdom. Extreme rainfall events which cause debris slides also occur in areas which have been recently glaciated, or subjected to cold climates and are covered with debris and weathered materials. Examples can be found throughout Europe, including less mountainous settings such as the UK. For example, on the 15th August 1952 a summer storm developed near Lynmouth, N. Devon where up to 200 mm of rain fell in 24 hours, (with most falling in a 5 hour period). A number of shallow landslides occurred on valley slopes of around 30° on Exmoor (Kidson and Gifford, 1955). Descriptions of the slides suggested they occurred at points of pre-existing saturation, i.e. at seepage faces, or springs, at slight inflexions of slope profiles, with a characteristic wet vegetation. It was also noted that water was trickling from the back of the landslide scars through underlying weathered slates.

Debris slides have been triggered by similar rainfall totals at other UK locations, for example in September 1983, 24 hour totals of between 170 and 200 mm occurred near Snowdonia, with intensities of between 16 mm/hour and 40 mm/hour. The A5 trunk road was blocked, causing substantial damage (Addison, 1987). The slides occurred at the junction between slope talus and rock on 25° slopes along pre-existing water courses and moved 600 m downslope. A similar summer daily total of 150 mm triggered a debris slide at Bilsdale, North York Moors in 1976 on a 17° slope, which consisted of weathered organic soils. Beven et al. (1978) described the presence of water at the head of this slip which was flowing through narrow cracks at the soil/rock boundary along distinct lines of seepage. In December 1986 a rainfall of 75 mm in 24 hours near Aberystwyth caused a debris slide on weathered Silurian slates on a 35° slope (Dowdeswell et al., 1988). The lower triggering threshold in this case (a 20 year event) may reflect winter antecedent conditions prior to the slide.

Intense rainfall in November 1984 caused several failures on steep slopes in the Ochil Hills, Scotland (Jenkins et al., 1988). Estimated rainfall totals were around 70 mm in 24 hours. The largest slide occurred on slopes of 32°-35° at Menstrie, which engulfed a house. The slip took place at the junction between soil and bedrock, with water flowing in a soil pipe at the head of the slide.

Ireland. On the 29th December, 1896 a peat bog, near Killarney in Ireland, circa 1,200 m^2 in area and 10 m thick failed after heavy rainfall. It flowed for nearly 8 kms, caused the deaths of 8 people, and destroyed many acres of agricultural land, roads and buildings (Sollas et al., 1897). It was situated directly on Middle Coal Measures, which were dipping towards a faulted contact with Carboniferous limestones and shales. The fault passed beneath the centre of the bog and probably influenced both the development of a spring, and the form of the depression in which the peat had accumulated. In the susbequent enquiry it was pointed out that nearly all previous peat-bog flows in Ireland and elsewhere, had occurred where springs and subterranean water were present. A quote from evidence given at the enquiry stated:

> "Springs are often met with in the deepest part of the bog, rushing up sometimes with much violence and often strongly impregnated with sulphate of iron, carbonic acid and earth".

The importance of springs was recognised even earlier by a report published by an Irish Government Commission on peat-bogs, dated 1811, (Sollas, 1897):

> "To ascertain whether the wetness of these bogs originates solely from rainwater falling on the surface, or from springs in the interior of the bogs, or from both, is an enquiry of very great importance."

Malawi. A similar triggering mechanism was put forward by Schroder (1976), in a different climatic zone, for mass movement activity on the Nyika plateau in Malawi (mean annual rainfall between 1000 mm and 2500 mm). Organic soils (peats), were shown to have developed on deeply weathered metamorphic granites, in slope depressions known locally as dambos. He described a ground failure of April 1960, involving 95,000 m^3 of material which moved 400 m on a 8° slope. A major spring was present at the edge of the slipped soils, and had clearly been responsible for repetitive periods of degradation.

8.6.2 Physical Models of Rainfall Infiltration and Mechanisms of Translational Failure

There is a widespread literature describing the movement of water from the ground surface through an unsaturated zone to an unconfined saturated zone, or water table. For example, the theory and controlling flow equations have been thoroughly reviewed and modelled one-dimensionally by Freeze (1969). His study has shown there are two separate mechanisms of effective stress change which could cause first-time failure,

(i) A loss of suction in the unsaturated zone;
(ii) A rise in the water table.

Several, interelated physical parameters and boundary conditions were shown to control the extent and timing of the changes, i.e. the attainment of ponding conditions at the ground surface, the antecedent moisture content, the rate of recharge, the depth to the saturated zone, rainfall intensity, and the soil hydraulic conductivity.

A number of publications have also considered two dimensional hydrological flow equation models of hillslopes consisting of a homogeneous soil (generally cohesionless) overlying an unspecified bedrock with a much lower hydraulic conductivity. Early examples which ignored unsaturated flow include those by Schmid and Luthin (1964), Henderson and Wooding (1964), Wooding and Chapman (1966), Wooding (1966), Childs (1971), Youngs (1971), Jaiswal and Chauhan (1975), and Towner (1975). The topic has continued to attract much attention and more recent analyses include those by Chapman (1980), Beven (1981), Sloan and Moore (1984), and Yates and Warrick (1985). Some of these have also included estimates of water pressure change during unsaturated flow, e.g. Beven (1982), and Hurley and Pantelis (1985). Reddi and Wu (1991), and Rhett-Jackson and Cundy (1992) also attempted to take into account topographic convergence and divergence. Iverson and Major (1986), Iverson and Reid (1992), and Vaughan, (1985) discussed applications to slope instability, and Buchanan and Savigny (1990), have attempted to use a two dimensional approach to model measured water pressures at the head of debris slides.

Whilst these models give further insights into the controlling variables and may have some practical applications on individual slopes, the examples described previously suggest that in real landscapes, the variations in topography, (slope), vegetation, anthropogenic interference, the geological heterogeneity, structure and weathering products, the spatial variability of soil properties, anisotropy, non-Darcian flow, and rainfall intensity, all contribute to hydrological systems of some complexity. The difficulties of accurately measuring and mapping these variables thus generally precludes the use of a single, accurate physical model which can be used to predict when and where shallow first-time landslides will occur. Therefore simplified empirical methods must be developed for this purpose.

A promising approach, which takes into account topography and rainfall, but not variations in the other physical properties has been put forward by Tsukamoto et al. (1982). They suggested that steep hillslopes can be classified into convergent, divergent and planar elements, and that translational failure and erosion are most active on convergent units or 'zero-order basins'. These basins are found at the upper parts of first order stream channels and are also the sites of soil piping. Dietrich et al. (1986), O'Loughlin (1986), Montgomery and Dietrich (1989), Dietrich et al. (1992) and Montgomery and Dietrich (1994) have subsequently developed the idea by combining digital terrain data with shallow saturated throughflow, and infinite slope stability models.

Their combined model attempts to predict areas of shallow soil saturation where a cohesionless soil overlies a relatively impermeable bedrock and takes a threshold of ground saturation of,

$$\frac{a}{b} \geq \frac{T}{q} \sin \Theta \cos \Theta' \tag{8.4}$$

where Θ = slope angle, a = area of upslope catchment, b = unit length across which the catchment is draining, T = saturated hydraulic conductivity, ($K \times$ depth of flow), q = effective rainfall. This derivation follows Darcy's law and assumes a groundwater discharge Q parallel to the slope surface is given by $K \sin \theta$ where $\sin \theta$ is the hydraulic gradient. The horizontal velocity V, (from one contour to another) is thus $Q \cos \theta$, and the discharge, for a depth of saturation h, is $hK \sin \theta \cos \theta$. The value for q is obtained by multiplying the catchment area by the rainfall. Saturation is thus assumed to occur when $q \times a$ is greater than the maximum flux the slope can conduct which is the product of the other terms. This equation has been coupled with the infinite slope stability equation, i.e.

$$\frac{a}{b} \geq \frac{\gamma_s}{\gamma_w}\left(1 - \frac{\tan \theta}{\tan \phi}\right) \frac{T}{q} \sin \theta \cos \theta, \tag{8.5}$$

where γ_s = bulk density of the saturated soil, γ_w = density of water, ϕ = friction angle of soil. In this model if a/b exceeds or equals the term on the right, slope instability occurs with the assumption that the pore pressure ratio r_u ($\gamma_w z/\gamma_s z$, where z is depth of failure) does not exceed 0.5. The minimum steady state rainfall for failure can thus be predicted for different slope elements by rearranging the equation. Descriptions of actual translational slides reviewed previously suggest that most do take place under saturated rather than unsaturated conditions. However, as shown by Tsukamoto et al. (1982) there is focused groundwater flow within the basins, usually with input from the bedrock, or soil pipes which is difficult to include in two dimensional slope models. Nevertheless, Montgomery and Dietrich (1994) have applied their method to a number of study areas in the USA and concluded that despite some limitations, the distribution of shallow landslide scars generally supported its use in regional hazard assessment studies.

Other approaches to water level prediction for shallow landslides have included in situ measurement, (Haneberg, 1991), combined water balance and probabalistic methods, (Sangrey et al., 1984, Fell et al., 1988) and the establishment of direct empirical links between rainfall and a maximum piezometric level in a given soil type and climate. One of the most thorough studies of this kind has been carried out in Alaska, (Sidle, 1992) where measurements were made in 10, continuously recording piezometers during 44 storms on a 46° slope over a 4 year period. The regression equation for piezometric head (p) was found to be

Landslides

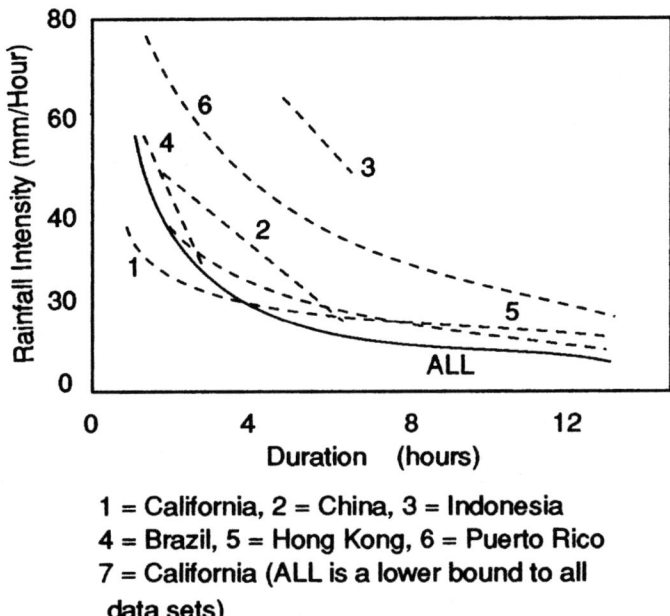

1 = California, 2 = China, 3 = Indonesia
4 = Brazil, 5 = Hong Kong, 6 = Puerto Rico
7 = California (ALL is a lower bound to all data sets)

Fig. 8.1. Summary of rainfall duration thresholds for first time translational/debris slides (after Jibson, 1989).

$$p = 0.149 \ln(\text{TOTPPT}) + 0.0398(\text{ANT2})^{1/3} + 0.0668 \ln(\text{INT1}), \qquad (8.6)$$

where TOTPPT is total storm rainfall (mm), ANT2 is antecedent 2-day rainfall (mm) and INT1 is maximum 1 hour rainfall intensity (mm/hour). Piezometric head is expressed in metres.

8.6.3 First-Time Translational Slides: Rainfall Triggers

An alternative to physical modelling is to plot rainfall intensity against duration to define an empirical threshold for landslide initiation. Figure 8.1, from Jibson (1989) summarises a number of these from different countries. The form of the thresholds is similar, but Figure 8.2 from Wieczorek and Sarmiento (1988), also illustrates the importance of antecedent rainfall, in this case on intensity-duration curves from landslide events in California, where a figure of at least 280 mm is required before the threshold becomes meaningful. These plots, with a time scale of hours require data from closely spaced and continuously recording gauges, which are not normally available in most areas. Daily, 24 hour rainfall figures are more readily obtained, and in Hong Kong extreme events have been plotted against the 15 day antecedent rainfall by Lumb (1975), who found that disastrous slides occurred when daily figures exceeded 100 mm, and antecedent totals had reached 200 mm. Crozier, (1985) has extended the approach by plotting the antecedent

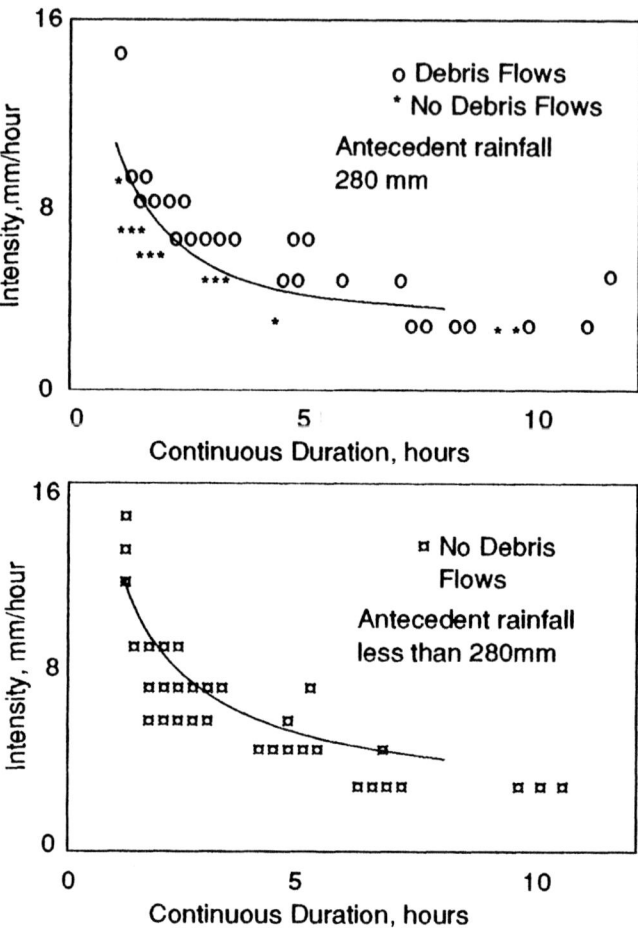

Fig. 8.2. Rainfall-intensity/duration relationship showing importance of antecedent rainfall in triggering first-time slides in California (after Wieczorek and Sarmiento, 1988).

soil moisture status, i.e. taking into account evapotranspiration (Figure 8.3) for a 10 day period against 24 hour rainfall in New Zealand, which defines a linear threshold between landsliding and non-landsliding events.

These empirical approaches can be used to calculate return periods of triggering variables and when accurately calibrated, provide some guidance to when landslides might occur, but they do not take into account the initial groundwater conditions, geology, topography etc, to predict the type of slide or where it will take place.

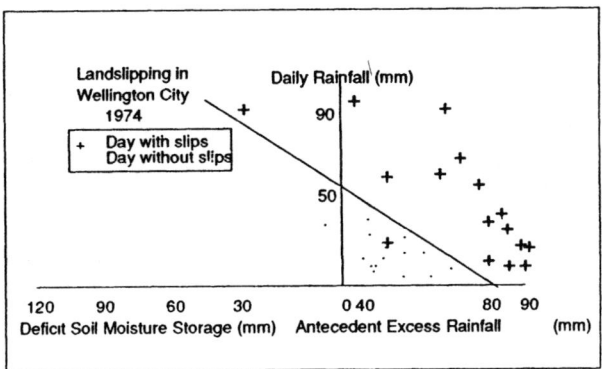

Fig. 8.3. First-time landslide thresholds in New Zealand, taking into account soil moisture status (after Sidel et al., 1985).

8.7 First-Time Rotational and Complex Deep-Seated Pre-Existing Landslide Movements

In situations where a soil layer is not constrained by a layer of lower permeability parallel to the slope surface more complex and deeper failure surfaces are able to develop during the process of groundwater recharge and discharge. The generation of positive pressures which control their position will depend on subsurface geological boundary conditions and variations in hydraulic conductivity. According to Tsukamoto et al. (1982) natural first time slides of this type are perhaps less common than those on cut and fill slopes, although in 'active' geomorphological areas such as river valleys and coasts, (Hutchinson, 1976, Eyles and Howard, 1988) they may occasionally follow extreme rainfall events. More usual are reactivations of large landslides or landslide complexes, with pre-existing shear surfaces which, like translational movements in zero-order basins, are often associated with groundwater discharge areas and springs. In this review a distinction is thus made between springs associated with the permanent water table, and those generated by a temporary saturated subsurface flow, described above.

8.7.1 Hydrogeology of Pre-Existing Landslides

8.7.1.1 General

A spring, springline, or seepage face is a surface expression of the water table, and can be defined following Bear (1979) as a 'point or small area through which groundwater emerges from an aquifer to the ground surface'. Seepage vectors are directed upwards and outwards, and water pressures, u, measured vertically, increase with depth at a rate greater than hydrostatic. Piezometric head fluctuations occur as the angles between seepage vectors and the ground surface change following recharge, and Iverson and Major, (1986) have reviewed their importance in relation to potential instability. In contrast, the catchments, aquifers, or landslide blocks supplying springs have seepage vectors orientated downwards, and piezo-

metric levels measured vertically, increase with depth, but at a rate which is less than hydrostatic. Effective shear strengths are thus generally lower in discharge areas and there is a wealth of published evidence to show they are associated with pre-existing landslide complexes, (e.g. Zaruba and Mencl, 1969). The hydrological conditions which are required to maintain a continuous groundwater spring outflow have been summarised by Toth (1971) as
1. A discharge rate in excess of the local rate of evapotranspiration.
2. A sufficiently high rate of precipitation to keep the flow system recharged.
3. A sufficiently steep hydraulic gradient.

If these conditions are not met, springs can cease to flow, but may emerge temporarily, if groundwater recharge occurs during extreme rainfall events. Bear (1979) divides springs into two broad types i.e.,
1. Depression;
2. Perched.

Depression springs are associated with a comparatively homogeneous geology, and are controlled by topographically driven gravitational flow. Perched springs are controlled by a geological heterogeneity and occur where an impervious layer, or aquiclude underlies an aquifer, (Rulon et al., 1985).

Rainfall falling on saturated, or tension saturated ground which extends to the surface, such as near a seepage face from both spring types will quickly raise piezometric levels, (Gillham, 1984) causing runoff, mass movement, and surface erosion.

8.7.2 Mass Movement Associated with Depression and Perched Springs

8.7.2.1 *Geologically Controlled (Perched) Groundwater Discharge*

A number of geological situations provide the conditions for springs, (Freeze and Cherry, 1979, Bear, 1979). In each of these there is a potential reduction in effective stress and instability in three groundwater zones which are related to aquifer/aquiclude discharge boundaries, i.e.
(i) Slope movements downslope of the spring discharge area.
(ii) Slope movements or superficial failure of the seepage face.
(iii) Slope movements involving failure of both the seepage face, the upslope aquifer, and downslope, or underlying aquiclude.

Landslides in geologically-controlled discharge areas which exhibit instability in all three of these groundwater zones often have a complex, or compound morphology and, by definition, may be found at elevations above, and away from stream channels.

8.7.2.2 *Topographically Controlled (Depression) Groundwater Discharge*

Because there is no single aquifer controlling discharge, the values of groundwater equipotentials are related solely to topography and climate. Hence areas with the greatest potential for instability are generally found where saturated ground coincides with steep slopes next to first order plus, stream channels, for example in hard rock, mountainous terrains. In some neotectonic regions, such as southern

Landslides

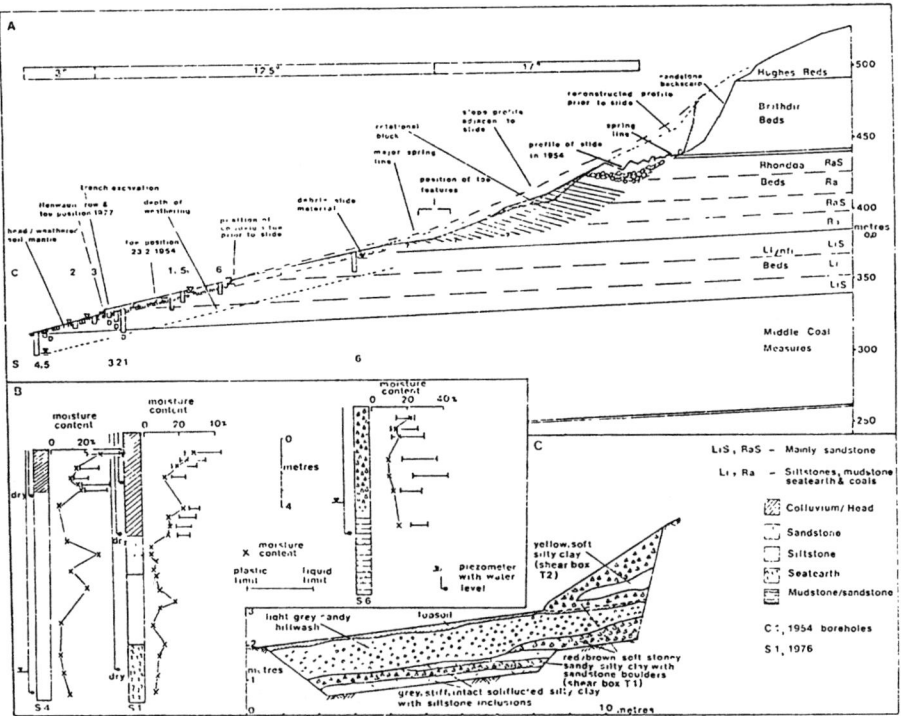

Fig. 8.4. Cross spectra through a valley slope in South Wales (UK), showing the relationship of a geologically controlled springline from an aquifer consisting of hard fractured rock, with mass movement both upslope and downslope of the discharge pont.

Italy (see below), ancient, deep-seated landslides, which perhaps fall into this category are also found next to uplifted river terraces.

8.7.3 Geological Susceptibility to Hydrological Landslide Disasters

Four examples, with different geomorphological and geological environments have been chosen to illustrate landslides and groundwater discharge areas which may be susceptibile to reactivation from hydrological triggers.

8.7.3.1 South Wales, UK

The Carboniferous Upper and Middle Coal Measures in South Wales consists of faulted, shallow dipping, alternating coals, argillaceous and arenaceous sediments (hard shales and mudstones, weathered at the ground surface and strong cemented sandstones). They outcrop on oversteepened, glaciated valley sides. Figure 8.4, from Gostelow (1977) is a cross section through a typical slope, at East Pentwyn, near Blaina in the NE part of the Coalfield. The section illustrates a major groundwater discharge from a sandstone aquifer with an upper elevation of 700 m. The seepage face lies at between 370 m–375 m and 350 m–355 m OD (possibly fault

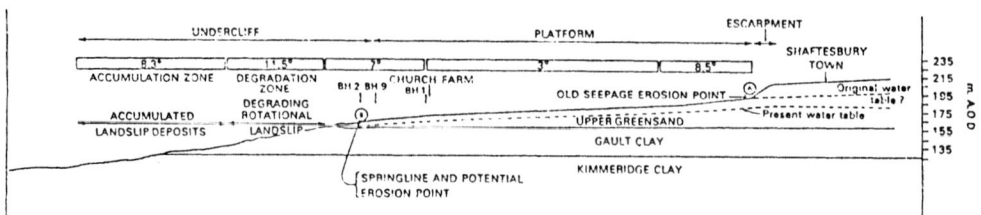

Fig. 8.5. Cross-section through an escarpment slope at Shaftesbury, Dorset, UK, showing the relationship of a geologically controlled springline from an aquifer consisting of a soft cohesionless rock-mass with landsliding both upslope and downslope of the discharge point.

controlled), with a rotational slide upslope of this level and a debris slide downslope. The hydraulic conductivity of the geological sequence has been affected by mining subsidence at this site, and the date of movement in 1954, may have reflected a change in subsurface water pressures associated with this. Nevertheless the spring, enhanced by rainfall caused slope movements upslope and downslope of the discharge point, and subsequent recharge in the sandstone scarp at the slope crest resulted in rockfalls. Photographs taken immediately after failure suggested a considerable quantity of water was released at the springline, although discharge measurements were not made at the time. Mean average annual rainfalls are c. 2000 mm, with a one day mean annual maximum of c. 70 mm. Approximate water balance calculations within the landslide and adjacent area carried out by Halcrow (1985) over a 6 month period shown below suggested that, 35% of the water was derived from the aquifer, with 100% occurring as groundwater. This can be compared to the adjacent area where discharge from the slope was mainly from runoff with only 22% occurring as groundwater.

Slope	Length of Hill (m)	Total discharge $\times 10^2$ m^3	Groundwater $\times 10^2$ m^3	
			Direct	From aquifer
E. Pentwyn	230	68	44	24
Adjacent area	240	45	- - - 10 - - -	

The figures illustrate the variability of groundwater movement in a slope length of only 470 m, and the dominant effect of springs which control the position of the landslides along the valley sides. Average winter discharge figures at E. Pentwyn (from the total), are circa 6 litres/sec.

Reactivations of other slope movements in South Wales are also largely related to the response of groundwater/springs to rainfall. The region is one of the most active areas in the UK with regard to mass movement, and Rouse and Bridges (1985) have found that between 1901 and 1925, there were 42 daily rainfalls in excess of 50 mm, compared to 164 in the period 1926 and 1980. This trend of increased storminess which developed between 1920 and 1930 also seemed to correspond with a known general increase in landslide activity.

8.7.3.2 Shaftesbury, Dorset, UK

Figure 8.5, from Gostelow (1991b) is a cross section through the escarpment slopes below the town of Shaftesbury, Dorset, UK. The geology consists of stiff fissured Gault clay of medium to high plasticity and undrained strengths of between 60–200 kPa, overlain by loosely cemented Upper Greensand (UGS), known locally as the Cann Sand with an upper elevation of c. 200 m above sea level. This fine sand is capped by a thin layer of cemented sandsone and chert which is also part of the UGS. Both formations are of Cretaceous age. The slope, consisting of escarpment, platform and undercliff (Figure 8.5) has developed through river downcutting during the Middle and Upper Pleistocene periods (500,000–10,000 years BP), but has not been covered by ice. A prominent springline occurs at the junction between the Gault and Cann Sands at the edge of the platform. There are active shallow rotational slides on the 10°–12° slopes of the Gault clay undercliff, downslope of the discharge area, and a series of "sand steps" towards the edge of the platform, upslope from the springline (Figure 8.5). Boreholes have proved that the steps, 5 m–10 m thick, are sliding on a preferred plane of weakness, probably a bedding plane, with a thin (0.5–1 mm thick), highly plastic smectitic clay dipping at only 1° (Gostelow, 1991b). The current slope angle and width of the platform varies from between 3° and 7°, and 350–400 m, respectively.

The geomorphological and geological evidence from the area suggested that the slope forms at Shaftesbury developed by processes of seepage erosion, Late-glacial periglacial activity, and recent Holocene landslide reactivation (Gostelow, 1991b). Mean average annual rainfall is c. 800 mm, and the area has a 1 day mean annual maximum of c. 35 mm. The slope forms thus contrast with those from the South Wales valleys because of a) their elevation, (they have not been glaciated), b) escarpment setting, c) a greater past susceptibility to freeze-thaw action, and d) the lower shear strengths of the overlying aquifer. These factors have, over a period of time resulted in removal of the sands, slope retreat and development of a springline in the manner outlined on Figure 8.6. Measurements of current spring discharge have not been made, but estimates, based on the geometry of the slopes and the known hydraulic gradient suggested it varies from 0.1 to 0.007 litres/sec, for an hydraulic conductivity, K, of the platform of 0.0001 m/sec and 0.00001 m/sec respectively.

Springs are a common feature of the UGS/Gault outcrop in the UK, and actively eroding slopes are often associated with landslides (Brunsden and Jones, 1976). Hutchinson et al. (1991) have recorded the large springs which have led to the coastal landslides at St Catherines Point, Isle of Wight. There, the slides and mass movement debris are downslope of the spring discharge points. The same paper presented a model for the long-term development of such slopes, which included a retreat of the UGS, upslope of the spring discharge area, at the top of the Gault, as outlined above for Shaftesbury.

These UK examples of a soft aquifer retreating over a clay base are comparatively small in size, but are not unique to that country. A much larger version, which is the result of similar processes, can be found in the Chuska mountains in New Mexico (Watson and Wright, 1962). There, a Miocene sandstone caprock some 300 m high with an upper elevation of 3000 m has retreated over 11 km,

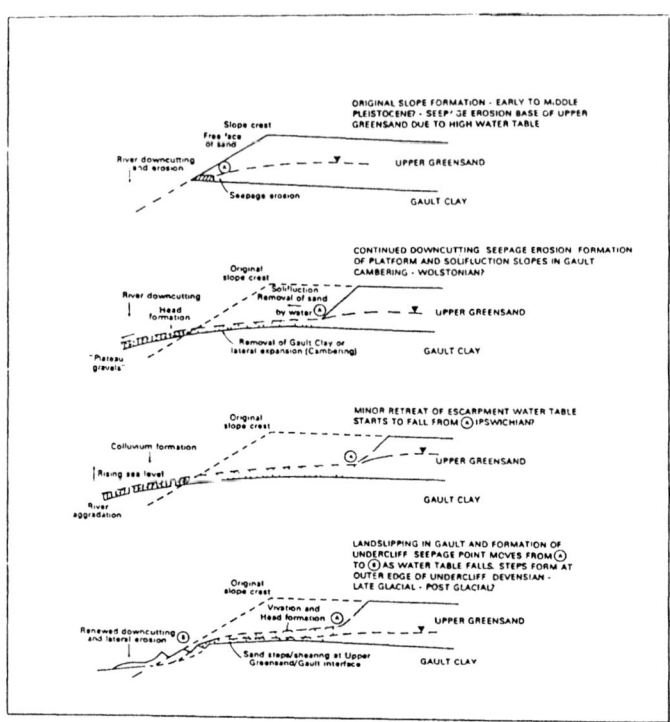

Fig. 8.6. Slope development at Shaftesbury, showing the role of Quaternary climate and springlines of the Gault Clay and Cann Sands in the evolution of the slope profile.

leaving a wide platform of landslide debris, 50 km long, which overlies Cretaceous shales. Relict mudflows (probably of Pleistocene age) are present beyond the platform downslope of spring discharge points, and this slide complex is generally considered to be the most distinctive area of mass movement in the United States. The current mean annual rainfall is only c. 250 mm, (1-day mean annual maximum unknown) and the slides are inactive.

Neither of the UK landslides have been associated with specific disasters. However, they have been chosen to illustrate the hydrogeological settings and ground conditions, which might be susceptible to renewed movements following anthropogenic interference and an unforeseen extreme, climatic event or climatic change. The common factors are an upland aquifer with local seepage faces, and a boundary with weak clayey materials. A similar geological succession is present in Southern Italy which is more geomorphologically active, and in this case there has been a long history of disastrous events.

8.7.3.3 *Campania and Basilicata, Italy*

These regions are in the Appennine mountain chain and settlements have followed young, uplifted valley slopes eroded into weak tectonised sediments. At the village of Senerchia, in the Sele valley (Campania region), Cotecchia et al. (1986a) and

Alexander and Coppola (1989) record that rainstorms directly after the November 1980 earthquake caused a large quantity of water to be released from an upland limestone aquifer at the crest of the valley side, into underlying Sicilide variegated scaley clays. A mudslide, downslope of the spring developed, which reached 2500 m in length, 500 m in width, 33 m in depth, and carried debris up to 2 km downstream. Alexander and Coppola (1989) recorded that high piezometric levels were found, with a sufficient hydraulic gradient to deliver groundwater at up to 200 litre/sec. During the same earthquake a similar mudslide developed at Buoninventre near Caposele (Cotecchia et al., 1986b). It moved on slopes of 6°–7° and was 3 km long and 1 km wide. Major springs, derived from a faulted carbonate aquifer to the north of the slide discharged groundwater at the unfaulted contact between an arenaceous member of the varicoloured clays and the clays themselves. Cotecchia (1987) noted the presence of other, seemingly identical mudslides in the valley, which had remained stable during the earthquake, and suggested this was because groundwater discharge was absent.

Similar geological situations exist in Calabria (Guerricchio et al., 1993) and the western parts of Basilicata. For example, in the valley sides, near Maratea (Figure 8.7), Cretaceous limestones have also been overthrust onto Liguride clayey flysch, and the hydrogeology of the resulting slides has been studied by Cotecchia et al. (1990). A number of large springs (South Basile, S.S. Maria and South Sorgimpiano Figure 8.8) issue from the limestones which act as an aquifer (Figure 8.7), and outcrop some 1200 m above sea level. In this case, (unlike Senerchia, although the aquifer reaches similar elevations) large sections of the limestone have become detached upslope of the discharge area which extends laterally for over 4 kms, (the town of Maratea is sited on one of these). The blocks have slid downslope for a distance of 1 to 1.5 km in a huge complex of mass movement, into the Maratea valley (Figure 8.7).

Winter spring discharge reaches 90 litre/sec, and there is a high maximum average of over 50 litre/sec from the slipped blocks, (mean annual rainfall 1734 mm, mean annual 1-day maximum c. 100 mm). The contrast with Senerchia reflects variations in the local climate, and topography, but suggests that effective stresses at the limestone/clay boundary are likely to be lower at Maratea, with lower frictional strengths. For example, there is evidence, (Guerricchio et al., 1988) that the slipped blocks at Maratea move periodically, ie, the overall factor of safety of the complex is low, and may be susceptible (and sensitive) to change following fluctuations in rainfall, climate or to climatic change.

The lower parts of the Basento valley lie in the Bradano foretrough of southern Italy, and consist of a sequence of Pliocene and Pleistocene marine sands and clays. Neotectonic uplift has resulted in downcutting through the sediments by rivers, such as the Basento, which drain SE to the Gulf of Taranto.

Ferrandina, Pisticci and Grassano are typical examples of small towns which have developed on the summits of the valley sides which are now 300–400 m above river floodplains. The slopes below the towns have continued to be subjected to erosion and landsliding, and historical evidence has suggested this has been the result of past agricultural practices (changing land-use), urbanisation, rainfall, and earthquakes (Cotecchia and Melidoro, 1974).

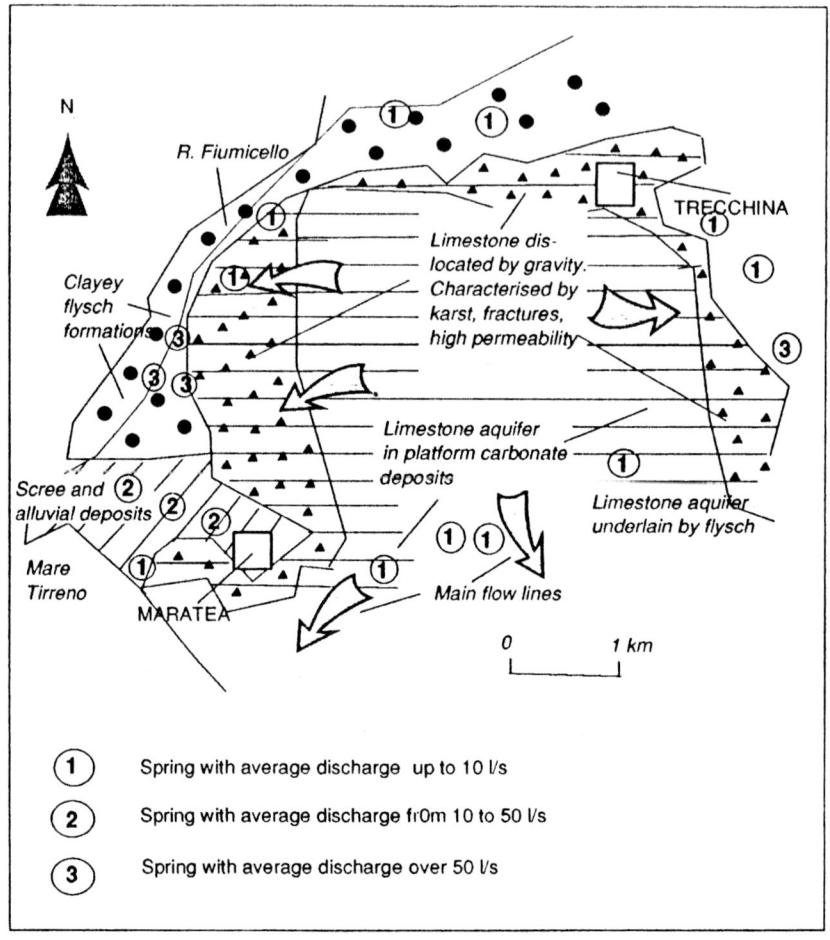

Fig. 8.7. Spring discharge from a limestone caprock, Mareta, Basilicata, Italy, showing aquifer (horizontal lines) springs, landslides and main directions of groundwater flow (after Cotecchia et al., 1990).

The geological succession beneath, and surrounding, the towns generally consists of a cemented conglomerate (the Irsinia formation), loose, fine sands (Monte Marano formation) and stiff, but weakly lithified, grey blue clays of medium plasticity, with undrained strengths (unweathered), of between 100–300 kPa, (the Apennine formation). The geology and thickness of these formations near Ferrandina is shown on Figures 8.9 and 8.10. The sequence can thus be likened to a younger version of the succession at Shaftesbury, UK, with comparable geotechnical properties. Figure 8.11 also illustrates that the Pleistocene M. Marano sands have nearly identical grading curves to the Cretaceous Cann sands, and fall within the range known to be susceptible to seepage erosion.

Subsurface investigations at Ferrandina, Pisticci, (Guerricchio and Melidoro, 1979) and Grassano (Cotecchia and Del Prete, 1986), have shown that the slopes

Landslides

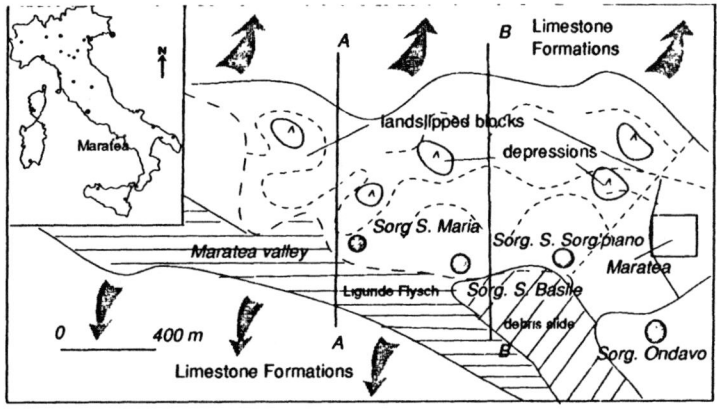

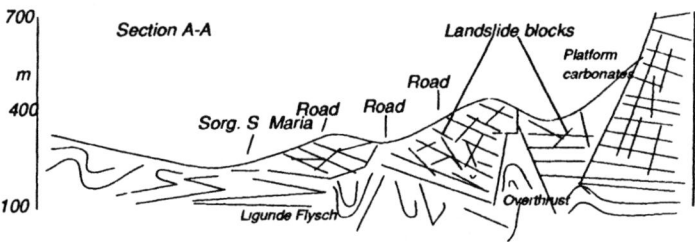

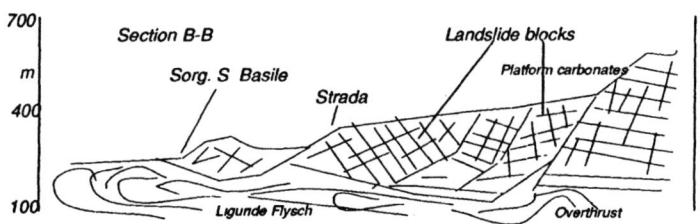

Fig. 8.8. Cross-sections through the Maretea landslides (after Cotecchia et al., 1990).

immediately surrounding the towns are covered with a layer of ancient cohesionless debris (colluvium), consisting of reworked conglomerates and sands, beween 2 and 15 m thick, (Figure 8.9). At Grassano, the debris occupies a platform, 400–500 m wide, comparable to that at Shaftesbury, which slopes at between 5° and 7°, steepening to nearly 14° immediately below the in situ slope crest which stands out as an escarpment on both sides of the valley divide. Beyond the platform edge slope angles on the blue clays average 11°–12°, but they have been heavily dissected by gulleying. The geometry of the slopes below Pisticci, and Ferrandina appear to be similar.

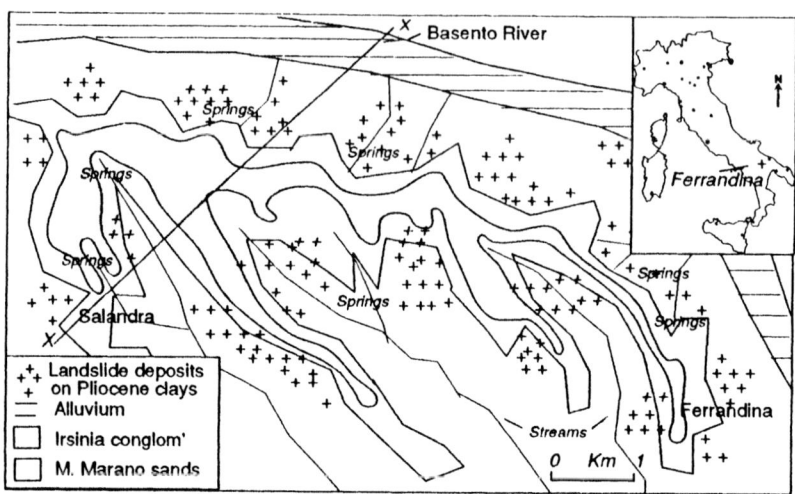

Fig. 8.9. Springs and areas of past mass movement on the debris platform in the area around Ferrandina in the Basento valley, Basilicata, Italy.

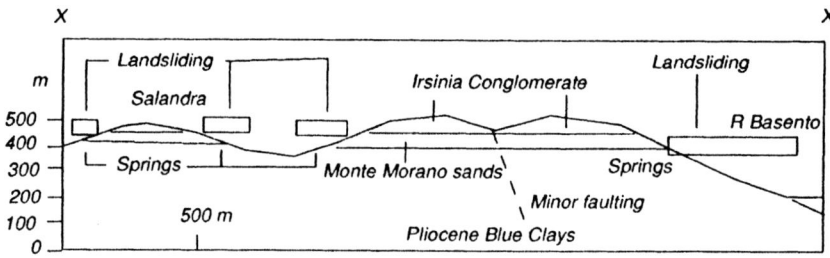

Fig. 8.10. Cross-sections showing the relationship of the geology with mass movement in the Basento valley area near Ferrandina, Basilicata, Italy.

The majority of the buildings at Grassano have been sited on the platform, whilst at Pisticci and Ferrandina they are on in situ ground behind the slope crest. In both towns construction has taken place close to the escarpment edge, and at Pisticci on the upper parts of one of the slides of 1688, where 400 lives were lost (Cotecchia and Melidoro, 1974).

A prominent springline has developed in the valley sides, at the junction of the M Marano sands and the underlying blue clays. This was recognised by Baldassarre and Radina (1971), who mapped their distribution and discharge characteristics in a rural area to the east of Gravina, in Puglia, which drains to the Bradano river, north of the Basento valley. They found that of the 130 which were surveyed, 8 had discharges greater than 1 litre/sec, half of the remainder were between 0.1 and 1 litre/sec and the rest between 0.01 and 0.1 litre/sec. The geomorphology of the areas surrounding the springs was not described in their paper, and it would be interesting to compare mass movement activity and slope development, between areas with different levels of discharge.

Landslides 205

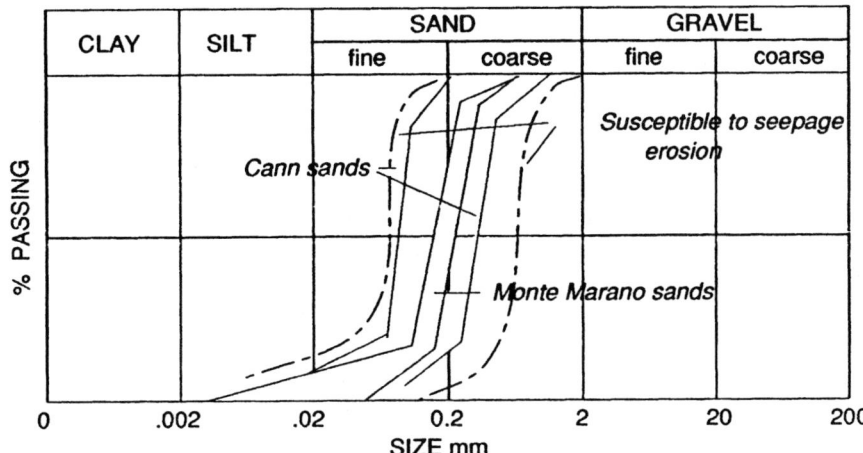

Fig. 8.11. Grading curves of Cann Sands and Monte Marano Sands in relation to particle sizes susceptible to seepage erosion.

At Grassano, two major springs, possibly fault controlled, discharge from the edge of the debris platform into large gullies which are eroding back across the platform towards the slope crest. The break of slope which occurs at the platform edge possibly reflects a period of renewed downcutting following uplift and this, together with past land-use practices has resulted in desiccated bare soils, slaking, piping and gulleying. The development of a calanchi (badland) zone on the blue clays is now a serious problem here and elsewhere in the Basento valley. Increased water pressures associated with rainfall, (1 day mean annual maximum lies between 50–60 mm) and the springs downslope of the discharge area appear to have been responsible for major instability at the edge of the platform after a storm on the 21st November 1976, which damaged the cemetery there, (Cotecchia and Del Prete, 1986). Other movements downslope of the platform edge have been recorded (Del Prete et al., 1992), but there have been no significant slides in historic time, of the conglomeritic escarpment slopes, areas upslope of these discharge points, or zones of groundwater saturation, (aquifers).

In contrast, Pisticci, with a greater catchment area, and a higher one day annual maximum rainfall (100 mm) has had a long history of instability, upslope of the spring area, (ie it involves the aquifer) which has been summarised by Guerricchio and Melidoro, (1979). They have referred to a large landslide in 1688 which destroyed the village and produced long fissures on the south side of the town. These fissures have since contributed to further instabilities. In one example, landsliding occurred in 1976 after the same heavy rainfall event mentioned above and was probably first-time. It also seems to have taken place at the escarpment edge, upslope of a spring discharge level.

Mean annual rainfall in this part of Basilicata varies from between 600–800 mm, but there is a comparatively high 1-day mean annual maxima, which varies between c. 50 mm and 100 mm with a coefficient of variation, (CV) of c. 25%. These figures can be compared with those further west at Maratea, (CV of the

1 day annual maximum is c. 10%) where the mean annual rainfall is much higher, but the one-day mean annual maxima are similar, but with a lower CV. In this area, mass movement and erosional activity are certainly present, but are less severe than in the eastern parts of Basilicata, suggesting that the relationship, (difference) between the mean annual rainfall, and the variability of 1 day maximum extremes, could be significant. The reason may be connected to the greater transient water pressures which will be experienced by a groundwater table and discharge area in a region with a low annual mean, but high, and variable 1-day annual maximum rainfall.

8.7.3.4 Papua New Guinea

In Papua New Guinea upland aquifers next to mountainous valley slopes occur at much higher altitudes than reviewed above, and are similarly associated with large scale mass movement. Blong and Goldsmith (1993) described karstified limestone aquifers in the Miocene Darai formation near Porgera in central Papua New Guinea which outcrop at between 3000 and 4000 m above sea level. Mean annual rainfall is up to 4000 mm. They overlie a weak Cretaceous mudstone and ancient colluvial complex. Spring + river discharge coupled with extreme rainfalls and anthropogenic interference from mining has caused numerous active mudslides on slopes as low as 10°. However, despite this recent activity, radiocarbon dating has also shown that climatically induced degradation associated with individual slides has been intermittent, but continuing for over 40000 years.

8.8 Landslide Caprocks and Their Response to Rainfall

8.8.1 GENERAL

The first common feature of these examples of deep-seated, geologically controlled landslides are caprocks which act as aquifers. The currently active slides are located where the aquifers are in an upland position, are subjected to orographic rainfall, a comparatively high, variable mean annual 1-day extreme value, (in relation to mean annual maxima) and have potentially, a high hydraulic gradient from the aquifers to points of discharge. The second common feature is a seepage face, in the form of a spring, (usually structurally controlled) and the third and fourth are the presence of underlying weak materials and steep slopes. The fifth factor is the effective rainfall trigger, its input to the aquifer and landslipped mass, and its links to transient water levels, groundwater discharge, and mass movement.

8.8.2 LANDSLIDE AND AQUIFER RESPONSE TO RAINFALL

The rate of increase or decrease in aquifer recharge, hydraulic gradient, groundwater levels (pore pressure ratio), and spring discharge from within a landslide or near a discharge point, following a rainfall input, varies between two extremes depending on the geological materials involved, i.e.

Landslides

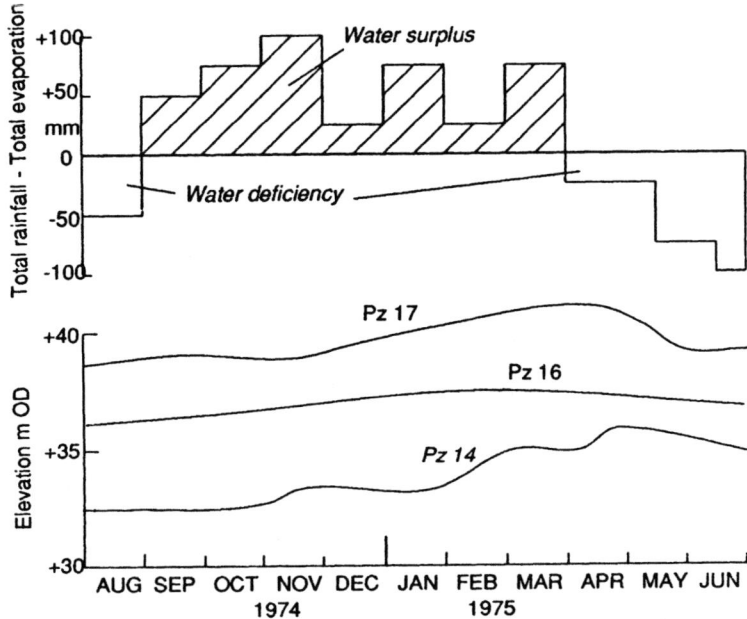

Fig. 8.12. Groundwater levels in relation to hydrometeorology in a landslipped, weathered, London Clay slope at Hadleigh, Essex, UK, illustrating the limited and slow response to rainfall (after Hutchinson and Gostelow, 1976).

8.8.2.1 Diffuse or Filtration (Atkinson, 1977, Khordikainen, 1982)

Diffuse groundwater flow is laminar, and takes place within a porous media or aquifer matrix where hydraulic conductivity (K) and storativity (S) are relatively low and high respectively. Seasonal effective strength changes are thus comparatively small, with extreme water table height variations rarely exceeding 1–2 metres. An example of seasonal groundwater variation in a series of piezometers below a weathered landslipped London Clay slope from Hadleigh, UK (Hutchinson and Gostelow, 1976), in Figure 8.12 illustrates typical fluctuations.

Within ancient slides it is possible for an exceptional storm rainfall to act as a trigger by temporarily raising the water pressures above the seasonal, or prevailing steady state values which are operating on the shear surface(s). Skempton et al. (1989) have called this a storm response, (Figure 8.13), obtained by the rise in water table (mm) divided by the rainfall (mm), and suggest this may be characteristic of a particular site or possibly material. At a large complex slide at Mam Tor in Derbyshire, UK the ratio was found to be about 4.0, but may be much greater in more fractured materials with lower storativity, (see below). The greater the storm response the greater the loss in factor of safety (F), which may decrease below 1.0.

A change in both the long term seasonal recharge and/or storm behaviour may therefore both trigger landslides either together or independently. Anthropogenic influences may also increase susceptibility to storm response movement.

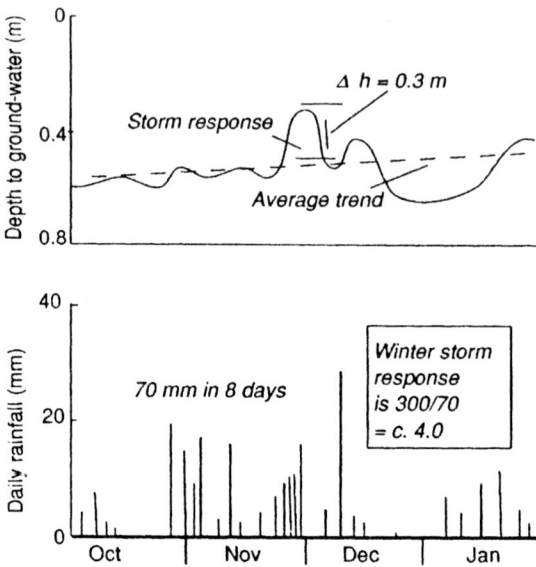

Fig. 8.13. Groundwater response to rainfall illustrating the concept of storm response (after Skempton et al., 1980).

In some geological situations diffuse aquifers may be weakly cemented, and susceptible to internal seepage erosion or mass transport, (Dunne, 1990). If there are alterations to the hydraulic gradient, for example by man-made activity, or climate, then reactivation of colluvial deposits upslope or downslope of a discharge point may occur.

Spring discharges, following recharge, generally have small variations in amplitude and are often perennial with quantities seldom exceeding 4 litre/sec. Aquifer, water level response is generally slow following a climatic, or individual rainfall event, and is followed by a long recession time. Extreme events of long duration, but short return period may thus be important in triggering landslide reactivations.

8.8.2.2 Conduit or Fluation (Atkinson, 1977, Khordikainen, 1982)

Conduit groundwater, as defined by Atkinson (1977), moves by turbulent flow, although here, this definition is extended, to include water movement in a largely fractured media such as rocks, or a soil with macropores. Groundwater may thus range from laminar to turbulent. Aquifers of this type tend to have a low storativity (the porosity is significantly less than within diffuse aquifers), and if fracture connectivity and hydraulic conductivity are high, a large potential effective stress reduction can take place following a rainfall (or snowmelt) event. In extreme cases, such as within conduits in karstic limestone aquifers, water table increases of up to 89 m in a 24 hour period have been measured (Milanovic, 1976). The water level hydrograph for this event is shown in Figure 8.14. Rapid rises in water level have also been recorded in UK carbonate aquifers which have been involved in deep

Landslides

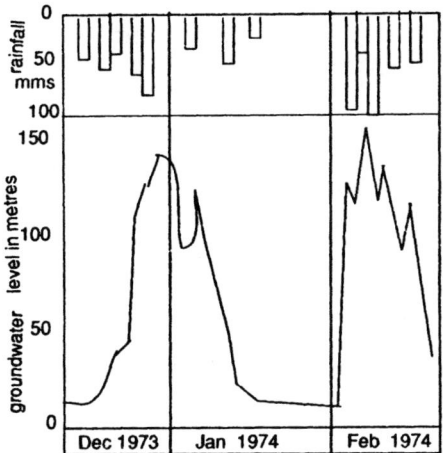

Fig. 8.14. Groundwater levels in relation to rainfall in a karstic aquifer, illustrating the large and rapid response plus the quick recession between events (after Milanchovich, 1976).

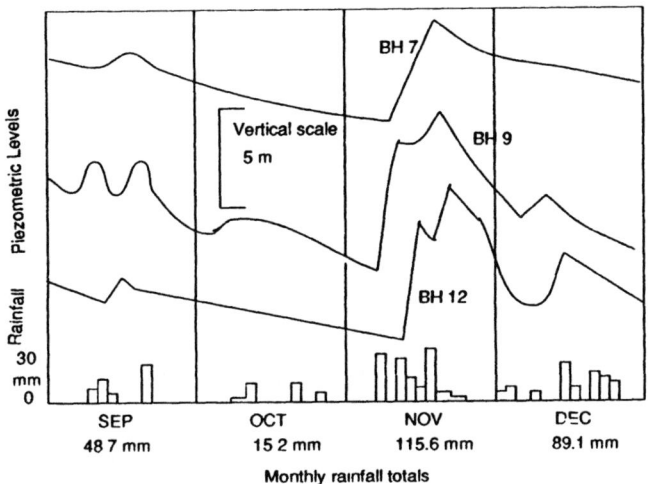

Fig. 8.15. Fluctuations in standing water levels with a series of rainfall events in a landslip consisting of fractured U. Carboniferous sandstones, mudstones and siltstones, Cullompton, Devon, UK. The figure illustrates a rapid response (and effective strength loss which is less than the conduit type behaviour but greater than found in 'diffuse' materials).

seated landslides, notably in the Chalk (Headworth, 1972, Atkinson and Smith, 1974), and the Lincolnshire limestone (Smith, 1979, Rushton and Rathod, 1979).

According to Khordikainen (1982), spring discharges in conduit aquifers are similarly variable, ranging from 2000 to 250 litre/sec (maximum) to 300 to 50 litre/sec (minimum) with many abrupt changes in discharge. Meinzer (1923) introduced the concept of spring variability, V where V = maximum − minimum discharge divided by the average (%). Figures greater than 100% were associated with extremes of climatic behaviour. The hydrograph recession time is usually shorter than with a diffuse response and hence instability may be related to greater rainfall extremes of shorter duration, higher intensity, and possibly longer return period.

Mixed-fluation and filtration responses are also possible (Khordikainen, 1982) with spring discharges showing more pronounced seasonal fluctuations than a purely filtration type aquifer. For example, Figure 8.15, from Sherrell (1971) illustrates a rapid water level response of up to 4.5 m in several piezometers following extreme rainfall in July 1969 in a landslide at Cullompton, Devon, UK. The geological setting included folded, mixed Carboniferous mudstones, siltstones and sandstones. The response and subsequent recession of water levels can be compared with the diffuse and conduit records shown in Figures 8.12 and 8.14. A mixed fluation response may also be present in some Hong Kong colluvial slopes where shallow piezometers have recorded water level increases of up to 8 m, (Brand et al., 1984). The controlling factor is the average specific storage and hence structure of the materials within these slopes.

8.8.3 Spring Discharge from Aquifers and Landslide Caprocks: Monitoring a Potential Disaster?

Most of the previous discussion has concerned unconfined water level fluctuations driving saturated groundwater discharge areas. Similar principles apply to aquifers which become confined, although in this case uplift pressures can be generated many kilometres from a rainfall intake area, (Werner, 1946). It also follows that with greater water level increases, i.e. in a conduit, or fractured material there is more opportunity for an aquifer to become confined within a heterogeneous lithological sequence. Aberfan, is an example of a landslide which occurred under confined flow conditions, and Misfeldt et al. (1991) have described a good example from Canada in thinly bedded sands and clays. However, in many landslide areas it is often unclear whether the groundwater conditions were confined or unconfined at the level of the principal failure surfaces.

In large deep-seated complexes, permanent spring, or channel courses containing surface water, are often present. These reflect aquifer hydrogeology, and their discharges are controlled by hydraulic gradient, climate and hence fluctuations in water pressures.

Figure 8.16 summarises idealised spring discharge hydrographs, (or water levels) for the two extremes described above, and a transient rainfall input. The different curves suggest it might be possible to classify preexisting landslides/susceptible areas, with respect to both variability, (V) and recession behaviour, in terms of a constant k.

This constant reflects the baseflow from the landslide and the shape of the curve thus depends on the geology and topography of the catchment area, and the hydraulic gradient driving the flow (Mero, 1963, Venetis, 1969). Its value over time may thus also reflect potential hazards from slope instability. The form of the recession equation is well known and given by

$$Q_t = Q_o e^{-kt}, \tag{8.7}$$

Landslides

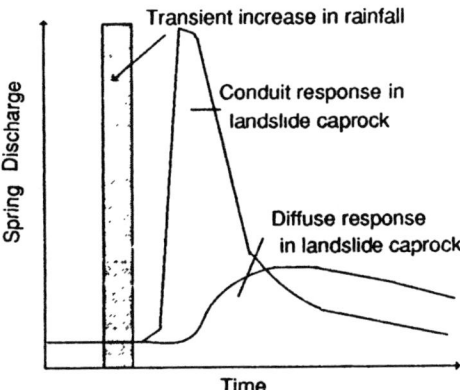

Fig. 8.16. Diagram illustrating the extremes in spring discharge of groundwater level behaviour between a diffuse porous media and a conduit type material following a transient rainfall event.

where Q_t is the discharge at time t, Q_o is the discharge at time zero, and k is the recession coefficient. The equation is derived from the same linearised solution to Boussinesq's groundwater flow equation for a uniform, unconfined homogeneous and isotropic aquifer (with no recharge). The constant k is given by

$$k = \frac{\pi^2 K h}{4a^2 S}, \qquad (8.8)$$

where h is the average saturated thickness, t is time, K is hydraulic conductivity, S is specific storage and a is the distance from the water table divide to the outlet. Recession coefficients from landslide springs have not been studied, but the hydraulic properties of the caprock/aquifer are included in the expression. The shape of the hydrograph, in relation to the magnitude and spacing of rainfall events, i.e. climate, is thus an important control on the susceptibility of a slope or landslide area to mass movement.

If a rise in groundwater level causes a pre-existing landslide to reach a state of limiting equilibrium, the factor of safety will fall below 1.0 and a deformation will occur. This magnitude depends on the length of time (duration) of the rainfall and the height of the piezometric level which Skempton et al. (1989) have called a storm-response movement. At the Mam Tor site, they suggested the residual strength of the shear surface increased with the rate of displacement, (i.e. there was a viscous effect) which restricted the total movement. However, this example, which included a clayey colluvial layer had a comparatively small storm increase in water level, and the strength loss/deformations were small. With a low storativity material, channeled flow, and a high increase in piezometric level, the loss of strength could be considerable, and catastrophic, particularly in heterogeneous aquifers and in materials which tend to decrease in volume during shear.

The shape of a hydrograph in a groundwater discharge area, or pre-existing landslide may thus also provide a measure of the degree of hazard from, a) reactivations, b) the extent of 'post-failure' displacements, and c) disastrous landslides.

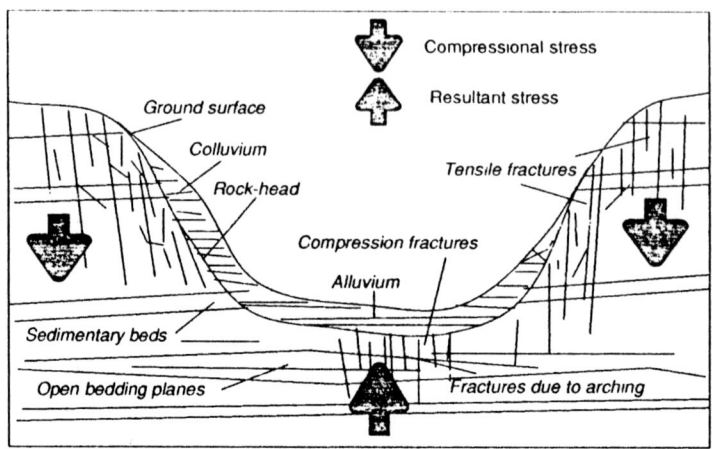

Fig. 8.17. Stress relief and the development of secondary permeability, valley side aquifers and landslides (after Ferguson, 1974).

This example of groundwater response to seasonal and storm rainfall, described by Skempton et al. (1989) applies to one geological setting, involving saturated weathered clay soils with a relatively high water table. However, the hydrogeology of landslide sites varies considerably, and in order to classify them for potential hazards from climatic events, it is necessary to review their behaviour.

8.8.4 Geological Structure, Aquifers, Valley Sides and Landslides

The most important hydrological characteristic of a caprock or aquifer, is its geological structure. Fault zones, jointing, bedding planes etc contribute to the degree of fracturing in valley sides and encourage conduit type conditions, rapid water level response times and spring development. Elastic rebound may be an important contributory process, which encourages the development of unfavourable hydraulic conditions, (Nichols, 1980).

Figure 8.17 from Ferguson (1974) illustrates the important mechanism of stress relief fracturing which may be significant in both weakening rock masses and creating local zones of high secondary hydraulic conductivity. Such fracturing may also be enhanced by underground mining which can lead to subsidence along pre-existing faults and the tilting of aquifers towards valleys. This has occurred in the Carboniferous Coal Measures in South Wales, UK, and was almost certainly instrumental in contributing to the Aberfan flow slide disaster, and the E. Pentwyn slide described previously, (plate 1).

The development of springs and their response to rainfall may be more complex in areas, such as in the UK, where the structure of weaker shales has been altered by glaciation or periglaciation. Figure 8.18 from Smith (1979) is a cross section through the Jurassic Lincolnshire Limestone escarpment near Spalding, UK which is underlain by impermeable clays and silts of the Lower Estuarine Series. Hydrogeological investigations of the limestone by Smith (1979) found

Landslides

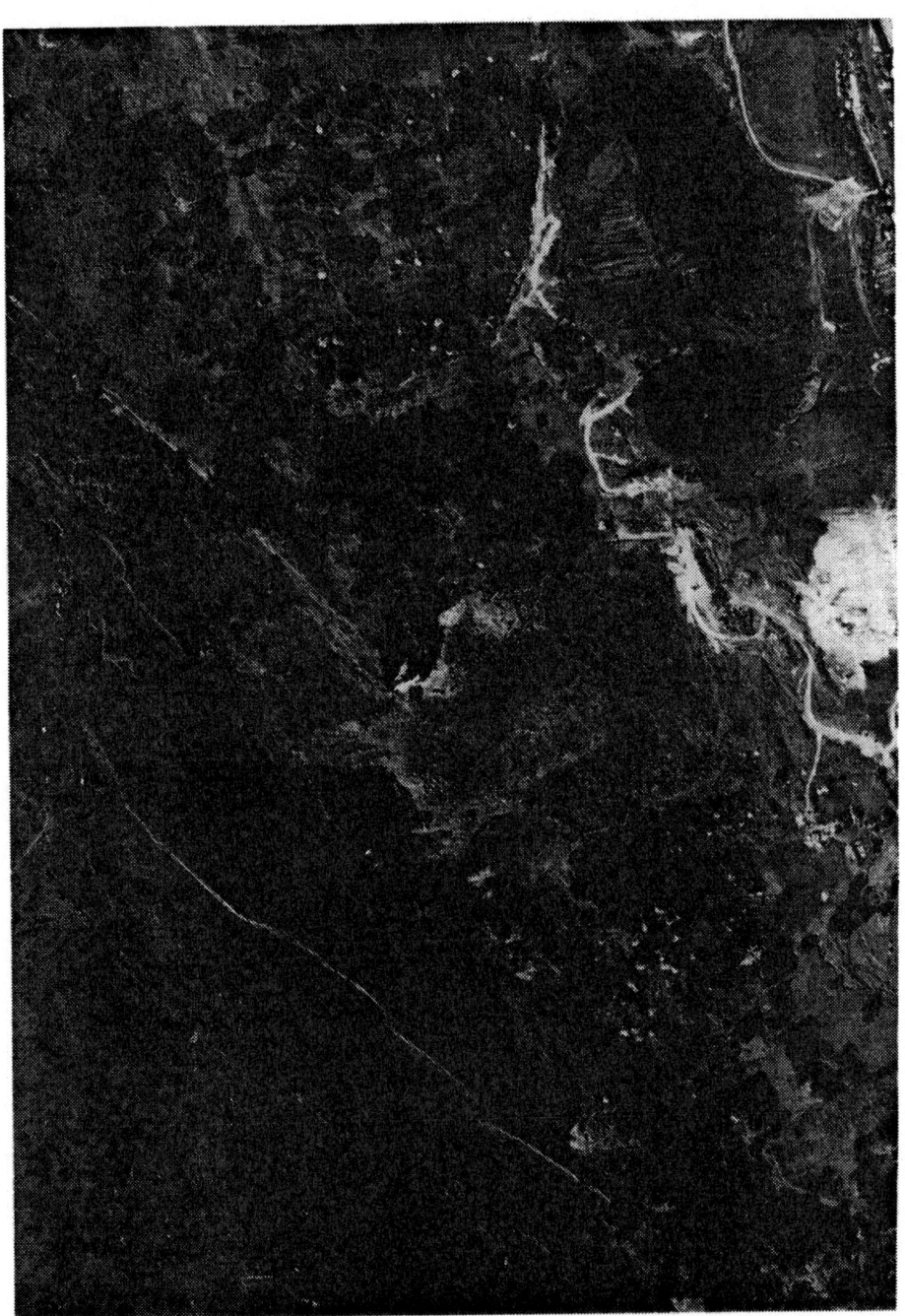

Plate 1. Faulting and stress relief fractures which have contributed to mass movements in Upper Coal Measure sandstones and shales near Blaina, South Wales, UK.

Holywell Brook catchment study area, UK

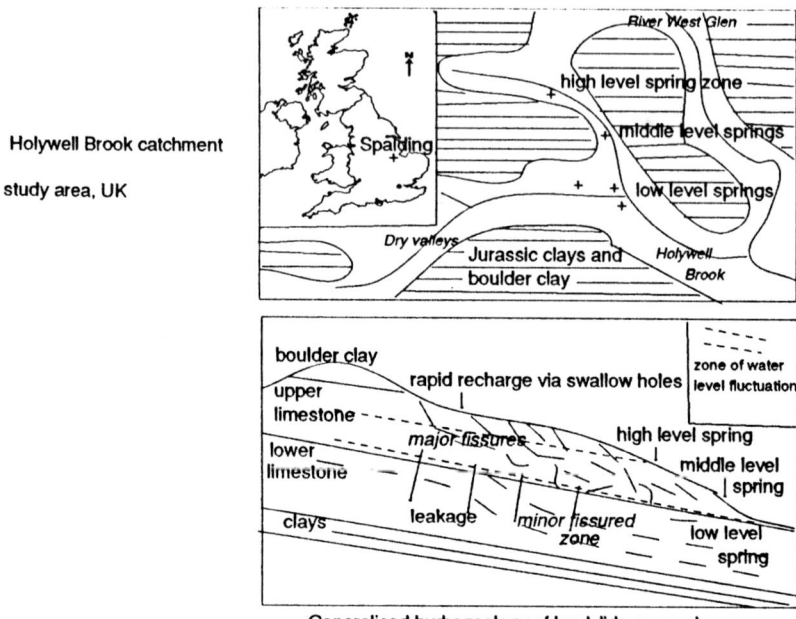

Fig. 8.18. The effects of geological structure on the response of groundwater levels and spring discharge in 'landslide caprocks' of the Lincolnshire Limestone, UK (after Smith, 1980).

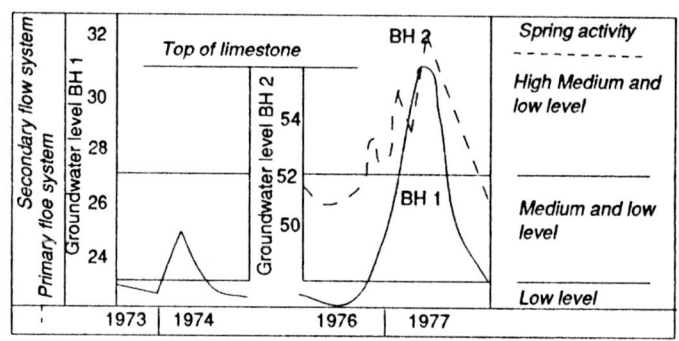

Fig. 8.19. Groundwater levels and spring discharge in the Lincolnshire Limestone, UK, illustrating the effects of geological structure on climatic response (after Smith, 1980).

that the storage coefficient was generally low (circa 10^{-4}), but there was a high transmissivity of between 2500 and 10000 m^2/day.

Three systems of springs were recognised (low, medium and high), based on their elevation above the valley floor (Figure 8.18). The low level springs are perennial, but it was found that during the onset of a period of high groundwater recharge (for example in 1976) the groundwater table rise reactivated the medium and high level areas of discharge. Hydrographs indicated that the low level perennial springs were associated with aquifer characteristics of relatively

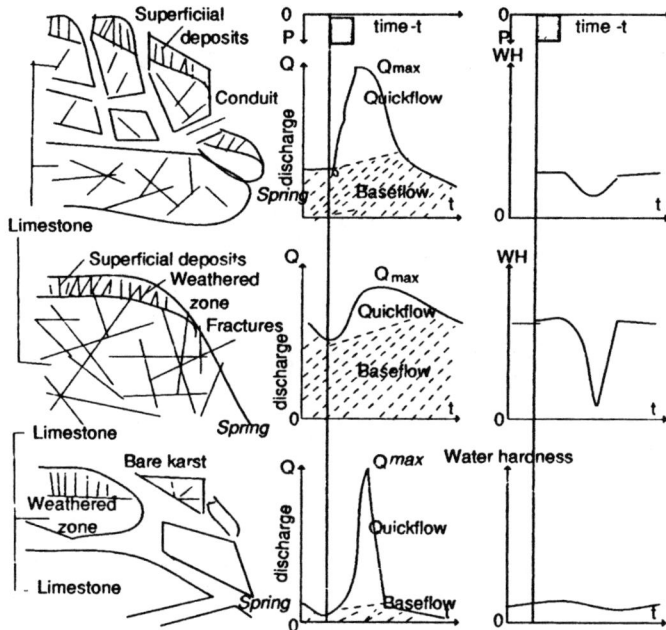

Fig. 8.20. Different forms of spring hydrograph from a landslide caprock (and aquifer) in relation to rock mass quality and structure (after Bonacci, 1993).

high storativity and low transmissivity (Figure 8.19), while the medium and higher level springs discharged from materials with relatively lower storativity and higher transmissivity (Figure 8.19). In other words, only certain climatic conditions activated the upper springs through a rapid rise in water table. The reason for this difference in behaviour is a vertical and spatial variation in aquifer properties. The low storativity zones were thought to have developed by solution, when the water table was generally higher at some time during the Quaternary.

These results have implications for landslide caprocks or other aquifers involved with slope instability which have spatial hydraulic variability, either upslope or downslope of spring discharge points. Comparatively small fluctuations in climate could exert a considerable influence on the volume of groundwater discharge, water table, (piezometric heights) and thus effective shear strengths.

Figure 8.20, from Bonacci (1993) develops these ideas regarding a) the shape of spring discharge hydrographs, b) the quality of the rock mass, c) the rise in the water table, d) the loss of strength following an extreme rainfall event and e) slope instability. In this figure the hydraulic behaviour of three examples of fractured and karstified limestone are shown, (a graph of water hardness against time is also included for each case, but other hydrochemical analyses could be included), and illustrates the kinds of discharge response (and strength loss) which might be expected after a rainfall event in a typical limestone cap-rock. The same principles apply to other fractured rock masses which include different lithologies, although as discussed above, limestones, (often karstified) are commonly found as aquifers supplying large landslides in the Mediterranean and elsewhere.

8.8.5 Development of Aquifer Properties over Time

Of general interest and importance, is the possibility that vertical zones, or regions of conduit behaviour, which have the highest potential strength losses, are present in diffuse aquifers/materials, and that these, and their properties develop over time. Processes which may be responsible include tectonism, volcanism, underground mining, chemical solution, glaciation, periglaciation, weathering, valley stress relief, and desiccation (Gburek and Urban, 1990, Gerhart, 1984, Wyrick and Borchers, 1981, Ruland et al., 1991, D'Astous et al., 1988, Keller et al., 1986, and Keller et al., 1989).

Potential groundwater conduit behaviour may thus be closely related to a) landslide reactivations following a long term trend in climatic change, and/or b) short term, first-time landslide disasters. For example, during the 1980 normal faulting earthquake in southern Italy, ground rupture took place along a 40 km length of a fault (the Irpinia fault), and further aftershocks occurred up to 15 km to the east, associated with an antithetic fault plane (Pantosti et al., 1993). Muir Wood (1994) has shown that such movements are associated with a recognisable hydrological signature, and that normal faults predominate in the shallow expulsion of meteoric water. He suggested that accompanying fractures, and their apertures also exert a considerable influence on subsequent readjustments to groundwater flow which might even oppose steady-state gravitational fluxes. A simple example, illustrating their importance, given by de Marsily, (1986) shows that a single fracture not wider than 0.2 mm conducts the same flow as a 100 m thick porous aquifer of unit width, with a hydraulic conductivity of 10^{-7} m/s. Considerable alteration to geomechanical and hydrogeological properties must have occurred in this large area of Basilicata, which probably now has an increased susceptibility to mass movement from rainfall events.

8.9 Groundwater Models, Pre-Existing Landslide Complexes and Regional Planning

8.9.1 Models, Planning and Geographic Information Systems (GIS)

Given the complexity of the hydraulic parameters and geological boundary conditions in landslides, it has become more usual to carry out borings with hydrogeophysical, packer, and shear strength testing coupled with in-situ water pressure monitoring for detailed engineering analysis. However, it is sometimes necessary to employ simpler models, perhaps for developing a set of rules for classification within a GIS so that critical sites can be identified and mapped for planning purposes. Maasland, (1959) has shown that for a hill which consists of an unconfined aquifer, or landslide caprock overlying an impermeable clay which is separated by 2 valleys L metres apart, that the maximum steady state water table height, h_m occurs at the midpoint, $L/2$, ie

$$h_\mathrm{m} = \sqrt{\left[\frac{RL^2}{4K}\right]},\qquad(8.9)$$

where R = effective rainfall (taking into account evapotranspiration, and runoff), and K = hydraulic conductivity. The water table height h, at a distance x, from the point of spring discharge on the slope is

$$h = \sqrt{\left[\frac{R}{K}(L-x)x\right]}.\qquad(8.10)$$

Without recharge the water table height h, declines according to a recession equation, (Jacob, 1943),

$$h = h_\mathrm{o}\exp\left(-\frac{\pi^2 K h_\mathrm{m} t}{L^2 S}\right),\qquad(8.11)$$

where h_o is the initial height, t is time, and S is storage. As with spring discharge, basic geometric and aquifer data can be obtained from DEM models and published results respectively. Freeze (1987) has used these equations with h_o equal to h_m to model climate/water levels, (ignoring evapotranspiration) on factors of safety in slopes for a variety of initial conditions. Transient effects were taken into account by adding a head component to h_m which was half the value obtained from the recession curve following individual rainfall events. Time, t, above is thus the average spacing between rainfalls. These become more important as the recession time increases, although extreme events are not taken into account. Nevertheless, his results confirm that hydraulic conductivity, storage, climate and the frequency of rainfall events represented by R/K and K/S control idealised slope form. Once upland aquifers and valley systems have been identified and mapped, this simple approach when coupled with extreme rainfall data may have further potential in regional classifications of areas susceptible to landsliding, (Gostelow and Gibson, 1993).

8.9.2 Recognition and Mapping Groundwater Discharge and Recharge Areas for Hazard Assessment

An initial guide to spring location, and hence slopes amenable to a more detailed modelling approach, can be obtained through geological and topographic maps, satellite imagery and aerial photographs. A number of geophysical, remote sensing techniques, such as infra-red linescan from low flying aircraft (Brereton and Hall, 1983) have also been used with varying success, particularly for defining the lateral extent of discharge, and catchments. Toth (1971) has reviewed and summarised the field characteristics of discharge areas as follows, i.e.

i) Spring seepages.
ii) Flowing (artesian) wells.
iii) Phreatophytic vegetation.
iv) Salt precipitations.
v) "Burnt" crops.

vi) Quicksand.
vii) Moist and dry depressions.
viii) Man-made drainage.

Ophori and Toth (1989) successfully used these mapping techniques for regional studies in Alberta, Canada, to distinguish springs, and broad areas of recharge and discharge. In Mediterranean Europe the CORINE land cover classes may also assist with differentiating these areas, (Gostelow and Gibson, 1994).

8.10 Landslides Associated with Snowmelt, Permafrost and Glaciers

This heading warrants a separate review and is not treated in detail here. However, melting snow and glacier movements are comparable in many ways to upland aquifers, i.e. they store and release water with a high potential energy, and are hence dangerous when combined with the other parameters discussed previously. The difference perhaps lies with the introduction of freeze-thaw processes, increased run-off, other erosional effects, and the timing, intensity and location of released water. An extensive literature exists on this subject, and good reviews of the hydrology of frozen ground and its influence on solifluction, landsliding, rock glaciers and thermokarst can be found in Williams and Smith (1989).

Disastrous events are more likely in areas of high relative relief, close to the snowline; Flerchinger et al. (1992) have shown that at this elevation the groundwater response in fractured basalts in Idaho, USA varied considerably from year to year depending on the extent of the accumulated snowpack. They found that flow might be unconfined during low snowmelt years and confined when above normal. The distribution of snowpacks in relation to susceptible slopes and materials is thus obviously important. For example a snowpack in a similar position triggered a slump earthflow in June 1974 at the edge of the Manti Canyon in Utah which turned into a debris slide. The canyon is incised into the Wasatch plateau which reaches up to 3000 m in altitude and is underlain by basalt and carbonate upland aquifers. These drain into a number of major springs, which discharge onto mixed sedimentary sequences at points around the aquifers, (Mundorff, 1971). One of these was present below the snowpack and because of adjacent steep slopes the initial failure was able to travel up to 1.2 km on to the head of the ancient, inactive Manti landslide, (Fleming et al., 1988). As a result of the undrained loading, the upper parts of this slide moved and propagated downslope, involving 19 million m^3 of debris 3 km long, and up to 800 m wide. More than 40 m of displacement occurred and nearly $2 million was subsequently spent on repairing damage and disaster preparedness.

Water released from the edge of glaciers has a similar effect on slopes; Grove (1987) has described the recession of tropical glaciers after 1932 in South America which have created a series of high level lakes in Peru. These were held back by weak morainic materials which when breached, caused sudden floods and alluviones (mud runs). Grove (1987) pointed out that the most destructive alluvione was caused by a huge ice-rock avalanche at 5000 m near Mount Huascaran in 1962. This descended 4000 m and drowned 4000 people. In 1970 a landslide from the same area caused 15000 deaths in Yungay. Similar, but less disastrous movements involving snowmelt and glacier fluctuations are known from the French Alps and have been documented by Pilot and Durville (1988).

High level melting of snow also takes place on active volcanoes, and at Ruiz in Colombia in 1985 these generated mudslides which killed 22000 people, (Herd, 1986).

8.11 Erosion, Rivers and Landslides

The effects of hyrologically-induced surface erosion, or slope undercutting by rivers or glaciers has also attracted a large literature, but is not reviewed in detail here. Changing river or lake levels during floods may also cause groundwater fluctuations in banks and adjoining slopes causing failure (Werner and Noren, 1951). Susceptible areas adjacent to first order streams and above can be identified comparatively easily by the position of glaciers, or rivers and their flood plains, (and past floods) with respect to the geology, groundwater conditions, land-use, existing landslides and slope angles of valley slopes. Remote sensing techniques may be particularly valuable in studies of this kind.

Of greater long-term interest perhaps, and more difficult to assess, are the effects of slow geomorphological processes and trends. For example river bed degradation in upland areas is particularly important in relation to slope undercutting and instability. It is caused by both natural river behaviour, i.e. through climatic factors, or by man-made changes, and can progress both upstream and downstream. The subject has been reviewed by Galay (1983), and Figure 8.21 summarises causes of degradation. A range of field effects, also described by Galay (1983) can be observed, again most easily by remote sensing techniqes, and used to assess the extent of change and areas most susceptible to increased erosion.

The use of historical data, related to rivers, especially land-use change, urbanisation, neotectonism, (including geodetic measurements) long-term river discharge fluctuations etc may be valuable for erosional studies. An example from Saskatoon, Canada has been described by Clifton et al. (1980) and the subject has been reviewed more generally in a book edited by Petts et al. (1989), using examples from European alluvial rivers. The spatial and temporal datasets which are described might benefit from a GIS approach so that the other variables described here can be included for combined analysis.

Occasionally landslide disasters occur without any obvious immediate cause or trigger and can perhaps be attributed to these slow geo-hydrological changes. An example described by Hutchinson (1975) took place on 25th April 1974 in the central Peruvian Andes at Mayunmarca on the Rio Mantaro, a tributary of the Amazon, where there had been no high rainfall or earthquakes. It consisted of mixed sandstone and siltstone rock debris, and moved on a 35° slope at an estimated 130 km/hr for over 8 km, killing more than 400 people. Its maximum thickness was 200 m, and the difference in elevation between the source area and river valley was 1900 m. A landslide lake was formed in the valley which was subsequently overtopped. Of interest also was the presence of several free flowing springs and high level lakes in the source area of the slide.

Type of Degradation	Primary Cause	Type of River Change or Engineering Works to Cause Degradation
1. Downstream Progressing	Decrease of bed material discharge	(1) construction of high dam (2) construction of low dam (3) excavation of bed material (4) diversion of bed material (5) change in land use (6) storage of bed material
	Increase water discharge	(1) diversion of flow (2) rare floods
	decrease in bed material size	(1) river processes
	other	(1) river emerging from lake (2) thawing of subsurface permafrost
2. Upstream Progressing	lower base level	(1) drop in lake level (2) drop in level of main river (3) excavation of bed material
	decrease river length	(1) cutoff (2) channelisation and regulation (3) horizontal shift of base level (4) stream capture
	removal of control point	(1) natural erosion (2) removal of dam

Fig. 8.21. Causes of river bed degradation (after Galay, 1983).

8.12 Storm-Induced Submarine Landslides

Cyclonic storms are more usually associated with subaerial landsliding, but Lee et al. (1993) described an additional instability mechanism which was caused by Hurricane Camille in 1969. At an oceanographic station on the Mississippi Delta, exceptional wave heights of 21–23 m were recorded, and it was found later that one of the drilling platforms had been displaced downslope by 30 m. The cause was almost certainly connected to the water waves which can create dynamic pressures on submarine slopes, increasing the pressure below crests and reducing it below troughs. In an extreme example associated with a storm this may result in wave-induced stresses, a progressive build up of pore pressure, loss of shear strength, deformation and slope failure. Bea, (1983) and Rahman and Jaber (1991) have provided good reviews of the mechanism, and also re-analyse the failure of the Mississippi platform in relation to the wave heights slope geometries, geotechnical properties and sea-bed deformations found there.

Critical offshore areas are obviously those in the path of cyclonic storm tracks. However, in common with subaerial slopes, initial conditions such as bathymetry, local geology, and sedimentological processes are also equally important. One aspect not often considered, is the initial pressure field in sub-sea pore waters, which may be influenced by steady state flow, (recharge) from adjacent land masses, (Robb, 1984), over-pressures, chemical dissolution, heat, and tectonism.

Storm induced runoff and sediment build-up on continental slopes can also trigger disasters. Rainfalls in the Peloponnese, Greece, were particularly severe during 1962–1963; apparently the worst since 1900, (Ambraseys, 1967). On February 7, 1963, during this period, a tsunami, with 3–6 m high waves occurred in the Gulf of Corinth, and flooded the coast from Patras to Aiyon. This was caused by a submarine landslide, which also removed part of the southern shoreline. Two people were drowned, 12 injured, and a number of fishing boats lost, (Comninakis et al., 1966). There were no major seismic shocks prior to the event and it was concluded that the landslide was triggered directly by the winter rains. Offshore cable breaks in this area have been recorded in 1907, 1921, and 1927, suggesting there may be an ongoing coastal stability problem in the Corinth area.

8.13 Conclusions

Hydrological landslide disasters involve a number of triggering mechanisms which either reduce effective shear strengths through groundwater recharge, or increase shear stresses by altering the geometry of subaerial and submarine slopes. Susceptibility to failure depends on a number of interelated factors, particularly topography, (slope), geology, land-use, and the initial pore water pressure field. The wide variety of hydrological landslides described here suggests that on subaerial slopes, the presence of pre-existing saturated ground, most often represented by certain geological conditions, springs, or seepage faces is closely associated with active landslide sites.

A large number of deterministic hydrogeological models, and empirical hydrological triggering factors related to the return periods of rainfall events of different durations have been proposed to predict slope instabilities. Several of these have been reviewed here. However, because of geological heterogeneity it is concluded that they are generally unable to predict when or where a first-time, or reactivated disastrous landslide will occur.

Hydrological landslide disasters rarely take place individually, and in this review emphasis has been placed on the importance of spatial factors which might influence their distribution. Of these, geological boundaries, records of historical movements, land-use, seismicity, topography, drainage basin characteristics, rainfall extremes, upland aquifers, and hydrogeological parameters have been shown to be of most importance.

Regional geomorphology and orographic effects on cyclonic rainfall influence the generation and movements of localised high intensity rainfall cells over slopes. Hydrometeorological modelling in relation to catchment geometry and landslide distribution from individual storm events may thus help to refine empirical rainfall triggers. Finally, it is considered that much of this controlling spatial information can be usefully entered into a GIS environment for management and further analysis.

Acknowledgements

The author would like to acknowledge the support of CEC DG X11 through their EPOCH R&D programme under contract PL 0029. Published with the permission of the Director of the British Geological Survey (NERC).

References

ADDISON K (1987) Debris flow during intense rainfall in Snowdonia, North Wales: A preliminary study. Earth. Surf. Proc. Land. 12, 561-566.
ALEXANDER D, COPPOLA L (1989) Structural geology and the dissection of alluvial fan sediments by mass movement: An example from the southern Italian Apennines. Geomorphology, 2, 341-361
ALEXANDER D (1992) On the causes of landslides: Human activities, perception, and natural processes. Environ. Geol. Water Sci. 20, 3, 165-179
AMBRASEYS NN (1967) The earthquakes of 1965-1966 in the Pelponnesus Greece: A field report. BSSA, 57, 5, 1025-1046
ATKINSON TC (1977) Diffuse flow and conduit flow in limestone terrain in the Mendip Hills, Somerset, Great Britain. Jnl. Hyd. 35, 93-110
ATKINSON TC, SMITH DI (1974) Rapid groundwater flow in fissures in the chalk: An example from South Hampshire. Q. Jnl. Eng. Geol. 7, 197-205
BAKER V (1990) Spring sapping and valley network development. Geol. Soc. Am. Spec. Pap. 252, 235-267
BALDASSARRE G, RADINA B (1971) Sorgenti di una parte del medio bacino del fiume Bradano. Geol. Applic. e Idrol. 6, 137-160
BALLANTYNE CK (1991) Late Holocene erosion in upland Britain:climatic deterioration or human influence. The Holocene, 1, 81-85
BARROWS AG, TAN SS, IRVINE PJ (1993) Damaging landslides related to the intense rainstorms of Jan-Feb 1993. Landslide News, No 7, 4-7
BEA RG, (1983) Wave induced slides in South Pass Block 70, Mississippi Delta. Jnl. Geotech. Eng. Div.; ASCE 109, 619-644
BEAR J (1979) Hydraulics of Groundwater. McGraw-Hill
BEVEN K, LAWSON A, MCDONALD A (1978) A landslip/debris flow in Bilsdale, North York Moors, Sept, 1976. Earth Surf. Proc. 3, 407-419
BEVEN K (1981) Kinematic subsurface storm flow. Wat. Res. Res. 17, 1419-1424
BEVEN K (1982) On subsurface stormflow: Predictions with simple kinematic theory for saturated and unsaturated flows. Wat. Res. Res. 18, 1627-1633
BISHOP AW, (1973) The stability of tips and spoil heaps. QJEG, 6, 335-376
BLONG RJ, GOLDSMITH RCM (1993) Activity of the Yakatabari mudslide complex, Porgera, Papua New Guinea. Engineering Geology, 35, 1-17
BONACCI O (1993) Karst springs hydrographs as indicators of karst aquifers. Hyd. Sci. Jnl. 38, 51-62
BRAND EW (1984) Landslides in Southeast Asia: A state of the art report. Proc. 4th Int. Symp. Landslides, Toronto, 1, 17-59
BRAND EW, PREMCHITT J, PHILLIPSON HB (1984) Relationship between rainfall and landslides in Hong Kong. Proc. 4th. Int. Symp. Landslides. Toronto, 1, 377-384

BRERETON, NR, HALL DH (1983) Groundwater discharge mapping at Altnabreac by thermal infra-red linescan surveying. BGS Report FLPU 83-7 51pp

BROWN WM (1988) Landslides floods and marine effects of the storm of 3-5 Jan 1982 in the San Francisco Bay region, California. Historical setting of the storm. USGS. Prof. Pap. 1434, 7-17

BRUNSDEN D, JONES DKC (1976) The evolution of landslide slopes in Dorset. Phil. Trans. Roy. Soc. A. 283, 605-631

BUCHANAN P, SAVIGNY KW (1990) Factors controlling debris avalanche initiation. Can. Geotech. Jnl. 27, 659-675

CAMPBELL RH (1975) Soil slips, debris flows and rainstorms in the Santa Monica mountains and vicinity, southern California, USGS. Prof. Pap. 851, 51pp

CHAPMAN TG (1980) Modelling groundwater flow over sloping beds. Wat. Res. Res. 16, 6, 1114-1118

CHILDS EC (1971) Drainage of groundwater resting on a sloping bed. Wat. Res. Res. 7,5 1256-1263

CLIFTON AW, KRAHN J, FREDLUND DG (1980) Riverbank instability and development control in Saskatoon. Can. Geotech. Jnl. 18, 95-105

COOKE RU (1984) Geomoprphological hazards in Los Angeles. G. Allen & Unwin. 206pp

COMNINAKIS N, DELIBASIS, N, GALANOPOULOUS A (1966) A tsunami generated by an earth slump set in motion without shock. Ann. Geol. des pays Hell. 16, 93-110

COSTA JE, SCHUSTER RL (1988) The formation and failure of natural dams. BGSA, 100, 1054-1068

COTECCHIA V, MELIDORO G (1974) Some principal geological aspects of the landslides of southern Italy. Bull. Int. Ass. Eng. Geol. 9, 23-32

COTECCHIA V, DEL PRETE M (1986) Some observations on stability of old landslides in the historic centre of Grassano after the earthquake of 23rd November, 1980 Geolo. Applic. e Idrogeol. 21, 4, 155-167

COTECCHIA V, DEL PRETE M, TAFUNI N (1989) Effects of earthquake of 23rd November 1980 on preexisting landslides in the Senerchia area of Southern Italy. Geol. App. e Idrogeol. 21, 4, 177-198

COTECCHIA V (1987) Earthquake-prone environments. in "Slope Stability" ed MG ANDERSON, KS RICHARDS, 287-330, J. Wiley & Sons Ltd.

COTECCHIA V, D'ECCLESIIS G, POLEMIO M (1990) Studio geologico e idro geologico dei monte di Maratea. Geol. App. e Idrogeol., 25, 139-178

CROZIER MJ (1985) Landslides: Causes, consequences and environment. Croom-Helm, London

D'ASTOUS AY, RULAND WW, BRUCE JRG, CHERRY JA, GILLHAM RW (1988) Fracture effects in the shallow groundwater zone in weathered Sarnia-area clay. Can. Geotech. Jnl. 26, 43-56

DEL PRETE M, GOSTELOW TP, PININSKA J (1992) The importance of historical observations in the study of climatically controlled mass movement on natural slopes, with examples from Italy, Poland and the UK. Proc. 6th. Int. Symp. Landslides. Christchurch New Zealand. (vol 3, in press)

DIETRICH WE, WILSON CJ, RENEAU SL (1986) Hollows, colluvium, and landslides in soil mantled landscapes. "Hillslope Processess", ed AD ABRAHAMS, Allen & Unwin, Boston, 461pp

DIETRICH WE, WILSON CJ, MONTGOMERY DR, McKEAN J, BAUER R (1992) Erosion thresholds and land surface morphology. Geology, 20, 675-679

DOWDESWELL JA, LAMB HF, LEWIN J (1988) Failure and flow on a 35° slope: Causes and threedimensional observations. Earth. Surf. Proc. land. 13, 737-746

DUNNE T, (1978) Field studies of hillslope flow processes. in 'Hillslope Hydrology', J Wiley and sons, N. York pp227-293

DUNNE T (1990) Hydrology, mechanics, and geomorphic implications of erosion by subsurface flow. Geol. Soc. Am. Soc. Am. Spec. Pap. 252, 1-29

EISBACHER GH, CLAGUE JJ (1981) Urban landslides in the vicinity of Vancouver, British Columbia with special reference to the December 1979 rainstorm. Can. Getech. Jnl. 18, 205-216

ELLEN SD, WIECZOREK GF (1988) Landslides floods and marine effects of the storm of January 3-5 1982, in the San Francisco Bay Region, california. USGS Prof. Pap. 1434

EYLES N, HOWARD KWF (1988) A hydrochemical study of urban landslides caused by heavy rain: Scarborough Bluffs, Ontario, Canada. Can. Geotech. Jnl. 25, 455-466

FELL R, MOSTYN GR, MAGUIRE P, O'KEEFE L (1988) Assessment of the probability of rain induced landsliding. 5th. Aust. NZ. Conf. Geomech. Sydney, 1, 72-77

FERGUSON HF (1974) Geologic observations and geotechnical effects of valley stress relief in the Allegheny Plateaus. Pap. Presented to Am. Soc. Civ. Eng. Wat Res. Meeting, L. A. 31pp

FLEMING RW, JOHNSON RB, SCHUSTER RL (1988) The reactivation of the Manti landslide, Utah. USGS Prof. Pap. 1311-A

FORSTER C, SMITH L (1988a) Groundwater flow systems in mountainous terrain. 1. Numerical modelling technique. Wat. Res. Res. 24, 7, 999-1010

FORSTER C, SMITH L (1988b) Groundwater flow systems in mountainous terrains. 2. Controlling factors. Wat. Res. Res. 24, 7, 1011-1023

FREDLUND DG (1987) Slope stability analysis incorporating the effects of soil suction in 'Slope Stability' J Wiley and sons Ltd, London

FREEZE RA (1969) The mechanism of natural groundwater recharge and discharge, 1. One dimensional, vertical, unsteady, unsaturated flow above a recharging or discharging groundwater flow system. Wat. Res. Res. 5, 1, 153-171

FREEZE RA (1971) Three dimensional, transient, saturated-unsaturated flow to a groundwater basin. Wat. Res. Res. 7, 2, 347-366

FREEZE RA (1987) Modelling interrelationships between climate, hydrology, and hydrogeology and the development of slopes, in "Slope Stability", ed MG ANDERSON, KS RICHARDS, J Wiley & Sons Ltd, 648pp

FREEZE RA (1980) A stochastic-coneptual analysis of rainfall-runoff processes on a hillslope. Wat. Res. Res. 16, 2, 391-408

FREEZE RA, CHERRY JA (1979) Groundwater. Prentice-Hall, New Jersey

FREEZE RA, WITHERSPOON PA (1967) Theoretical analaysis of regional groundawater flow. 2. Effect of water table configuration and subsurface permeability variation. Wat. Res. Res. 3, 623-634

GALAY VJ (1983) Causes of river bed degradation. Wat. Res. Res. 19, 5, 1057-1090

GBUREK WJ, URBAN JB (1990) The shallow weathered fracture layer in the near stream zone. Groundwater. 28, 6, 875-883

GERHART JM (1984) A model of regional groundwater flow in secondary permeability terrain. Groundwater, 22, 2, 168-175

GOSTELOW TP (1977) The development of complex landslides in the Upper Coal Measures at Blaina South Wales. Ass.Geotec.Ital. Symp. on 'The Geotechnics of Structurally Complex Formations', Capri, 1, 255-268

GOSTELOW TP (1990) An investigation into the origin and engineering geology of periglacial slope deposits in Wenlock Shale, Ironbridge, UK, using interactive digital modelling. Proc. 6th Int. IAEG Conf. Amsterdam, 1, 111-118

GOSTELOW TP (1991a) Rainfall and Landslides. in "Prevention and Control of Landslides and other Mass Movements". CEC Report EUR 12918 EN, Luxembourg, 37-53

GOSTELOW TP (1991b) Geological processes and their effect on the engineering behaviour of the Gault/UGS at Shaftesbury, Dorset. British Geological Survey Technical Report WN/91/9, 50pp
GOSTELOW TP, LOUCAIDES G (1988) Slope stability in allocthonous sediments: An example from the Moni Melange, Cyprus. Proc. 5th. Int. Symp. Lands. Lausanne, 1, 161-168
GOSTELOW TP, GIBSON JR (1993) Rainfall induced landslides in selected Mediterranean mountainous zones of Italy, Spain and Greece: The application of Geographic Information Systems to hazard mapping. BGS Tech Rep WN/93/36
GOSTELOW TP, GIBSON JR (1994) CORINE land cover data: Its application to regional landslide susceptibility mapping in Basilicata, Italy. Proc. Int. Conf. Vegetation and Slopes, Oxford. Pub T. Telford, ICE, (in Press)
GROVE JM (1987) Glacier fluctuations and hazards. The Geog. Jnl. 153, 351-369
GRYTA JJ, BARTHOLOMEW MJ (1989) Factors influencing the distribution of debris avalanches associated with the 1969 Hurricane Camille in Nelson County, Virginia. Geol. Soc. Am. Spec. Pap. 236, 15-28
GUERRICCHIO A, MELIDORO G (1979) Fenomeni franosi e neotettonici nelle argille grigio-azzure Calabriane di Pisticci (Lucania) con saggio di cartografia. Geol. Applic. e Idrol. 14,1 105-138
GUERRICCHIO A, MELIDORO G, RIZZO V (1988) Instrumental observations of the slope deformations and deep phenomena in Maratea Valley. Proc. 5th Int Symp. landslides, Lausanne, 1, 415-422
GUERRICCHIO A, MELIDORO G (1988) Franosita nei territori comunali di Gorgoglione e Cirigliano (Basilicata). CNR, Conv. Cartog. Monit. dei Moviment. Fran., Bologna, 65-87
GUERRICCHIO A, MELIDORO G (1991) Deformazioni gravitative profonde prodotte dall'erosione fluviale in Basilicata. Convegno Univ. Degli Studi, di Ancona, Fenomeni di Erosione e alluvionamenti degli alvei fluviali. Ancona, 63-77
GUERRICCHIO A, MASTRMATTEI R, BRUNO F (1993) Centri abitati instabili in Calabria deformazioni gravitative profonde di verante e grandi frane nel territorio comunale di lungro (Calabria settentrionale). Proc. 3rd. Conv. Naz. dei Giov. Ric. di Geol. Applic. Potenza. (in Press)
HACK JT, GOODLETT JC (1960) Geomorphology and forest ecology of a mountain region in the Central Appalachians. Geol. Surv. Prof. Pap. 347, 66 p
HALCROW (1985) East Pentwyn and Bournville landslips research project. DOE, Welsh Office, 132pp
HANEBERG WC (1991) Observation and analysis of pore pressure fluctuations in a thin colluvium landslide complex near Cincinatti, Ohio. Eng. Geol. 31, 159-184
HENDERSON FM, WOODING RA (1964) Overland flow and groundwater flow from a steady rainfall of finite duration. Jnl. geophys. Res. 69,8 1531-1540
HERD DG (1986) The 1985 Ruiz volcano disaster. Trans. Am. Geophys. Un. 67,457-460
HODGE RAL, FREEZE RA (1977) Groundwater systems and slope stability. Can. Geotech. Jnl. 14, 466-476
HOEK E, BRAY J (1981) 'Rock Slope Engineering' Inst. Min. Met. 309pp
HOUSNER GW (1989) Coping with natural disasters: The international decade for natural disaster reduction. SECED. Inst. Civ. Eng. London.
HURLEY DG, PANTELIS G (1985) Unsaturated and saturated flow through a thin porous layer on a hillslope. Wat. Res. Res. 21, 821-824
HUTCHINSON JN (1975) The Mayunmarca landslide of 25 April 1974. UNESCO, Rep No. 3124/RMO.RD/ SCE, 23pp

HUTCHINSON JN (1976) Coastal landslides in cliffs of Pleistocene deposits between Cromer and Overstrand, Norfolk, England. Laurits Bjerrum Memorial Vol. NGI, Oslo, 155-182

HUTCHINSON JN, GOSTELOW TP (1976) The development of an abandoned cliff in London Clay at Hadleigh Essex. Phil. Trans. Roy. Soc. Lond. A, 283, 557-604

HUTCHINSON JN, BROMHEAD EN, CHANDLER MP (1991) Investigations of landslides at St Catherines Point, Isle of Wight. "Slope Stability Engineering" ICE, T. Telford, London, 169-179

INNES JL (1983) Stratigraphic evidence of episodic talus accumulation on the Isle of Skye. Earth Surface Proc. Land. 8, 399-403

IVERSON RM, MAJOR JJ (1986) Groundwater seepage vectors and the potential for hillslope failure and debris flow mobilisation. Wat. Res. Res. 32, 1543-1548

IVERSON RM (1990) Groundwater flow fields in slopes. Geotechnique, 40, 1, 139-143

IVERSON RM, REID ME (1992) Gravity-driven groundwater flow and slope failure potential. 1. Elastic effective stress model. Wat. Res. Res. 28, 3, 925-938

JACOB CE (1943) Correlation of groundwater levels and precipitation on Long Island, New York (Part 1.). Trans. Am .Geophys. Un. 564-573

JACOBSEN RB, CRON ED, McGEEHIN JP (1989) Slope movements triggered by heavy rainfall, Nov 3-5, 1985 in Virginia and West Virginia, USA. Geol. Soc. Am. Spec. Pub. 236, 1-13

JACOBSEN RB (1993) Geomorphic studies of the storm and flood of November 3-5, 1985, in the U. Potomac and Cheat river basins in West Virginia and Virginia. USGS Bull, 1981

JAISWAL CS, CHAUHAN HS (1975) A Hele Shaw model study of steady state flow in an unconfined aquifer resting on a sloping bed. Wat. Res. Res. 11, 595-600

JENKINS A, ASHWORTH PJ, FERGUSON IC, GRIEVE P, ROWLING P, STOTT TA (1988) Slope failures in the Ochill hills, Scotland, November 1984. Earth. surf. Proc. and Land. 13, 69-76

JIBSON RW (1989) Debris flows in southern Puerto Rico. Geol. Soc. Am. Bull. Spec. Pap. 236, 29-55

JOHNSON KA, SITAR N (1990) Hydrologic conditions leading to debris flow initiation. Can. Geotech. Jnl. 27, 789-801

KELLER CK, VAN DER KAMP G, CHERRY JA (1986) Fracture permeability and groundwater flow in clayey till near Saskatoon, Saskatchewan. Can. Getech. Jnl. 23, 229-240

KELLER CK, VAN DER KAMP G, CHERRY JA (1989) A multiscale study of a thick clayey till. Wat. Res. Res. 25, 11, 2299-2317

KHORDIKAINEN MA (1982) Assessment of exploitable groundwater resources on the basis of measurements of spring flow. in 'Groundwater Models and Studies and Reports in Hydrology, No 34', UNESCO Press, 131135

KIDSON C, GIFFORD J (1955) The Exmoor storm of 15th August 1952, Geography, 38, 1-17

LEACH B, HERBERT R (1982) The genesis of a numerical model for the study of the hydrogeology of a steep hillside in Hong Kong. QJEG. 15, 243-259

LEE HJ, SCHWAB WC, BOOTH JS (1993) Submarine landslides: An introduction. USGS Bull. 2002, 1-22

LERNER DN (1986) Predicting piezometric levels in steep slopes. "Groundwater in Engineering Geology", Geol. Soc. Eng. Geol. Spec. Pub. No 3, 327-333

LIPPINCOTT DK, BREDEHOEFT JD, MOYLE WR (1985) Recent movement on the Garlock fault as suggested by water level fluctuations in a well in Fremont valley, California. Jnl. Geophys. Res. 90, 1911-1924

LUMB P (1975) Slope failures in Hong Kong. Q. Jnl. Eng. Geol. 8, 31-65
MAASLAND M (1959) Water table fluctuations induced by intermittent recharge. Jnl. Geophys. Res. 64, 549559
MALONE AW (1988) The role of government in landslide disaster prevention in Hong Kong and Indonesia. Geotech. Eng. 19, 227-252
MARSILY G de (1986) Quantitative hydrogeology. Academic Press, London, 440 pp
MEGAHAN WF (1983) Hydrologic effects of clearcutting and wildlife on steep granitic slopes in Idaho. Wat. Res. Res., 19, 811-819
MEINZER OE (1923) Outline of groundwater hydrology. USGS Wat. Supp. Pap. 494
MERO F (1963) Application of the groundwater depletion curves in analysing and forecasting spring discharges influenced by well fields. Symp. Int. Ass. Sci Hyd. IUGG Pub. No 63, California. 107-117
MILANOVIC V (1976) Water regime in deep karst. Case study of the Ombla spring drainage area. Proc. USYugoslavian Symp. Karst Hydrology and Water Resources, Dubrovnik, 1, 165-191
MILLER SM (1988) A temporal model for landslide risk based on historical precipitation. Math. Geol. 20, 529542
MISFELDT GA, SAUER EK, CHRISTIANSEN EA (1991) The Hepburn landslide: An interactive slope stability and seepage analysis. Can. Geotech. Jnl. 28, 556-573
MONTGOMERY DR, DIETRICH WE, (1989) Source areas, drainage density and channel initiation. Wat. Res. Res. 25, 1907-1918
MONTGOMERY DR, DIETRICH WE (1994) A physically based model for the topographic control on shallow landsliding. Wat Res. Res. 30, 1153-1171
MOORE R, LEE EM, LONGMAN F, (1991) The impact, causes and management of landsliding at Luccombe village, Isle of Wight. 'Slope Stability Engineering' Thomas Telford, Ltd, London, 225-230
MOORE LR (1969) Geological report on the tipping site and its environs at Merthyr Vale and Aberfan. "A Selection of Technical Reports submitted to the Aberfan Tribunal. Welsh Office, HMSO, 147-187
MOSER M, HOHENSIN F (1982) Geotechnical aspects of soil slopes in alpine regions. Eng. Geol. 19, 185-211
MUIR-WOOD R (1994) Earthquakes, strain-cycling and the mobilisation of fluids. Geological Soc. Lond. Spec. Pub. No 78, 85-98
MUNDORFF JC (1971) The non-thermal springs of Utah. Utah. Geol. Min. Surv. Wat. Res. Bull. 16, 70pp
NEELY MK, RICE RM (1990) Estimating risk of debris slides after timber harvest in northwestern California. Bull. Ass. Eng. Geol. 27, 281-289
NEUZIL CE (1986) Groundwater flow in low permeability environments. Wat. Res. Res. 22, 1163-1195
NICHOLS TC (1980) Rebound, its nature and effect on engineering works. Q. Jnl. Eng. Geol. 13, 133-152
O'LOUGHLIN EM (1986) Prediction of surface saturation zones in natural catchments by topographic analysis. Wat. Res. Res. 22, 794-804
OPHORI D, TOTH J (1989) Characterisation of groundwater flow by field mapping and numerical simulation. Ross Creek Basin, Alberta, Canada. Groundwater, 27, 2, 193-201
PANTOSTI D, SCHWARTZ DP, VALENISE G (1993) Paleoseismology along the 1980 surface rupture of the Irpinia fault:Implications for earthquake recurrence in the Southern Apennines, Italy. Jnl. Geophys. Res. 98, B4, 6561-6577
PETTS GE, MOLLER H, ROUX AL (1989) Historical change of large alluvial rivers: Western Europe, J Wiley and Sons, 355pp

PHILLIPS CJ (1988) Geomorphic effects of two storms on the Upper Waitahaia river catchment, Raukumara peninsula, New Zealand. Jnl. Hyd. (NZ) 27,2, 99-112

PILOT G, DURVILLE JL (1988) Landslides in the French Alps. Proc. 5th Int. Conf. Lands. Lausanne, 3, 15151537

POMEROY JS (1980) Storm induced debris avalanching and related phenomena in the Johnstown area, Pennsylvania, with reference to other studies in the Appalachians. USGS. Prof. Pap. 1191, 24pp

RAHMAN MS, JABER (1991) Submarine landslides: Elements of analysis. Mar. Geotech. 10, 97-124

REDDI LN, WU TH (1991) Probabilistic analysis of groundwater levels in hillside slopes. ASCE, Jnl. Geotech. Eng. 117, 872-891

REID ME, IVERSON RM (1992) Gravity-driven groundwater flow and slope failure potential. 2. Effects of slope morphology, material properties, and hydraulic heterogeneity. Wat. Res. Res. 28, 3, 939-950

RENEAU SL, DIETRICH WE, DORN RI, BERGER CR, RUBIN M (1986) Geomorphic and paleoclimatic implications of latest Pleistocene radiocarbon dates from colluvium-mantled hollows, California. Geology, 14, 655-658

RHETT-JACKSON CR, CUNDY TW (1992) A model of transient, topographically driven, saturated subsurface flow. Wat. Res. Res. 28, 1417-1427

ROUSE WC, BRIDGES EM (1985) Landslide susceptibility in the South Wales coalfield. Proc. Symp. Pontypridd. South Wales. Pontypridd, 1, 189-201

RULON JJ, RODWAY R, FREEZE RA (1985) The development of multiple seepage faces on layered slopes. Wat. Res. Res. 21, 11, 1625-1636

RULAND WW, CHERRY JA, FEENSTRA S (1991) The depth of fractures and active groundwater flow in a clayey till plain in SW Ontario. Groundwater, 29, 3, 405-417

RUSHTON KR, RATHOD KS (1979) Modelling rapid flow in aquifers. Groundwater, 17, 4, 351-358

SANGREY DA, HARROP-WILLIAMS KO, KLAIBER JA, (1984) Predicting groundwater response to precipitation. ASCE, Jnl. Geotech. Eng. 110, 957-975

SCHMID P, LUTHIN J (1964) The drainage of sloping lands. Jnl. Geophys. Res. 69, 8, 1525-1529

SHERRELL FW (1971) The Nag's Head landslips, Cullompton By-pass Devon. Quart. Jn. Eng. Geol. 4, 37-73

SCHRODER JF (1971) Landslides of Utah. Utah. Geol. Min. Surv. Bull. 90, 1-51

SCHRODER JF (1976) Mass movememt on the Nyika Plateau, Malawi. Z. Geomorph. N. F. 20, 1, 56-77

SIDLE RC, PEARCE AJ, O'LOUGHLIN C (1985) Hillslope stability and land use. Am Geophys. Un. Wat. Res. Mono. 11 140pp

SIDLE RC, TERRY PKK (1992) Shallow landslide analysis in terrain with managed vegetation. "Erosion, debris flows and environment in mountain regions", IAHS Publ. 209, 289-298

SIDLE RC (1992) A theoretical model of the effects of timber harvsesting on slope stability. Wat. Res. Res. 28, 7, 1897-1910

SKEMPTON AW, HUTCHINSON JN, (1969) Stability of natural slopes and embankments. Proc. 7th Int. Conf. S. M. and Found. Eng. State of the Art Vol, 291-334

SKEMPTON AW, LEADBETTER AD, CHANDLER RJ (1989) The Mam Tor landslide, North Derbyshire. Phil. Trans. Roy. Soc. A, 329, 503-547

SLOAN PG, MOORE ID (1984) Modelling subsurface storm flow on steeply forested watersheds. Wat. Res. Res. 20, 1815-1822

SMITH EJ (1979) Spring discharge in relation to rapid fissure flow. Groundwater, 17, 4, 346-350

SOLLAS WJ, PRAEGER RL, DIXON AF, DELAP A (1897) Report of the committee on the recent bog-flow in Kerry. Sci. Proc. Royal. Dublin Soc. 8, 5, 475-507

STARKEL L (1972) The role of catastrophic rainfall in the shaping of the relief of the Lower Himalaya (Darjeeling Hills). Geographica Polonica, 21, 103-147

TERZAGHI K (1950) Mechanisms of landslides. Geol. Soc. Am. Application of Geology to Engineering Practice, Berkey Volume, 83-123

TOWNER GD (1975) Drainage of groundwater resting on a sloping bed with uniform rainfall. Wat. Res. Res. 11, 144-147

TOTH J (1963) A theoretical analysis of groundwater flow in small drainage basins. Jnl. Geophys. Res. 68, 16, 4795-4812

TOTH J (1971) Groundwater discharge: A common generator of diverse geologic and morphologic phenomena. Bull. Int. Sci. Hyd. 16, 1, 7-24

TOTH J (1978) Gravity-induced cross-formational flow of formation fluids, Red Earth region, Alberta, Canada: Analysis, patterns, and evolution. Wat. Res. Res. 14, 5, 805-843

TOTH J, MILLAR RF (1983) Possible effects of erosional changes of the topographic relief on pore pressures at depth. Wat. Res. Res. 19, 6, 1585-1597

TSUKAMOTO Y, OHTA T, NOGUCHI (1982) Hydrological and geomorphological studies of debris slides on forested hillslopes in Japan. IAHS Pub. No 137, 89-98

VAUGHAN PR (1985) Pore pressures due to infiltration into partly saturated slopes. Proc. 1st. Int. Conf. Geomech. Trop. Soils. Brasilia, 2, 61-71

VENETIS C (1969) A study on the recession of unconfined aquifers Bull. Int. Ass. Sci. Hyd. 14, 119-125

WARRICK AW, LOMEN DO (1974) Seepage through a hillside: The steady water table. Wat. Res. Res. 10, 279-283

WATSON RA, WRIGHT HE (1962) Landslides on the east flank of the Chuska mountains, NW New Mexico. Am. Jnl. Sci. 525-548

WERNER PW (1946) Notes on flow-time effects in the great artesian aquifers of the earth. Trans. Am. Geophys. Un. 27, 5, 687-708

WERNER PW NOREN D (1951) Progressive waves in non-artesian aquifers. Trans. Am. Geophys. Un. 32, 2, 238-244

WIECZOREK GF, HARP EL, MARK RK (1988) Debris flows and other landslides in San Mateo, Santa Cruz, Contra Cosata, Alameida, Napa, Solano, Sonoma, Lake, and Yolo counties, and factors influencing debris flow distribution. USGS Prof. Pap. 1434, 133-161

WIECZOREK GF, SARMIENTO J (1988) Landslides, floods and marine effects of the storm of Jan 3-5, 1982 in the San Francisco bay region, California. Rainfall and piezometric levels and debris flows between 1975 and 1983 near La Honda, in storms between 1975 and 1983. USGS Prof. Pap. 1434, 43-63

WILLIAMS GP, GUY HP (1973) Erosional and depositional aspects of Hurricane Camille in Virginia, 1969. USGS Prof. Pap. 804, 80pp

WILLIAMS PJ, SMITH MW (1989) The frozen earth: Fundamentals of geocryology. Camb. Univ. Press. 306pp

WILSON EM (1990) Engineering Hydrology. Macmillan, London, 348pp

WOO B (1992) Major landslides and associated rehabilitation measures in urban areas of the Republic of Korea. IAHS Pub No 209, 347-355

WOODING RA (1966) Groundwater flow over a sloping impermeable layer (2). Jnl Geophys. Res. 71, 12, 29032910

WOODING RA, CHAPMAN TG (1966) Groundwater flow over a sloping impermeable layer. Jnl.Geophys. Res. 71, 12, 2895-2902

WYRICK GG, BORCHERS JW (1981) Hydrologic effects of stress-relief fracturing in an Appalachian valley. Geol. Surv. Wat. Supp. Pap. 2177 36pp.

YATES SR, WARRICK AW (1985) Hillside seepage: An analytical solution to a non-linear Dupuit-Forchheimer problem. Wat. Res. Res. 21, 331-336

YOUNGS EG (1971) Seepage through unconfined aquifers with lower boundaries of any shape. Wat. Res. Res. 7, 624-631

YOUNGS EG (1990) An examination of computed steady-state water table heights in unconfined aquifers: Dupuit-Forchheimer estimates and exact analytical results. Jnl. Hyd. 119, 201-214

ZARUBA Q, MENCL V (1969) Landslides and their control. Elsevier, Amsterdam, 205pp

CHAPTER 9

Land Subsidence

Giuseppe Gambolati, Mario Putti and Pietro Teatini

ABSTRACT. The increasing exploitation of subsurface fluids, such as water, oil, or natural gas, from basins filled with unconsolidated deposits of alluvial, lacustrine, or shallow marine origin, may cause the sinking or settlement of the land surface, a phenomenon known as land subsidence. Observed ground settlements of about 9 m have been reported from Wilmington, Los Angeles County, due to oil extraction, and Mexico City, due to extensive groundwater withdrawal. When land subsidence takes place in coastal areas, huge damages to the natural environment, and to man-made structures as well, can be experienced.

A pore pressure decline in a pumped aquifer, or reservoir, produces an increase of the effective stress in that part of the overburden supported by the depressurized sediments. As a major result, these sediments shorten, or compact, and the land surface subsides. Horizontal displacements may also occur, but in a lesser amount. The flow and stress-strain fields, which develop because of fluid withdrawals, are intimately connected and theoretically coupled. However, coupling is weak in most real hydrogeological settings, and uncoupled prediction models can be used. Furthermore, in large aquifer or multi-aquifer systems subject to distributed pumping, horizontal soil displacements may be negligible, and one-dimensional vertical compaction can be assumed.

This chapter presents a summary of the most well-known subsiding sites in the world, and provides a review of the basic linear theory used to build the mathematical models for the prediction of land subsidence due to fluid withdrawal. The coupled and uncoupled approaches are compared and shown to yield similar head drawdown and land settlement in realistic geological settings. Three well-known examples of land subsidence due to water and gas extraction are reviewed and discussed, together with the related models and model predictions. The examples include the sites of Ravenna, Italy, where extraction of both water and natural gas caused a ground settlement of more than 1 m, and the site of Mexico City, where land subsidence of around 9 m was caused by the extensive groundwater withdrawal from a shallow, highly compressible, semi-confined aquifer system.

9.1 Introduction

The first scientific report on land subsidence due to fluid withdrawal is due to Pratt and Johnson (1926). They write (page 578): "beginning in 1918 it became apparent that Gaillard Peninsula (mouth of Goose Creek, Texas), near the center of the (oil) field, and other nearby low land, was becoming submerged". They recognize the localized nature of the phenomenon (page 585): "contours for the 1-year period, for which we have ample data, do prove conclusively that the continuing subsidence is purely a local phenomenon", and its strict relation with fluid removal (page 588): "in our opinion the cause of the subsidence is to be found in the extensive extraction of oil, water, gas, and sand from beneath the affected area". They envisage an explanation to the ground movement in the fact that (page 590): "the pore spaces are occupied by water draining in more slowly from the adjacent clays, and it is a well-known fact that the draining of clays causes them to become more compact. This in turn, would permit subsidence of the overlying surface".

Although Pratt and Johnson were unable to mathematically describe the mechanism connecting land settlement to pore pressure decline, nevertheless they were among the first to recognize that fluid extraction from the subsurface may produce land subsidence. Since their contribution, ground surface settlement due to fluid removal has been experienced worldwide, causing big damages to the environment and to man-made structures. Table 9.1 summarizes the information related to some of the most important sites where large land subsidence occurred, possibly causing major damages to the surrounding environment. Inspection of this table reveals that the observed historical maximum subsidence amounts to almost 9 m (Mexico City, S. Joachin Valley and Wilmington, California) and the depth of the extraction wells ranges from very shallow water table aquifers to very deep (4000-5000 m) gas/oil bearing formations. Most of land settlement has taken place in densely populated areas close to seas or oceans. The overall extent of the sinking bowl may also be very large, e.g. 13500 km^2 in the S. Joachin Valley, California. Usually, groundwater withdrawal produces larger subsidence because it affects the most superficial ground layers where the most compressive soils are located. However, large settlements are reported from oil and gas fields as well, as is shown in Table 9.1, depending on the mechanical properties of the specific reservoir and the magnitude of the pore pressure decline, which may be orders of magnitude larger than that experienced by the aquifer systems. Large subsidence is the result of large drawdowns in shallow, thick, and highly compressible units. However, even small subsidence records can also be of great concern, especially in those areas that lie at sea level and are densely populated, (e.g. Venice, Italy).

Subsidence develops mainly under two different ground environments. The first is characterized by carbonate rocks overlain by unconsolidated deposits. When water is pumped and hydraulic gradients form, the unconsolidated material may move downward into openings in the underlying carbonate rocks, causing therefore the collapse of the ground surface. We do not address this type of subsidence, which is not frequently encountered. The second scenario, and by far the most common, is characterized by recent unconsolidated clastic sediments of high porosity laid down in alluvial, lacustrine, or shallow marine environments.

Among the subsiding areas listed in Table 9.1, almost all those related to water withdrawal are formed by a succession of semiconfined or confined aquifers, made up of sand or gravel with high permeability and low compressibility, interbedded

TABLE 9.1
Selected areas of major land subsidence worldwide

Location	Depth of pumping (m)	Maximum subsidence (m)	Time of main occurrence	Selected references
Latrobe Valley, Australia	10–300 *	1.6	1961-1978	Gloe (1984)
Shangai, China	3–300 *	2.6	1921-1965	Luxiang and Manfang (1984)
Tokyo, Japan	0–400 * 800–2000 o	4.6	1918-1978	Yamamoto (1984)
Mexico City, Mexico	0–50 *	9	1891-1978	Figueroa-Vega (1984)
Wairakei, New Zealand	250–800 ◁	6-7	1952-1978	Bixley (1984)
San Joaquin Valley, California	60–900 *	9	1930-1975	Poland and Lofgren (1984)
Houston-Galveston area, Texas	60–900 *	2.8	1943-1978	Gabrysch (1984)
Venezia, Italy	70–350 *	0.15	1952-1970	Gambolati and Freeze (1973) Gambolati et al. (1974) Teatini et al. (1995)
Ravenna, Italy	80–450 * 1700-4000 o	1.30	1955-1985	Gambolati et al. (1991, 1996)
Goose Creek, Texas	150-1250 •	1	1918-1926	Pratt and Johnson (1926)
Wilmington, Long Beach Beach, California	600-1200 •	9	1926-1968	Allen and Mayuga (1969) Allen (1969)
Bolivar coast, Venezuela	500-1000 •	3	1928-1960	Finol and Sancevic (1995) Nuñez and Escojido (1976)
Groningen, Netherlands	2800-3000 o	0.2	1959 to present	Boot (1973) Geertsma (1989)
Ekofisk, North Sea	3000-3200 •	6	1970 to present	Jones (1985) Chierici (1992) Zaman et al. (1995)
Po river Delta, Italy	0-600 ◇	3.5	1938-1961	Schiesano (1980) Gambardella et al. (1991)

Legend: * water; ◁ hot water; o gas; • oil; ◇ gas-bearing water with clayey aquitards of low vertical permeability and high compressibility under virgin stress conditions. These deposits are usually normally loaded (for a few exceptions, see Holzer (1981)). They compact along the virgin compression curve in response to increased effective stress due to artesian-head decline, in coarse-grained aquifers, and excessive pore-pressure dissipation, in fine-grained confining or semi-confining beds, causing therefore the subsidence of the ground.

By distinction, natural gas and oil are extracted from much deeper reservoirs, and only occasionally significant land settlement occurs (Martin and Serdengecti, 1984). In this case, large subsidence records are related to very large pressure drawdown, and, therefore, to very large increase of the effective intergranular stress in the fluid bearing formations, and are associated with possible rock failure, which may involve the fracturing, crushing, and rearranging of grains. The resulting strong reduction in porosity may induce large shear stresses in the rock surrounding the reservoir, which may in turn cause a second type of in situ failure: fractures and faults.

In principle the mechanism relating land subsidence to fluid removal is well understood. The total geostatic load, acting on the aquifer, confining beds or reservoir, is balanced by the pore pressure p and the effective vertical and horizontal stresses. As p declines, the pore pressure cannot bear the entire total stress, and, therefore, there is a stress transfer from the fluid to the solid phase and the effective stress increases in both the reservoir and the adjacent formations, which are progressively drained. The resulting compaction extends its effect to the ground surface, which therefore subsides. In case of a deep reservoir, the behavior is like an inclusion in a semi-infinite structured body. The surrounding rocks absorb part of the loss of support caused by a local pressure decline, and the actual land subsidence depends mostly on the depth, volume, and compressibility of the reservoir. Typically, land settlement above a gas/oil field is smaller than the reservoir compaction, and spreads over a larger area than the reservoir horizontal section. By contrast, aquifer and multi-aquifer systems are generally much shallower, and are characterized by larger areal extents (regional basin). For these systems it is easy to conceive the existence of a central area where soil compaction is not contrasted by the overlying rocks and is simply transferred to the ground surface with a subsidence spreading factor equal to one. These stratified systems behave as if they were one-dimensional structures, and, although flow may be fully three-dimensional (e.g. vertical in the confining beds, or aquitards, and horizontal in the aquifers) soil movement occurs mostly in the vertical direction.

In addition to dimensionality, there are other factors that differentiate gas and oil field compaction from the compaction of multi-aquifer systems. Usually both consist of a sequence of sands and clays (aquifers) and sands and shales (reservoirs). Shales are surface clays that have undergone extensive mineralogical changes in the burial process, which associated them with the oil or gas bearing strata. These mineralogical changes may have a strong influence on the compaction properties of the shales. Moreover, at larger depths, where reservoirs are normally found, a preconsolidation effect may exist. The reservoir rocks may therefore behave as though they experienced, in the geological past, higher effective stresses than those existing at the beginning of the fluid production. As a major consequence, we observe that the sediments respond to a large increase of loading (i.e. pressure decline) with a small amount of deformation as long as the applied stress does not exceed the preconsolidation stress. Thereafter, the rate of incremental deformation per unit pressure decline suddenly increases. This occurrence may pose severe difficulties in the prediction of the ultimate subsidence at the production inception phase. Furthermore, the difference between the compressibility of sand and clay (shale) decreases with depth, and, for the deepest reservoirs, soil compressibility may be the same irrespectively of the sediment nature (Martin and Serdengecti, 1984; Gambolati et al., 1991).

In addition to anthropic land subsidence caused by fluid withdrawal, there are other types of man-induced subsidence. However they are generally much less important in terms of their environmental impact. A good review of these types of problems can be found in Allen (1984), and includes various activating mechanisms such as underground salt, gypsum and carbonate rock solution, subsurface erosion, lateral flow (*fluage*), surface loading, land drainage and reclamation, biochemical oxidation of organic soils, vibration, and water application.

In this chapter, we first provide a review of the general mathematical theory of coupled flow and stress, which correctly relates ground motion to pore pressure drawdown caused by fluid pumping. The development framework is constituted by the coupled linear theory of poroelasticity. The uncoupled approach is then presented and justified on the basis of the physical characteristics of the settings where the land subsidence occurs. Next we present an uncoupled prediction model that relies on a multi-dimensional flow model and a one-dimensional soil compaction model. This approach is extensively used by hydrogeologists for the simulation of the compaction of large, shallow aquifer systems. Results obtained from coupled and uncoupled three-dimensional finite element models are then compared. The differences in terms of aquifer drawdown and horizontal and vertical ground surface displacements show that the uncoupling assumption is acceptable when dealing with large stratified hydrogeological settings. Finally, one of the most important and complex subsidence events experienced in Italy (Ravenna) and the largest subsidence case occurred in the world (Mexico City) are reviewed, and the results from recent numerical predictions are presented and discussed.

9.2 Review of Mathematical Theory of Land Subsidence due to Fluid Withdrawal

When an aquifer, a reservoir, or a confining bed experiences a variation of the internal flow and stress fields (due, for instance, to pumping or external loading), the incremental effective stress and the hydraulic gradients that develop are intimately connected. This complex inter-relationship was first described by Biot (1941), who used a linear mathematical model for a three-dimensional porous medium. A model of flow and stress based on the Biot equations is said to be a *coupled* model.

Groundwater hydrologists and petroleum engineers, however, are primarily interested on the fluid dynamics of the system, and therefore are interested in solving the *uncoupled* flow equation (Theis, 1935), a diffusion type equation incorporating the soil mechanical reaction into a single distributed parameter, called the elastic storage coefficient. By contrast, soil engineers tend to give a limited credit to the uncoupled approach. They are mainly concerned with small scale problems (surface loading, pile foundations, embankments, fill dams, etc.) in systems characterized by low permeability coefficients, where a coupled consolidation process is likely to occur, and also the solid skeleton stress field may influence the transient hydraulic head before steady state is achieved (Lewis and Schrefler, 1987; Lewis et al., 1991). However, in natural hydrogeological basins, such as those involved in land settlement caused by extensive fluid pumping, coupling is less important (Gambolati, 1977; Debbarh, 1988; Gambolati, 1992), as will be shown later with some practical examples.

The Biot equations represent the most elegant approach at the theoretical level, but their analytical solution is available only for extremely simplified systems. The numerical solution of these equations is expensive also on modern powerful computers, and is frequently affected by instability and ill-conditioning, which make the simulation results often unreliable. Finite element solutions to the coupled equations in two-dimensional porous systems were originally obtained by Sandhu and Wilson (1969), Christian and Boehmer (1970), Hwang et al. (1971), Desai (1975), Sandhu (1976), Smith and Hobbs (1976), and Verruijt (1977). Numerical difficulties related to instability were reported by Booker and Small (1975), Sandhu et al. (1975) and Vermeer and Verruijt (1981).

9.2.1 Coupled (Biot) Model of Land Subsidence

Denote by E and ν the Young and Poisson moduli of the porous matrix, respectively, and by E_r, ν_r the same elastic constants for the individual grain. The equations shown below rely on the following fundamental assumptions:

1. Terzaghi's principle of effective intergranular stress holds;

2. the solid grains change their volume exclusively because of the pore (or neutral) pressure variation p, i.e. a variation of the effective stress σ does not induce, by itself, any volume variation in the solid grains.

If we indicate by ε_{xx}, ε_{yy} and ε_{zz} the principal components of the incremental volumetric strain ε, and both p and σ are incremental quantities, with the aid of the elastic theory and the three-dimensional form of Terzaghi's effective stress principle, we can write for a mechanically isotropic porous medium:

$$\varepsilon_{xx} = \frac{1}{E}[\sigma_{xx} - \nu(\sigma_{yy} + \sigma_{zz})] - \frac{1 - 2\nu_r}{E_r}p$$
$$\varepsilon_{yy} = \frac{1}{E}[\sigma_{yy} - \nu(\sigma_{xx} + \sigma_{zz})] - \frac{1 - 2\nu_r}{E_r}p \quad (9.1)$$
$$\varepsilon_{zz} = \frac{1}{E}[\sigma_{zz} - \nu(\sigma_{xx} + \sigma_{yy})] - \frac{1 - 2\nu_r}{E_r}p$$

In eq. (9.1) σ_{xx}, σ_{yy} and σ_{zz} are the principal components of the (incremental) effective stress tensor σ, and p is taken to be positive when increasing. The equilibrium (or Cauchy) equations for a porous medium are:

$$\frac{\partial(\sigma_{xx}-p)}{\partial x} + \frac{\partial \sigma_{xy}}{\partial y} + \frac{\partial \sigma_{xz}}{\partial z} = 0,$$
$$\frac{\partial \sigma_{xy}}{\partial x} + \frac{\partial(\sigma_{yy}-p)}{\partial y} + \frac{\partial \sigma_{yz}}{\partial z} = 0, \qquad (9.2)$$
$$\frac{\partial \sigma_{xz}}{\partial x} + \frac{\partial \sigma_{yz}}{\partial y} + \frac{\partial(\sigma_{zz}-p)}{\partial z} = 0.$$

Recalling that:

$$\sigma_{xy} = \frac{G}{2}\left(\frac{\partial u_x}{\partial y} + \frac{\partial u_y}{\partial x}\right),$$
$$\sigma_{xz} = \frac{G}{2}\left(\frac{\partial u_x}{\partial z} + \frac{\partial u_z}{\partial x}\right), \qquad (9.3)$$
$$\sigma_{yz} = \frac{G}{2}\left(\frac{\partial u_y}{\partial z} + \frac{\partial u_z}{\partial y}\right),$$

where $\mathbf{u} = (u_x, u_y, u_z)^T$ is the soil displacement vector and G is the shear modulus of the porous skeleton, and replacing in (9.2) the components of σ as calculated from eqs. (9.1) and (9.3), we obtain to the final equations, written in terms of $\mathbf{u}$ and p:

$$(\lambda + G)\frac{\partial \varepsilon}{\partial x} + G\nabla^2 u_x = \left(1 - \frac{E}{E_r}\right)\frac{\partial p}{\partial x},$$
$$(\lambda + G)\frac{\partial \varepsilon}{\partial y} + G\nabla^2 u_y = \left(1 - \frac{E}{E_r}\right)\frac{\partial p}{\partial y}, \qquad (9.4)$$
$$(\lambda + G)\frac{\partial \varepsilon}{\partial z} + G\nabla^2 u_z = \left(1 - \frac{E}{E_r}\right)\frac{\partial p}{\partial z},$$

where $\lambda = \nu E/[(1 - 2\nu)(1 + \nu)]$ is the Lamè constant of the porous matrix and ∇^2 is the Laplace operator.

Writing the continuity equation for the compressible fluid contained within the porous volume, and applying Darcy's law, yields (Verruijt, 1969; Gambolati and Freeze, 1973):

$$\frac{1}{\gamma}\nabla \cdot (K\nabla p) = n\beta\frac{\partial p}{\partial t} + \frac{\partial \varepsilon}{\partial t}, \qquad (9.5)$$

where ∇ is the gradient operator, γ is the specific weight of the fluid, K is the hydraulic conductivity tensor, n is the medium porosity, and β is the volumetric fluid compressibility.

Eq. (9.5) is based on the assumption that the single grains are incompressible. This equation must be slightly modified to take into account the compressibility of the grain, always keeping in mind assumption 2. In this case, in eq. (9.5), $\varepsilon = \Delta V/V$ must be replaced by $\varepsilon_p = \Delta V_p/V$, where V_p is the pore volume of the representative elementary volume V of the porous medium. Obviously $\varepsilon = \varepsilon_p$ only if the volumetric compressibility of the grains c_{br} is zero. If $c_{br} \neq 0$ we can develop the ε_p-ε relationship as follows.

Denote by V_s the volume of the solid grains contained in V. Then:

$$V = V_p + V_s \quad \text{and} \quad \Delta V = \Delta V_p + \Delta V_s$$

On the basis of the previous definition we can write:

$$\Delta V_s = -c_{br} V_s p$$

Hence:

$$\frac{\Delta V_p}{V} = \frac{\Delta V}{V} - \frac{\Delta V_s}{V} = \varepsilon + c_{br} \frac{V_s}{V} p = \varepsilon + c_{br}(1-n)p \tag{9.6}$$

Therefore, the flow equation for a compressible fluid moving in a deformable porous medium, whose single grains are considered compressible, is:

$$\frac{1}{\gamma} \nabla \cdot (K \nabla p) = [n\beta + c_{br}(1-n)] \frac{\partial p}{\partial t} + \frac{\partial \varepsilon}{\partial t} \tag{9.7}$$

Eq. (9.7) is similar, although not identical, to Geertsma's (1966) equation.

Eqs. (9.4) and (9.7) are coupled and represent a straightforward extension of the original Biot (1941) equations. They predict land subsidence (u_z at $z = 0$) employing a coupled model of flow and stress. For stratified multi-aquifer systems, eqs. (9.4) are easily extended to include soil transversal anisotropy (Gambolati et al., 1984) using five distinct elastic parameters for the porous skeleton.

Solution to the coupled model is a difficult task even on modern powerful computers. Typically, large uncertainties in the assessment of the soil parameters are likely to offset the expected greater accuracy of the coupled formulation. It is therefore apparent that the simpler uncoupled approach is often used in real applications.

9.2.2 Uncoupled Model of Land Subsidence

Differentiating with respect to x, y and z the first, second and third of equations (9.4), respectively, and summing up the results, we obtain:

$$(\lambda + 2G)\nabla^2 \varepsilon = (1 - \beta')\nabla^2 p \tag{9.8}$$

where

$$\beta' = \frac{E}{E_r} = \frac{(1-2\nu)}{(1-2\nu_r)} \frac{c_{br}}{c_b} \simeq \frac{c_{br}}{c_b} \tag{9.9}$$

is the ratio between the volumetric compressibility of the solid grain c_{br} and of the porous medium c_b.

Integration of (9.8) yields:

$$(\lambda + 2G)\varepsilon = (1 - \beta')p + f(x,y,z,t) \tag{9.10}$$

Land Subsidence

where $f(x, y, z, t)$ satisfies Laplace equation $\nabla^2 f = 0$ for every value of t, i.e. f is a harmonic function. For the expression of f in some specific problems see Verruijt (1969). If we discard f, we obtain an explicit relation between ε and p:

$$\varepsilon = \frac{1 - \beta'}{\lambda + 2G} p = \alpha(1 - \beta')p \tag{9.11}$$

where $\alpha = 1/(\lambda + 2G) = (1 - 2\nu)(1 + \nu)/E(1 - \nu)$ is the vertical soil compressibility, i.e. the rate of volume change of a soil sample subject to a unit change of effective stress along the vertical axis with lateral expansion precluded. Replacing (9.11) into (9.7) provides the uncoupled flow equation in the only variable p. This equation can now be solved independently of the equilibrium eqs. (9.4):

$$\frac{1}{\gamma} \nabla \cdot (K \nabla p) = [n\beta + c_{br}(1 - n) + \alpha(1 - \beta')] \frac{\partial p}{\partial t} \tag{9.12}$$

If c_{br} is negligibly small compared to β and α, as is usually the case in real porous media, the above equation becomes the classical flow equation as is generally used by hydrologists and petroleum engineers:

$$\frac{1}{\gamma} \nabla \cdot (K \nabla p) = (n\beta + \alpha) \frac{\partial p}{\partial t} = S_s \frac{\partial p}{\partial t} \tag{9.13}$$

where we have indicated with S_s the specific elastic storage coefficient.

The solution of the uncoupled model of land subsidence consists of the following two step procedure. First eq. (9.13) is solved for the incremental pore pressure, or pore pressure drawdown p, in the aquifer system or reservoir. At any given time t, the pressure gradients are computed and the equilibrium eqs. (9.4) solved for the displacement vector u. The component u_z at $z = 0$ (assuming that the origin of the vertical axis is at the ground surface) gives the land subsidence.

In the mathematical analysis of land sinking due to groundwater pumping, it is quite common to employ an uncoupled approach formed by a two or three-dimensional flow model and a one-dimensional vertical settlement model (see, for instance, Gambolati et al. (1974), Helm (1975), Brutsaert and Corapcioglu (1976), Helm (1976), Harris Galveston Coastal Subsidence District (1982), Narasimhan and Goyal (1984), Rivera (1990) and Gambolati et al. (1991)). The main assumption in the one-dimensional subsidence model is that $u_x = u_y = 0$ and $p(x, y, t) = \sigma_{zz}(x, y, t)$. The solution of eq. (9.13) is used in a one-dimensional consolidation model to produce the pointwise settlement prediction of the ground surface:

$$u_z(x, y, t) = \int_0^{b(x,y)} p(x, y, z, t) \alpha(z) \, dz, \tag{9.14}$$

where $b(x, y)$ is the thickness of the compacting aquifer system.

Some authors (Bear and Corapcioglu, 1981; Lewis and Schrefler, 1987) have questioned the validity of the assumption of one-dimensional vertical compaction. In fact, close to a single pumping well, the horizontal soil displacement may achieve a magnitude of some relevance. Debbarh (1988) has shown that Terzaghi's approach of one-dimensional consolidation may be safely used at a distance from the well exceeding five times the wellbore depth. Due to the superposition of effects, this limit can be significantly reduced if pumping is distributed over the subsiding area.

Helm (1987) noted that "in spite of its conceptually unsatisfactory nature, one-dimensional analysis has been successful in modeling complex aquifer problems", and one reason for this success was found to lie in the stratified nature and the anisotropic properties of the vast majority of the sedimentary basins, where the most important and productive aquifers are located.

9.2.3 COMPARISON OF COUPLED AND UNCOUPLED LAND SUBSIDENCE PREDICTIONS

In the present section we solve the coupled and uncoupled models of land subsidence by finite elements in realistic hydrogeologic settings and compare the corresponding results in terms of aquifer pressure drawdown and horizontal and vertical displacement at the ground surface. Eqs. (9.4) and (9.5), and eqs. (9.4) and (9.13) are solved in a three-dimensional porous system consisting of a pumped aquifer overlain and underlain by confining beds. Other unpumped formations are incorporated into the model (Figure 9.1). The mean aquifer depth and thickness are denoted by h and b, respectively, while H indicates the depth of the rigid basement or bedrock (Figure 9.1). It is assumed that the aquifer experiences an instantaneous unit pressure drawdown at time $t = 0$ in the pumping well ($r = 0$). The unit decline is kept constant until final equilibrium (steady state) is achieved. Because of the linearity of both the coupled and uncoupled approaches, the results from the comparison are representative of any drawdown or pumping rate, and of any arbitrary distribution of extraction wells in space and time.

We assume an aquifer average thickness of $b = 20$ m and study the influence of depth h by setting $h = 100, 200$ and 500 m. The following boundary conditions are prescribed: $p = \sigma_{zz} = \sigma_{rz} = 0$, for $z = 0$ (ground surface), and $p = \sigma_{rr} = \sigma_{rz} = 0$ for $r = R_e$, $R_e = 30000$ m being the outer radius of the simulated system; $\partial p/\partial z = 0$ and $u_r = u_z = 0$ at $z = H$; $u_r = 0$ at $r = 0$; $\partial p/\partial r = 0$ at $r = 0$ except in the aquifer, where $p = -1$ kg/m^2 is set.

Two different geological configurations are simulated:

Problem 1: a mechanically homogeneous porous medium with $H = 1000$ m and a uniform compressibility taken from Figure 9.2 at the average depth of the pumped aquifer.

Problem 2: a non-homogeneous porous medium with $H = 20000$ m where α decreases with z according to the experimental profile of Figure 9.2, i.e. the compressibility profile of the Quaternary and Pliocene sediments that was used at Ravenna to predict land subsidence caused by water and gas removal (Gambolati et al., 1991; Bertoni et al., 1995). Below 5000 m, α is kept constant and equal to the value corresponding to $z = 5000$ m ($\alpha_{z=5000} = 0.42 \times 10^{-8}$ m^2/kg).

Land Subsidence

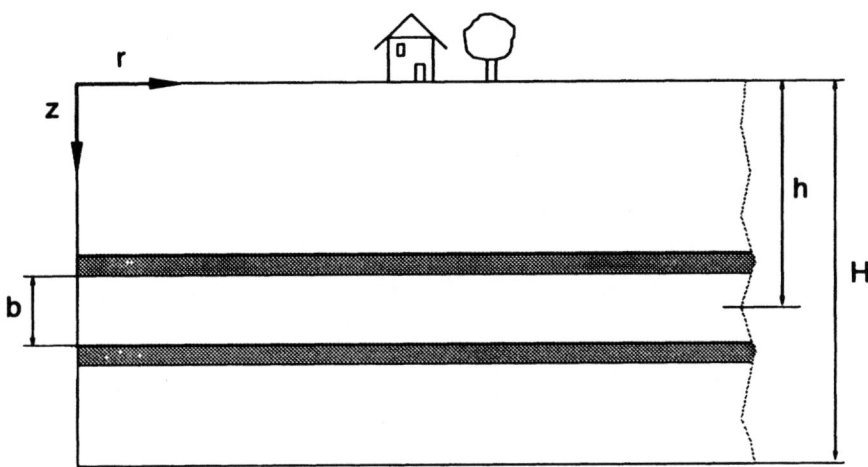

Fig. 9.1. Scheme of a three-dimensional aquifer system.

Above and below the pumped aquifer and the confining beds, the domain is constituted by an alternating sequence of sandy and clayey units, which we call unpumped aquifers and aquitards, respectively. For both problems, we assume a permeability of $K_{\text{aquifer}}=10^{-4}$ m/s in the pumped and unpumped aquifers, and $K_{\text{aquitard}}=10^{-7}$ m/s in the aquitards. The entire domain is assumed saturated with water.

Figure 9.3 shows a vertical cross-section of the triangular finite elements adopted in both the coupled and uncoupled models for test problem 1 with $h = 100$ m, and also gives the layering used in the simulation. Similar scenarios are used in the case of $h = 200$ m and $h = 500$ m and in problem 2. In each lithologic unit compressibility is considered constant, with its value obtained from Figure 9.2 as a function of the mid-depth of the layers. The Poisson ratio ν is set to 0.25 in problem 1, while in problem 2 we set $\nu = 0.25$ in the aquifers and $\nu = 0.4$ in the aquitards. Porosity n is constant and equal to 0.4. The compressibility of the grains in the aquifers and in the aquitards is assumed equal to that of quartz and calcite, respectively, i.e. $c_{br,\text{quartz}} = 0.16 \times 10^{-9}$ m^2/kg and $c_{br,\text{calcite}} = 0.8 \times 10^{-10}$ m^2/kg (Geertsma, 1973), and the compressibility of water is $\beta = 0.43 \times 10^{-8}$ m^2/kg.

Figures 9.4 and 9.5 provide the behavior of p versus r at the aquifer mid-depth for the three h values and different times for problems 1 and 2, respectively, as calculated from the solution of the numerical model. The solid and dotted curves represent the coupled and the uncoupled solutions, respectively. Note that the aquifer pressure drawdown is practically the same irrespective of the model

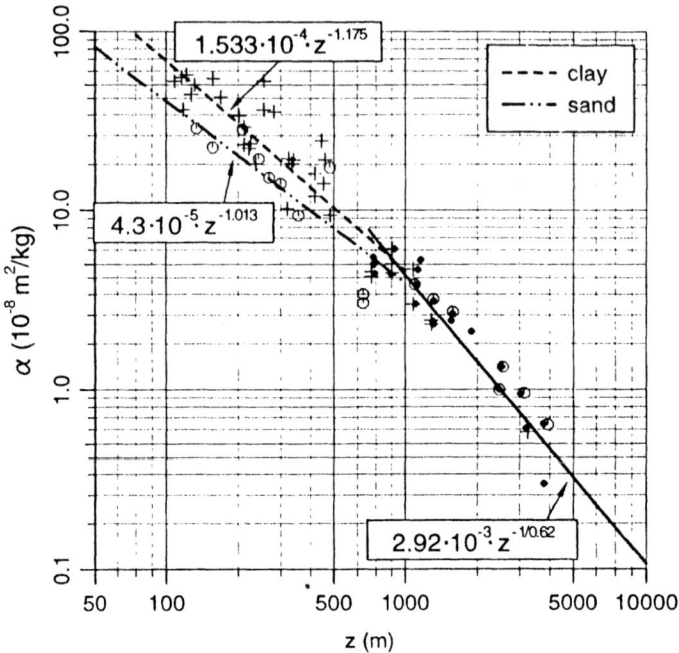

Fig. 9.2. Vertical soil compressibility α versus depth z for the subsurface system underlying the Ravenna area (after Bertoni et al. (1995)).

(coupled or uncoupled) used, for both the mechanically homogeneous and the non-homogeneous systems, and the various depths of stress source. Response in time is, however, different for different h, since α diminishes with z (Figure 9.2).

A similar result is obtained for the land subsidence, as may be seen in Figures 9.6 and 9.7. There, we can see that the ground settlement predicted by the uncoupled approach is slightly larger than the coupled settlement for intermediate time values. The difference, however, is not such as to require the use of the coupled model even for the worst case (Figure 9.7(c)), where it amounts to about 10%. In fact, typical uncertainties in the soil parameters may lead to much larger errors.

Finally the horizontal displacement u_r at $z = 0$ is given in Figures 9.8 and 9.9. Note in these figures, and especially in Figure 9.9(a), that the uncoupled u_r may exhibit a difference of some relevance from the coupled u_r during an intermediate time interval, at the distance r where u_r achieves a peak value. This difference does not persist in time, and is much less pronounced far away from the well. However, Figure 9.9 indicates that, for problem 2, the use of a coupled model is required for the accurate time prediction of the maximal u_r.

From this analysis we can conclude that the uncoupled model of flow and stress predicts the aquifer pressure drawdown and the related land subsidence quite satisfactorily in hydrogeological basins of regional size.

Land Subsidence

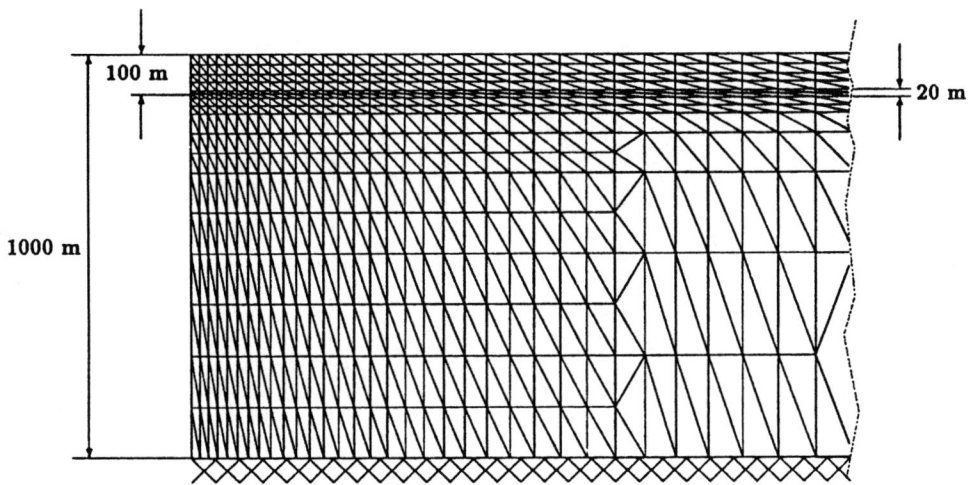

Fig. 9.3. Detail of the finite element mesh of the coupled and uncoupled models to simulate a pumped aquifer system resting on a 1000 m deep bedrock ($h = 100$ m).

9.3 Illustrative Case Studies

We will illustrate three important case studies where land subsidence has contributed to a significant modification of the environment, and has been successfully simulated and predicted with the aid of numerical models. The first two examples are concerned with Ravenna (Italy), where both groundwater and gas withdrawal have produced (and are still producing) a large scale surface settlement, which has markedly increased the exposure of the city to the offense of periodic flooding by the Adriatic Sea (Gambolati et al., 1991, 1996). The third case study deals with the land subsidence of Mexico City, one of the most impressive anthropic events ever observed in the world (Rivera, 1990; Rivera et al., 1991). The realistic simulation of the sinking of Mexico City has required the consideration of non linear effects, and hence the development of a relatively complex model.

9.3.1 Land Subsidence at Ravenna due to Groundwater Withdrawal

Ravenna is situated close to the Adriatic coast, approximately 120 km south of Venice (Figure 9.10(a)), another city affected by anthropic land subsidence (Gambolati and Freeze, 1973; Gambolati et al., 1974; Teatini et al., 1995). In the last two decades, a hydrogeologic reconnaissance study was carried out, for the development of which a number of shallow and deep boreholes, pumping tests, soil

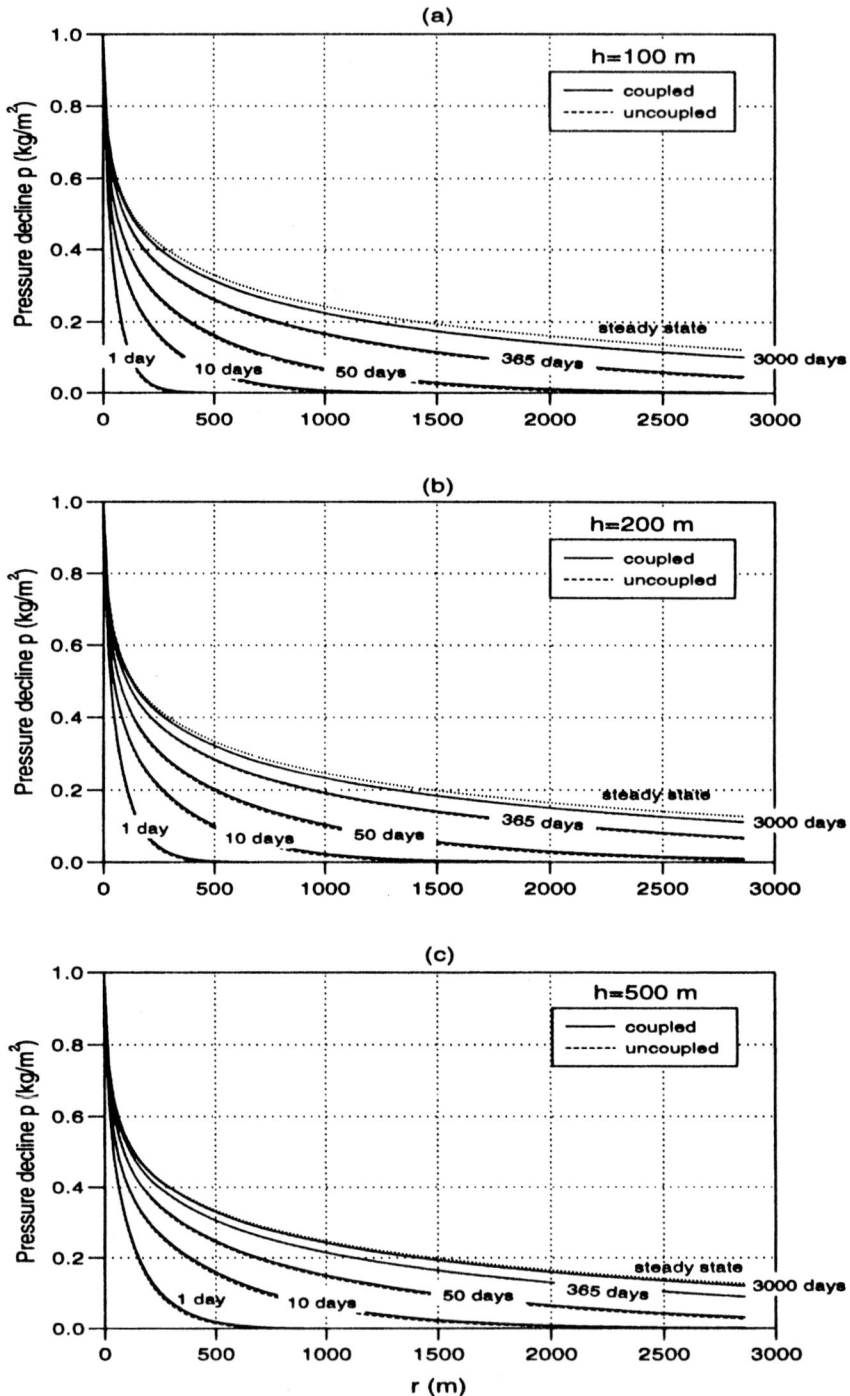

Fig. 9.4. Comparison of coupled and uncoupled aquifer pressure decline p vs. distance r from pumping well at different values of time for problem 1.

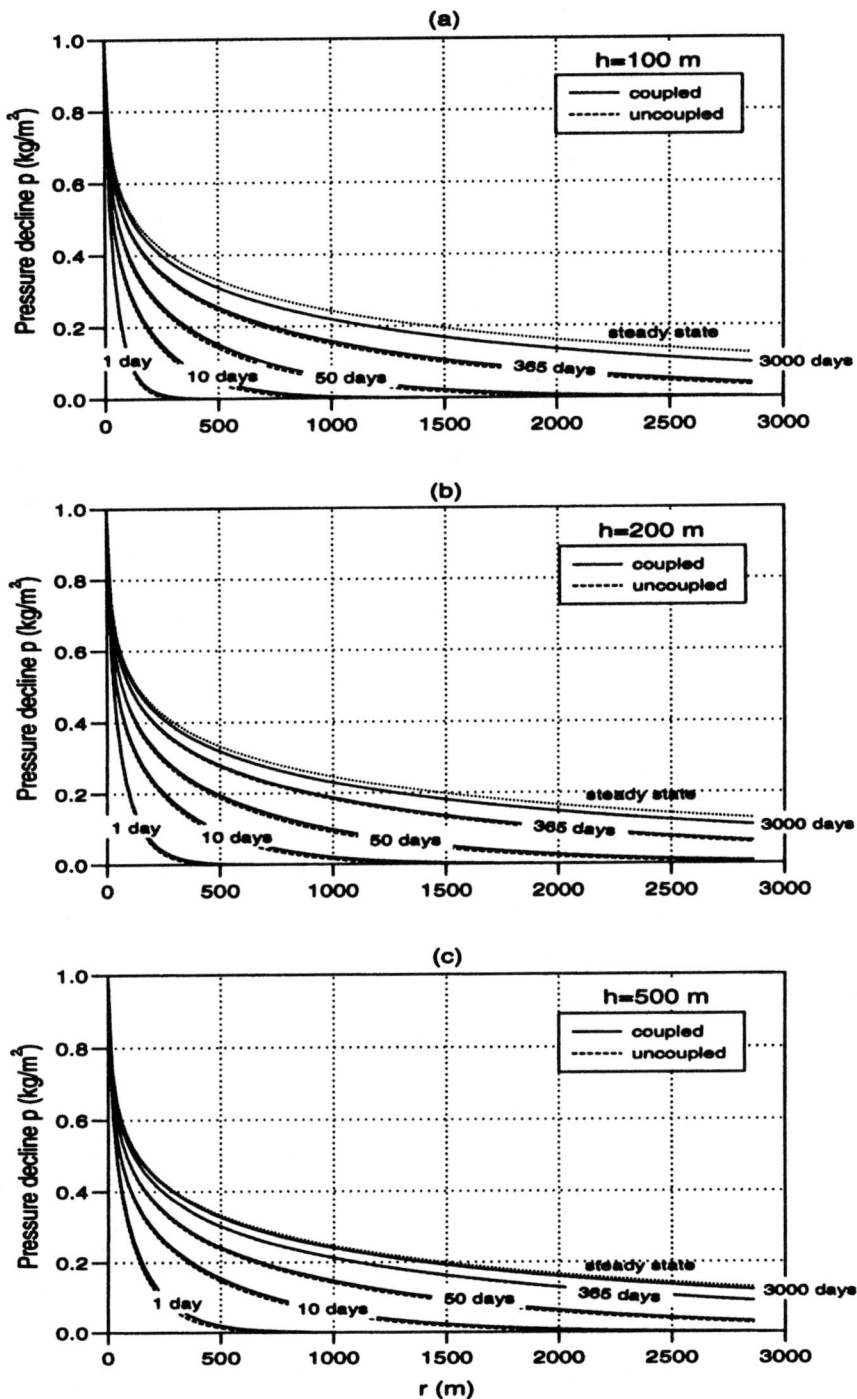

Fig. 9.5. Comparison of coupled and uncoupled aquifer pressure decline p vs. distance r from pumping well at different values of time for problem 2.

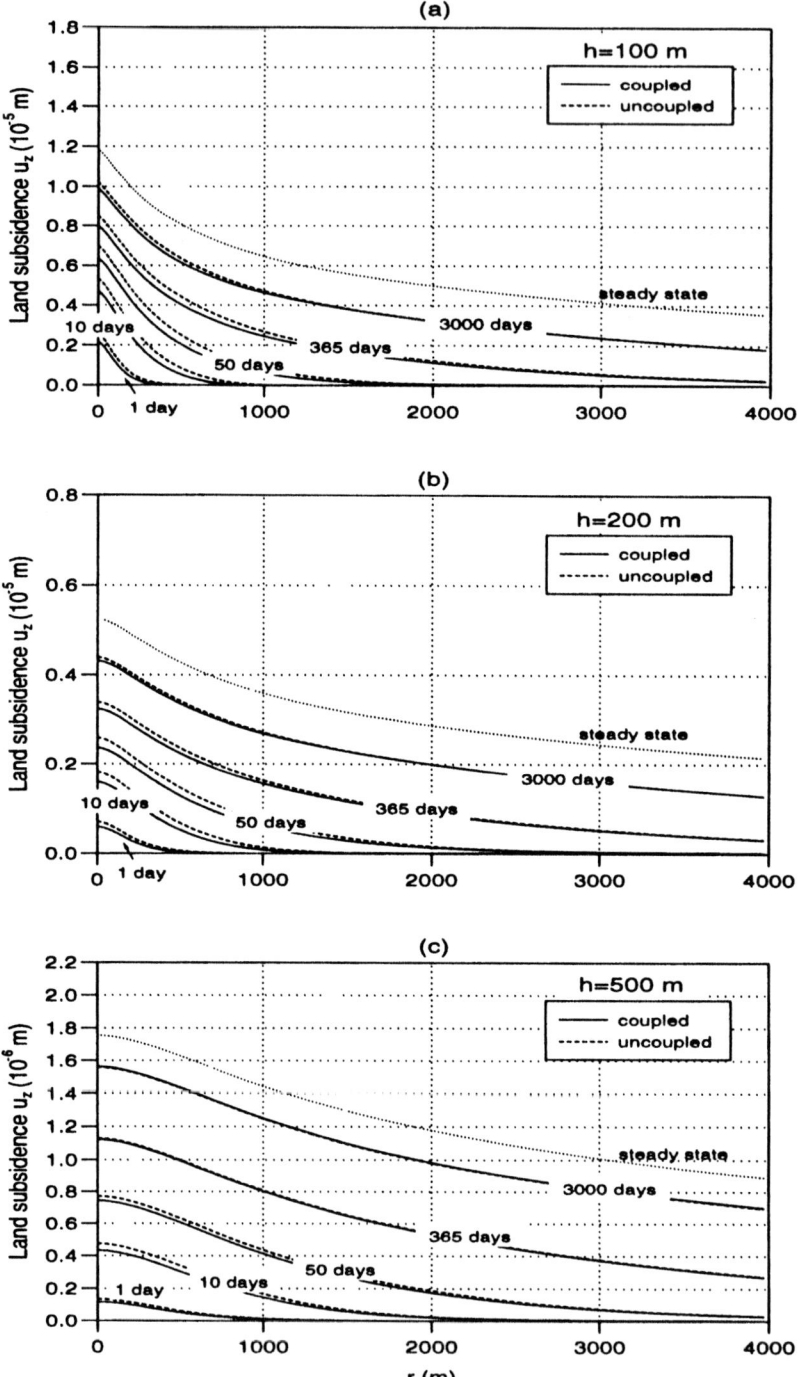

Fig. 9.6. Comparison of coupled and uncoupled land subsidence u_z ($z = 0$) vs. distance r from pumping well at different values of time for problem 1.

Land Subsidence

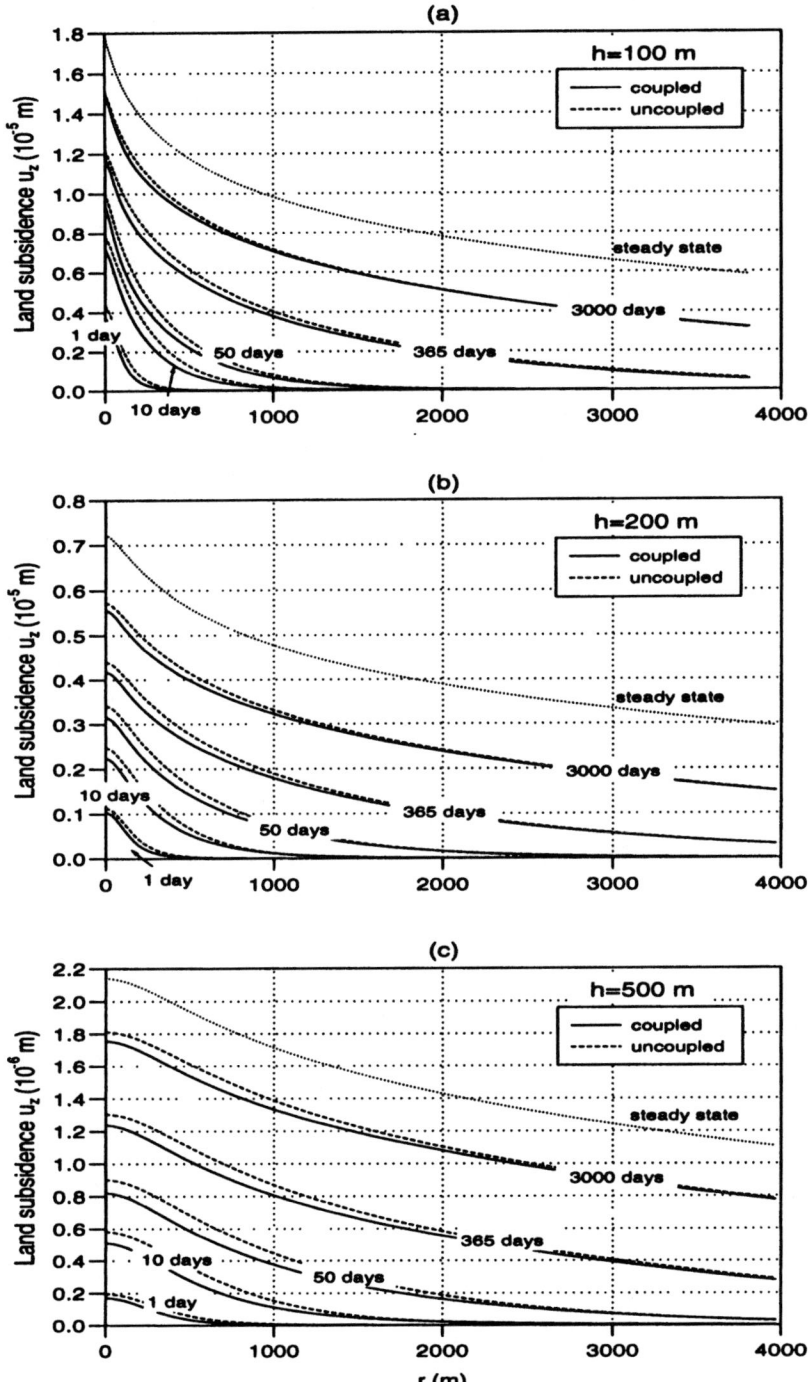

Fig. 9.7. Comparison of coupled and uncoupled land subsidence u_z ($z = 0$) vs. distance r from pumping well at different values of time for problem 2.

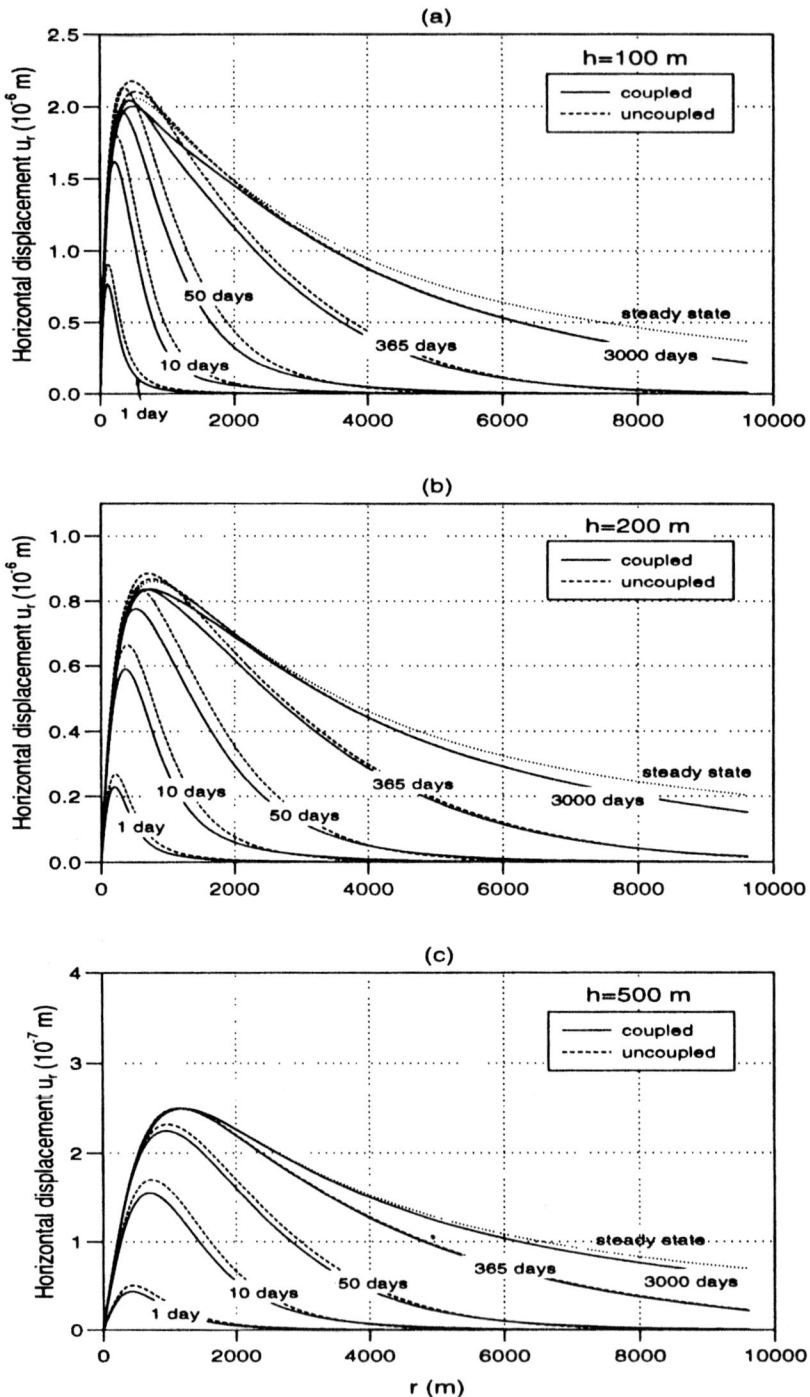

Fig. 9.8. Comparison of coupled and uncoupled horizontal displacements u_r at ground surface ($z = 0$) vs. distance r from pumping well at different values of time for problem 1.

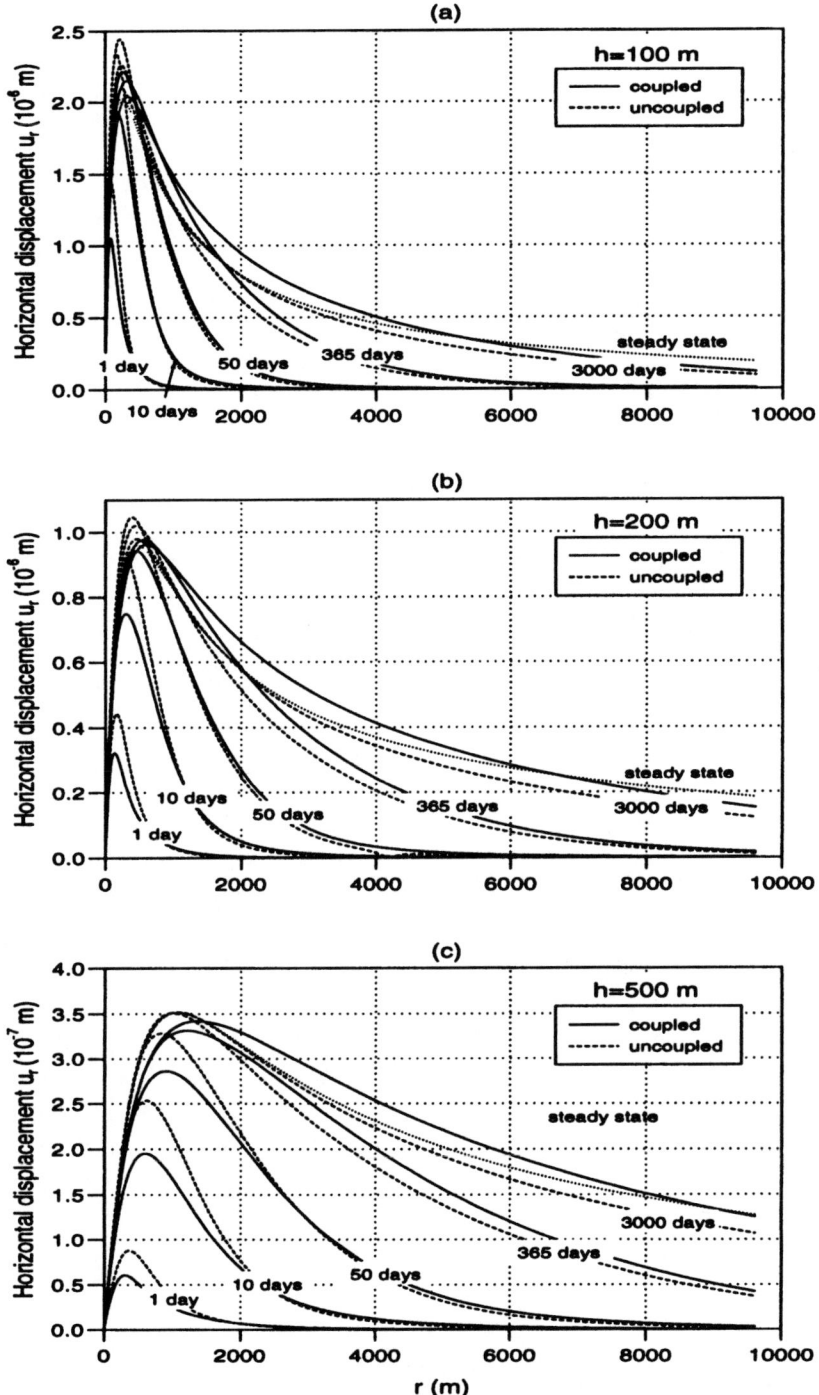

Fig. 9.9. Comparison of coupled and uncoupled horizontal displacements u_r at ground surface ($z = 0$) vs. distance r from pumping well at different values of time for problem 2.

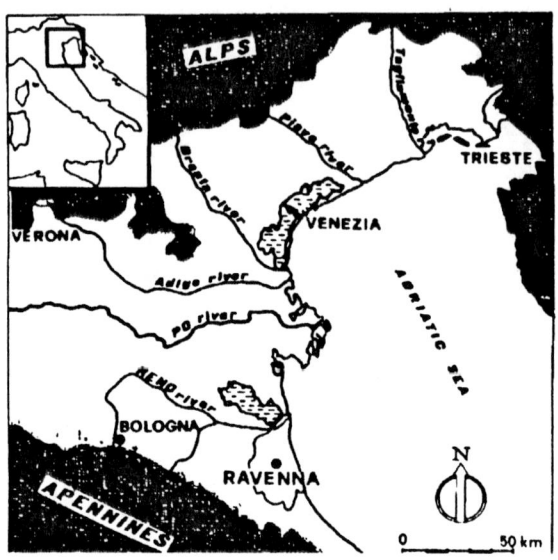

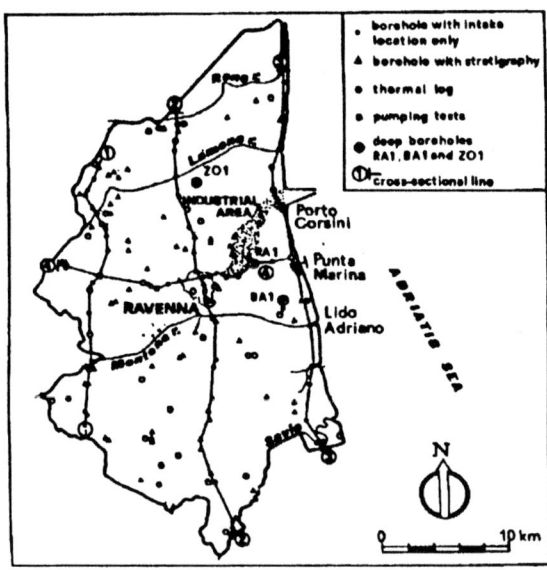

Fig. 9.10. Map of the eastern part of the Po river basin, (a), and the Ravenna area with the location of several boreholes, test wells and the trace of four hydrologic cross-sectional profiles, (b), (after Gambolati et al., (1991)).

Land Subsidence

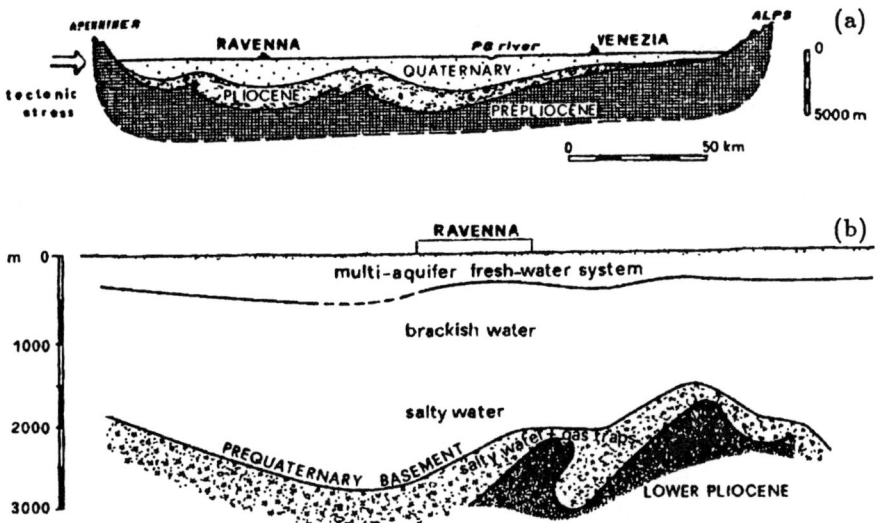

Fig. 9.11. Schematic geological section of the subsurface system between Venice and Ravenna, (a), (after AGIP Mineraria (1969)); cross-section of the Ravenna subsurface system, (b), (after Gambolati et al. (1991)).

laboratory analyses was employed, together with four stratigraphic profiles, as shown in the borehole map of Figure 9.10(b).

The Ravenna underground system consists of a sequence of stratified deposits laid down during the Quaternary and the upper Pliocene in different environments, of continental, lagoonal and deltaic type in the upper zone, and of littoral and marine type in the lower one (Figure 9.11(a)). Gas pools are mostly located in the pre-Quaternary basement, as is shown in Figure 9.11(b), which provides a schematic cross-section of the Ravenna system down to a depth of 3000 m. Fluid removal has occurred, and still occurs as far as gas is concerned, from the upper fresh-water multi-aquifer system and the gas bearing strata.

The finite element mesh used to simulate and predict the multi-aquifer pressure drawdown is displayed in Figure 9.12, where the six confined aquifers and the intervening aquitards forming the multi-aquifer system are visible. Figure 9.13, which provides the piezometric levels recorded in Ravenna from 1950 to 1990, shows the pronounced flow field recovery in the deepest aquifers observed after the middle seventies as a consequence of the drastic reduction in the pumping rate (Figure 9.14). The land subsidence recorded in the Ravenna area from 1949 to 1986 is shown in Figure 9.15. The largest recorded value is about 1.3 m in the industrial area, a few kilometers north-east of the Ravenna historical center. The land subsidence of Figure 9.15 includes the geologic settlement ($2 \div 3 \times 10^{-3}$ m/year) and the settlement due to both water pumping and gas production.

The mathematical model of land subsidence at Ravenna due groundwater withdrawal is based on an uncoupled approach making use of a quasi three-dimensional flow model combined with a one-dimensional consolidation model. On account of the distributed nature of the withdrawals, it is believed that the

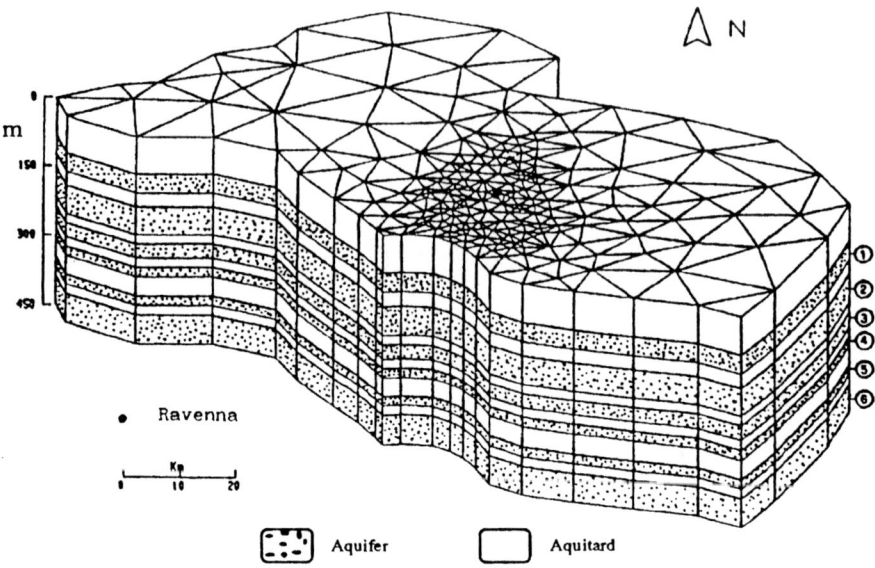

Fig. 9.12. Perspective three-dimensional view of the regional finite element-convolution integral model used to simulate the Ravenna multi-aquifer system (after Gambolati et al. (1991)).

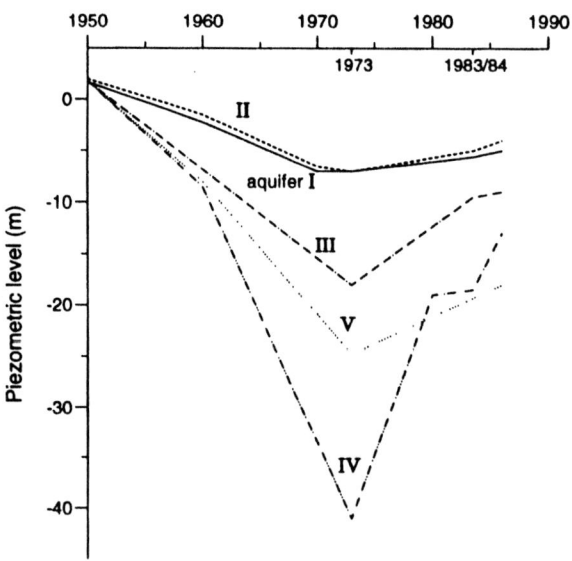

Fig. 9.13. Piezometric levels above mean sea level vs. time in the various aquifers underlying the Ravenna area (after Gambolati et al. (1991)).

Land Subsidence

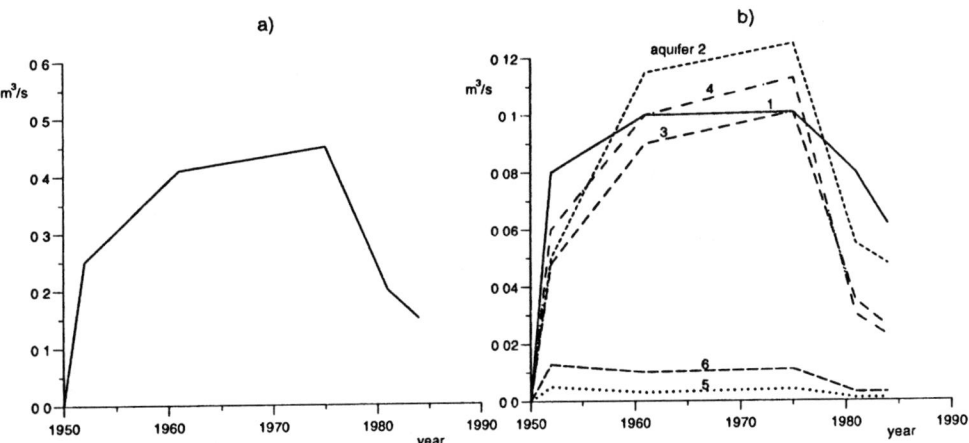

Fig. 9.14. Cumulative pumping rate in the Ravenna area, (a), and pumping rate from the various aquifers from 1950 to 1985, (b), (after Gambolati et al., (1991)).

Fig. 9.15. Contour lines of equal land subsidence (10^{-2} m) in the Ravenna area between 1949 and 1972, (a), 1949 and 1977, (b), and 1977 and 1986, (c), (after Gambolati et al. (1991)).

error introduced by the one-dimensional vertical approximation remains small at the scale of the present analysis. The hydrologic model involves triangular finite elements in the aquifers (Figure 9.12), and convolution integrals in the aquitards, to correctly represent leakage into each aquifer from above and below. For a description of the model, its computer implementation with the available data and boundary conditions, together with the numerical solution, the reader is referred to Gambolati et al. (1986) and (1991).

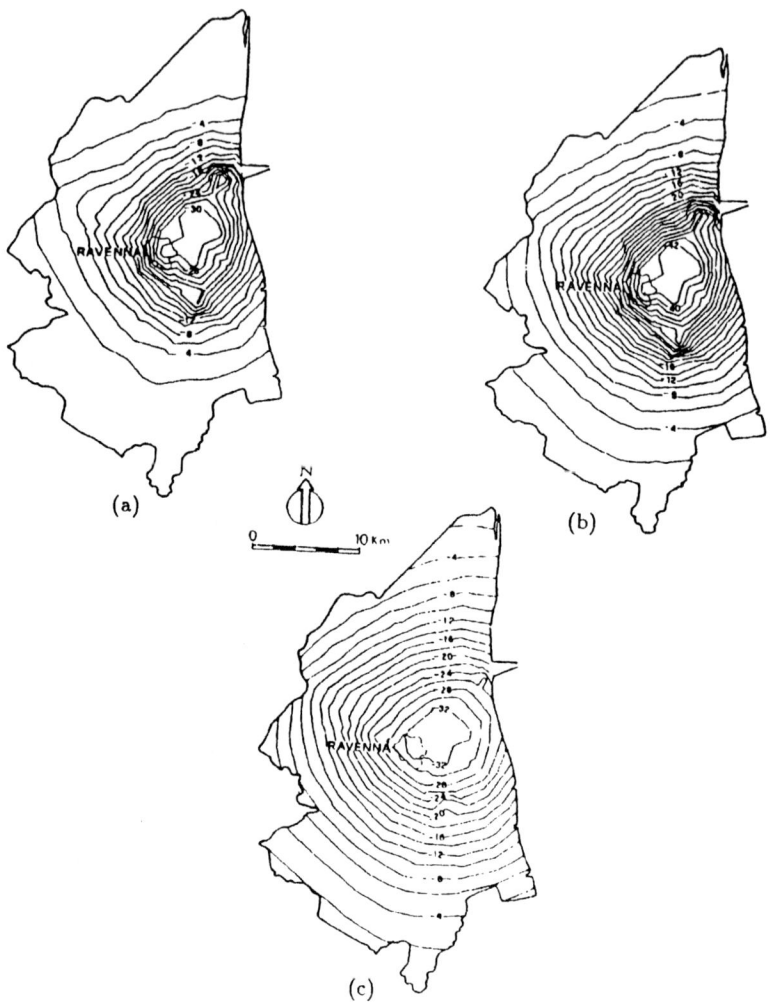

Fig. 9.16. Contour lines of equal drawdown (m) in the Ravenna deepest aquifers (4, 5 and 6) as provided by the three-dimensional model of flow in 1965, (a), 1975, (b), and 1985, (c), (after Gambolati et al., (1991)).

Solution of the regional flow model (Fig. 9.12) provides the regional head field (Fig. 9.16) as well as the aquifer drawdowns at Ravenna (Fig. 9.17). The latter are used as specified boundary conditions in the refined consolidation models of each aquitard. The one-dimensional vertical form of eq. (9.13) is solved by finite differences using a realistic soil representation of each aquitard below Ravenna. If expansion occurs, the clay compressibility α in rebound it taken to be ten times less than the corresponding value in compression (Gambolati et al., 1974).

Figure 9.18 shows the land subsidence behavior at Ravenna caused by water pumping as predicted by the three-dimensional multi-aquifer flow and the one-

Land Subsidence 255

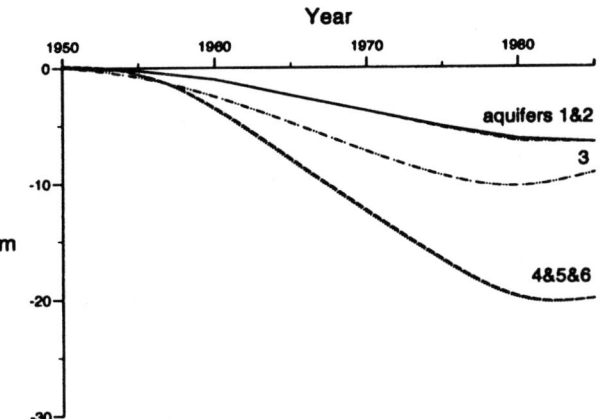

Fig. 9.17. Potential drawdown in the various aquifers at Ravenna from 1950 to 1985 as provided by the three-dimensional model of flow (after Gambolati et al., (1991)).

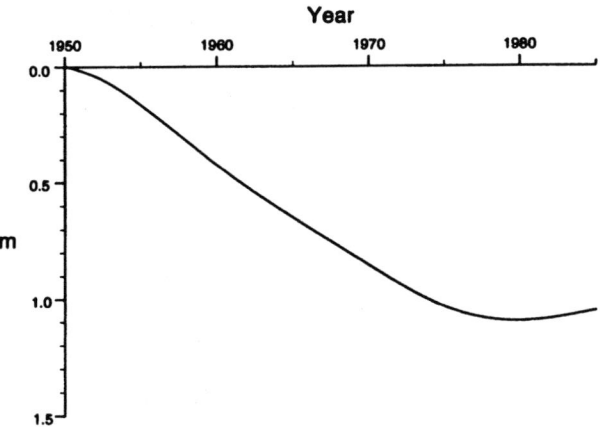

Fig. 9.18. Land subsidence due to groundwater withdrawal at Ravenna from 1950 to 1985 as predicted by the numerical model (after Gambolati et al. (1991)).

dimensional vertical consolidation models, making use of the stratigraphic profile derived from a deep borehole drilled at Ravenna. Note that the largest predicted settlement is 1.1 m in the late 1970s with a slight rebound thereafter. This value is consistent with the contour map of Figure 9.15 and provides evidence that water pumpage is likely to account for most of the land sinking occurred in the city of Ravenna.

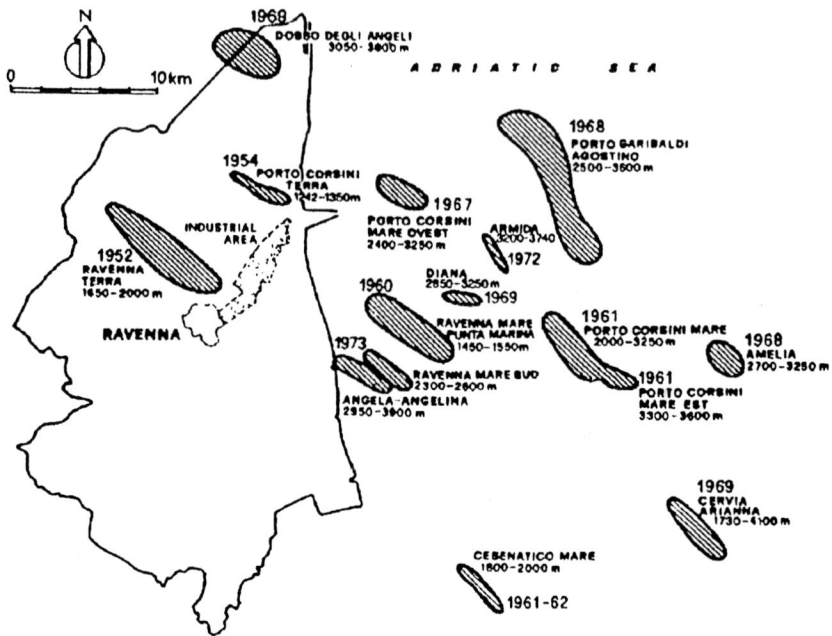

Fig. 9.19. Location of the main gas fields in the Ravenna area and neighboring Adriatic sea (after Gambolati et al. (1991)).

9.3.2 LAND SUBSIDENCE IN THE RAVENNA AREA CAUSED BY GAS PRODUCTION

Most of the gas fields discovered in the Ravenna area after 1952 are located in the Pliocene basement (Figure 9.11(b)). They are most frequently situated offshore, but some are located inland near the coastline (Figure 9.19). Only a limited amount of data is available for the reconstruction of the mechanical properties of the subsurface in this area. The existing information is summarized in Figure 9.2. Note that the number of data points becomes smaller as z increases, and that no data are available below $z = 4000$ m. Also note that α_{clay} is larger than α_{sand} in the upper sediments ($z \leq 1000$ m), with two distinct regression lines for clay and sand in this depth interval. For $z > 1000$ m the soil compressibility is the same irrespective of the lithology. The mean value of Poisson's ratio has been found to be equal to 0.25. The other important physical parameters are the vertical permeability K_z of the clay overlying the reservoir and the hydraulic transmissivity T of the lateral aquifer. For the Ravenna Terra gas field, K_z is evaluated at an order of magnitude of 10^{-12} m/s, i.e. a much smaller value than that of the permeability of the sandy formations (Gambolati et al., 1991). No direct measurements are available for T, which was therefore calculated, using the aquifer model, by requiring that the cumulative water inflow into the reservoir be consistent with the estimate independently made by the gas company (Gambolati, 1996).

In what follows, we will show the results (Gambolati, 1996) from the simulation of the land subsidence over the Dosso degli Angeli field (Figure 9.19), one of the

Land Subsidence

largest and most productive reservoirs in the area. The analysis has been performed with the collaboration of the Municipality of Ravenna and the gas company.

The most salient features of the numerical approach are summarized below. The subsidence model relies on the three-dimensional form of eqs. (9.4). The pressure drawdown p in the reservoir is made available by the gas company, while in the lateral aquifer p is computed by finite elements using the two-dimensional form of eq. (9.13). Dosso degli Angeli is made up of three major pools (P_1, P_2, P_3) located at a mean depth of 3033, 3160 and 3232 m, respectively. Eqs. (9.4) have been solved by the *strain nucleus* or *tension center* approach (Gambolati et al., 1991). Following this procedure, a fundamental solution u_z^* for the land subsidence due to a unit pressure decline in a unit volume reservoir is first computed using a three-dimensional axi-symmetrical finite element model. The overall subsidence is next computed by the formula:

$$u_z(P,t) = \sum_{k=1}^{N} u_z^*(r_k, P) A_k b_k p_k(t) \tag{9.15}$$

where P is an arbitrary point belonging to the ground surface, and N is the number of triangular elements into which P_1, P_2, and P_3, and the respective lateral aquifers have been discretized. The pore pressure drop over element k at time t is indicated by $p_k(t)$, and we have denoted with A_k the area of element k, b_k its thickness, considered constant over the whole element, and r_k the horizontal distance between the centroid of element k and point P. Figure 9.20 provides the fundamental solution u_z^* for the Ravenna subsurface system for a varying depth z in the interval 2900-4000 m. As was anticipated, u_z^* decreases with z for two main reasons: i) the compressibility of the tension center reduces with z (Figure 9.2); ii) the vertical displacement at the ground surface is smaller for a deeper tension center.

Figure 9.21 shows the finite element triangulation of the lateral aquifer surrounding each pool. On the inner aquifer boundary, which coincides with the outer reservoir boundary, $p(t)$ is prescribed, while on the outer aquifer boundary zero pressure is imposed. Pore pressure decline in the reservoir amounts to almost 300 kg/cm^2 over the period 1971-1992, and behaves linearly with time. Since the α values are kept constant over the entire range of the very large pressure variation, land subsidence is probably overestimated by our model. Actually, it is known that α decreases for increasing effective stress. However, no information is at present available about the non linear behavior of the in situ rock system.

The land subsidence over the gas field Dosso degli Angeli, as of 1992, is shown in Figure 9.22. The contributions to the total subsidence of the reservoir and of the lateral aquifer are provided in Figures 9.22(a) and 9.22(b), respectively. Note that the land subsidence due to the aquifer compaction is almost equal to the subsidence caused by the reservoir compaction. The total land settlement over the central area of the reservoir reaches a maximum value of 1 m. The influence of the aquifer is indicated by the spreading of subsidence over a much larger area than the reservoir horizontal section, as can be seen in Figure 9.22(c). Figure 9.22 points out the importance of the lateral aquifer and reveals that the maximum subsidence due to

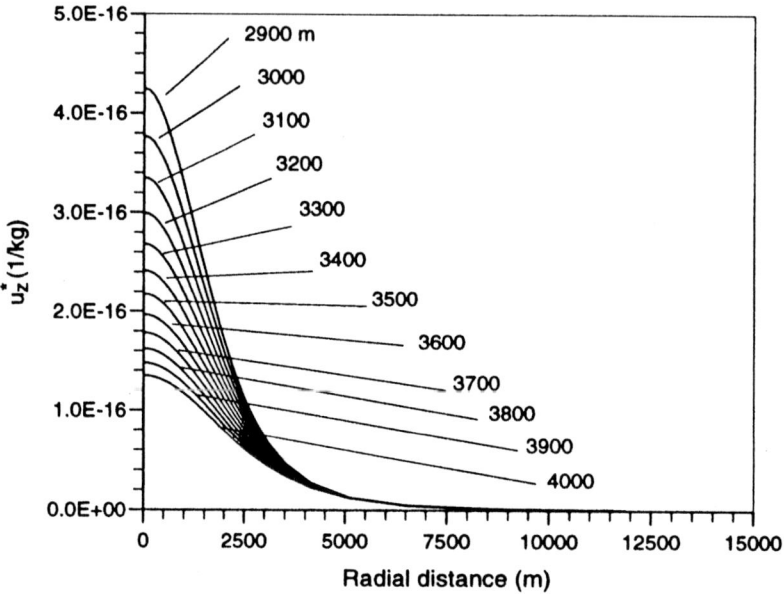

Fig. 9.20. Fundamental solution u_z^* at Ravenna in the depth interval 2900–4000 m.

gas pumping can achieve a very high value, despite the fact that the reservoir is very deep and quite limited in areal extent. Since Dosso degli Angeli is overlain by a lagoon, measured subsidence records over this gas field are missing, and thus a direct comparison of predicted and observed values is not possible. The present analysis, although crude, suggests that land sinking over gas or oil fields can be significant, and its environmental impact can be of great concern for the preservation of coastal areas.

9.3.3 LAND SUBSIDENCE PREDICTION AT MEXICO CITY

Land subsidence at Mexico City is a worldwide known event, thoroughly studied for a long time (Figueroa Vega, 1984). Early analyses started in the late sixties, and important field and theoretical investigations were contributed by Cruickshank et al. (1989), Durazo and Farvolden (1989), Herrera et al. (1989), Ortega and Farvolden (1989) and Rudolph and Frind (1991). The most recent and advanced numerical model of the land subsidence of Mexico City has been developed by Rivera (1990) and Rivera et al. (1991) and is described below.

The numerical simulation relates to the period 1934-1986. Starting in 1934 several wells were drilled in the downtown area and in the northern and western surroundings of the city. Later in the fifties, new wells were drilled in the southern neighborhoods. Since 1934, groundwater withdrawal in Mexico City steadily increased, as is shown in Figure 9.23, and exceeded 21 m^3/s in the early eighties,

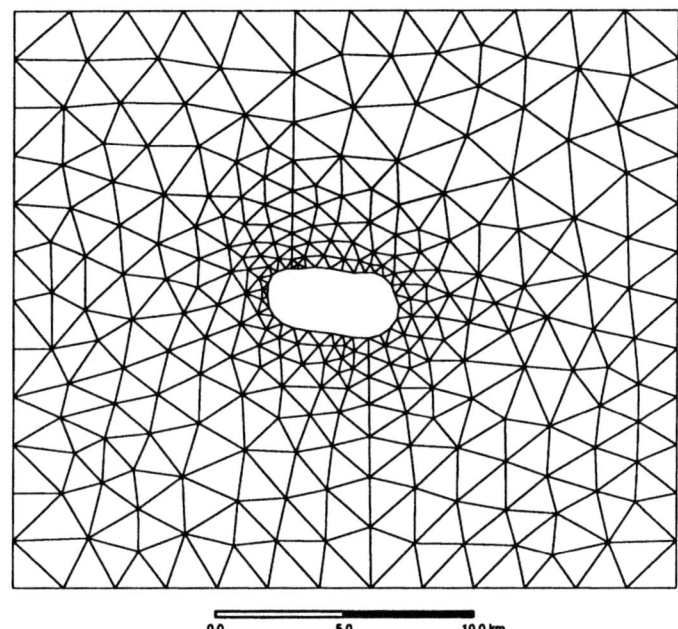

Fig. 9.21. Finite element triangulation of the lateral aquifer. The inner boundary is the outer boundary of the gas reservoir Dosso degli Angeli.

from more than 600 wells. The largest head drawdown, recorded in some of the artesian wells, was over 70 m in 1986. The maximum land settlement amounted to as much as 9 m in some sites, with more than 6 m in the old Mexico City center (Figure 9.24).

The simulated aquifer system comprises a confined aquifer, with the bottom located at 50 m below ground surface, that becomes unconfined when the water level is lower than the aquifer top, and an overlying 44 m thick aquitard made up of lacustrine, highly compressible sediments. The modeling approach is very similar to the one developed for the Ravenna site, and makes use of an uncoupled quasi three-dimensional flow model, combined with a one-dimensional vertical consolidation model. The hydrologic model is solved over a number of regularly nested finite difference grids by the code NEWSAM (for the basic theory underlying NEWSAM see de Marsily et al. (1978)). A perspective view of the modeled one aquifer-one aquitard system discretized into finite differences is provided in Figure 9.25. The aquifer mesh is made up of squares with edges of size varying between 500 and 2000 m, while the vertical aquitard layering is formed by up to 23 nodes at some locations. The subsidence model uses the same aquitard discretiza-

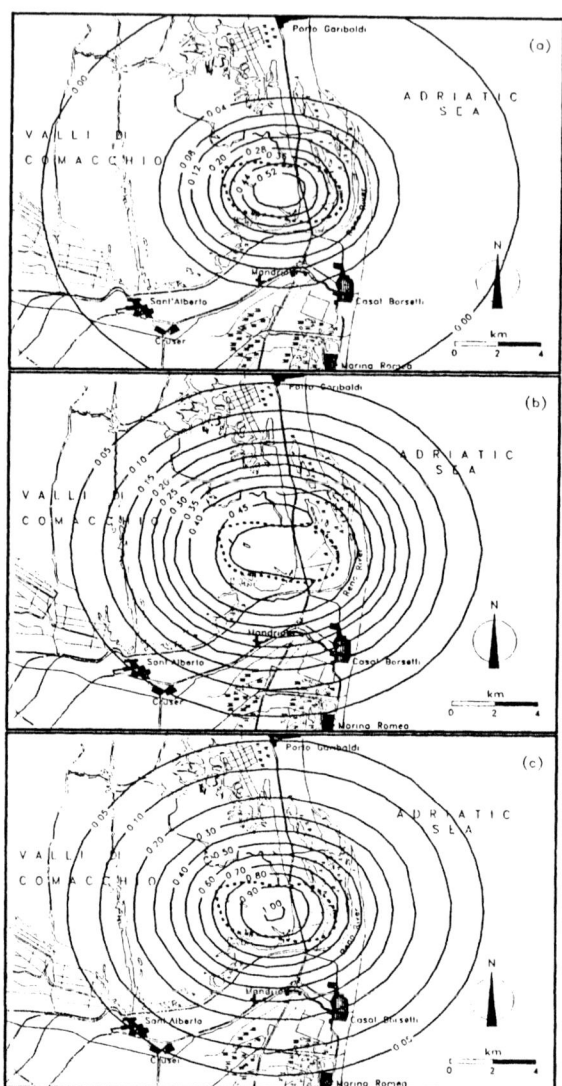

Fig. 9.22. Land subsidence (m) over Dosso degli Angeli gas field in 1992: land subsidence due to the compaction of the reservoir alone, (a); contribution from the lateral aquifer, (b); overall cumulative subsidence, (c). The trace of the reservoir is also shown (after Gambolati et al., 1996).

tion as the flow model, and predicts the land settlement on the basis of eq. (9.14).

The most interesting and innovative feature of the model is its ability to handle non-linear behavior of the soil parameters. Actually, the observed large porosity, soil compressibility, and vertical permeability variations over the range of the aquifer pore pressure drawdown, make the subsidence event of Mexico City a highly non-linear phenomenon. At the end of the consolidation process, α may be

Land Subsidence

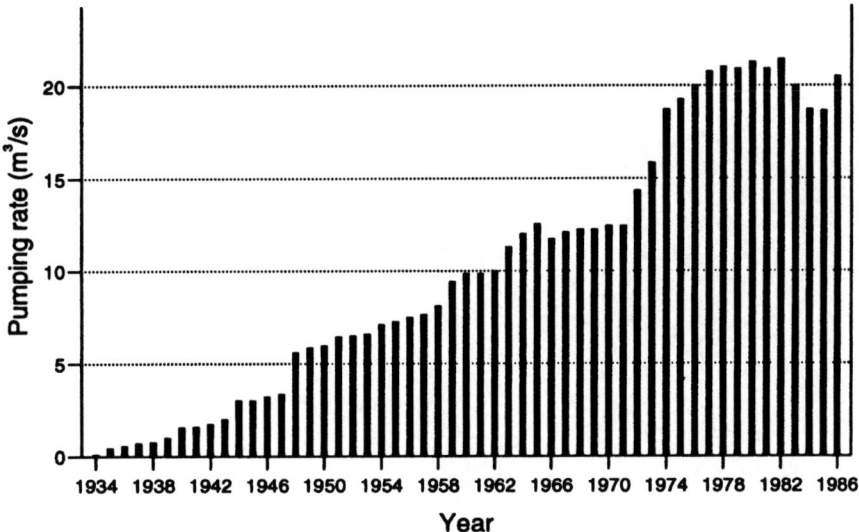

Fig. 9.23. Groundwater pumping in Mexico City for the period 1934-1986 (after Rivera, (1990)).

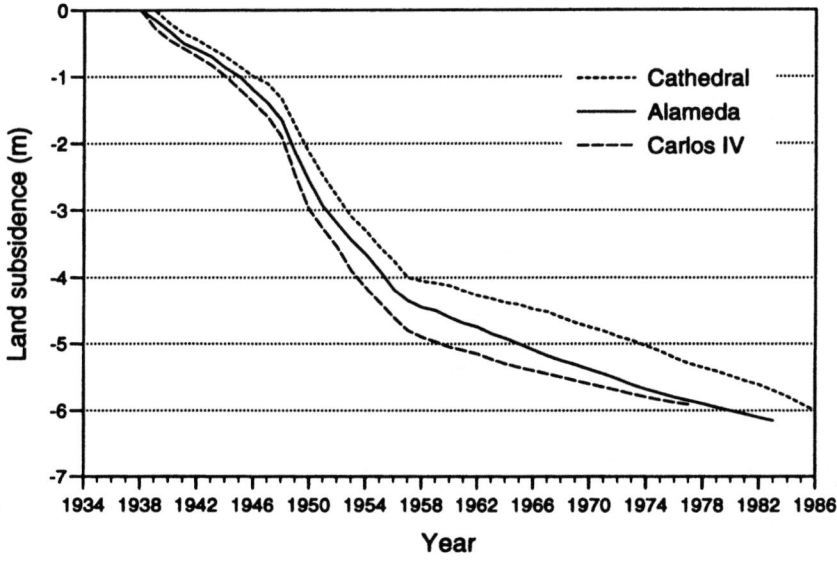

Fig. 9.24. Observed land subsidence at three selected sites in downtown Mexico City (after Rivera, (1990)).

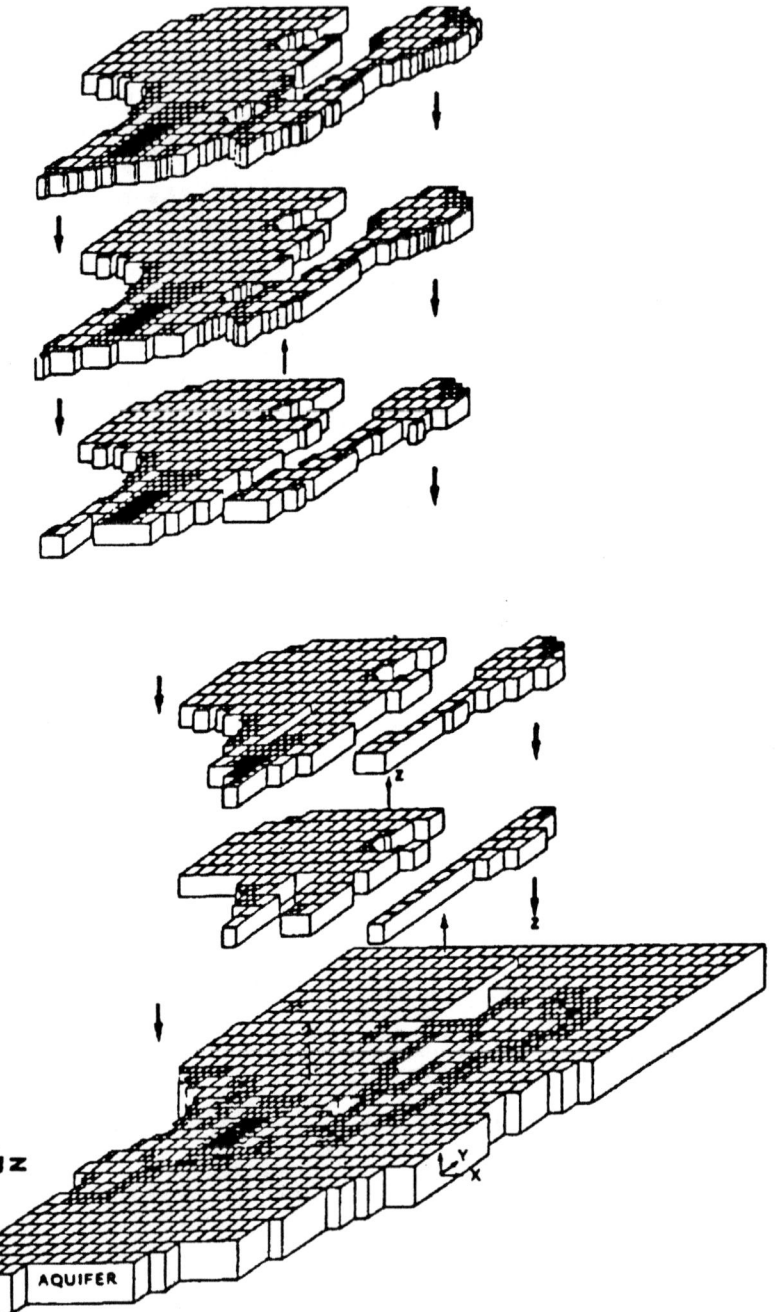

Fig. 9.25. Perspective view of the numerical discrètization by finite differences of the one aquifer - one aquitard system at Mexico City (after Rivera (1990)).

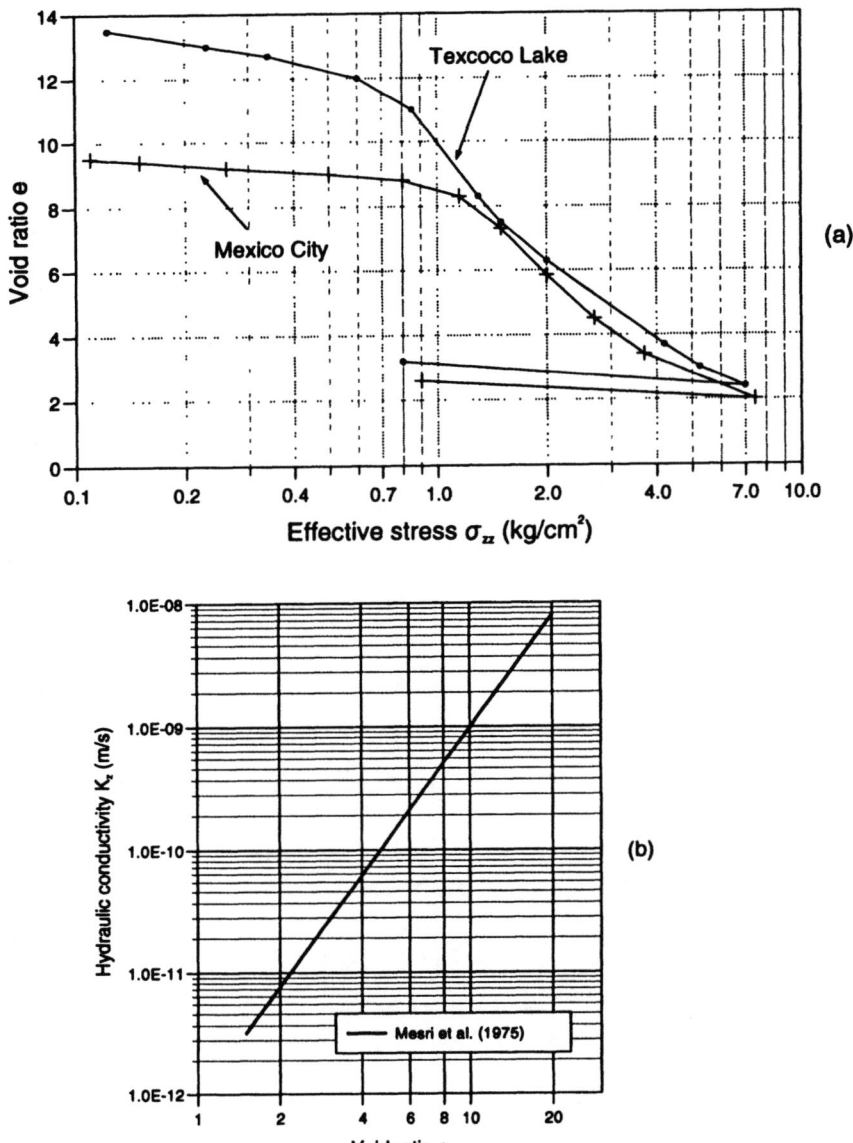

Fig. 9.26. Typical consolidation curve (void ratio–$\log\sigma_{zz}$) of Mexico City clay pointing out a large α-variation, (a), (after Zeevaert, (1953)), and behaviour of K_z versus void ratio $e = n/(1 - n)$ for Mexico City clay, (b), (after Mesri et al., (1975)).

reduced by a factor of 2 or 3, while porosity n is decreased from a maximum value of 0.87 to 0.42, and K_z has varied by 2 orders of magnitude (Figure 9.26). The

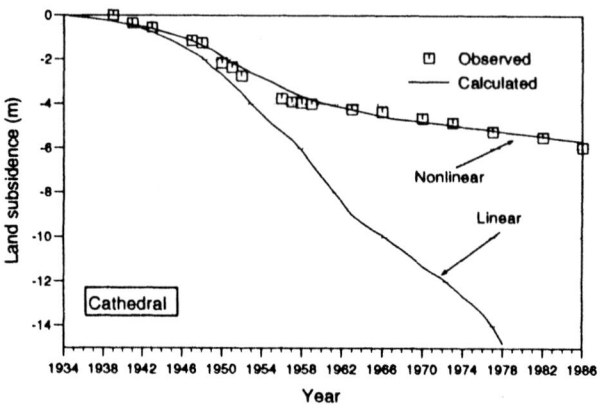

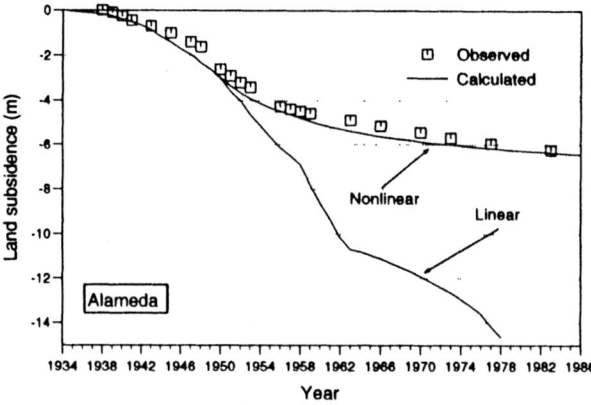

Fig. 9.27. Comparison of simulated and measured land subsidence in Mexico City. The results from the linear model are also shown (after (Rivera, 1990)).

non-linear relationships for α and K_z used by Rivera (1990) in the simulations are:

$$\alpha(e, \sigma_{zz}) = 0.434 \frac{C_c}{(1+e_0)\sigma_{zz}} \qquad \sigma_{zz} > \sigma_{zz,pre}$$

$$K_z = K_{0z} \left(\frac{e}{e_0}\right)^3$$

where $e = n/(1-n)$ is the void ratio, $\sigma_{zz,pre}$ is the preconsolidation effective stress, C_c is the compression index, i.e. the slope of the virgin consolidation straight line-segment of the semi-log plot ($e - \log \sigma_{zz}$) obtained from oedometer tests on clay samples (Figure 9.26a), and e_0 is the initial void ratio. If σ_{zz} is less than $\sigma_{zz,pre}$, or a clay expansion occurs due to a pore pressure increase, C_c in the above equation is replaced by the swelling index C_s, which is smaller than C_c by one order of magnitude (Figure 9.26(a)).

The non-linear flow model is integrated in time with an implicit-explicit scheme. The α coefficient is updated explicitly after each time step Δt ($\Delta t =$ 30 days) while porosity (or void ratio) and vertical hydraulic conductivity are modified after each phase (1 year). Land subsidence is also computed by eq. (9.14) at the end of each phase.

Figure 9.27 shows the observed and the predicted land subsidence in downtown Mexico City obtained with the non-linear model. The agreement is excellent and emphasizes that non-linearity plays an important role in the mechanism of clay compaction in Mexico City. By distinction, Figure 9.27 reveals that the linear model (using the initial compressibility, porosity and permeability) would lead to a settlement prediction which is overestimated by a factor of 2. A smaller discrepancy could be obtained from the linear model with a better choice (e.g. intermediate values) of the controlling parameters. Nevertheless, Rivera's (1990) analysis proves that a non-linear approach is most suited for a reliable prediction of land subsidence due to groundwater withdrawal at Mexico City.

Acknowledgements

This research was supported by the EC Environment Research Programme (contract EV5V-CT94-0498, Climatology and Natural Hazards).

References

AGIP Mineraria, La pianura Padano-Veneta, In: Italia, Geologia e Ricerca Petrtolifera, Enciclopedia Petrologia e Gas Naturali. ENI. Colombo Ed., Milan, 1969.
Allen, D. R., The mechanics of compaction and rebound, Wilmington oil field, Long Beach, California. Technical Report, Dept. of Oil Properties, City of Long Beach, CA, 1969.
Allen, D. R. and M. N. Mayuga, The mechanics of compaction and rebound, Wilmington Oil Field, Long Beach, California, U.S.A., In: Tison, L. J. (ed.) Land Subsidence, Vol.2, pp. 410-423,. IASH Publ. no. 89, 1969.
Allen, S. A., Types of land subsidence, In: Poland, J. (ed.) Guidebook to Studies of Land Subsidence Due to Ground-Water Withdrawal. UNESCO, Paris, pp. 133-142, 1984.
Bear, J. and M. Y. Corapcioglu, Mathematical model for regional land subsidence due to pumping. 2. Integrated aquifer subsidence equations for vertical and horizontal displacement, Water Resour. Res. 17(4), 947–958, 1981.
Bertoni, W., G. Brighanti, G. Gambolati, G. Ricceri and F. Vuillermin, Land subsidence due to gas production in the on-offshore natural gas fields of the Ravenna area, Italy, in: F.B.J. Barends et al. (eds.), Land Subsidence, pp. 13-20. Wallingford, UK, IAHS Publ. No. 234, 1995.
Biot, M. A., General theory of three-dimensional consolidation, J. Appl. Phys. 12, 155–164, 1941.

Bixley, P. F., Case history no.9.9: The Wairakei Geothermal Field, New Zeland, In: Poland, J. (ed.) Guidebook to Studies of Land Subsidence Due to Ground-Water Withdrawal. UNESCO, Paris, 1984.
Booker, J. R. and J. C. Small, An investigation of the stability of numerical solutions of Biot's equations of consolidation, Int. J. Solids Struct. 11, 907–917, 1975.
Boot, R., Level control surveys in the Gröningen Gasfield, Verhandelingen Kon. Ned. geol. mijnouwk. Gen. 28, 105–109, 1973.
Brutsaert, W. and M. Y. Corapcioglu, Pumping of aquifer with visco-elastic properties, J. Hydraul. Div. ASCE. 102(11), 1663–1675, 1976.
Chierici, G. L., La subsidenza, In: Subsidenza naturale e da estrazione di fluidi dal sottosuolo. AGIP, Milan, 1992.
Christian, J. J. and J. W. Boehmer, Plain strain consolidation by finite elements, J. Soil. Mech. and Found. Div. ASCE. 96, 1435–1457, 1970.
Cruickshank, V. C., I. Herrera, R. Yates, J. Hennart, D. Balazero and R. Magana, Modelos de prediccion del hundimiento del subsuelo de Valle de Mexico. Proyecto 9138, Inst. de Ingenieria, UNAM, Mexico City, 1989.
de Marsily, G., E. Ledoux, A. Levassor, D. Poitrinal and A. Salem, Modeling of large multiaquifer systems: Theory and application, J. Hydrol. 36, 1–33, 1978.
Debbarh, A., Consolidation of Elastic Saturated Soil due to Water Withdrawal by Numerical Laplace-Fourier Inversion Methods. PhD thesis, Dept. of Civil and Min. Eng., Univ. of Minn., Minneapolis, 1988.
Desai, C. S., Analysis of consolidation by numerical methods, In: Proc. Symp. on Recent Developments in the Analysis of Soil Behavior and Application to Geotechnical Structures. Univ. of New South Wales, Sidney, 1975.
Durazo, J. and R. Farvolden, The groundwater regime of the Valley of Mexico from historical evidence and field observation, J. Hydrol. 112, 171–190, 1989.
Figueroa Vega, G. E., Case history no.9.8: Mexico, D.F., Mexico, In: Poland, J. (ed.) Guidebook to Studies of Land Subsidence Due to Ground-Water Withdrawal. UNESCO, Paris, pp. 217-232, 1984.
Finol, A.S. and Z.A. Sansevic, Subsidence in Venezuela, in: G. Chilingarian et al. (eds.), Subsidence due to Fluid withdrawal, Developments in Petroleum Science, 41, 337-378, Elsevier, Amsterdam, 1995.
Gabrysch, R. K., Case history no.9.12: The Huston-Galveston Region, Texas U.S.A., In: Poland, J. (ed.) Guidebook to Studies of Land Subsidence Due to Ground-Water Withdrawal. UNESCO, Paris, pp. 253-262, 1984.
Gambardella, F., S. Bortolotto and M. Zambon, The positioning systems GPS for subsidence control of the terminal reach of the Po river, In: IV Int. Symposium on Land Subsidence. IASH Publ. no. 200, pp 433–441, 1991.
Gambolati, G., Deviations from Theis solution in aquifers undergoing three-dimensional consolidation, Water Resour. Res. 13(1), 62–68, 1977.
Gambolati, G., Comment on "Coupling versus uncoupling in soil consolidation", Int. J. Numer. Analytic. Methods Geomech. 16(11), 833–837, 1992.
Gambolati, G. and R. A. Freeze, Mathematical simulation of the subsidence of Venice. 1. Theory, Water Resour. Res. 9(3), 721–733, 1973.
Gambolati, G., P. Gatto and R. A. Freeze, Mathematical simulation of the subsidence of Venice. 2. Results, Water Resour. Res. 10(3), 563–577, 1974.
Gambolati, G., A. M. Perdon and G. Ricceri, A coupled finite element model of flow in porous layered media, In: Laible, J. P., C. A. Brebbia, W. G. Gray and G. Pinder (eds.) 5th Int. Conf. Finite Elements Water Resources. Springer-Verlag, Berlin, 1984.
Gambolati, G., F. Sartoretto and F. Uliana, A conjugate gradient finite element model of flow for large multiaquifer systems, Water Resour. Res. 22(7), 1003–1015, 1986.

Gambolati, G., G. Ricceri, W. Bertoni, G. Brighenti and E. Vuillermin, Mathematical simulation of the subsidence of Ravenna, Water Resour. Res. 27(11), 2899–2918, 1991.

Gambolati, G., P. Teatini and W. Bertoni, Numerical prediction of land subsidence over Dops degli' Angli gas field, Ravenna, Italy, in: J. Borchers (ed.), Current Research and Case Studies on Land Subsidence, (Proc. J.F.Poland Symp. on Land Subsidence, Sacramento, CA, October 1995, AZG Publ., 1996.

Geertsma, J., Problems of rock mechanics in petroleum production engineering, In: Proc. 1st Cong. Int. Soc. Rock Mechan. Lisbon, 1966.

Geertsma, J., Land subsidence above compacting oil and gas reservoir, J. Pet Technol. 25, 734–744, 1973.

Geertsma, J., On the alert for subsidence, AGIP Review. 6, 39–43, 1989.

Gloe, C. S., Case history no.9.1: Latrobe Valley, Victoria, Australia, In: Poland, J. (ed.) Guidebook to Studies of Land Subsidence Due to Ground-Water Withdrawal. UNESCO, Paris, pp. 145-153, 1984.

Harris Galveston Coastal Subsidence District, Houston, Texas, 68 pp., Water Management Study: Phase 2 and Supplement 1. 1982.

Helm, D. C., One-dimensional simulation of aquifer system compaction near Pixley, California. 1. Constant parameters, Water Resour. Res. 11(3), 465–478, 1975.

Helm, D. C., One-dimensional simulation of aquifer system compaction near Pixley, California. 2. Stress-dependent parameters, Water Resour. Res. 12(3), 375–391, 1976.

Helm, D. C., Three-dimensional consolidation theory in terms of the velocity of solids, Geotechnique. 37, 369–392, 1987.

Herrera, I., R. Martinez and G. Hernandez, Contribución para la adimistración científica del agua subterránea de la Cuenca de México, In: Symposium: El systema acuífero de la Cuenca de México. Geofisica International, Mexico City. Volume 28, pp 297–334, 1989.

Holzer, T. L., Preconsolidation stress of aquifer systems in areas of induced land subsidence, Water Resour. Res. 17(3), 693–704, 1981.

Hwang, C. T., N. R. Morgenstern and D. T. Murray, On solutions of plane strain consolidation problem by finite element methods, Can. Geotech. J. 8, 109–119, 1971.

Jones, M. E., Compaction analysis of the Ekofisk Field. Technical Report, Sediment Deformation Group, University College of London, Dept. of Geological Sciences, 1985.

Lewis, R. W. and B. A. Schrefler, The Finite Element Method in the Deformation and Consolidation of Porous Media. J. Wiley, New York, 1987.

Lewis, R. W., B. A. Schrefler and L. Simoni, Coupling versus uncoupling in soil consolidation, Int. J. Numer. Analytic Methods Geomech. 15(8), 533–548, 1991.

Luxiang, S. and B. Manfang, Case history no.9.2: Shangay, China, In: Poland, J. (ed.) Guidebook to Studies of Land Subsidence Due to Ground-Water Withdrawal. UNESCO, Paris, pp. 155-160, 1984.

Martin, J. C. and S. Serdengecti, Subsidence over oil and gas field, In: Holzer, T. (ed.) Man-Induced Land Subsidence, Rev. Eng. Geol., vol. VI. Geol. Soc. of America, Boulder (CO), 1984.

Mesri, G., A. Rokhsar and B. Bohor, Composition and compressibility of typical samples of Mexico City clay, Geotechnique. 25(3), 527–554, 1975.

Narasimhan, T. N. and K. P. Goyal, Subsidence due to geothermal fluid withdrawal, In: Holzer, T. (ed.) Man-Induced Land Subsidence, Rev. Eng. Geol., vol. VI. Geol. Soc. of America, Boulder (CO), pp. 35-66, 1984.

Nuñez, O. and D. Escojido, Subsidence in the Bolivar Coast, In: 2nd Int. Symp. on Land Subsidence. IASH Publ. no. 121, Anaheim (CA), pp. 257-266, 1976.

Ortega, A. and R. Farvolden, Computer analysis of regional groundwater flow and boundary conditions in the Basin of Mexico, J. Hydrol. 110, 271–294, 1989.

Poland, J. F. and B. E. Lofgren, Case history no.9.13: San Joaquin Valley, California, U.S.A., In: Poland, J. (ed.) Guidebook to Studies of Land Subsidence Due to Ground-Water Withdrawal. UNESCO, Paris, pp. 263-277, 1984.
Pratt, W.E. and D.W. Johnson, Local subsidence of the Goose Creek oil field, J. Geol. XXXIV(7, Part.1), 577–590, 1926.
Rivera, A., Modèle hydrogéologique quasi-tridimensionnel non-linéaire pour simuler la subsidence dans les systèmes aquifères multicouches. Cas de Mexico. PhD thesis, Ecole des Mines de Paris-CIG, 1990.
Rivera, A., E. Ledoux and G. de Marsily, Non-linear modeling of groundwater flow and total subsidence of the Mexico City aquifer-aquitort system, in: Land Subsidence, pp. 44-58, IAHS Publ. no. 200, 1991.
Rudolph, D. and E. O. Frind, Hydraulic response of highly compressible aquitards during consolidation, Water Resour. Res. 27(1), 17–30, 1991.
Sandhu, R. S., Finite element analysis of soil consolidation. Geotech. Engng. Rep. to NSF 6, The Ohio State University, 1976.
Sandhu, R. S. and E. L. Wilson, Finite element analysis of seepage in elastic media, J. Eng. Mech. Div. ASCE. 95, 641–652, 1969.
Sandhu, R. S., H. Lin and K. J. Singh, Numerical performance of some finite element schemes for analysis of seepage in porous elastic media, Int. J. Numer. Analytic. Methods Geomech. 1, 177–194, 1975.
Schiesano, G., Il bacino gassifero polesano: una indagine- una proposta, L'Industria Mineraria. 5, 15–23, 1980.
Smith, I. M. and R. Hobbs, Biot analysis of consolidation beneath embankments, Geotechnique. 26, 149–161, 1976.
Theis, C.V., The relationship between the lowering of the piezometric surface and the rate and duration of discharge of a well using groundwater storage, Eos. Trans. AGU. 16, 519–524, 1935.
Teatini, P., G. Gambolati and L. Tosi, A new 3-D non-linear model of the subsidence of Venice, in: F.J.B. Barends et al. (eds.), Land Subsidence, pp. 353-361, Wallingford, UK. IAHS Publ. No. 234, 1995.
Vermeer, P. A. and A. Verruijt, An accuracy condition for consolidation by finite elements, Int. J. Numer. Analytic. Methods Geomech. 5, 1–14, 1981.
Verruijt, A., Elastic storage of aquifers, In: De Wiest, R. (ed.) Flow Through Porous Media. Academic Press, New York, pp 331–376, 1969.
Verruijt, A., Generation and dissipation of pore water pressure, In: Gudehus, G. (ed.) Finite Elements in Geomechanics. John Wiley, London, pp. 293-317, 1977.
Yamamoto, S., Case history no.9.4: Tokyo, Japan, In: Poland, J. (ed.) Guidebook to Studies of Land Subsidence Due to Ground-Water Withdrawal. UNESCO, Paris, pp. 175-184, 1984.
Zaman, M.M., A. Abdulraheem and J.C. Roegiers, Reservoir compaction and surface subsidence in the North Sea Ekofisk field, in: G. Chilingarian et al. (eds.), Subsidence due to Fluid Withdrawal, Developments in Petroleum Science, 41, pp. 373-423, Elsevier Amsterdam. 1995.
Zeevaert, L., Outline of the stratigraphical and mechanical characteristics of the unconsolidated sedimentary deposits in the basin of the Valley of Mexico, In: IV INQUA Congress. Rome, pp. 976-987, 1953.

CHAPTER 10

Saltwater Intrusion

M.M. Sherif and V.P. Singh

ABSTRACT. Saltwater intrusion is a serious menace to the groundwater quality of several coastal aquifers around the world. Fundamental to controlling or reducing this intrusion is the determination of its spatial and temporal distribution. The sharp-interface and the density-dependent approaches are utilized for modeling saltwater intrusion. This chapter reviews these approaches, discusses the mechenisms of saltwater intrusion, and formulates the governing equations along with initial and boundary conditions. With an intoduction of finite-difference and finite-element methods, a finite-element formulation for density-dependent problems is presented. The chapter is concluded with a discussion of three case studies.

10.1 Hydrological Aspects

A coastal aquifer is bounded at least from one end by an extensive saltwater body such as a salt lagoon or a sea or an ocean. Due to the direct hydraulic link between the freshwater in the aquifer and the former saltwater body, such an aquifer is usually threatened by the encroachment of saline water and, therefore, degradation of the water quality in the aquifer. A two to three percent mixing with seawater renders freshwater inadequate for human consumption. A five percent mixing is enough to destroy a freshwater resource.

In many coastal areas, especially in arid and semi-arid regions, groundwater is the only source of freshwater. The flow of the freshwater towards the sea limits the land-ward intrusion of seawater. The growth of population in these areas and hence the conjugate increase in human activities, agricultural and industrial, has imposed an increasing demand for freshwater. This increase in demand is often covered by extensive pumping of fresh groundwater, causing subsequent lowering of the water table or piezometric head and upsetting the dynamic balance between the freshwater body and the saline water body. The classical result of such a development is saltwater intrusion. This phenomenon has been reported in several parts of the globe. At least in 20 of the coastal states in the United States, the invasion of saltwater has resulted in degradation of the freshwater in the aquifers

(Newport, 1977). The best reported cases in the United States include many parts of California (California Department of Water Resources, 1958), Long Island (Lusczynski and Swarzenski, 1966 and Pinder, 1973), and Miami (Kohout, 1961 and Segol and Pinder, 1976). Other cases reported are in The Netherlands (Ernest, 1969), Israel (Schmorak, 1967), Senegal (Debuisson, 1970) and India (Sherif et al., 1990). A disaster case of saltwater intrusion is reported in the Nile Delta aquifer in Egypt where the seawater intruded inland to a distance of 35.0 km from the shore line (Sherif et al., 1988).

The shape and degree of saltwater intrusion in a coastal aquifer depend on several factors, including the type of the aquifer (confined, Phreatic, leaky, or multi-layer) and its geology and geometry, water table and/or piezometric head, seawater concentration and density, natural rate of flow, capacity and duration of water withdrawal or recharge, rainfall intensities and frequencies, rates of evaporation, physical and geometric characteristics of the porous media, geometric and hydraulic boundaries, tidal effects, variations in barometric pressure, earth tides, earthquakes and other vibrational effects, water wave actions, and chemical changes (Walton, 1970). The depth of the aquifer at the seaside through which the saltwater intrudes inland ultimately affects the degree of intrusion (Sherif et al., 1990).

Pumping programs in coastal areas must be carefully studied and established to prevent disasters of freshwater degradation. On the other hand, it is also recommended to provide productive coastal aquifers with an adequate dynamic monitoring network along the shore line to monitor any variation in groundwater quality. Monitoring should include water level, water abstraction and electrical conductivity as well as chloride, sulphate, sodium and calcium ion concentration. As possible vertical flow in the observation wells may alter the vertical distribution of salinity, it is recommended to prefer separate observation wells with short screens in the monitoring network. Intrusion of saltwater due to intensive pumping is not recognized timely, effects may lag behind the cause. These records should be analyzed on a regular basis to prevent any hazardous situation. In some cases, it may be necessary to stop pumping and relax it to allow a freshwater recharge of the aquifer.

10.2 Sharp Interface and Density Dependent Approaches

When dealing with saltwater intrusion problems, two different approaches can be employed. The sharp interface approach and the density dependent approach. The basic concept and even the governing equations are totally different in the two approaches. Each one is appropriate under certain conditions.

10.2.1 SHARP INTERFACE APPROACH

In this approach, the freshwater in the aquifer and the saltwater body are assumed to be two immiscible fluids, like water and oil. Since the density of the freshwater is 1.0 gm/cm^3 and that of the seawater is 1.025 gm/cm^3 on average, the freshwater body would float above the saltwater body. The boundary between the two water

Saltwater Intrusion

bodies is known as the interface or the wedge, as shown in Figure 10.1. An interface is mainly developed in an undisturbed aquifer which has not yet been used and unaffected by artificial causes (Kashef, 1975). If the freshwater-saltwater interface has a very steep slope, then it is called a sharp front. The point at which the interface intersects the bottom of the aquifer is defined as the toe point (T.P) of the interface. If the interface does not intersect the bottom, the fresh water will have a shape of an optical lens above the saline water. The upper point on the interface is defined as top tip point (T.T.P). The interface will be an exterior one if the T.T.P. lies on the sea boundary, as shown in Figure 10.1a. Otherwise it is an interior interface, as shown in Figure 10.1b.

An interface in a confined coastal aquifer must be an exterior one to allow for freshwater discharge to the open sea. The segment on the sea boundary through which the freshwater discharges into the sea is called the window or the seepage face, below which the saline water intrudes the aquifer, as shown in Figure 10.1a. In semi-confined (leaky) coastal aquifers, the location of the interface depends mainly on the free water table and aquifer piezometric head. If the water table is higher than the piezometric head, the interface must be exterior, otherwise it may be either exterior or interior. An interior interface may exist in semi-confined coastal aquifers where the piezometric head is higher than the free water table at the sea boundary, as shown Figure 10.1b.

In phreatic aquifers, the interface may be either exterior or interior depending on the hydraulic parameters and freshwater discharge to the sea. Discontinuous interface would exist in a multilayered coastal aquifer system, as shown in Figure 10.1c.

The Ghyben-Herzberg Relation. At the turn of the century, it was generally thought that saltwater in coastal aquifers occurred at a depth approximating sea level (Domenico and Schwartz, 1990). Badon-Ghyben (1888) and Herzberg (1901) independently from each other investigated the shape and position of the sharp interface under static equilibrium conditions. The weight of a column of freshwater extending from the interface to the free water table is simply balanced by the weight of a column of seawater extending from the same point to the sea level. With the nomenclature of Figure 10.2, we can write:

$$\gamma_s h_s = \gamma_f (h_s + h_f)$$

from which

$$h_s = \left(\frac{\gamma_f}{\gamma_s - \gamma_f}\right) h_f = \delta h_f, \tag{10.1}$$

where h_s is the depth of the interface below the sea level (L), γ_f and γ_s are the specific weights of freshwater and saltwater ($ML^{-2}T^{-2}$), respectively, and h_f is the height of the free water table above the sea level (L). For example, for $\gamma_s = 1.025$ gm/cm^3, $\delta=40$, and $h_s = 40h_f$. In other words, at any distance from the shore line, a one meter height of the water table above the sea level ensures a 40.0 m of freshwater below the sea level. The slope of the interface is 40 times greater than that of the water table. If extractions of groundwater from a coastal

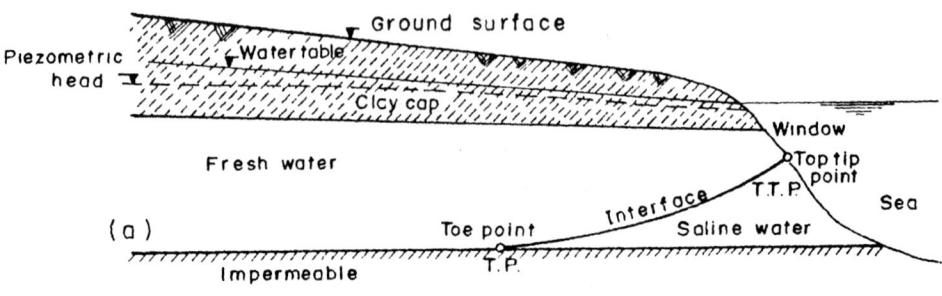

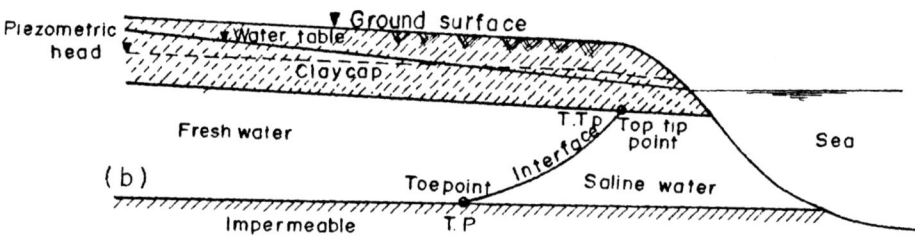

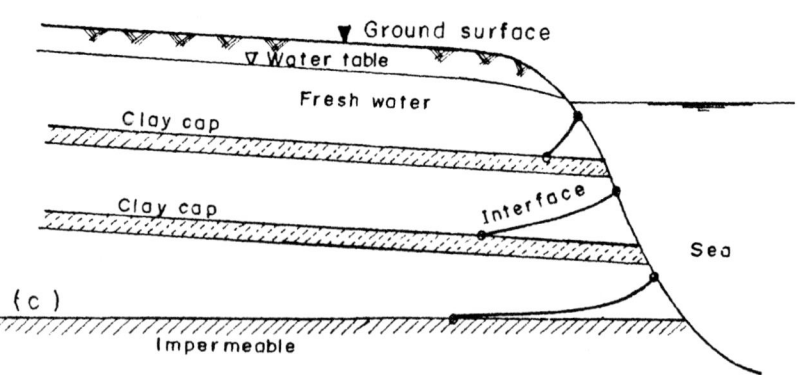

Fig. 10.1. Seawater–freshwater sharp interface. (a) Exterior interface. (b) Interior interface. (c) Discontinuous interface.

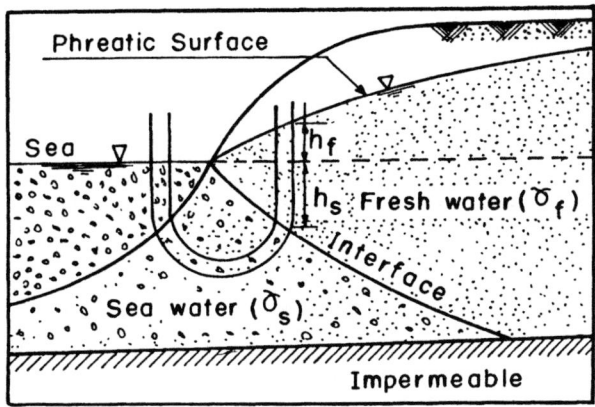

Fig. 10.2. Ghyben-Herzberg relation (Bear, 1979).

aquifer cause a drop of 10.0 cm in the free water table, the saltwater interface will rise 4.0 m.

The flux, q, of freshwater per unit width of a confined aquifer can be calculated from Darcy's law as $q = Kib$, where K is the hydraulic conductivity (LT^{-1}), i is the hydraulic gradient (dimensionless), and b is the thickness of the confined aquifer (L). Therefore (Domenico and Schwartz, 1990),

$$q = K(\frac{\psi_f - 0}{x})b, \qquad (10.2)$$

where ψ_f is the freshwater piezometric head above the toe point, and x is the distance between the shore line and the toe point. From Ghyben-Herzberg relation, we have

$$\psi_f = (\frac{\rho_s - \rho_f}{\rho_f})b = \rho_r b, \qquad (10.3)$$

where $\rho_r = (\rho_s - \rho_f)/\rho_f$ is the relative density (dimensionless). Substituting into equation (10.2), one can write

$$x = \frac{\rho_r K b^2}{q}, \qquad (10.4)$$

From the above equation, it can be concluded that the intrusion length, x, is directly proportional to the relative density, hydraulic conductivity, and saturated thickness squared and inversely proportional to the flux of freshwater to the sea. For the case of unconfined aquifers, the Dupuit-Ghyben-Herzberg model of one-dimensional flow yields the following expression for the intrusion length, x

$$x = \frac{\rho_r K b^2}{2q}. \qquad (10.5)$$

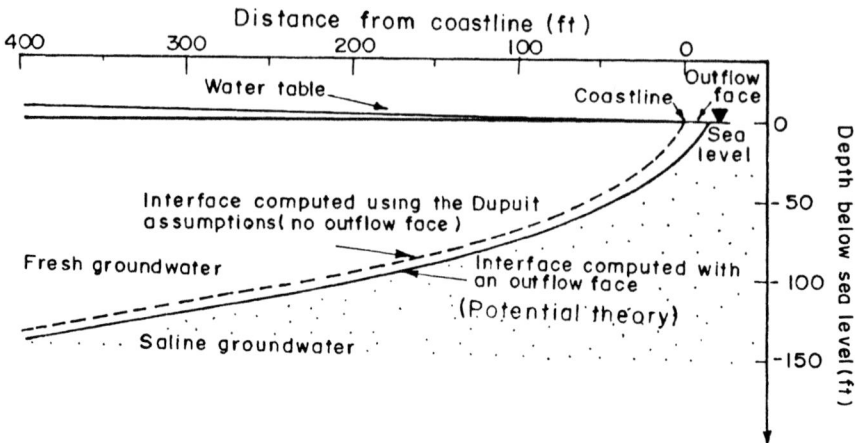

Fig. 10.3. Position of the interface as computed from potential theory versus Ghyben-Herzberg relation (Fetter, 1972).

where b is the saturated thickness at the toe point.

One of the weaknesses of the Ghyben-Herzberg relation is that the saltwater interface intercepts the water table at the shore line. The zero freshwater and saltwater heads at this line, along with the rigid interface assumption through which no flow exist, closes the system at its discharge point. In other words, the freshwater flux to the sea is not allowed. However, in a study of saltwater intrusion in eastern Long Island, Fetter (1972) showed that even at the coast line where the greatest deviations from the Dupuit assumptions may exist, the difference in the position of the interface as computed from potential theory versus Ghyben-Herzberg relation was very small, as shown in Figure 10.3. Bear and Dagan (1962) investigated the accuracy of the Ghyben-Herzberg relation. For a confined horizontal aquifer with constant thikness D, the error in the location and length of the interface was less than 5%, provided that $\pi K'D/Q$ is greater than 8, where $K' = K(\gamma_s - \gamma_f)/\gamma_f$ and Q is the fresh water discharge to the sea. Generally, as the coast is approached, the depth of the interface is more than that given by the Ghyben-Herzberg relation (Bear, 1972).

Glover (1964) presented the following approximate equation for the shape of interface considering a seepage face at the sea boundary to allow for freshwater flux, as shown in Figure 10.4:

$$h_s^2 = \frac{2qx}{K\rho_r} + \left(\frac{q}{K\rho_r}\right)^2, \tag{10.6}$$

where x is the distance measured positive inland from the shore. Point A on the interface (Figure 10.4) indicates the actual depth of the interface at that distance from the shore line. Point B corresponds to the Ghyben-Herzberg approximation. The length of the seepage face, W, through which freshwater seeps into the sea can be obtained by setting h_s to equal zero in equation (10.6),

Saltwater Intrusion

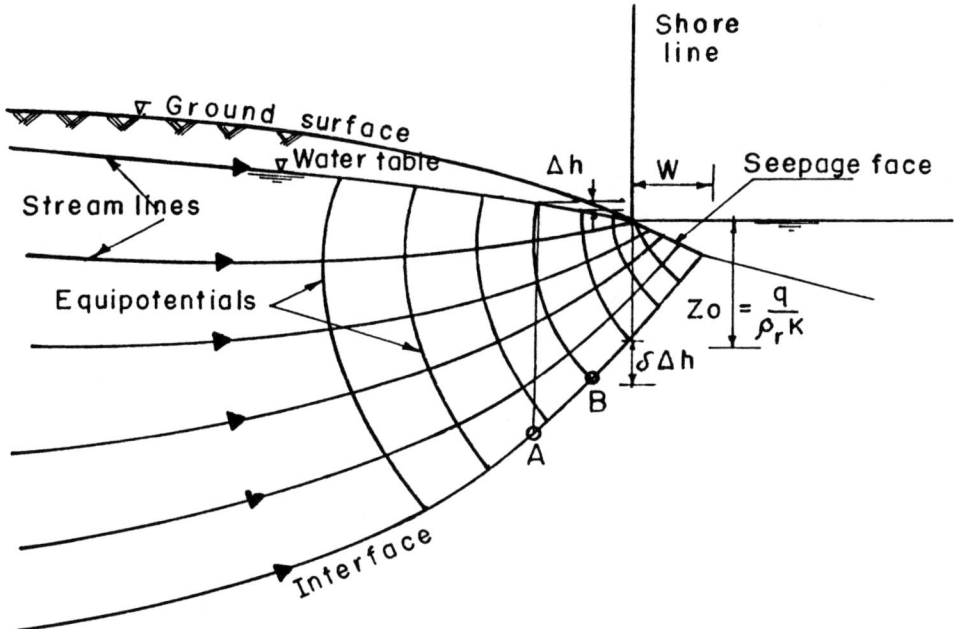

Fig. 10.4. Interface and seepage face in coastal aquifers (Glover, 1964).

$$W = -\frac{q}{2\rho_r K}. \tag{10.7}$$

If x is set equal to zero in equation (10.6), one can obtain the depth of the interface, Z_o, at the shore line as

$$Z_o = \frac{q}{\rho_r K}, \tag{10.8}$$

i.e. at the shore line the depth of the interface is twice the seepage face.

To predict the intrusion length, x, we replace h_s in equation (10.6) by the saturated thickness of the aquifer, b. Therefore,

$$x = \frac{1}{2}\left(\frac{\rho_r K b^2}{q} - \frac{q}{\rho_r K}\right). \tag{10.9}$$

The first term in the brackets is identical to equation (10.5) based on the hydrostatic condition, whereas the second term is equal to Z_o. The role of freshwater flux to the sea as a major control on retarding the advance of saltwater is very clear in equation (10.9).

When applying the sharp interface approach, one should notice that the freshwater head should be measured above the local mean sea level. It may differ substantially from the topographic datum. Even if it coincides in some places with the mean sea level, there are latitudinal effects, marine currents, and coastal effects that may produce deviations of several centimeters at a given coastal point

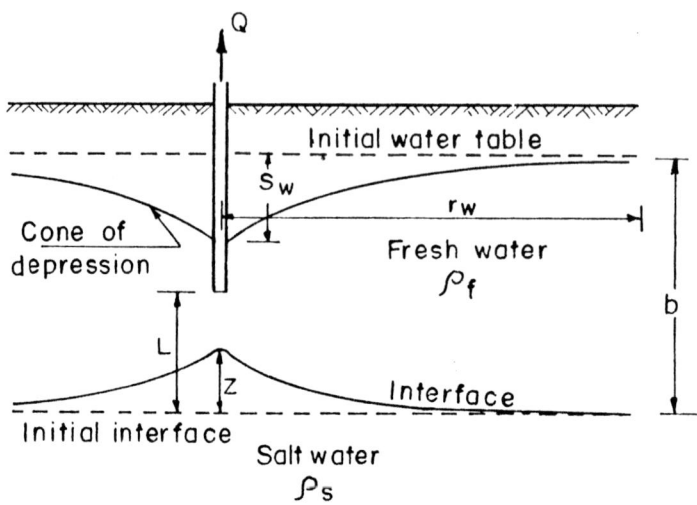

Fig. 10.5. Saltwater upconing.

of a country. Poor levelling, modification of the references and naturally induced subsidence add to this effect. In areas of wide sea tide it is common to place the datum at the lowest stage of the mean equinoccial tide, which sometimes may be more than 1.0 m lower than local mean sea level (Custodio, 1993).

Saltwater Upconing. In coastal areas, productive wells may pump from a fresh water strata which is underlain by saline water strata. As pumping is initiated, the water table or the piezometric head decreases, forming the cone of depression. The horizontal interface between freshwater and saline water rises (in the form of mound) at first slowly and then faster toward the pumping well, as shown in Figure 10.5. Under certain conditions, a state of dynamic equilibrium will be attained before the upconed interface reaches the pumping well. When the pumping rate is raised from one certain (steady) value to another higher (steady) one, yet below the critical pumping rate, a new higher interface is established following a transition period. If the bottom of the well is relatively close to the saline water strata or the well discharge exceeds a certain critical value, the saltwater cone will reach the pumping well, causing the well discharge to be brackish. The exact shape of the upconed interface and the critical pumping rate beyond which saltwater will enter the pumping well are dependent on many parameters, including the rate of accretion, aquifer permeability, freshwater flux above the toe of the interface, distance between well bottom and initial interface, and aquifer thickness (Bear, 1972). When pumping stops, the upconed interface undergoes decay toward the initial steady state horizontal interface. Mean time freshwater flux will displace the upconed interface as it rises and decays (Bear, 1972).

The upconing is in response to the pressure reduction due to drawdown around the well. If the pumping rate exceeds the critical level, the saline water cone reaches the bottom of the well with a sudden jump, indicating conditions of instability. One should notice that saltwater intrusion may occur without any pumping, and the

Saltwater Intrusion

dispersion phenomenon will take place; whereas saltwater upconing exists only under severe pumping conditions.

In a steady horizontal fresh water flow field, the height of the cone below the well center Z_w (Figure 10.5) can be calculated from the Ghyben-Herzberg relation as:

$$Z_w = \frac{\rho_f}{\rho_s - \rho_f}(S_w) = \delta S_w, \tag{10.10}$$

where S_w is the drawdown of the water table at the well. Bear and Dagan (1968) presented the following equation to evaluate the height of the center of the cone (Z_w) at any pumping time t:

$$(Z_w)_t = \frac{\rho_f Q}{2\pi(\rho_s - \rho_f)K_x L}\left(1 - \frac{2\rho_f n L}{2\rho_f n L + (\rho_s - \rho_f)K_z t}\right), \tag{10.11}$$

where L is the depth of the initial interface below the well bottom, K_x and K_z are the horizontal and vertical hydraulic conductivities, and n is the porosity of the aquifer. The maximum height of the saltwater cone can be obtained by setting $t = \infty$ in equation (10.11), therefore,

$$(Z_w)_\infty = \frac{\rho_f Q}{2\pi(\rho_s - \rho_f)K_x L} = \frac{\delta Q}{2\pi K_x L} \tag{10.12}$$

Heights of saltwater cones obtained from equation (10.10) agreed with field measurements up to some critical height, which was generally between 0.4 L and 0.6 L (Schmorak and Mercedo, 1969 and Haubold, 1975). As Z_w exceeded 0.6 L, saline water entered the well with a sudden jump. As a practical matter, the interface is apparently stable for upconed heights less than or equal to about one third L, as shown in Figure 10.5 (McWhorter and Sunada, 1977). The steady discharge for a well is given as (McWhorter, 1972)

$$Q = \frac{\pi K}{4(r_e/r_w)}[2bS_w - (1+\delta)S_w^2], \tag{10.13}$$

where r_e is the radius of the depression cone, r_w is the radius of the well, and b is the initial freshwater thickness.

The maximum sustainable safe discharge Q_m can be obtained by setting S_w to equal $(L/3)(\gamma_s - \gamma_f)/\gamma_f$ in equation (10.13):

$$Q_m = \frac{\pi K L}{3\delta \ln r_e/r_w}\left[2b - \frac{L}{3}\left(1 + \frac{1}{\delta}\right)\right]. \tag{10.14}$$

The assumption of horizontal flow beneath the well overestimates the upconed interface height and, hence reduces Q_m. Bear and Dagan (1968) accounted for the vertical components of flow below wells, and gave the following approximate formula to calculate the maximum safe sustained discharge:

$$Q_m = \frac{2}{3\delta}\pi L^2 K. \tag{10.15}$$

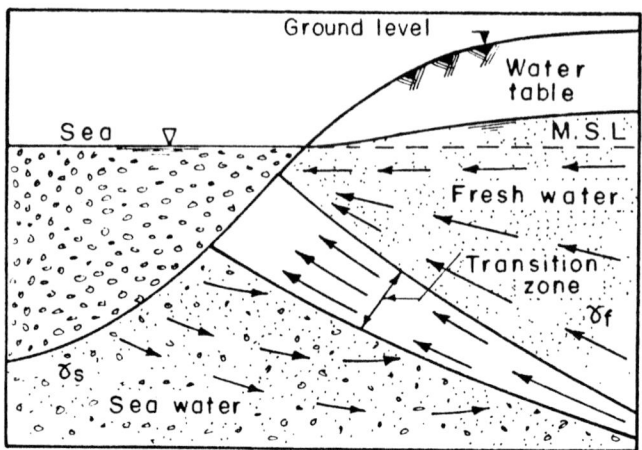

Fig. 10.6. Dispersion zone in coastal aquifers.

The above discussion and equations are based on the sharp interface assumption which does not hold in many cases. The density dependent approach may provide better simulation for concentration distribution and flow pattern beneath pumping wells. Governing equations are the same as those describing density dependent saltwater intrusion problems presented in Section 10.5.2. Initial and boundary conditions differ from one case to the other.

10.2.2 Density Dependent Approach

In many situations, especially when the depth of the aquifer at the sea boundary is relatively big, the sharp interface assumption does not provide appropriate simulation for the saltwater intrusion phenomenon. In reality, freshwater in the aquifer and saline water in the sea are two miscible fluids. The two water bodies merge in a transition or dispersion zone in which concentration and the fluid density vary gradually from those of freshwater at the land side to those of seawater at the sea side, as shown in Figure 10.6. In the dispersion zone the flow of water and the transport of salt ions are coupled. Flux of water is under hydraulic gradient, while transport of salt ions is under concentration gradient. An extensive zone of dispersion may alter the flow pattern. Sherif et al. (1988) reported a case in the Nile Delta aquifer in Egypt, where the dispersion zone extended for tens of kilometers. Because of its higher density, the seawater migrates to the bottom of the aquifer and mixes with the freshwater. This mixed water is of lower density, so it finds its way back again to the sea through the upper part of the sea boundary. Pandit and Anand (1984) reported the existence of cyclic flow near the sea boundary when the velocity at this boundary was bigger than the longitudinal dispersivity. Sherif et al. (1990) among others confirmed the cyclic flow pattern near the shore boundary in many cases. The width of the dispersion zone is a function of the hydraulic and transport parameters of the porous medium, aquifer geometry, natural recharge, hydraulic gradient, and relative density. The aquifer depth that comes

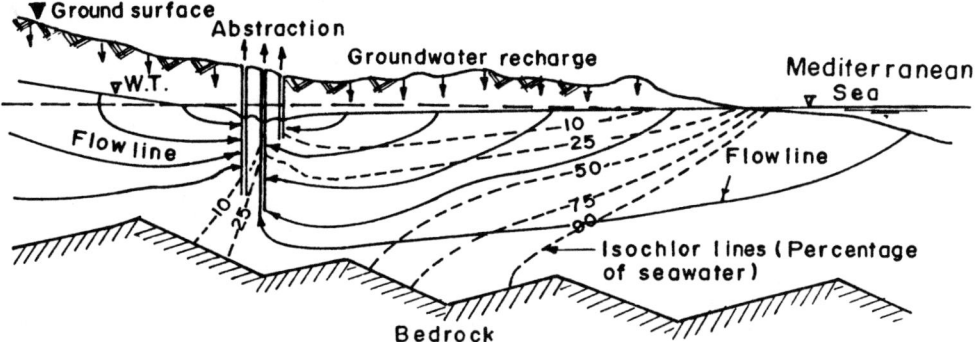

Fig. 10.7. Groundwater flow conditions and mixing with seawater in the Garraf carbonate massif, Catalonia, Spain (Pascual and Custodio, 1990).

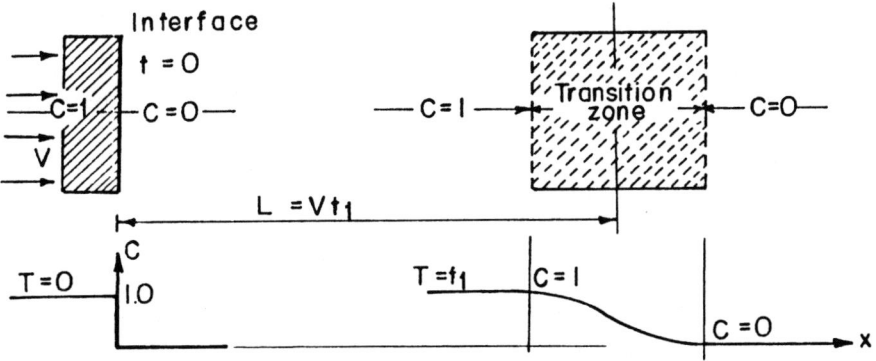

Fig. 10.8. Longitudinal spreading of a sharp interface (Bear, 1979).

into direct hydraulic contact with sea determines the degree of migration of saline water into the aquifer. Tidal effects, recharge and discharge events, and artificial actions enlarge the dispersion zone, especially when combined with the effects of medium heterogeneities. In double porosity media, water moving through the high permeability parts may be easily salinized from less permeable parts.

There is no general way to determine the thickness of the dispersion zone, which is varied from the toe to the tip. A thick zone may be expected where the freshwater flux to the sea and the recharge rate are small. Displacement of groundwater fronts during the transient stages greatly enhances mixing (Custodio, 1993). An extreme situation is when the freshwater flow is reversed and both the freshwater and saltwater flow towards an abstraction point, as shown in Figure 10.7.

10.3 Dispersion in Porous Media

Vital to the study of any saltwater intrusion problem (using density dependent approach) is the understanding of the dispersion phenomenon and solute movement

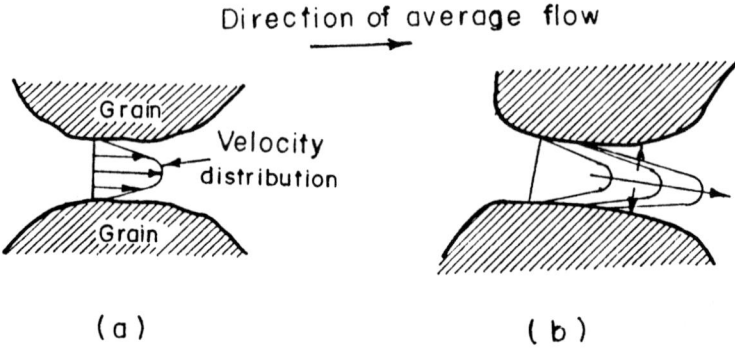

Fig. 10.9. (a) Velocity distribution inside pores. (b) Molecular diffusion (Bear and Verruijt, 1987).

through porous media. To that end, consider separated flow through a porous medium, as shown in Figure 10.8. An initial vertical interface separates the domain occupied by seawater (of concentration $C = C_{sea}$) from the one occupied by freshwater (of concentration $C = C_f$) at time $t = 0$. According to Darcy's law, if the flow is uniform (normal to the interface), the two water bodies should continue to occupy two separate domains and the interface should always be vertical. The new position of the interface at time $t = t_1$ is at $x = Vt_1$, as shown in Figure 10.8. In reality, if we measure the salinity via a system of observation wells, a gradual transition from seawater concentration C_{sea} to freshwater concentration, C_f, will be detected. The density of the water will also vary in this zone. The width of the dispersion zone will increase with time. Mixing and spreading of the seawater cannot be explained by the averaged water movement. The spreading process described above is called hydrodynamic dispersion or miscible displacement (Bear and Verruijt, 1987). It is a nonsteady irreversible process (in the sense that the initial concentration distribution or the initial vertical interface cannot be obtained by reversing the direction of the uniform flow) in which the salt ions of the seawater mix with the freshwater.

At the microscopic level, a velocity variation in both the magnitude and direction is found inside any pore. The maximum water velocity is encountered at some internal point and a zero velocity is usually assumed at the solid surface, as shown in Figure 10.9a. The velocity distribution is dependent on the shape and the size of the pore. It is anticipated that the spreading of any plume will take place mainly in the average flow direction. However, a considerable spreading will also be recognized normal to the direction of flow. As the flow continues, the plume will occupy an ever increasing space of the flow domain.

Mechanical dispersion is responsible for the spreading caused by the velocity variations at the microscopic level. Molecular diffusion is responsible for the spreading caused by the random movement of molecules in a fluid and also by molecular movement under concentration gradient. Molecular diffusion tends to equalize the concentrations along the stream tube as well as between adjacent stream lines across stream tubes, as shown in Figure 10.9b (Bear and Verruijt,

Saltwater Intrusion

1987). Hydrodynamic dispersion combines mechanical dispersion and molecular diffusion.

Molecular diffusion is mainly dependent on the chemical properties of the flowing fluid and time. It takes place even in the absence of motion. Therefore, the hydrodynamic dispersion in purely laminar flow field is an irreversible phenomenon. The effect of mechanical dispersion on the spreading process is usually much more significant. However, spreading due to molecular diffusion may predominate in very low velocity fields. In other words, hydrodynamic dispersion results from the simultaneous action of both a purely mechanical dispersion and a purely physicochemical dispersion or molecular diffusion. Mechanical dispersion results from the nonuniform velocity distribution of a fluid flowing through a porous medium. Molecular diffusion results from the chemical potential gradient which is correlated to the fluid concentration and takes place even in a fluid at rest (Fried, 1975). In flow fields of higher velocities, molecular diffusion is always overshadowed by the mechanical splitting created by the porous matrix. In particular, two pollutant particles starting from two close points in the same passageway of a porous matrix, in general, travel through two entirely different paths to arrive at the same downstream cross section. These two particles will arrive at this cross section at two different times due to the effect of longitudinal dispersion and at entirely two different points in the cross section due to the effect of lateral dispersion (Hunt, 1983).

10.3.1 Experimental Investigations on Longitudinal and Lateral Dispersion

Many experimental studies have been conducted to evaluate longitudinal dispersion in unconsolidated porous media. Results are presented in a dimensionless form on two types of graphs: D_L/D^* versus P_e, as shown in Figure 10.10a, and D_L/Vd versus P_e, as shown in Figure 10.10b, where D_L is the coefficient of longitudinal dispersion, P_e is the Peclet number of dispersion ($P_e = Vd/D^*$), D^* is the coefficient of molecular diffusion in fluid continuum (equals about 10^{-5} cm^2/sec in dilute systems), and d is some characteristic length of the pores. Five dispersion regimes can be distinguished from Figure 10.10b (Fried, 1975):

– **Regime a**: Molecular diffusion is the only component of the dispersion process and D_L/D^* is then constant (Figure 10.10a). This regime occurs when the mean velocity is very small (Peclet number less than 0.4). The porous medium slows down the diffusion processes (Fried and Combarnous, 1971). For a homogeneous medium made up of identical spheres, the ratio $(D_L/D^*)_o = 0.67$.

– **Regime b**: Corresponds to Peclet number between 0.4 and 5.0. Both the mechanical and molecular diffusion contribute significantly to hydrodynamic dispersion.

– **Regime c**: Mechanical dispersion predominates. However, molecular diffusion reduces the effects of mechanical dispersion. Taking into account the linear relationship observed between $\log_{10}(D_L/D^*)$ and $\log_{10} P_e$, a formula can be developed as:

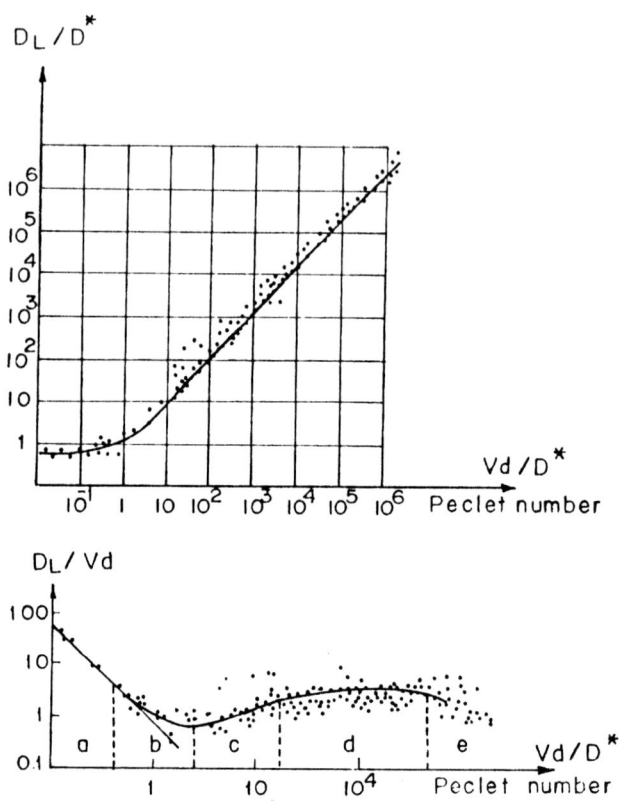

Fig. 10.10. Experimental investigations on longitudinal dispersion. (a, *Top*) D_L/D^* versus Peclet number. (b, *Bottom*) D_L/Vd versus Peclet number (Fried, 1975).

$$\frac{D_L}{D_*} = \left(\frac{D_L}{D^*}\right)_o + \alpha_1 \left(\frac{Vd}{D^*}\right)^m. \tag{10.16}$$

For the same regime, Bear and Verruijt (1987) introduced the formula:

$$\frac{D_L}{D^*} = \alpha_1 (Pe)^m, \tag{10.17}$$

where $\alpha_1 = 0.5$, and $1 < m < 1.2$.

– **Regime d**: The effect of molecular diffusion is totally over-shadowed by the mechanical effects. The following formula holds:

$$D_L = \left(\frac{D_L}{D^*}\right)_o + \alpha_2 (Vd), \tag{10.18}$$

where α_2 is equal to 1.8 ± 0.4 (Fried, 1975).

$(D_L/D^*)_o$ in equation (10.18) introduces the effect of molecular diffusion which can be neglected in this regime; therefore,

$$D_L = \alpha_3 V, \qquad \alpha_3 = \alpha_2 d. \tag{10.19}$$

Saltwater Intrusion

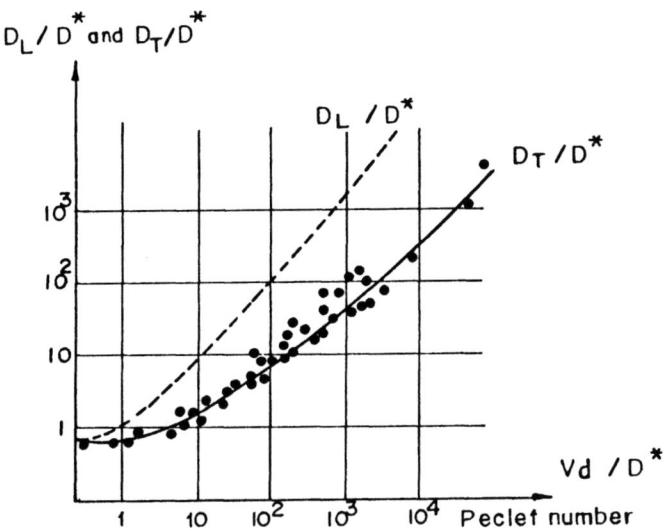

Fig. 10.11. Experimental investigations on lateral dispersion, D_L/D^* and D_T/D^* versus Peclet number (Fried, 1975).

Bear and Verruijt (1987) presented equation (10.18) in the form:

$$\frac{D_L}{D^*} = \beta(P_e), \qquad \beta = 1.8. \tag{10.20}$$

For practical purposes, equations (10.18) and (10.19) can be used in Regime c.

– **Regime e**: Mechanical dispersion predominates. Darcy's law is not applicable in this regime and turbulent flow is found. Few results are available for this regime. A considerable change in the dispersion pattern is observed, as shown in Figure 10.10b.

Few experiments were conducted to investigate lateral dispersion for Peclet numbers ranging from 10^{-2} to 10^{+4} and have shown the existence of four regimes of lateral dispersion, as shown in Figure 10.11 (Fried, 1975).

– **Regime a**: Molecular diffusion predominates and the velocities are very low. For a homogeneous medium made up of identical spheres $(D_T/D^*)_o$ is roughly equal to 0.7.

– **Regime b**: Both the molecular diffusion and mechanical dispersion contribute to the lateral hydrodynamic dispersion. Neglecting the effects of molecular diffusion will result in an overestimation of lateral dispersion.

– **Regime c**: Mechanical dispersion predominates and the following formula holds:

$$\frac{D_T}{D^*} = \left(\frac{D_T}{D^*}\right)_o + \alpha\left(\frac{Vd}{D^*}\right)^m, \tag{10.21}$$

where $\alpha = 0.025$, and $m = 1.1$.

– **Regime d**: Pure mechanical dispersion. Neglecting the effect of molecular diffusion in equation (10.21), i.e., $(D_T/D^*)_o = 0$, and setting $m = 1$, then

$$D_T = \alpha(Vd). \tag{10.22}$$

10.3.2 Hydrodynamic Dispersion Equation

The mass balance equation of a solute or a pollutant, expressed in terms of $C = C(x, y, z, t)$ which is often called the hydrodynamic dispersion equation or the advection–dispersion equation, can be written as:

$$\frac{\partial nC}{\partial t} = -\nabla \cdot (qC - nD \cdot \nabla C) + RC^* - PC, \tag{10.23}$$

where n is the porosity of the porous medium (dimensionless), C is the solute concentration (ML^{-3}), t is the time (T), q is the specific discharge (LT^{-1}), D is the hydrodynamic dispersion tensor (L^2T^{-1}), R and P are the recharge and discharge rate per unit volume of aquifer medium (T^{-1}), C^* is the concentration of the recharged water (ML^{-3}), and ∇ is an operator ($\partial/\partial x \; \partial/\partial y \; \partial/\partial z$) of dimension ($L^{-1}$).

Hydrodynamic dispersion coefficient. Many studies have been conducted to identify the relationship between the hydrodynamic dispersion coefficient D and porous matrix configuration, flow velocity and molecular diffusion. The following equation has been obtained (Nikolaevskii, 1959; Scheidegger, 1961; Bear and Bachmat, 1967):

$$D_{ij} = \alpha_{ijkm} \frac{\bar{V}_k \bar{V}_m}{\bar{V}} f(P_e, \delta), \tag{10.24}$$

where V is the average velocity, δ is the ratio of the length characterizing the individual pores of a porous medium to the length characterizing their cross-section, and $f(P_e, \delta)$ is a function to introduce the effect of molecular diffusion on hydrodynamic dispersion. For practical applications the function $f(P_e, \delta)$ is assumed equal to 1 (Bear and Verruijt, 1987).

The coefficient α_{ijkm} is the dispersivity of the medium (L). It is a parameter which accounts for the loss of information about the pore velocity fluctuation when passing from the microscopic to the macroscopic scale. It is a fourth-rank tensor which expresses the microscopic configuration of the solid liquid interface.

For an isotropic medium, the number of non-zero components of the dispersivity tensor is 21. All components are related to two parameters only, the longitudinal dispersivity, α_L and the lateral dispersivity, α_T. Laboratory experiments have shown that α_L is of the order of magnitude of the average sand grain. Transversal dispersivity α_T is estimated as 10 to 20 times smaller than α_L. Volker and Rushton (1982) have indicated that values of longitudinal dispersivity that have been reported in field investigations range from 6 to 150 m. Lateral dispersivity, α_T, has usually been taken to be 10–30% of the longitudinal value (Pickens and Grisak, 1980). In laboratory experiments, longitudinal dispersivities of 0.1–10 cm have been found for different types of materials (Klotz, 1973). Longitudinal dispersivities in the field are usually much larger than those in laboratory experiments

Saltwater Intrusion

(Lenda and Zuber, 1970). The main reason for this is the influence of small-scale inhomogeneities in the permeability of the aquifer, which means the onset of macro dispersion. Owing to this scale behavior of dispersion, no generally valid values for longitudinal dispersivities can be given (Kinzelbach, 1986). Values between 0.1 and 500 m can be found in the literature. The components of the dispersivity for an isotropic porous medium can be expressed in the form (Bear and Verruijt, 1987):

$$\alpha_{ijkm} = \alpha_T \delta_{ij}\delta_{km} + \frac{\alpha_L + \alpha_T}{2}(\delta_{ik}\delta_{jm} + \delta_{im}\delta_{jk}), \tag{10.25}$$

where δ_{ij} denotes the Kronecker delta (with $\delta_{ij} = 0$ for $i \neq j$ and $\delta_{ij} = 1$ for $i = j$). For an isotropic medium, α_{ijkm} do not change with the rotation of the coordinate system.

Substituting equation (10.25) into equation (10.24) and for $f(P_e, \delta) = 1$, we get

$$D_{ij} = \alpha_T V \delta_{ij} + (\alpha_L - \alpha_T)V_i V_j / V. \tag{10.26}$$

In cartesian coordinates, equation (10.26) can be written as:

$$\begin{aligned}
D_{xx} &= \alpha_T V + (\alpha_L - \alpha_T)V_x^2/V = [\alpha_T(V_y^2 + V_z^2) + \alpha_L V_x^2]/V, \\
D_{xy} &= (\alpha_L - \alpha_T)V_x V_y/V = D_{yx}, \\
D_{xz} &= (\alpha_L - \alpha_T)V_x V_z/V = D_{zx}, \\
D_{yy} &= \alpha_T V + (\alpha_L - \alpha_T)V_y^2/V = [\alpha_T(V_x^2 + V_z^2) + \alpha_L V_y^2]/V, \\
D_{yz} &= (\alpha_L - \alpha_T)V_y V_z/V = D_{zy}, \\
D_{zz} &= \alpha_T V + (\alpha_L - \alpha_T)V_z^2/V = [\alpha_T(V_x^2 + V_y^2) + \alpha_L V_z^2]/V,
\end{aligned} \tag{10.27}$$

where V_x, V_y, and V_z are the velocity components in the x, y, and z directions. If one of the axes of the cartesian coordinate system coincides at a point with the direction of the average uniform velocity V, then equation (10.27) at that point reduces to

$$D_{xx} = \alpha_L V, \qquad D_{yy} = D_{zz} = \alpha_T V, \qquad D_{ij} = 0, \quad i \neq j, \tag{10.28}$$

which can be written in a matrix form as

$$D_{ij} = \begin{bmatrix} \alpha_L V & 0 & 0 \\ 0 & \alpha_T V & 0 \\ 0 & 0 & \alpha_T V \end{bmatrix}. \tag{10.29}$$

The coefficients D_{xx}, D_{yy}, and D_{zz} in equation (10.29) are the principal values of the mechanical dispersion coefficient. The coefficient D_{xx} is the coefficient of longitudinal dispersion, whereas D_{yy} and D_{zz} are the coefficients of transversal dispersion.

Molecular diffusion coefficient. In stagnant or slowly moving fluids, molecular diffusion is the main dispersive transport mechanism. The rate of macroscopic or

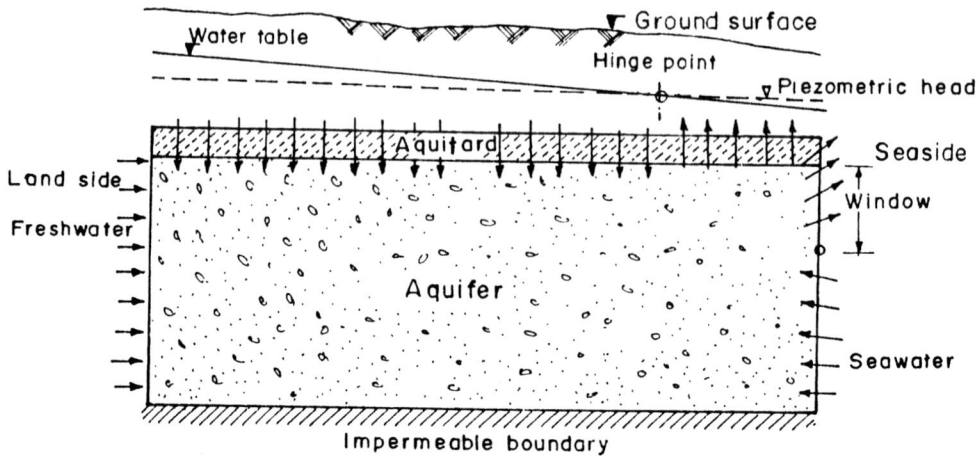

Fig. 10.12. Idealized leaky coastal aquifer.

overall molecular diffusion in a porous medium is related to the rate of diffusion in a free fluid (Bear, 1972):

$$D_p^* = \sigma D^*/\lambda, \tag{10.30}$$

where D_p^* is the diffusion coefficient in a porous medium (L^2T^{-1}), σ is the free space (dimensionless), D^* is the diffusion coefficient in free fluid (L^2T^{-1}), and λ is the tortuosity (dimensionless). Tortuosity is defined as the ratio of the average path length followed by a solute molecule to the straight line distance between the initial and final locations. As it would be expected, the reduction of free space, σ, and the increase in path length, λ, reduces the overall diffusion rate.

Fick's law says that the flux of solute mass, that is, the mass of a solute crossing a unit area per unit time in a given direction, is proportional to the gradient of solute concentration in that direction. It can be written as:

$$q = -D_p^* \cdot \nabla C, \tag{10.31}$$

where q is the mass flux vector with components (q_x, q_y and q_z) in a cartesian coordinate system, and C is the mass concentration of diffusing solute.

Empirical studies have found the tortuosity to be related to the formation factor, F, which is the ratio of the resistivity of the porous medium saturated with electrolyte to the resistivity of the electrolyte over an identical distance. Bear, (1972) presented the formula:

$$\lambda = (F\sigma)^{3/5}. \tag{10.32}$$

10.4 Mechanism of Saltwater Intrusion into Coastal Aquifers

Consider the idealized leaky system as shown in Figure 10.12. The aquifer is recharged by freshwater entering from the landward boundary and by leakage through the upper clay cap (aquitard) in parts where the free water table is higher than the aquifer piezometric head (before the point of the intersection, hinge point, between the free water table and the aquifer piezometric head). In some cases, the free water table may be higher than the piezometric head all over the upper clay cap;therefore, the system will be recharged by the freshwater throughout the entire upper boundary as well as the landward boundary. An unconfined (phreatic) aquifer will be recharged by freshwater through infiltration of precipitation. A confined aquifer will be recharged only by freshwater from the boundary at the land side. At the seaward boundary, there will be an influx of seawater into the system which, because of its greater density, will migrate to the bottom of the aquifer and displace the freshwater. Upward leakage of mixed water will also take place through the aquitard in parts where the free water table is lower than the aquifer piezometric head (near the seaside). The rest of the mixed water will find its way out of the system through the window at the seaside. This discharge through the window and the upward flux through the aquitard cause a loss of salt from the system which is replenished from the new seawater moving in from the seaward boundary.

If the boundary conditions remain constant, a state of dynamic equilibrium will eventually be attained by the system. At equilibrium, the total fluid mass entering at both ends of the aquifer plus the leakage influx, will be balanced by the upward leakage through the aquitard plus the flux out through the window at the sea side. Likewise, the salt mass entering at the seaward side will be balanced by the salt mass swept out from the system with the mixed water through the aquitard and the window at the seaward boundary.

10.5 Governing Equations

The basic concept and the governing equations describing saltwater intrusion under the sharp interface assumption are entirely different from those describing the phenomena via density dependent approach.

10.5.1 SHARP INTERFACE APPROACH

The mathematical statement of saltwater intrusion into coastal aquifers under the assumption of sharp interface gives rise to the problem of two-phase flow. Each flow phase occupies a separate portion of the entire flow domain, as shown in Figure 10.13. The interface, separating freshwater body from saline water body, is subject to movement as sources and sinks may exist in each region of the two water bodies. In the absence of any external sources or sinks, the interface will be stationary.

Assuming that the two water bodies are compressible, the piezometric head in the two regions can be defined as (Bear, 1979):

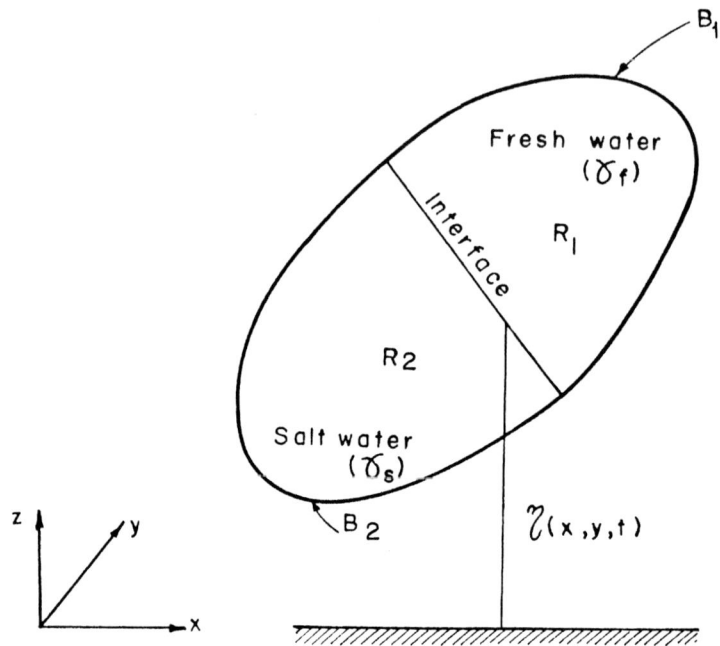

Fig. 10.13. Sharp interface between two-phase flow (Bear, 1972).

$$\psi_f = z + \int_{p_o}^{p_f} \frac{dp_f}{\gamma_f p_f} \tag{10.33}$$

and

$$\psi_s = z + \int_{p_o}^{p_s} \frac{dp_s}{\gamma_s p_s}, \tag{10.34}$$

where ψ is the hydraulic head (L), z is the elevation of a point in the flow domain above a specified datum (L), and p is the pressure ($ML^{-1}T^{-2}$). The subscripts f, s, and o refer to freshwater, saltwater, and initial value, respectively.

The objective is to determine the hydraulic head, ψ, in the two regions such that the continuity requirements are satisfied, i.e,

$$S_f \frac{\partial \psi_f}{\partial t} = \nabla \cdot (K_f \cdot \nabla \psi_f) - Q_f \quad \in R_1 \tag{10.35}$$

and

$$S_s \frac{\partial \psi_s}{\partial t} = \nabla \cdot (K_s \cdot \nabla \psi_s) - Q_s \quad \in R_2, \tag{10.36}$$

where $K_f = k\gamma_f/\mu_f$, $K_s = k\gamma_s/\mu_s$, k is the permeability (L^2), μ is the dynamic viscosity ($ML^{-1}T^{-1}$), S is the specific storativity (L^{-1}), Q is the source/sink term (T^{-1}), R_1 and R_2 refer to the freshwater and saline water regions, respectively.

Saltwater Intrusion

In many problems, the specific storativity is assumed the same in both regions. Equations (10.35) and (10.36) constitute the main equations under the sharp interface assumption. For practical purposes, the elastic storage of the medium can be neglected. Therefore, the main equations are reduced to

$$\nabla \cdot (K_f \cdot \nabla \psi_f) - Q_f = 0 \quad \in R_1 \tag{10.37}$$

and

$$\nabla \cdot (K_s \cdot \nabla \psi_s) - Q_s = 0 \quad \in R_2. \tag{10.38}$$

10.5.2 Density Dependent Approach

The governing equations describing the dispersion zone and the flow pattern in coastal aquifers subjected to saltwater intrusion under unsteady state conditions are:

1. The general Darcy equation for groundwater flow

$$q = -\frac{k}{\mu}(\nabla p + \rho g \nabla z), \tag{10.39}$$

where q is the specific discharge vector (LT^{-1}), k is the permeability tensor (L^2), μ is the dynamic viscosity ($ML^{-1}T^{-1}$), p is pressure ($ML^{-1}T^{-2}$), ρ is the fluid density (ML^{-3}), g is the gravitational acceleration (LT^{-2}) and z is a space coordinate (L). Substitution of

$$\psi = \frac{p}{\rho_f g} + z, \quad K = \frac{k \rho_f g}{\mu}, \quad \text{and} \quad \rho_r = \frac{\rho}{\rho_f} - 1$$

into equation (10.39) yields

$$q = -K(\nabla \psi + \rho_r \nabla z), \tag{10.40}$$

where K is the hydraulic conductivity tensor (LT^{-1}), ψ is the equivalent hydraulic head (L), and ρ_r is the relative density (dimensionless).

2. The basic fluid continuity equation or the mass balance equation for the fluid can be written as

$$\frac{\partial n \rho}{\partial t} = -\nabla \cdot \rho q + R \rho^* - P \rho, \tag{10.41}$$

where n is the effective porosity (dimensionless), R and P are the recharge and pumping rates per unit volume of the aquifer medium, respectively (T^{-1}), and ρ^* is the density of the recharged water (ML^{-3}).

3. The hydrodynamic dispersion equation or the mass balance equation for the salt ions can be written as

$$\frac{\partial n C}{\partial t} = -\nabla \cdot (qC - nD \cdot \nabla C) + RC^* - PC. \tag{10.42}$$

In the above equation, the effect of adsorption on the dispersion process and solute transport is neglected. For two-dimensional vertical cross-section problems, the components of the hydrodynamic dispersion tensor D can be calculated as follows (Bear, 1979):

$$D_{xx} = \alpha_L \frac{V_x^2}{|V|} + \alpha_T \frac{V_z^2}{V} + D^*_p$$

$$D_{zz} = \alpha_T \frac{V_x^2}{|V|} + \alpha_L \frac{V_z^2}{V} + D^*_p, \tag{10.43}$$

$$D_{xz} = D_{zx} = (\alpha_L - \alpha_T)\frac{V_x V_z}{V}.$$

4. A constitutive equation relating fluid density to solute concentration can be expressed as:

$$\rho = \rho_f + a(C - C_f), \tag{10.44}$$

where C_f is the freshwater (reference) concentration (ML^{-3}), and a is a constant (dimensionless) given by

$$a = \frac{\rho_s - \rho_f}{C_s - C_f}, \tag{10.45}$$

where ρ_s is the seawater density (ML^{-3}), and C_s is the seawater concentration (ML^{-3}). A linear relationship between the density and concentration is assumed in equation (10.44). Baxter and Wallace (1916) developed an empirical relation which relates the salt concentration to fluid density as

$$\rho = \rho_f + (1 - E)\rho_s, \tag{10.46}$$

where E is a constant (dimensionless) and has a value of 0.3 for seawater concentration.

An examination of the main equations (10.40), (10.41), (10.42), and (10.44) reveals that there are four unknowns (ψ, V, C and ρ) in four equations. However, these equations can be combined into two nonlinear partial differential equations in only two variables, namely, the hydraulic head ψ and the concentration C.

10.6 Initial and Boundary Conditions

The governing equations which describe the saltwater intrusion in coastal aquifers are partial differential equations. Each equation has an infinite number of possible solutions for the flow field and solute concentration, each of which corresponds to a particular problem. Only one particular solution is correct with respect to a certain case. Additional information and constrains are needed to define an individual problem. These are (Bear, 1979):

(a) Domain geometry in which the problem is under consideration and possibly parts of the boundary are at infinity.

Saltwater Intrusion

(b) Data about all relevant physical parameters and fluid properties (for example n, K, α_L, γ_f and C_s).

(c) Initial conditions which describe the initial state of the unknowns in the study domain.

(d) Boundary conditions which describe how the fluid in the study domain interacts with its surroundings.

Any mathematical problem that corresponds to a physical reality must be a well-posed problem. A problem is well posed under the following conditions:
- The solution must exist (existence).
- The solution must be unique (uniqueness).
- The solution must be dependent on the data (stability).

10.6.1 Initial Conditions

To model the saltwater intrusion phenomenon by either of the sharp interface approach or the density dependent approach, we have to specify some initial and boundary conditions. Once the initial freshwater levels and seawater heads are specified, the initial location of the interface can be estimated (from the Ghyben-Herzberg relation). Initial assumptions for the hydraulic head and the concentration are required if the density dependent approach is utilized. The final solution should not be dependent on these initial conditions under any circumstances. However, unrealistic assumptions of the initial conditions may diverge the solution, whereas good estimates will accelerate the convergence. Realistic initial conditions can be obtained from field observations.

10.6.2 Boundary Conditions

Sharp interface approach. Equations (10.37) and (10.38) are the main equations to be solved to locate the interface between the freshwater body and the saline water body. Other equations may be used when necessary, depending on the assumptions made with respect to the porous medium and the flowing fluids. The location of the interface is not known a priori, hence it is impossible to write the boundary conditions directly. However, in three dimensions, the location of the interface may be expressed as:

$$F(x, y, z, t) = 0, \tag{10.47}$$

the elevation z of points on the interface (Figure 10.13) is $\zeta = \zeta(x, y, t)$; therefore $z = \zeta(x, y, t)$, or

$$F \equiv z - \zeta(x, y, t) = 0. \tag{10.48}$$

Since the pressure on both sides of the interface must be the same, then

$$\zeta(x, y, t) = \psi_s \frac{\gamma_s}{\gamma_s - \gamma_f} - \psi_f \frac{\gamma_f}{\gamma_s - \gamma_f}, \text{ or}$$
$$\zeta(x, y, t) = \psi_s(1 + \delta) - \psi_f \delta, \tag{10.49}$$

where δ is equal to $\gamma_f/(\gamma_s - \gamma_f)$. Substituting into equation (10.48)

$$F \equiv z - \psi_s(1 + \delta) + \psi_f\delta. \tag{10.50}$$

The boundary conditions on the interface are
1. The same specific discharge on both sides of the interface

$$(q_n)_f = (q_n)_s, \tag{10.51}$$

and
2. The same pressure on both sides of the interface

$$\gamma_f(\psi_f - \zeta) = \gamma_s(\psi_s - \zeta). \tag{10.52}$$

Since the interface is a material surface, one can write

$$\frac{dF}{dt} \equiv \frac{\partial F}{\partial t} + V_f \cdot \nabla F = 0, \tag{10.53}$$

also

$$\frac{dF}{dt} \equiv \frac{\partial F}{\partial t} + V_s \cdot \nabla F = 0, \tag{10.54}$$

where $V_f = -(K_f/n) \cdot \nabla \psi_f$ and $V_s = -(K_s/n) \cdot \nabla \psi_s$. Combining equation (10.50) with equations (10.53) and (10.54), we get

$$n\delta \frac{\partial \psi_f}{\partial t} - n(1+\delta)\frac{\partial \psi_s}{\partial t} - K_f \cdot [\nabla z - (1+\delta)\nabla \psi_s + \delta \nabla \psi_f] \cdot \nabla \psi_f = 0 \tag{10.55}$$

and

$$n\delta \frac{\partial \psi_f}{\partial t} - n(1+\delta)\frac{\partial \psi_s}{\partial t} - K_s \cdot [\nabla z - (1+\delta)\nabla \psi_s + \delta \nabla \psi_f] \cdot \nabla \psi_s = 0. \tag{10.56}$$

Equations (10.55) and (10.56) constitute the boundary conditions on the interface. They are nonlinear partial differential equations in ψ_f and ψ_s. It is, therefore, practically impossible to solve the governing equations with consideration of such boundary conditions. Some approximations should be considered such as employment of the hydraulic approach and averaging the three dimensional balance equation separately for each region over the vertical. There are many different ways for expressing the boundary conditions at the interface. Each involves some approximation which may fit for certain problems. Several cases are reported by Bear (1979).

Density dependent approach. The discussion in this part will be limited to the case of two-dimensional vertical cross-section. Of course, the same procedure and concept can be followed when dealing with three-dimensional problems and two-dimensional areal problems.

Saltwater Intrusion

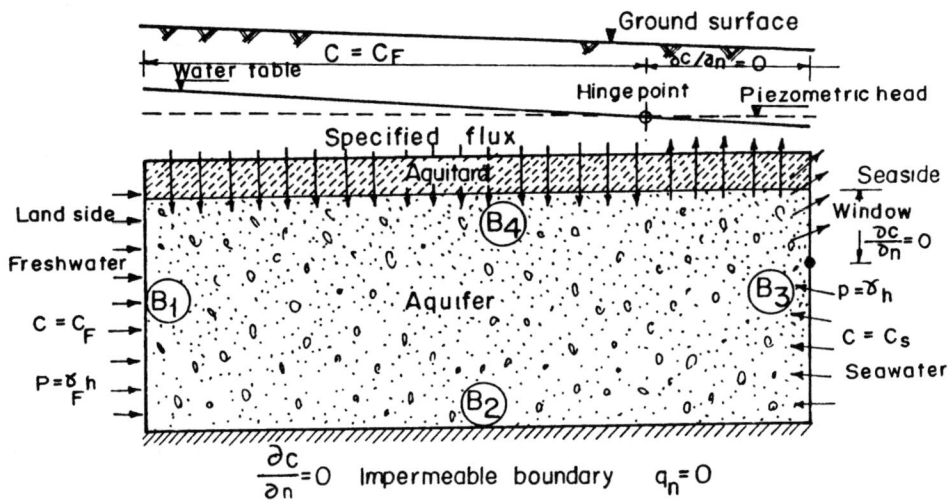

Fig. 10.14. Boundary conditions.

Consider an idealized leaky aquifer system as shown in Figure 10.14. The lateral boundary (B1) at the landward side should be located at a point where either the concentration is constant and equal to the freshwater concentration C_f, or where the change in concentration across the boundary is negligible, i.e., the concentration gradient is equal to zero. The former condition of prescribed concentration is preferable because it accelerates the convergence. The pressure at this boundary is hydrostatic and the freshwater flux through it is defined and can be calculated from Darcy's equation. The bottom of the aquifer (B2) is impermeable, i.e., the normal flux through the bed for both fluid and salt ions is equal to zero.

The seaward boundary (B3), where the aquifer is exposed to the open sea, is a very important boundary. It has been customary (Huyakorn et al., 1987; Sherif et al., 1988; Galeati et al., 1992) to deal with the seaside boundary as the prescribed concentration boundary when flow across it is directed inward, and as the zero dispersive flux boundary when the flow direction is outward. Since conditions along this boundary vary with time, it may not be adequately prescribed on the basis of preliminary simulations as done by Huyakorn et al. (1987). We follow the approach utilized by Sherif et al. (1988) and Galeati et al. (1992). During each iteration the directions of velocities at all the nodal points on the boundary are checked, then the appropriate boundary conditions are assigned accordingly, i.e., the boundary conditions at the sea side are updated after each iteration. The concentrations are prescribed and equated to seawater concentration C_s over the segment of inward flow. Over the window (Figure 10.14), where the flow is outward, the concentration gradient is equal to zero. The pressure is hydrostatic at this boundary and should also be updated after each iteration according to the new nodal concentration.

For the upper leaky boundary (B4), as shown in Figure 10.14, in the segment where the free water table is higher than the aquifer piezometric head, i.e., there is a leakage of freshwater into the system, the concentration is known and is equal to that of freshwater, C_f. Otherwise, where the piezometric head is higher than the water table, the concentration along any vertical stream tube will be constant, that is, the concentration gradient is zero. The direction of flow through the boundary should be checked after each iteration and consequently the boundary conditions should be updated. The flux in the aquitard is vertical and is governed by Darcy's equation and may be downward or upward depending on whether the free water table is higher or lower than the piezometric head.

In the case of confined aquifers, the upper boundary is impermeable, then the normal flux through it for both fluid and salt ions will be equal to zero. For phreatic aquifers, where the pressure at the upper boundary is equal to zero and the location of the boundary is not known a priori, the phreatic surface with accretion boundary should be utilized (Bear and Verruijt, 1987, p. 72). A similar strategy to that of the seaside boundary should be employed for the concentration of the phreatic surface. A Cauchy boundary condition with zero total solute mass is prescribed when flow is directed inward, and a Newmann boundary condition with zero diffusive flux when the flow is directed outward. This way, no artificial boundary conditions need to be prescribed anywhere and the salt content ensues directly from the simulation (Galeati et al., 1992).

For the idealized leaky domain shown in Figure 10.14, the boundary conditions can be written as follows:

At the landward side (B1):

$$\psi = \psi_L,$$
$$C = C_f,$$
(10.57)

where ψ_L is the piezometric head at the landside.

At the bottom boundary (B2), impermeable:

$$q_n = 0,$$
$$\partial C / \partial n = 0.$$
(10.58)

At the seaside boundary (B3):

$$\psi = \psi_s,$$
$$\partial C / \partial n = 0, \quad \text{over the window, and}$$
$$C = C_s, \quad \text{below the window,}$$
(10.59)

where ψ_s is the equivalent freshwater hydraulic head at seaside.

At the upper boundary (B4):

$$q_z = K'_z \left(\frac{\partial \psi}{\partial Z} + \rho_r \right) = \frac{K'_z}{b'}(\psi_t - \psi) + K'_z \rho_r,$$

where K'_z is the vertical hydraulic conductivity for the aquitard (LT^{-1}), b' is the aquitard thickness (L), and ψ_t is the free water table elevation (L). Therefore,

$$q_z = \frac{K'_z}{b}(\psi_t - \psi) + \frac{aK'_z}{\rho_f}(C - C_f).$$

Saltwater Intrusion

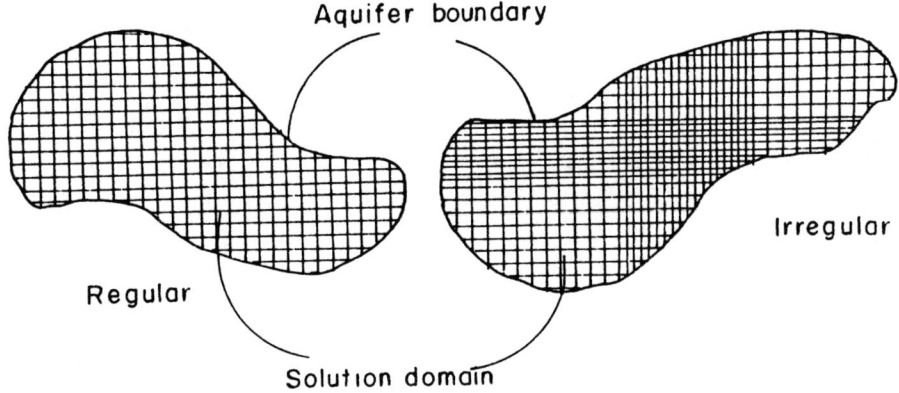

Fig. 10.15. Regular and irregular finite difference grid.

$$C = C_f, \quad \text{for downward flux,}$$
$$\partial C/\partial n = 0, \quad \text{for upward flux.} \tag{10.60}$$

10.7 Numerical Methods

Analytical methods have failed to provide solutions when the density dependent approach has been utilized in modeling saltwater intrusion. Few approximate solutions for idealized domains may be obtained analytically for some simplified problems using the sharp interface approach. Numerical methods are used almost exclusively to model saltwater intrusion problems encountered in practice. Numerical methods replace the partial differential equations by a system of linear or nonlinear equations to be solved to yield numerical values for the unknowns at predetermined points inside the study area. Two numerical methods, the finite difference method and the finite element method, are very powerful and extensively used. The method of characteristics has been used to solve the hydrodynamic dispersion equation (Konikow and Bredefoeft, 1978). Sanford and Konikow (1985) employed the method of characteristics in modeling density dependent groundwater flow. Generally, this method has some stability criteria which, when adhered to, yields accurate solutions. There are some other methods which can be employed in modeling saltwater intrusion problems, such as the boundary element method (Liggett and Liu, 1983) and the analytical element method (Strack, 1987).

10.7.1 Finite Difference Method

The finite difference method is the oldest and most widely used numerical method for solving differential equations (Forsythe and Wasow, 1960). However, it has some limitations when used to solve the hydrodynamic dispersion equation because of a phenomenon known as numerical dispersion which may result in an overestimation of solute migration than is physically possible.

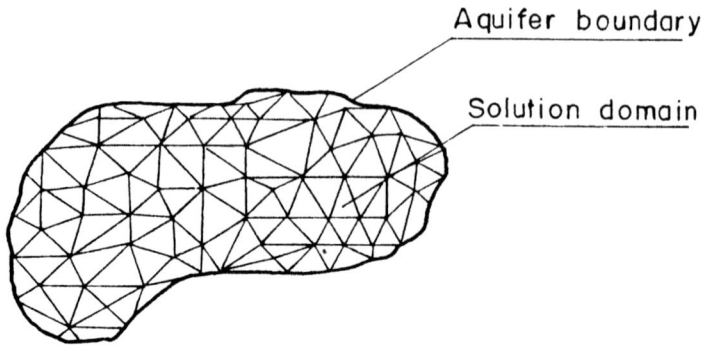

Fig. 10.16. Typical triangular finite element grid.

A system of nodal points is superimposed over the study domain which is covered by either a regular or an irregular rectangular grid, as shown in Figure 10.15. Governing differential equations are replaced by a set of difference equations, one for each grid point or block. The resultant is $n \times m$ simultaneous sets of equations, where n is the number of rows and m is the number of columns of the grid. Solution of these equations yields numerical values at the specified nodal points. The choice of a node centered or block centered scheme is mainly dependent on the user. However, the block-centered node is preferred for large scale problems.

The stability criterion of the finite difference method is well defined. To minimize numerical oscillations due to overshooting or undershooting when solving the hydrodynamic dispersion equation, the Peclet number ($P_e = \Delta x / \alpha_l$) should be less than 2, and Courant number ($C = V \Delta t / \Delta x$) should be less than 1, where Δx is some characteristic grid size, α_l is the dispersivity, V is the velocity of flow and Δt is the time step. In other words, to satisfy the Courant number requirement, the time step Δt should be less than the time needed by a solute particle to advance one grid distance.

10.7.2 Finite Element Method

The finite element method (Zienkiewicz, 1977) is an integral approach in which the study domain is represented by an irregular polygonal mesh, therefore providing more flexibility in representation of a natural boundary. Triangular elements are used extensively in modeling groundwater flow and pollution. Occasionally, quadrilateral elements are also employed. A typical finite element grid is shown in Figure 10.16. In the standard formulation of the finite element method, each element can have its own physical and hydraulic parameters. Therefore, heterogeneous domains are naturally represented.

In this method, the variables (piezometric head and concentration) are approximated by a series of small triangular surfaces (when triangular elements are used) which can be flat or curved. If the chosen basis functions are linear then the surfaces will be flat and variation of the variables within each element will be linear. If the elements are not too big then this is a perfectly reasonable assumption to make

Saltwater Intrusion

(Goodwill, 1993). Furthermore this assumption makes the ensuing analysis much simpler than it would otherwise be. The equations are solved by a weighted residual technique of which the most popular one for groundwater flow and solute transport problems is the Galerkin technique (Pinder, 1973; Grove, 1977). In this technique the weighting functions are made equal to the basis functions and the integration is then performed over each element and summed to yield the contribution from all the elements that make up the solution.

The finite element method is a powerful and mathematically elegant technique but it is difficult to program. A major advantage of this method is its high flexibility to refine the grid wherever needed. In saltwater intrusion problems, an intensive grid is essential near the sea boundary where the flow is rotational in character and the concentration gradient is relatively high. Unfortunately, it is not possible to set up stability criteria for the finite element method in the same way as for the finite difference method and data needed to define the hydraulic parameters and the grid is much more time consuming.

10.8 Finite Element Formulation for Density Dependent Problems

The mathematical development of saltwater intrusion problems under the sharp interface assumption is well established and reported widely in literature (Bear, 1979; Mercer et al., 1980a and 1980b; Wilson and Sa da Coasta, 1982; Bear and Verruijt, 1987; Essaid, 1987 and Sakr, 1992). Thereupon, the attention is focused here on solving saltwater intrusion problems using the density dependent approach.

Equations (10.40)–(10.42) and (10.44) constitute the governing equations for three-dimensional flow fields. Considering two-dimensional cross sectional (x-z plane) domain, the basic fluid continuity equation (10.41) can be written as

$$\frac{\partial(n\rho)}{\partial t} = -\frac{\partial}{\partial x}\rho q_x - \frac{\partial}{\partial z}\rho q_z + R\rho^* - P\rho. \tag{10.61}$$

For an incompressible medium and neglecting the compressibility of the fluid,

$$\frac{\partial(n\rho)}{\partial t} = na\frac{\partial C}{\partial t}. \tag{10.62}$$

Applying Galerkin's technique to fluid continuity equation

$$\int_A [N]^T [(\nabla \cdot \rho q_x) + (\nabla \cdot \rho q_z) - R\rho^* + P\rho + na\left(\frac{\partial C}{\partial t}\right)] \, dA = 0, \tag{10.63}$$

where, $[N] = [N_i \ N_k]$ is the shape function. Using Gauss' theorem and assuming unit thickness, equation (10.63) can be written as:

$$\int_{A^e} \frac{\partial [N]^T}{\partial x}(\rho q_x) \, dA^e + \int_{A^e} \frac{\partial [N]^T}{\partial z}(\rho q_z) \, dA^e + na \int_{A^e} [N]^T \left(\frac{\partial C}{\partial t}\right) dA^e$$
$$= \int_{L^e} [N]^T (\rho q_n) \, dL^e. \tag{10.64}$$

Let

$$\rho = \lambda_0 + aC, \tag{10.65}$$

where λ_0 is a constant (ML^{-3}) and equal to $(\rho_f - aC_f)$ and a is another dimensionless constant given by equation (10.45).

Substituting equations (10.40) and (10.65) into equation (10.64), one gets

$$K_{xx} \frac{\partial [N]^T}{\partial x} \frac{\partial [N]}{\partial x} \int_{A^e} (\lambda_0 + a[N][C^*]) \, dA^e [\psi^*]$$

$$- K_{zz} \int_{A^e} \frac{\partial [N]^T}{\partial z} (\lambda_0 + a[N][C^*]) \left(\frac{\partial [N]}{\partial z} [\psi^*] + \frac{1}{\rho_f} (\lambda_0 + a[N][C^*]) - 1 \right) dA^e$$

$$- \int_{A^e} [N]^T (R\rho^*) \, dA^e + \int_{A^e} [N]^T (P\rho) \, dA^e + na \int_{A^e} [N]^T \left(\frac{\partial C}{\partial t} \right) dA^e$$

$$= \int_{L^e} [N]^T (\rho q_n) \, dL^e, \tag{10.66}$$

where C^* and ψ^* are the concentration and the equivalent freshwater piezometric head, respectively, at the nodal points of the finite element grid. The integration on the right hand side of equation (10.66) must be evaluated over all the boundaries.

The final form, after integration, of equation (10.66) can be written in a matrix form as:

$$[W_\psi]\{\psi^*\} + [P_\psi]\{\partial C/\partial t\} = \{F_\psi\}, \tag{10.67}$$

where $[W_\psi]$ is the condutance matrix which has been developed for the steady-state case, $\{\psi^*\}$ is the column matrix of the unkown nodal piezometric heads, $[P_\psi]$ is a square matrix which accounts for the storge term in the transient flow condition, $\{\partial C/\partial t\}$ is a column matrix of the time derivatives $\partial C/\partial t$, and $\{F_\psi\}$ is a column matrix which represents the boundary conditions.

Similary, applying Galerkin's technique to the hydrodynamic dispersion equation (10.42), one gets

$$\int_V [N]^T [\nabla \cdot (nD \cdot \nabla C - qC) + RC^* - PC - n \frac{\partial C}{\partial t}] \, dV = 0. \tag{10.68}$$

Using Gauss' theorem and assuming unit thickness, equation (10.68) can be written for two-dimensional vertical cross sections as:

$$\int_{A^e} \frac{\partial [N]^T}{\partial x} \left(n \left(D_{xx} \frac{\partial C}{\partial x} + D_{xz} \frac{\partial C}{\partial z} \right) - q_x C \right) dA^e$$

$$+ \int_{A^e} \frac{\partial [N]^T}{\partial z} \left(n \left(D_{zx} \frac{\partial C}{\partial x} + D_{zz} \frac{\partial C}{\partial z} \right) - q_z C \right) dA^e$$

$$+ \int_{A^e} [N]^T (RC^*) \, dA^e - \int_{A^e} [N]^T (PC) \, dA^e - n \int_{A^e} [N]^T \left(\frac{\partial C}{\partial t} \right) dA^e$$

$$= \int_{L^e} [N]^T (nD \cdot \nabla C - qC)_n \, dL^e. \tag{10.69}$$

Saltwater Intrusion

Substituting equation (10.40) into equation (10.69) yields

$$\int_{A^e} \frac{\partial [N]^T}{\partial x}\left(n\left(D_{xx}\frac{\partial [N]}{\partial x} + D_{xz}\frac{\partial [N]}{\partial z}\right)\right) dA^e [C^*]$$

$$+ K_{xx} \int_{A^e} \frac{\partial [N]^T}{\partial x}\frac{\partial [N]}{\partial x} dA^e [\psi^*]$$

$$+ \int_{A^e} \frac{\partial [N]^T}{\partial z}\left(n\left(D_{xz}\frac{\partial [N]}{\partial x} + D_{zz}\frac{\partial [N]}{\partial z}\right)\right) dA^e [C^*]$$

$$+ K_{zz} \int_{A^e} \left(\frac{\partial [N]}{\partial z}[\psi^*] + \frac{1}{\rho_f}(\lambda_0 + a[N][C^*]) - 1\right) dA^e$$

$$+ \int_{A^e} [N]^T (RC^*) dA^e - \int_{A^e} [N]^T (PC) dA^e - n \int_{A^e} [N]^T \left(\frac{\partial C}{\partial t}\right) dA^e$$

$$= \int_{L^e} [N]^T (nD \cdot \nabla C - qC)_n dL^e. \quad (10.70)$$

As before, the integration on the right hand side of equation (10.70) must be evaluated over all the boundaries. The final form, after integration, of equation (10.70) can be written in matrix form as:

$$[W_c]\{C^*\} + [P_c]\{\partial C/\partial t\} = \{F_c\}, \quad (10.71)$$

where $[W_c]$ is the coefficient matrix which has been developed for the steady-state case, $\{C^*\}$ is the column matrix of the unknown nodal concentration, $[P_c]$ is a square matrix which accounts for the storage term in the transient conditions, $\{\partial C/\partial t\}$ is a column matrix of the time derivatives $\partial C/\partial t$, and $\{F_c\}$ is a column matrix which represents the boundary conditions.

Equations (10.67) and (10.71) constitute the main equations for density dependent groundwater flow and solute transport which when solved with the boundary conditions presented before yield the concentration distribution and the flow pattern in coastal aquifers. The same equations can be employed with appropriate boundary conditions to model saltwater upconing beneath pumping wells.

10.9 Study Cases

10.9.1 Hypothetical Case

Kawatani (1980) considered the case of an unconfined coastal aquifer with a channel carrying fresh water at the land side, as shown in Figure 10.17a. The water infiltration from the channel creates a constant water table above the sea level. If the infiltration from the channel is sufficient, a water divide will be formed somewhere beneath the channel, as shown in Figure 10.17a. The study region was 500×20 m. Three cases of different geological structure were investigated, as shown in Figure 10.17b. In the first case, the whole study region was homogeneous and isotropic. In the second case, the aquifer was divided into two strata by

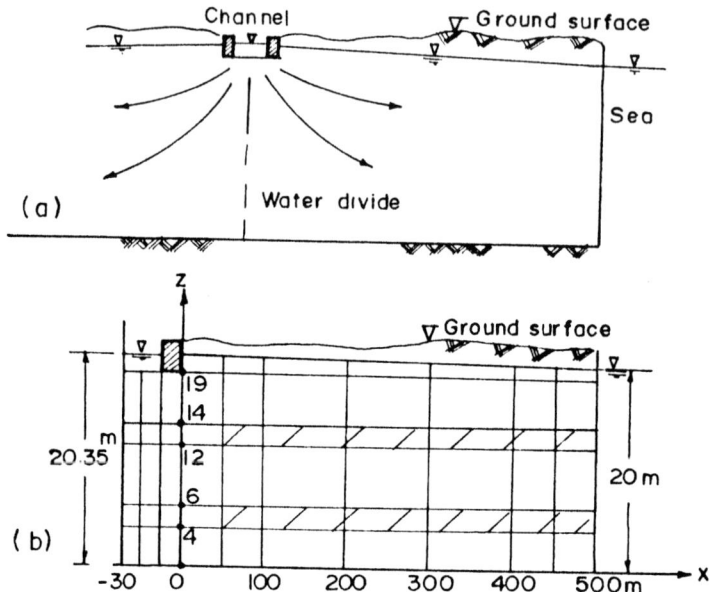

Fig. 10.17. Hypothetical case (Kawatani, 1980), (a) Water divide beneath an open channel. (b) Study region and location of semipervious layers.

interlaying an isotropic semipervious layer extending from $x = 50$ to 500 m between $Z = 4$ and 6 m. In the last case, the aquifer was divided into three strata by two isotropic semipervious layers. An additional semipervious layer extended from $x = 50$ to 500 m between $Z = 12$ and 14 m was introduced, as shown in Figure 10.17b. Eight-node quadrilateral isoparametric elements were used.

The water table at the land side (fresh water level in the channel) was 0.35 m above the sea level. The hydraulic conductivity in the x-direction in the pervious strata, K_{xx} was assumed to be constant and equal to 0.06 cm/min, while in the Z-direction, K_{zz} was set to equal 0.006 cm/min. The hydraulic conductivity for the semipervious layers was set to equal 0.0006 cm/sec. An effective porosity of 0.25 was chosen over the entire flow domain.

While solving the transport equation by finite difference or finite element methods, instability occurs if the convection terms exceed a certain value related to the dispersion coefficients and the size of the grid, namely, when the local Peclet number is greater than 2.0 (Heinrich et al., 1977). To avoid this instability problem, Kawatani assumed constant dispersion coefficients. He set D_{xx}, D_{zz} and D_{xz} to be equal to 5, 0.5, and 0.05 cm^2/min, respectively.

The flow pattern for the three different cases is presented in Figure 10.18. A strong and clear cyclic flow was found in the first case, whereas in the second and third cases, the semipervious strata interfered with the occurrence of the cyclic flow and the saltwater intruded further inland particularly in the lower strata near the bottom of the aquifer where the piezometric head at the sea boundary was relatively high. The rotational character of the flow in the first case has altered the

Saltwater Intrusion

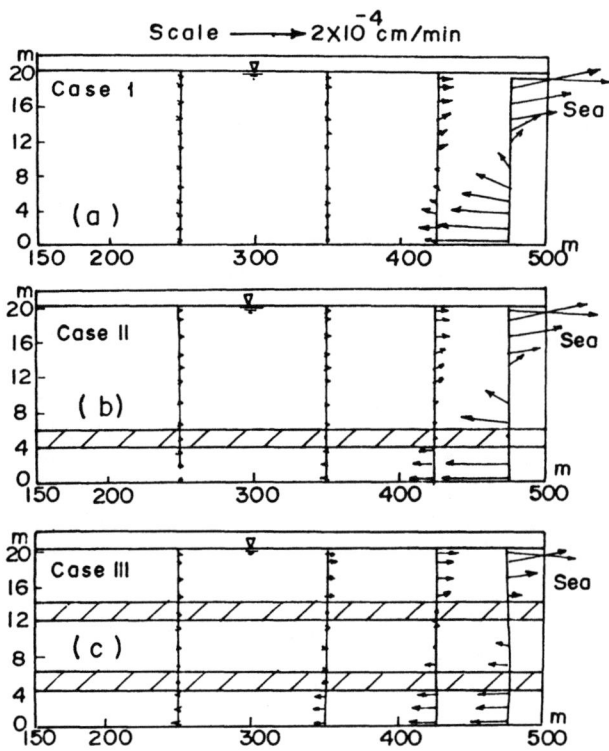

Fig. 10.18. Flow pattern for different cases. (a) Homogeneous medium. (b) One semipervious layer. (c) Two semi-pervious layers (Kawatani, 1980).

direction of the saltwater back to the sea through the window (upper part of the sea boundary). No seawater crossed the section at $x = 350$ m, where the velocity vector was found to be directed to the seaside (Figure 10.18a). In the last case, the saltwater crossed the section at $x = 250$ m in the lowest stratum (Figure 10.18c).

Equiconcentration lines for the three different cases are presented in Figure 10.19. Equiconcentration line 0.05 may be regarded as the front of the saltwater intrusion. In the first case, where the aquifer was homogeneous, equiconcentration line 0.05 advanced to a distance of 142 m measured from the sea boundary on the bottom boundary (Figure 10.19a). In case two, the same equiconcentration line advanced to a distance of 250 m measured from the same boundary (Figure 10.19b). In the last case (Figure 10.19c), where two semipervious strata were placed, the saltwater front intruded the bottom layer to a distance of 400 m from the seaside.

Sherif et al. (1990) considered the same domain with the same hydraulic parameters and boundary conditions. The homogeneous domain was represented by a uniform mesh with 1280 triangular elements with 697 nodes. The dispersion coefficients were taken to be velocity dependent. Longitudinal and lateral dispersivities, α_l and α_t, were taken as 0.5 m and 0.25 m, respectively. The piezometric head adjusted quickly and was nearly independent of the concentration. On the other hand,

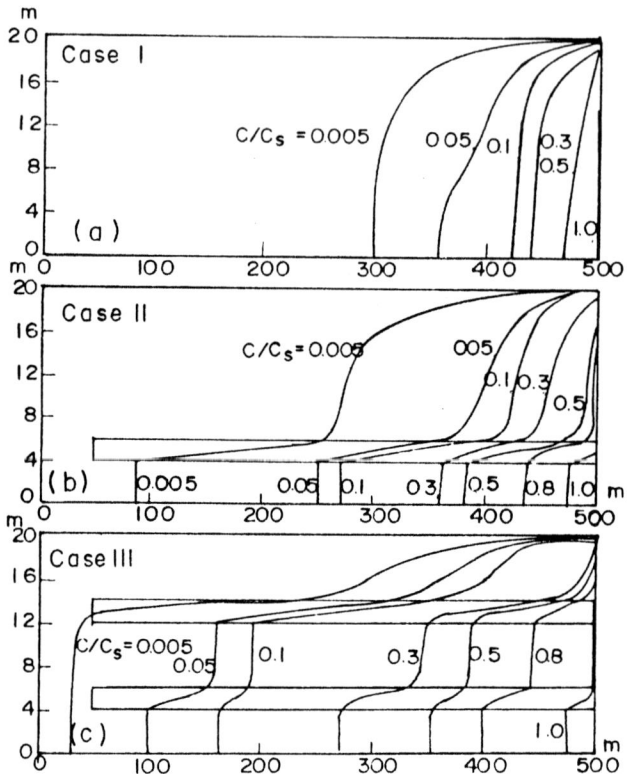

Fig. 10.19. Equiconcentration lines for different cases. (a) Homogeneous medium. (b) One semipervious layer. (c) Two semipervious layers (Kawatani, 1980).

the concentration was very sensitive to any variation in the piezometric head and required more iterations to achieve the convergence criterion. Figure 10.20a compares the equiconcentration lines obtained by Sherif et al. with those of Captain. A possible argument for the discrepancies between the two results is that different numerical schemes and different grid systems were used. Figure 10.20b presents the equipotential lines. Cyclic flow is evident at the sea boundary where the flow is rotational in character. This result is also consistent with the investigations of Huyakorn and Taylor (1976), Anand and Pandit (1982), Pandit and Anand (1984), among others.

10.9.2 THE MADRAS AQUIFER

The city of Madras is the capital city of Tamil Nadu in India and is situated on the coast of Bay of Bengal, as shown in Figure 10.21a. The climate is tropical with maximum mean temperature of 38.5°C. Madras is the fourth largest city in India with a population of over 4.0 million. Many big industries have been established in and around the city. Demand for water for drinking, industrial and

Saltwater Intrusion

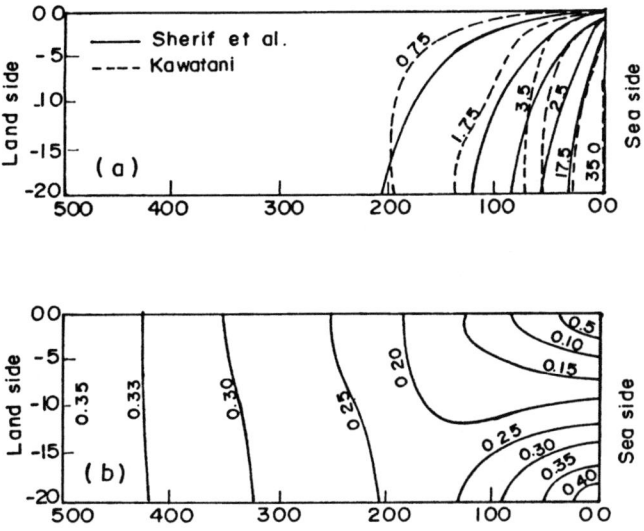

Fig. 10.20. (a) Comparison between equiconcentration lines obtained by Sherif et al. and Kawatani for a homogeneous medium. (b) Equipotential lines for homogeneous medium (Sherif et al., 1990).

agricultural purposes is satisfied by three major well fields, viz. Minjur, Panjetty and Tamaraipakkam (Rouve and Stoessinger, 1980). Water is pumped from these wells and supplied to industrial and drinking purposes by a pipe line system. The well field at Tamaraipakkam is far away from the coast and therefore not subject to the intrusion problem. The wells in the Minjur and Panjetty are relatively close to the coast (Figure 10.21a) and under uncontrolled pumping it may be subject to saltwater intrusion.

Geological formation of the aquifer along the east coast consists of unconsolidated and semiconsolidated formations with sand, silt and alluvial gravel. The geological cross section of the aquifer, taken through the Minjur and Panjetty fields approximately in the East-West direction, is shown in Figure 10.21b. The aquifer is confined, nearly homogeneous and isotropic in nature with granular sandy strata of average thickness of about 30.0 m. Upper and lower boundaries of the aquifer are impermeable clay layers (Rouve and Stoessinger, 1980). The aquifer has an effective porosity of 0.35 and an average hydraulic conductivity of 3.0×10^{-3} m/sec. No measurements were available about dispersivities. Based on similar studies (Pinder and Gray, 1977) in California, longitudinal and transversal dispersivities were set equal to 66.6 m and 6.6 m, respectively. Molecular diffusion D_{ij}^* was taken as 1.0×10^{-6} m^2/sec. It was assumed that no rainfall could infiltrate directly into the confined aquifer. Actual recharge takes place over a region far away from the pumping region. There was no correlation between the time variation of rainfall and groundwater fluctuations in the Minjur area. Based on some numerical simulations, it was found that pumping fields at Panjetty and Minjur were independent of each other. Thereupon only Minjur, which is near to the shore, was considered. Because the depth of the aquifer which is in direct hydraulic contact with the sea

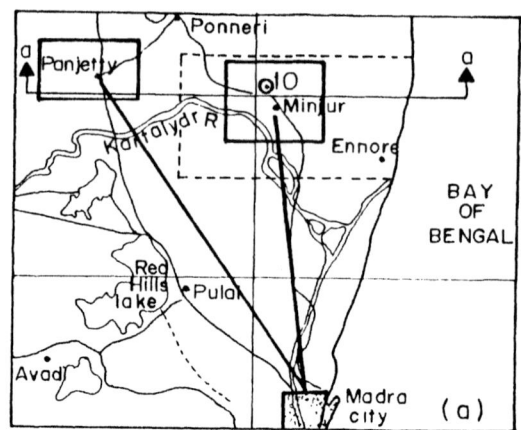

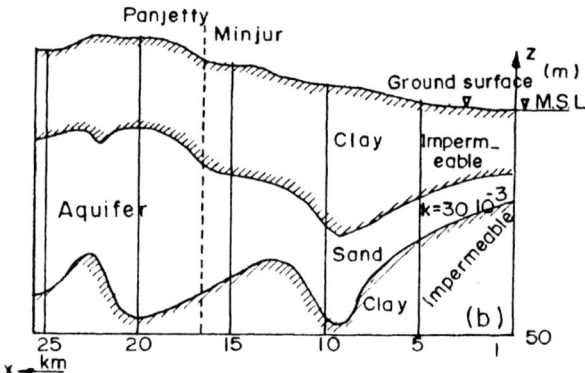

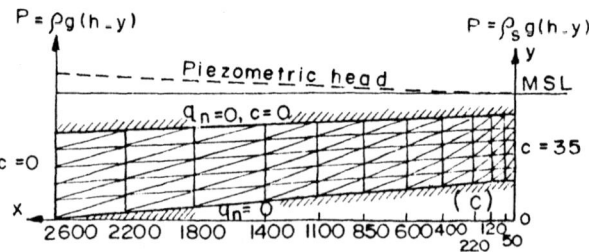

Fig. 10.21. The Madras Aquifer (Rouve and Stoessinger, 1980). (a) Minjur and Panjetty well field. (b) Geological cross section in the Madras aquifer. (c) Study domain and finite element grid.

Saltwater Intrusion

TABLE 10.1
Concentration distribution (kg/m^3) in Madras aquifer (Stoessinger, 1979)

Y (m)	Shoreline, X (m)											
	0	50	120	220	440	600	850	1000	1400	1800	2200	2600
0	12.4	0	0	0	0	0	0	0	0	0	0	0
5	18.3	12.0	3.6	1.6	0.4	0.1	0	0	0	0	0	0
10	27.9	19.3	9.4	3.7	1.0	0.2	0.1	0	0	0	0	0
15	35.0	22.5	13.9	6.7	1.6	0.4	0.1	0	0	0	0	0
20	35.0	24.4	15.8	9.1	2.4	0.5	0.1	0	0	0	0	0
25	35.0	25.8	17.1	9.9	2.9	0.6	0.1	0	0	0	0	0
30	35.0	26.7	18.0	10.4	3.2	0.6	0.1	0	0	0	0	0

TABLE 10.2
Piezometric head (m) in Madras aquifer (Stoessinger, 1979)

Y (m)	Shoreline, X (m)											
	0	50	120	220	440	600	850	1000	1400	1800	2200	2600
0	40.2	40.5	40.5	40.6	40.7	40.7	40.7	40.8	40.8	40.9	40.9	41.0
5	40.2	40.4	40.5	40.6	40.7	40.7	40.8	40.8	40.8	40.9	40.9	41.0
10	40.1	40.3	40.4	40.6	40.6	40.7	40.7	40.8	40.8	40.9	40.9	41.0
15	40.0	40.2	40.4	40.5	40.7	40.7	40.8	40.8	40.8	40.9	40.9	41.0
20	40.0	40.2	40.3	40.5	40.6	40.7	40.7	40.8	40.8	40.9	40.9	41.0
25	40.0	40.1	40.3	40.5	40.6	40.7	40.8	40.8	40.8	40.9	40.9	41.0
30	40.0	40.1	40.3	40.5	40.6	40.7	40.7	40.8	40.8	40.9	40.9	41.0

is only 30.0 m, the limit up to which the influence of hydrodynamic dispersion can be extended is about 3.0 km at which a boundary of freshwater concentration was assumed.

A study domain of 2600 m×30 m was considered. The domain was represented by 134 elements with 84 nodes, as shown in Figure 10.21c. For the actual groundwater pumping situation the piezometric height at the land side was 1.0m above the mean sea level. The resulting concentration distribution and piezometric head in the Madras aquifer are given in Tables 1 and 2 (Stoessinger, 1979).

In order to investigate the effect of any additional pumping at Minjur pumping field on the intrusion length, the piezometric head at the land side (2600 m from the sea boundary) was reduced by 0.2 m. In this case the equiconcentration line 17.5 kg/m^3 (maximum concentration = 35 kg/m^3) advanced inland 23.0 m more than in the first case (Figure 10.22). Rouve and Stoessinger (1980) concluded that the landward movement of the intrusion front would be of the order of 500 m if the piezometric head at the far end was further reduced to 40.7 m.

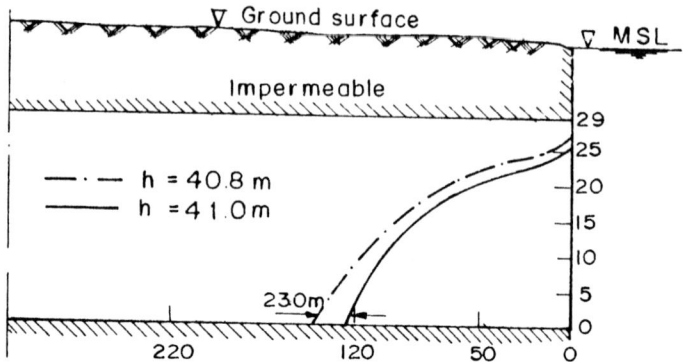

Fig. 10.22. Movement of equiconcentration line 0.5 due to lowering the piezometric head (Rouve and Stoessinger, 1980).

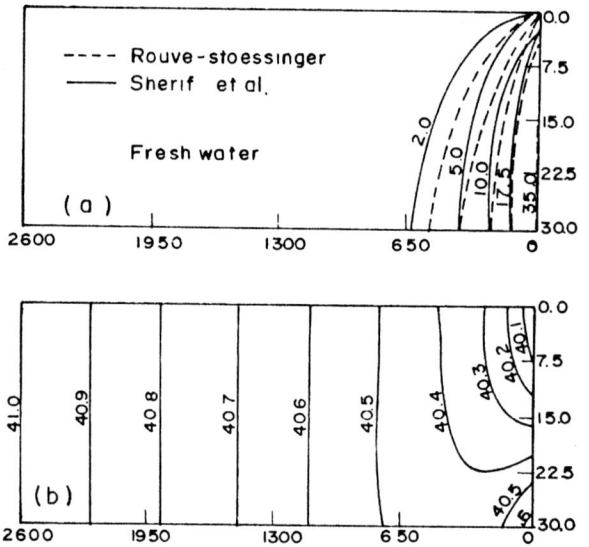

Fig. 10.23. Dispersion zone and flow pattern in Madras aquifer, case one (a) Equiconcentration lines. (b) Equipotential lines (Sherif et al., 1990).

Sherif et al. (1990) studied the same problem and used the same hydraulic parameters and boundary conditions. The domain was represented by 400 triangular elements and 246 nodes. The two cases for the piezometric head at the landside were considered. In the first case the piezometric head at the land side was set equal to 41.0 m (one meter above the sea level). It was found that the width of the dispersion zone, which can be defined between equiconcentration lines 35 and 2, measured at the lower boundary was about 610 m. In the Rouve–Stoessinger model, the width of the dispersion zone measured at the same boundary was about 520 m as shown in Figure 10.23a. The equiconcentration line 10 intruded inland

Saltwater Intrusion

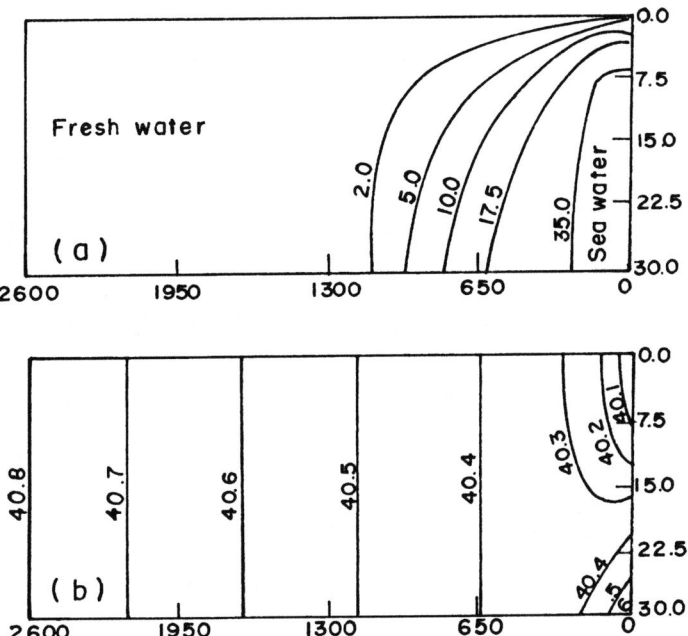

Fig. 10.24. Dispersion zone and flow pattern in Madras aquifer, case two (a) Equiconcentration lines. (b) Equipotential lines (Sherif et al., 1990).

250 m from the sea boundary, while in the Rouve–Stoessinger model, it intruded 232 m. Cyclic flow was predicted near the shore line, as shown in Figure 10.23b. The Rouve–Stoessinger model did not exhibit any cyclic flow as seen from Table 2. Mixed water swept out to the sea all over the sea boundary but under a different hydraulic gradient.

In the second case, the piezometric head at the landside was lowered by 0.2 m due to excessive pumping. Saltwater (equiconcentration line 35) intruded inland to a distance of 260 m measured at the bottom boundary from the sea side. The width of the dispersion was about 840m as shown in Figure 10.24a. Lowering the piezometric head at the land boundary allowed more saltwater to intrude the Madras aquifer through the lower part of the seaside boundary. Equipotential lines for this case are shown in Figure 10.24b.

10.9.3 The Nile Delta Aquifer in Egypt

Egypt lies between latitudes 22° and 32° north, and longitudes 25° and 35° east. The north boundary of Egypt is the Mediterranean sea, and the east is the Red sea. The south and west are political boundaries with Sudan and Libya, respectively. Land area of Egypt is equal to about one million km^2. The Nile Delta and its fringes (22000 km^2), as shown in Figure 10.25, lie between latitudes 30°05' and 31°30' north, and longitudes 29°50' and 32°15' east. At a distance of 20 km north-west of Cairo (Delta Barrage), and at an elevation of 17 m above the sea level, the Nile

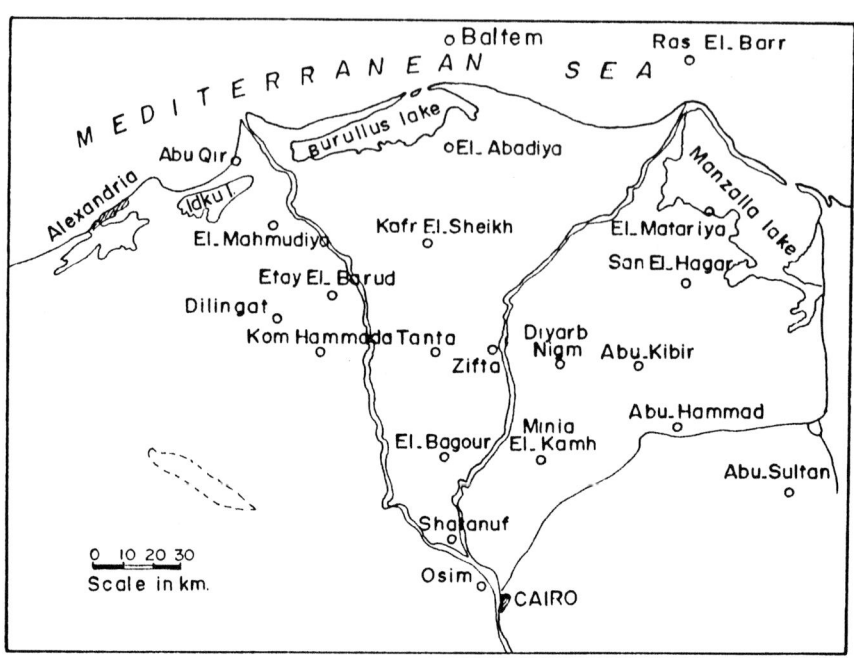

Fig. 10.25. The Nile Delta aquifer and its fringes.

Valley begins to open out into a triangular alluvial Delta with a base length of 275 km along the Mediterranean Sea joining Alexandria and Port-Said, as shown in Figure 10.25. The level of the Delta land ranges between +17 m above the sea level at the south to less than one meter at the north boundary (Farid, 1985).

The Nile Delta region lies within the temperature zone which is a part of the great desert belt. It also occupies a portion of the arid belt of the Southern Mediterranean region. The desert fringes on both sides of the Delta cause a rise in temperature and affect the changes in daily temperature. The average temperatures in January and July at Cairo are 12°C and 31°C, respectively. Minimum and maximum temperatures at Cairo are 3°C and 48°C, respectively.

The rainfall over the Nile Delta is rare and occurs in Winter. Maximum average rainfall along the Mediterranean Sea shore, where most of the rain occurs, is about 180 mm. This amount decreases very rapidly as one proceeds inland to about 26 mm at Cairo.

The Nile Delta aquifer is among the largest underground freshwater reservoirs in the world and it constitutes an important water resource in the future of Egyptian economy. It extends over six million acres, and is naturally bounded Northward by the Mideterranean Sea and eastward by the Suez Canal. The Western boundary extends well into the desert. At the south, the aquifer demises and seems to be isolated from the aquifer of upper Egypt by an aquiclude approaching the clay cap near Cairo at El-Manawat, as shown in Figure 10.26. It fills a vast underground bowl situated between Cairo and the Mediterranean Sea. The Nile Delta aquifer system is a complex groundwater system. It is a leaky one, with an upper semi-permeable

Saltwater Intrusion

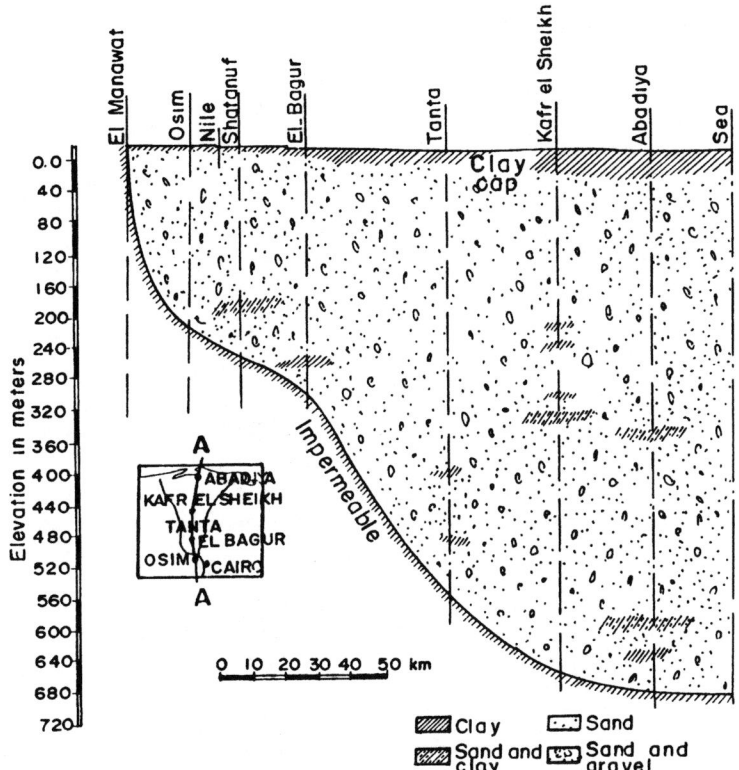

Fig. 10.26. Geological cross section (A-A) in the Nile Delta aquifer.

boundary and a lower impermeable boundary. The aquifer is recharged by the infiltration from irrigation networks, excess of irrigation water and precipitation through the upper clay layer. It may also be recharged by any possible flow comming from the upper Egypt aquifer.

There are three dependent sources of water in Egypt: Nile water, drainage water, and groundwater. The Egyptian share from the Nile water is limited to 55.5×10^9 m^3 per year. Reuse of drainage water is not recommended because of its low quality and high contents of chemicals and pesticides. The growth of population in the Nile Delta area and, hence, the increase in human activities, agricultural and industrial, has imposed an increasing demand for freshwater. This increase in demand in the Delta area was covered by intensive pumping of fresh groundwater, causing subsequent lowering of the piezometric head and upsetting the dynamic balance between freshwater body and saline water body in the aquifer. Like any coastal aquifer, an extensive saltwater flux has intruded the Nile Delta aquifer forming the major constraint against aquifer exploitation.

Few investigations have been carried out to study the saltwater intrusion in the Nile Delta aquifer. Farid (1985) modified AQUIFEM-1 (Wilson et al., 1979) to model areal intrusion in the Nile Delta aquifer under the sharp interface assumption.

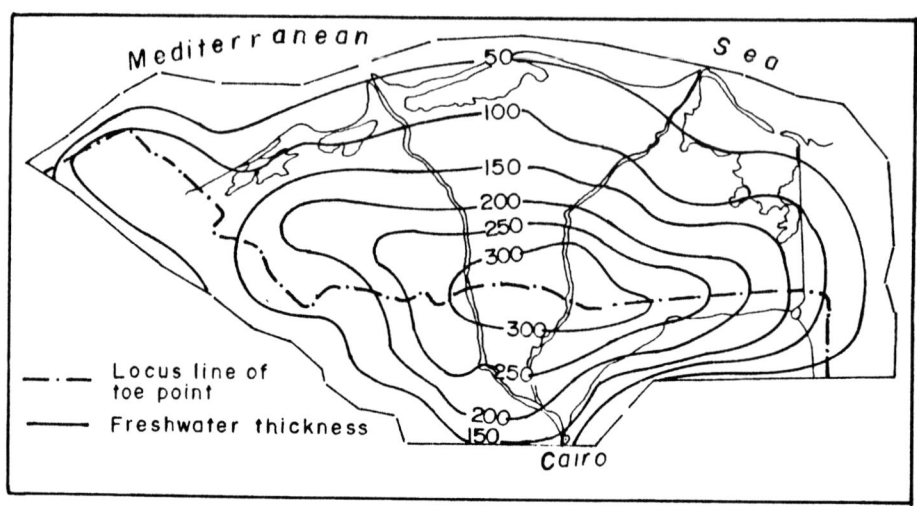

Fig. 10.27. Locus line of the toe point and contour lines of freshwater thickness in the Nile Delta aquifer (Farid, 1985).

Vertical equipotential lines (horizontal flow) were assumed and hence the solution was neither accurate nor justified. Figure 10.27 presents the locus line of the toe point (T.P.) and the contour lines of freshwater thickness in the aquifer.

Sherif et al. (1988) applied a two-dimensional finite element model (2D-FED) to identify the dispersion zone and the flow pattern in the Nile Delta aquifer using the density dependent approach. The 2D-FED model was applied to the longitudinal cross section (A–A) given in Figure 10.26. Based on some field data, it was assumed that saltwater would not migrate inland up to Shatanuf, which is 150.0 km from the Mediterranean Sea. Under this assumption a domain of 150.0 km in length was considered; the depth of the domain was varied from 680.0 m at the seaside to 240.0 m at the landside. The thickness of the upper confining semipervious layer was taken as 40.0 m from the sea boundary to Om-Sin which is about 17.0 km inland. From Om-Sin to Kafr-Elsheikh, 52.0 km from the sea, it was varied from 40.0 m to 15.0 m, after which the thickness of the semipervious layer is constant. The bottom boundary of the Nile Delta aquifer is impermeable.

Due to the uncertainty of the field data concerning the hydraulic parameters, a calibrated value for the hydraulic conductivities K_{xx} and K_{zz}, of 100.0 m/day was considered representative of the aquifer medium. The vertical hydraulic conductivity for the upper semipervious layer, K'_{zz}, was set equal to 0.05 m/day. The piezometric head at the land boundary was 14.0 m above sea level. At the sea boundary the piezometric head was 0.60 m. The free water table level was given for some stations, and between these stations it was assumed linear. There was no data available about the dispersivities. Based on similar studies, the longitudinal and transversal dispersivities, α_l and α_t, were assumed 100.0 m and 10.0 m, respectively.

Saltwater Intrusion

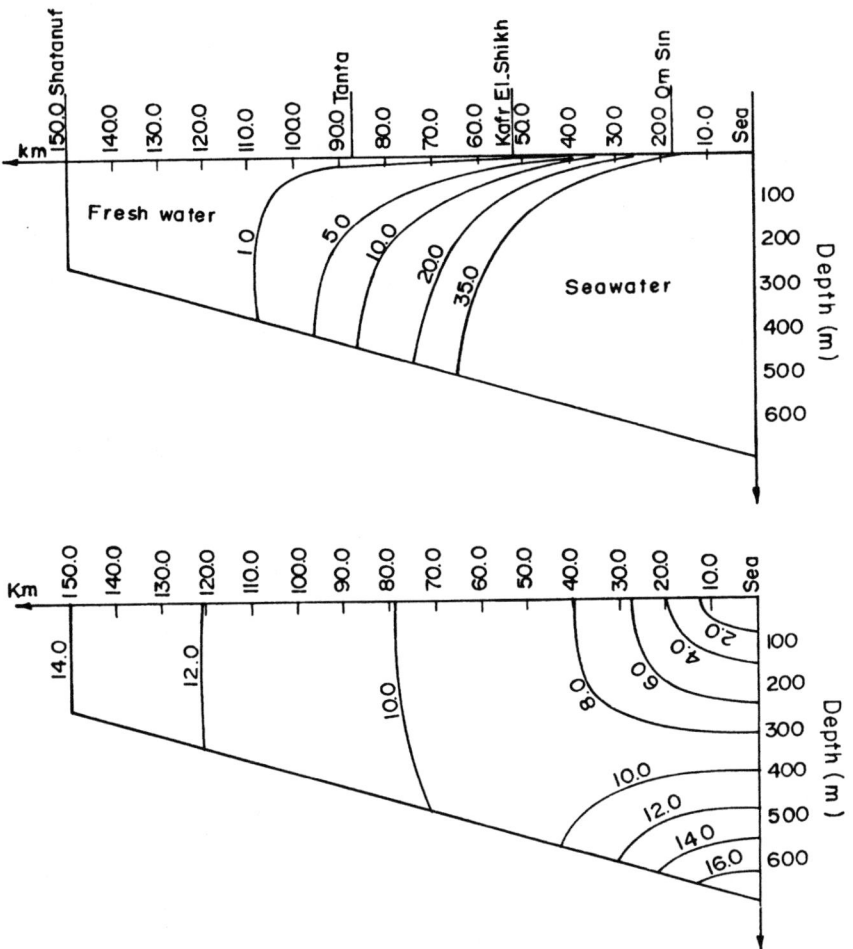

Fig. 10.28. Saltwater intrusion and flow pattern in the Nile Delta aquifer (Sherif et al., 1988). (a) Equiconcentration lines. (b) Equipotential lines.

The domain was subdivided into five subdomains; each subdomain was divided into a number of triangular elements with smaller areas in the regions where the variation in concentration gradient was relatively high. An intensive grid was also required near the shore boundary. The domain was finally represented by a nonuniform grid with 4020 nodes and 7600 triangular elements. The convergence criterion was set equal to 10^{-5}.

Unfortunately, due to deep and wide opening of the Nile Delta aquifer to the Mediterranean Sea, the seawater intruded the aquifer bottom under a high potential head, even more than that at the land side, and found its way back again to the sea through the window. At the last 22.0 km from the sea boundary, there was some upward flux of the mixed water through the upper semipervious layer. The

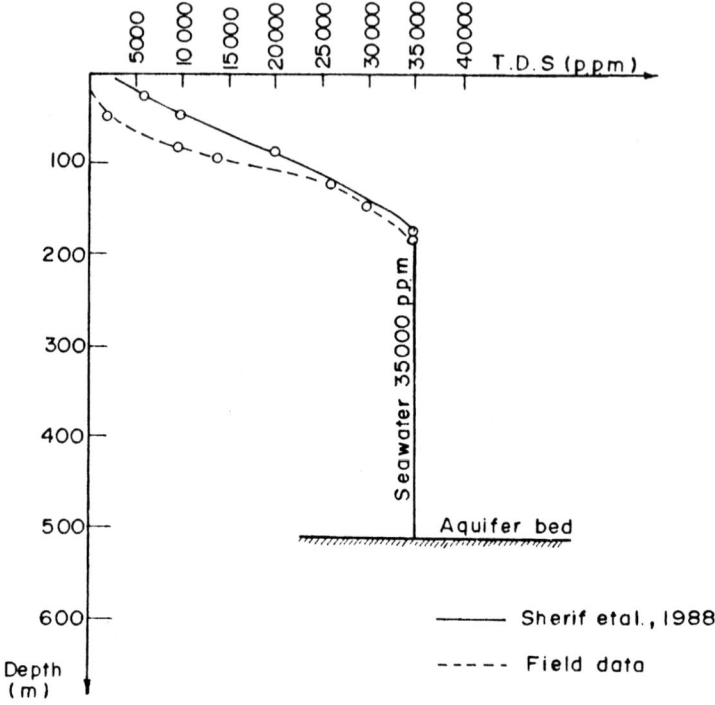

Fig. 10.29. Salinity variation with depth at Kafr-Elsheikh.

equiconcentration line 35.0 (seawater) intruded the Nile Delta aquifer to a distance of 63.0 km from the sea side measured at the bottom boundary, as shown in Figure 10.28a. The equiconcentration line 1.0 intruded the aquifer to a distance of about 108.0 km measured at the bottom boundary. The width of the dispersion zone is about 45.0 km, bigger than expected ever before (Figure 10.28a). It can be concluded from the shape of equipotential lines (Figure 10.28b) that the depth of the window at the seaside was about 350.0 m. Strong cyclic flow at the sea boundary was detected. Figure 10.29 presents a comparison of the results obtained by the 2D-FED model with the data available for the salinity distribution with depth at Kafr-Elsheikh.

10.10 Concluding Remarks

Saltwater intrusion into coastal aquifers is the classical result of groundwater exploitation and development in coastal areas. Saltwater upconing beneath pumping wells results from intensive pumping from freshwater strata overlying saltwater strata. Invasion of saltwater into coastal aquifers depends on many factors. The aquifer depth that is in direct hydraulic contact with the open sea determines the severity of intrusion. Two approaches can be utilized for modeling saltwater intrusion: the sharp interface approach and the density dependent approach. The sharp interface approach provides a quantitative evaluation for the problem under certain conditions, especially when the aquifer depth is relatively small and discharge events and artificial activities are absent. The density dependent approach provides a qualitative evaluation of the problem and accurately represents the flow pattern and the intrusion processes under different conditions.

Attention should be given to the sea side boundary when modeling saltwater intrusion. The problem is highly sensitive to this boundary. The reference density (freshwater density) concept should be used inside the study domain as well as on the boundaries. Cyclic flow is evident near the shore boundary where the seawater intrudes the bottom of the aquifer and migrates inland. The dispersion zone may vary from a few meters to tens of kilometers.

Establishment of dynamic monitoring networks in coastal areas is essential when groundwater exploitation activities are considered. Short screened observation wells are recommended. Monitoring should include water levels, abstraction rates, electrical conductivities and chemical characteristics of pumped water. Scattered productive wells are preferred and freshwater heads should be maintained as high as possible above mean sea level. Artificial recharge from surface water, lakes, and lagoons in coastal areas minimizes intrusion progress.

References

Anand, S.C. and Pandit, A., 1982. Finite element solution of coupled groundwater flow and transport equations under transient conditions including the effect of selected time step size. Proc. 4th Int. Conf. Finite Elements in Water Resources, Univ. of Hannover.
Badon-Ghyben, W., 1888. Nota in verband met de voorgenomen putboring nabij Amsterdam. Tijdschrift Koninklijk Instituut van Ingenieurs, the Hague, Netherlands, pp.8-22.
Baxter, G.P. and Wallace, C.C., 1916. Changes in volume upon solution in water of the halogen salts of alkali metals. J. Amer. Chem. Soc. 38: 70-104.
Bear, J., 1972. Dynamics of Fluids in Porous Media. American Elsevier, New York.
Bear, J., 1979. Hydraulics of Groundwater. McGraw-Hill, New York.
Bear, J. and Bachmat, Y., 1967. A Generalized theory on hydrodynamic dispersion in porous media. I.A.S.H. Symp. Artificial Recharge and Management of Aquifers, Haifa, Israel, IASH 72, PP.7-16.
Bear, J. and Dagan, G., 1968. Solving the problem of local interface upconing in a coastal aquifer by the method of small perturbations. J. Hydraul. Res., 6(1): 16-44.
Bear, J. and Dagan, G., 1962. The transition zone between fresh and salt water in coastal aquifers. Rep. 1. Hydraulic lab., Technion, Haifa, Israel.
Bear, J. and Verruijt, A., 1987. Modeling Groundwater Flow and Pollution. Reidel Publishing Co., Holland.
Bouwer, H., 1978. Groundwater Hydrology. McGraw-Hill, New York.
California Dept. Water Resources, 1958. Seawater intrusion in California. Water Res. Dept., Bull. 63, California, USA.
Custodio, E., 1993. Specific methodologies to identify and monitor seawater intrusion, especially in its early stages. Expert Consultation Meeting on Saltwater Intrusion into Coastal Aquifers in the Mediterranean Basin and the Near East, Cairo, Egypt.
Debuisson, J., 1970. La nappe aquifère du cordon dunaire de Malika (Sénégal). Bull. Bur. Rech. Géol. Min., 3-3:149-161.
Domenico, P.A. and Schwartz, F.W., 1990. Physical and Chemical Hydrogeology. John Wiley & Sons, Inc., New-York, 824p.

Ernest, L.F., 1969. Groundwater flow in the Netherlands delta area and its influence on the salt balance of the future Lake Zeeland J. Hydrol., V.8:137-172.

Essaid, H.I., 1987. Freshwater-saltwater flow dynamics in coastal aquifer systems: Development and application of a multi-layer sharp interface model. Ph.D. Thesis, Stanford Univ., Stanford, Calif.

Farid, M.S., 1985. Management of groundwater system in the Nile Delta. Ph. D. Thesis, Faculty of Engineering, Cairo Univ., Egypt.

Fetter, C.W., 1972. Position of the saline water interface beneath Oceanic Islands. Water Resour. Res., 8:1307-1314.

Fried, J.J., 1975. Groundwater Pollution. Elsevier Scientific Publication Company, Amsterdam, pp.5-46.

Fried, J.J. and Combarnous, M.A., 1971. Dispersion in porous media. In V.T. Chow (Editor), Advances in Hydroscience, Academic Press, New York, PP.169-282.

Forsythe, G.E. and Wasow, W.R., 1960. Finite Difference Methods for Partial Differential Equations. Wiley, New York.

Galeati, G., Gambolati, G. and Neuman, S.P., 1992. Coupled and partially coupled Eulerian-Lagrangian model of freshwater-saltwater mixing. Water Resources Research, 28(1): 149-165.

Glover, R.E., 1964. The pattern of fresh-water flow in coastal aquifer. In sea water in coastal aquifers, U.S.Geological Survey Water Supply paper 1613-C, pp.32-35.

Goodwill, I., 1993. Mathematical modeling of saline intrusion in coastal aquifers. Expert Consultation Meeting on Saltwater Intrusion into Coastal Aquifers in the Mediterranean Basin and the Near East, Cairo, Egypt.

Grove, D.B., 1977. The use of Galerkin finite-element methods to solve mass transport equations. U.S.G.S., Water Resources Investigations, 77-79.

Haubold, R.G., 1975. Approximation for steady interface beneath a well pumping fresh water overlying salt water. Ground Water, 13(3): 254-259.

Heinrich, J.C., Huyakorn, P.S., Zienkiewicz, O.C., and Mitchell, A.R., 1977. An upwind finite element scheme for two dimensional convection transport equation. Int. J. Num. Meth. Engng., 10:131-143.

Herzberg, A., 1901. Die wasserversorgung einiger Nordseebaden, Z. Gasbeleuchtung and Wasserversorgung. 44:815-819.

Hunt, B., 1983. Mathematical Analysis of Groundwater Resources. Butterworth & Co.pp.271.

Huyakorn, P.S., Andersen, P.F., Mercer, J.W. and White, H.O., 1987. Saltwater intrusion in aquifers: Development and testing of a three-dimensional finite element model. Water Resour. Res., 23(2): 293-312.

Huyakorn, P.S. and Taylor, C., 1976. Finite element model for coupled groundwater flow and convective dispersion. Proc. 1st Int. Conf. Finite Elements in Water Resources, Princeton Univ., Princeton, N.J.

Kashef, A.I., 1975. Management of Retardation of Saltwater Intrusion in Coastal Aquifers. Office of Water Res. and Techn., U.S. Dept., of Interior, 281 P.

Kawatani, T., 1980. Behavior of seawater intrusion in layered coastal aquifer. Proc. 3^{rd}Int. Con. Finite Elements in Water Resources, Univ. of Mississippi, Oxford, Mississippi.

Kinzelbach, W., 1986. Groundwater Modeling. Elsevier, New York, 333 PP.

Klotz, D., 1973. Untersuchunger Zur dispersion in porosen medien. Z Deutsch. Geol. Ges., 124 : 523-533.

Kohout, F.A., 1961. Case history of saltwater encroachment caused by a storm sewer in Miami. J. Amer. Water., Water Works Assoc.,V. 53, P. 1406-1416.

Konikow, L.F. and Bredehoeft, J.D., 1978. Computer model of two- dimensional solute transport and dispersion in groundwater. Techniques of Water Resources Investigations of the U.S.G.S., Chapter C2, Book 7, 90p.

Lenda, A. and Zuber, A., 1970. Tracer dispersion in groundwater experiments. Isotope Hydrol. IAEA-SM-129137, PP. 129-134.

Liggett, J.A. and Liu, P.L.F., 1983. The Boundary Integral Equation Method for Porous Media Flow. Allen & Wiley, London.

Lusczynski, N.J. and Swarzenski, W.V., 1966, Saltwater encroachment in Gnaws and Southeastern Queens Counties, Long Island, New-York. U.S. Geol. Survey Water Paper 1613-F.

McWhorter, D.B., 1972. Steady and unsteady flow of freshwater in saline aquifers. Water Management Techn. Rep. 20., Colorado State Univ., Fort Collins, Colorado.

McWhorter, D.B. and Sunada, D.K., 1977. Ground-Water Hydrology and Hydraulics. Water Resources Publications, Fort Collins, Colorado.

Mercer, J., Larson, S., and Faust, C., 1980a. Finite difference model to simulate the areal flow of saltwater and freshwater separated by an interface. U.S.G.S., Open File Rep., 80-407.

Mercer, J., Larson, S., and Faust, C., 1980b. Simulation of saltwater interface motion. Ground Water, 18(4): 374-385.

Newport, B.D., 1977. Saltwater intrusion in the Unites States. Rep. no. 60018-77-010, U.S. Environ Prot. Agency, Washington, D.C.

Nikolaevskii, V.N., 1959. Convective diffusion in porous media. Journal of Appl. Math. Mech. (PMM), 23(6), 1042-1050.

Pandit, A. and Anand, S.C., 1984. Groundwater flow and mass transport by finite elements- a parametric study. Proceedings of the 5th International Conference on Finite Elements in Water Resources, University of Vermont.

Pascual, J.M. and Custodio, E., 1990. Geochemical observation in a continuously saltwater intruded area: Garraf, Catalonia (Spain). 11th Saltwater Intrusion Meeting. Gdansk: 308-330.

Pickens, F.J. and Grisak, E.G., 1980. Scale-dependent dispersion in stratified granular aquifers. Water Resour. Res., 17(4):1191-1211.

Pinder, G.F., 1973. A Galerkin-finite element simulation of groundwater concentration on Long Island. Water Resour Res., 9(6):1657-1669.

Pinder, G.F. and Gray, W.C., 1977. Finite Element Simulation in Surface and Subsurface Hydrology. Acad. Press, New York.

Rouve, G. and Stoessinger, W., 1980. Simulation of the transport position of the saltwater intrusion in coastal aquifer near Madras coast. Proc. 3rd Int. Con. Finite Elements in Water Resources, Univ. of Mississippi, Oxford, Mississippi.

Sakr, S.A., 1992. Vertically integrated two dimensional finite element model of sea water intrusion in aquifers. M.Sc. Thesis, Colorado State University, Fort Collins, Colorado.

Sanford, W.E. and Konikow, L.F., 1985. A two-constituent solute-transport model for groundwater having variable density. U.S.G.S. Water Resources Investigations Report 85-4279, 88p.

Scheidegger, A.E., 1961. General theory of dispersion in porous media. Journal of Geophys. Res., 66, 3273-3278.

Schmorak, S. and Mercado, A., 1969. Upconing of fresh water-sea water interface below pumping wells, field study. Water Resour. Res., 5: 1290-1310.

Segol, G. and Pinder, G.F., 1976. Transient simulation of saltwater intrusion in south eastern Florida. Water Resour. Res., 12(1):65-70.

Sherif, M.M., Singh, V.P. and Amer, A.M., 1988. A two-dimensional finite element model for dispersion (2D-FED) in coastal aquifers. J. Hydrol., 103:10-36.

Sherif, M.M., Singh, V.P. and Amer, A.M.,1990. A sensitivity analysis of '2D-FED', a model for seawater encroachment in leaky coastal aquifers. J. Hydrol., 108:343-356.

Sherif, M.M., Singh, V.P. and Amer, A.M., 1990. A note on saltwater intrusion in coastal aquifers. Water Resour. Management, 4:123-134.

Stoessinger, W., 1979. Beschreibung der Hydrodynamic Dispersion mit der Methode der Finiten Elementen am Beispiel d. instationaren Interface zwischen Sub-U. Salzwasser in Grundwasser-Leitern. Mitt. Inst. f. Wasserbau u. Wasser- wirtschaft der RWTH Aachen, H. 28.

Strack, O.D.L., 1987. Groundwater Mechanics. Prentice-Hall, Englewood Cliffs, N.J.

Volker, R.E. and Rushton, K.R., 1982. An Assessment of the importance of some parameters for seawater intrusion in aquifers and a comparison of dispersive and sharp-interface modelling approaches. J. Hydrol., 56:239-250.

Walton, W.C., 1970. Groundwater Resources Evaluation. McGraw-Hill Co., 664p.

Wilson, J., Townley, L.R. and Sa Da Costa, A., 1979. Mathematical Development and Verification of a Finite Element Aquifer Flow Model AQUIFEM-1, Technoloy Adaptation Program, Report No. 79-2, M.I.T., Cambridge, Massachusetts.

Wilson, J. and Sa da Costa, A., 1982. Finite element simulation of saltwater/freshwater interface with indirect toe tracking. Water Resources Research, 18(4): 1069-1080.

Zienkiewicz, O.C., 1977. The Finite Element Method. 3rd edn., McGraw-Hill, London.

CHAPTER 11

Avalanche Dynamics

Kolumban Hutter

ABSTRACT. This article reviews recent work on avalanche, landslide and rockfall dynamics. Two *limiting cases* of these flows exist, the so-called *flow avalanche*, i. e., the dense gravity driven "laminar type flow" in which the role of the solid particles dominates, while that of the interstitial fluid is negligible — these flows are typical for most sturzstroms, debris flows, landslides, rockfalls and snow avalanches — and the less dense *powder avalanche*, i. e., the turbulent flow of air borne particles in a mixture, in which the role of the fluid dominates, while that of the particles is less significant — these flows are typical for density and turbidity currents such as dust clouds occuring in the desert, in pyroclastic volcanic eruptions, in submarine slope instabilities and in snow and ice avalanches. The latter application will be our focus. The state of the art in the description of both these phenomena is given. In the Introduction, after some historical remarks, we turn our attention to the characterization of the physical behaviour of the two limiting flow types and then discuss the laws of similitude and model theory relevant to modelling the flows in the laboratory.

Flow avalanches, landslides and rockfalls are discussed first. It is argued that these flows can be described as a continuum consisting of a cohesionless granular material. Such materials exhibit dilatancy effects and large energy dissipation, and under quasistatic or dynamic shear deformation, they exhibit the property that the ratio of the shear stress to the normal stress on any interior plane is nearly constant, a fact reflected in the constancy of the internal angle of friction. In shear cell tests under quasistatic and rapid deformation at constant normal load, the internal stress is practically independent of the rate of shear; when the shear deformation is performed at constant volume however, the dependence of the stress on the rate of shear is quadratic. Three different flow regimes which partly interact can be distinguished: (i), Dry Coulomb, rubbing frictional behaviour, typical when particles are in contact and ride one over the other; (ii), collisional interactions when particles bounce against each other, contact is short, and the mean free path is of the order of the particle diameter; (iii), translational transport when the mean free path is large and particle concentrations correspondingly small. For the second and the third regime, statistical theories along the lines of the kinetic theory of a dense gas have been developed, but some of these show ill behaviour in steady chute flow problems. An adequate formulation must incorporate quasistatic *and* collisional contributions; the emerging theories, however, are patched together from alien components. Furthermore, simple chute flow problems turn out to be very difficult to solve. In the flow regime where most of the moving granular mass rides more or less passively on a fluidized bed, Savage and Hutter (1989) have proposed to treat the material as being of the rate independent Coulomb type, both in the interior and at the bed, with constant internal and bed friction angle ϕ and δ, respectively. In this formulation, the bed friction angle δ is a measure of the resistance of the bed to a fluidized thin layer, a first approximation to a more general bed friction law that may also include viscous components. The depth averaged field equations of Savage and Hutter reproduce laboratory chute flow of a cohesionless granular material quite well even in cases when a localized bump in an otherwise convexly curved bed separates an initially single mass into two separate piles. The equations also permit similarity solutions, i.e., solutions of permanent shape (but not size). These solutions are not quantitatively corroborated in general by laboratory experiments, but they permit better analytic insight and thus help in understanding the model equations physically. Extensions of both theory and experiments to avalanche flows that spread in two dimensions indicate promising results.

Powder avalanches have been treated in the literature by essentially two different concepts: (i), a simple binary mixture of turbulent air and suspended particles with two balance laws for each constituent (or the mixture as a whole and one constituent: the particles) and a momentum balance for the mixture as a whole and (ii), a two-phase mixture with balance laws of mass and linear momentum for each constituent.

The first class of models concentrates on integral, or global representations. Long turbulent gravity currents from a steady continuous source on inclined planes and short, finite mass "clouds" or "thermals" are studied from the viewpoint of depth integrated mass, buoyancy and momentum balance laws. Typical mean velocity and averaged density variations are determined for idealized conditions, and corresponding results compared or matched with data obtained from laboratory experiments. Air entrainment through the upper free surface of the turbulent gravity current is accounted for, but entrainment of the particles from the bed or deposition of particles at the bed are often ignored when asymptotic behaviour for large time or large distances from the source are considered. Such earlier and simpler models of powder avalanches do not quantify the level of turbulence and cannot, therefore, properly describe the physical conditions of autosuspension or re-sedimentation of the particles. We present, however, a brief summary of formulations for long and short gravity currents in which the balance laws of mass, buoyancy and momentum are complemented by a balance of the averaged turbulent fluctuation energy that is coupled with these by closure statements on basal friction and snow entrainment. Such models can be computationally handled, but have not been verified in detail by laboratory experiments, even though an ad hoc application of the model equations to a real avalanche event indicates promising results. Proposals of theoretical formulations for simple mixtures of a turbulent fluid carrying suspended particles have also been put forward on the basis of k-ε closure and using *local* equations; very little experience, however, seems to have been done with this model.

Two phase turbulent mixtures that use the balance laws of mass and momentum for both constituents have also been proposed; they also use second order (k-ε) closure conditions. The novel theoretical feature in these models is, however, not so much the set of field equations than the handling of the free boundary as a singular surface, a concept that is briefly touched upon here. Finally we also present experimental results on the distribution of the streamwise particle velocity and density, both in the tail and the head of a laboratory powder avalanche of polystyrene particles suspended in water and moving down a straight and curved chute.

11.1 Introduction

Avalanches belong to the kind of natural phenomena to which the population of alpine regions is exposed during wintertime and spring. Research in avalanche prevention and the study of their dynamics are important responsibilities of most mountainous nations. In recent years, this responsibility has even increased due to the effects of the dying forest and the Greenhouse phenomenon. In the 48 winters from 1940/41 to 1987/88, Switzerland has had 7191 avalanches with damaged property, leading to 960 injured persons and 1269 deaths and an approximate total insurance refund for elementary damage of more than 300 Mio Swiss francs. The worst year was the winter 1950/51, with 1300 avalanches causing 42 Mio Swiss francs damage and 98 deaths. The yearly average over the 48 years period is 20 injured persons and 26 casualties and a refund of 6–8 Mio Swiss francs for elementary damage. Interesting in these figures is that the number of deaths always exceeds the number of injured people. Furthermore, the capital that is invested for direct and indirect *prevention* of damage and casualties due to avalanches is much more than the avaerage money spent for insurance claims (approximately 50 Mio Swiss francs per year).

11.1.1 SOME HISTORICAL NOTES

There exist a great number of historical reports on snow avalanches. Among the first who mentions them is the Greek geographer Strabon (63–23 BC), another one is the historian Titus Livius (59 BC–14 AD) who describes Hannibal's crossing of the Alps (218 BC); their statements are as follows, see Scheiwiller (1986):

> ".. ice avalanches which often carry away with them whole tourist parties and throw them into the abyss. For numerous are the layers lying one above the other. The snow develops layer by layer to ice and the uppermost detouches from time to time before it has been molten by the sun (Strabon).
>
> "The snow cover was caused to glide down by the weight of men and animals" (Livius).

The first appearance of the term Lavina is likely in Isodoros of Sevilla's encyclopedia of the sixth century. It traces the word Lavina to the Latin "Labes" which stands for "to tumble" or "to fall" and may express the fact that people were likely to tumble or fall when being caught by an avalanche.

Interesting are also the early pictures of avalanche events, because they tell us how people in the Middle Ages imagined avalanche events would occur. The oldest document known to me appears in Theuerdank's (1517) epos of the Emperor Maximilian, a love story, which contains the wood engraving shown in Figure 11.1. The presumption that avalanches are huge snow balls continued through the 18th and 19th century; the long persistence of this belief is probably due to the fact that mountainous regions in the Middle Ages were not populated or visited by intellectuals — the exception being the Swiss savant Johann Jakob Scheuchzer (1652–1733) — so pictures of avalanches turned out to be the result of imagination rather than observation. Serious research did not start before winter mountaineering and skiing became fashionable, i. e. well in this century, when the Swiss in 1936 founded their own snow and avalanche research institute; similar centres were subsequently also established in France, Canada, USA, Russia and other countries.

Fig. 11.1. Wood engraving of H. Schäufelein in the *Theuerdank* (1517)

These institutes are now leading centres of fundamental and applied avalanche research.

11.1.2 Physical behaviour

Landslides, rockfalls, sturzstroms, and ice and snow avalanches - in general gravity driven boundary layer shear flows — may be crudely classified into two types of limiting behaviour, the so-called *flow avalanches* and the *powder (snow) avalanches*. The former may be regarded as a fluid-like flow of a dense assemblage of particles in which the interstitial fluid plays a small to insignificant role. Most snow and ice avalanches, when they form from a fractured snow pack, develop as flow avalanches; the size of their particles depends upon the thermodynamic state of the snow pack, see Figure 11.2; fresh dry snow tends to form small granules of perhaps, 2–3 mm diameter, wet snow develops into hard snow balls (of several tens of mm diamter), and old snow in the spring which has undergone several metamorphoses (so-called "greasy snow") consists of ice grains of 5–10 mm diameter. In all these cases, a granular structure to the body of snow is suggested, akin to the motion of rock or sand down an inclined surface on a mountainside.

Avalanches that form from fresh, dry, unsettled, non-metamorphosed snow sometimes do not remain flow avalanches for very long. Either because of Kelvin-Helmholtz instabilities or because of impact due to bumps in the terrain, they may suddenly develop as powder snow avalanches, the second of the above mentioned two forms, see Figure 11.3. These avalanches are very similar to gravity, turbidity currents or heavy gas dispersion, and may also arise in submarine shore-shelf

Fig. 11.2. (a) *Left*: Snow deposit of a dry flow avalanche (Photo: E. Wengi, Swiss Federal Institute of Snow and Avalanche Research, Weißfluhjoch-Davos, Switzerland. (b) *Right*: Deposit of a wet flow avalanche. (Photo: E. Wengi, Swiss Federal Institute of Snow and Avalanche Research, Weißfluhjoch-Davos, Switzerland.

erosion, or in dust flows from volcanic eruptions. In these powder snow avalanches, the dynamic effects of the interstitial fluid are dominant, and the particles' role is that of a suspension; nevertheless, they are often treated as turbulent density currents. These avalanches are, however, turbulent two-phase (boundary layer) flows, in which both air entrainment *from* above, and snow deposition *to* the base are important dynamic mechanisms. These avalanches are much less frequent than flow avalanches; they are, however, generally large events which ought to, but cannot always, be avoided when avalanches are artificially released by detonations from a helicopter.

The two limiting situations just described actually represent idealized situations that are, if at all, seldom observed in pure form. By far the most often ocurring avalanche is of mixed type, e. g. a flow avalanche overlain by some snow dust. Despite this recognition, already mentioned by Seligman in 1936, the two limiting cases are so far in one way or another the only "rationally" treated cases. Their study is helpful because the relevant physics can be more easily described and corresponding physical processes therefore understood.

Typical characteristic velocities and flow heights (Table 11.1) of powder snow avalanches are much larger than for flow avalanches (80 to 120 m s^{-1} versus 30–60 m s^{-1} and 50–100 m versus 1–5 m); bulk densities, however, are smaller by a factor of between 50 and 100. Consequently, stagnation pressures ($\rho v^2/2$) are of similar magnitude, lever arms of resulting forces to the ground on obstacles, however, are much larger for powder snow avalanches than for flow avalanches; this explains their devastating effect. Flow avalanches are affected by the bed topography and, consequently, by basal friction; their track follows valley troughs. Powder snow avalanches are not influenced by basal friction except in the runout

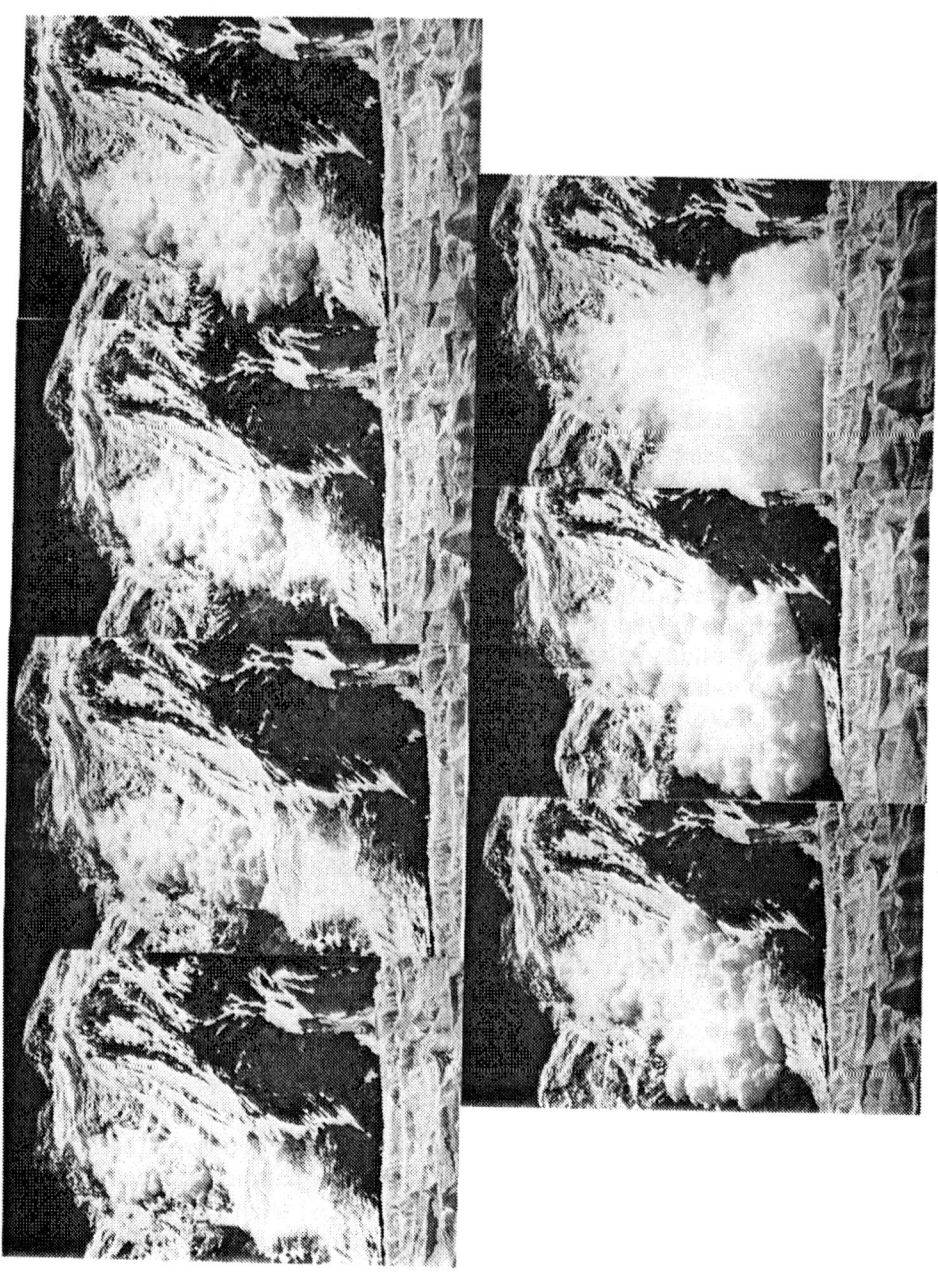

Fig. 11.3. A sequence of snapshots of a powder snow avalanche in the Himalaya (Photos: F. Tschirky).

TABLE 11.1
Characteristic parameters of the two limiting cases of avalanche flow

	Flow avalanche	Powder avalanche
Typical velocity	30–60 m s^{-1}	80–120 m s^{-1}
Flow height	1–5 m	50–100 m
Density	100–300 kg m^{-3}	5 kg m^{-3}
Effect by topography	Avalanche is channeled by small valleys	None, following the steepest descent
Basal effects	Bed roughness important	Bed roughness unimportant
Flow regime	Grain flow	Turbidity current

zone where snow settlement is important. Topography is irrelevant except for the direction of steepest descent which defines the flow direction.

11.1.3 Laws of Similitude

Physical variables that govern the turbulent processes of *powder snow avalanches* are

$$g, \Delta\rho, \rho, u, v, c, H, d, \nu. \tag{11.1}$$

These are the symbols typically used for the acceleration due to gravity, g, the density difference between the avalanche and the interstitial fluid (air), $\Delta\rho$, the density of the avalanche, ρ, the streamwise velocity, u, the particle settling velocity, v, the speed of sound of the suspension, c, the avalanche depth, H, a particle diameter, d, and a typical viscosity, ν. Since these flows are gravity driven, and because settling processes in the runout zone are significant, *dimensional analysis* indicates that the laws of similitude require the *densimetric Froude number* (Table 11.2) and the ratio of *longitudinal to depositional velocities* to be invariant under transformations to the laboratory scale

$$\mathbb{F}_{\Delta\rho} = \frac{u^2}{(\Delta\rho/\rho)gH}, \quad V = \frac{v}{u}. \tag{11.2}$$

Scale effects enter through the non-invariance of the remaining dimensionless parameters:

$$\mathbb{R}e = \frac{ud}{\nu}, \quad M = \frac{u}{c}, \quad \mathbb{B} = \frac{\Delta\rho}{\rho}, \quad A = \frac{d}{H}. \tag{11.3}$$

These dependences are masked by turbulence; a Reynolds number dependence is insignificant in fully turbulent motions; furthermore, because the aspect ratio, A, is in general small, both in the laboratory and in nature, it need not be invariant in transformations from the prototype to the laboratory scale. Scale effects due to the

TABLE 11.2
Typical dimensions and dimensionless numbers in nature and in the laboratory experiments of Scheiwiller et al. (1986), with changes

	Nature	Laboratory
Downslope velocity u	100 m s^{-1}	0.1 m s^{-1}
Flow height, H	50 m	0.1
Free fall velocity of the particles,	1 m s^{-1}	10^{-3} m s^{-1}
Speed of sound*	150 m s^{-1}	1400 m s^{-1}
Material density of the fluid phase $\hat{\rho}_1$	1 kg m^{-3}	1000 kg m^{-3}
Material density of the particle phase $\hat{\rho}_2$	917 kg m^{-3}	1250 kg m^{-3}
Volume fraction of the particle phase $c_{(2)}$	5×10^{-3}	5×10^{-3}
Viscosity of the fluid phase ν	1.7×10^{-5} m^2 s^{-1}	10^{-6} m^2 s^{-1}
Froude number $\mathbb{F}_{\Delta r}$	~5	~5
Velocity ratio $\mathbb{V} = v/u$	10^{-2}	10^{-2}
Boussinesq number $\mathbb{B}$	~4	$\sim 2 \times 10^{-4}$
Mach number $\mathbb{M}$	~0.6	$\sim 10^{-4}$
Reynolds number $\mathbb{R}e$	10^8	10^4
Aspect ratio $\mathbb{A}$	10^{-5}	10^{-3}

* According to Prandtl, Oswatitsch, Wieghardt (1984), the speed of sound in a suspension is, roughly, $c^2 = (\hat{\rho}_f/\hat{\rho}_s)(1/(a(1-a)))c_f^2$, where $\hat{\rho}_f$ and $\hat{\rho}_s$ are the peculiar densities of the fluid and solid, respectively, a is the volume fraction for the solid in the suspension and c_f the speed of sound of the fluid alone.

non-invariance of the Mach number $\mathbb{M}$, cannot be very dramatic in powder snow avalanches because possible shock waves damp out very quickly in suspensions. Somewhat critical is the Boussinesq number, $\mathbb{B}$; it is large or of order unity ($\mathbb{B} \geq 1$) for powder snow or dust avalanches (in air), but moderate ($\mathbb{B} < 1$) for subaquatic turbidity currents and small ($\mathbb{B} \ll 1$) for density currents (heavy gas, salt water in fresh water) and for our turbulent suspensions of polystyrene particles in water. This non-invariance of $\mathbb{B}$ in present day laboratory simulations of powder snow avalanches is indeed a critical point. Maintaining the invariance of the densimetric Froude number under model simulations is, however, more important.

For *flow avalanches* the physical parameters that describe the dynamic processes might be

$$g, u, H, d, \nu, \phi, \delta. \tag{11.4}$$

Thus, of the parameters appearing in Eq. (11.1) $\Delta\rho$ is insignificant (since the density of the granular material is very much larger than that of the interstitial fluid), and the settling velocity as well as the speed of sound, are irrelevant. However, two additional parameters have entered the list (11.4), namely ϕ, the internal angle of friction, and the bed friction angle δ; the former describes a Coulomb-type plastic yield between the moving particles. Similarly, the bed friction angle is a measure

Avalanche Dynamics

of the Coulomb-type friction between the particles and the bed. These variables will be explained in greater detail below. Here it may suffice to state that, besides geometric similarity, the parameters

$$\mathbb{F} = \frac{u^2}{gH}, \qquad \phi, \delta \tag{11.5}$$

are the dimensionless quantities that must be kept invariant in transformations from the prototype to the laboratory scale, while under most situations

$$A = \frac{d}{h}, \qquad \mathbb{R}e = \frac{ud}{\nu} \tag{11.6}$$

need not necessarily be, because the particle size is very small, and the interstitial fluid has a very small mass in comparison to the granules, which can in most cases be ignored (ν is the kinematic viscosity of the fluid and $\mathbb{R}e$ is large). The above variables cannot explain the size effects that are evidenced in large landslides and become significant at volumes of 10^6 to 10^{11} m^3. Obviously, the set of variables (11.4) is too small to cover this effect, which today is still an enigma.

In ensuing developments, we shall separately review the dynamics of granular avalanches (which are akin to flow avalanches) and those of powder snow avalanches.

I. DYNAMICS OF GRANULAR AVALANCHES

11.2 Some Distinctive Characteristics of Granular Flows

We list in this paragraph a number of properties that are treated in greater detail by Savage (1989).

11.2.1 DILATANCY, INTERNAL FRICTION, RATE DEPENDENCE OF STRESS, LARGE ENERGY DISSIPATION

The concept of *dilatancy* was defined by Reynolds (1885) as the property of a granular material to alter its volume in accordance with a change in the arrangement of its grains. For instance, if an arrangement of identical spheres (in which the solid volume fraction is 0.741) is sheared, then this deformation is necessarily accompanied by a reduction of the solid volume fraction. During *quasistatic* deformation, dilatancy is the consequence of kinematic constraints. Under rapid shearing when particles may collide and bounce, the many rapid collisions generate a so-called dispersive stress. Much like in a dense gas of molecules the bigger the fluctuation energy of the colliding particles, the larger the pressure; this dispersive pressure is responsible for dilatancy during *dynamic* deformation. It follows that a continuum theory of granular materials may need to incorporate compressibility. When a finite

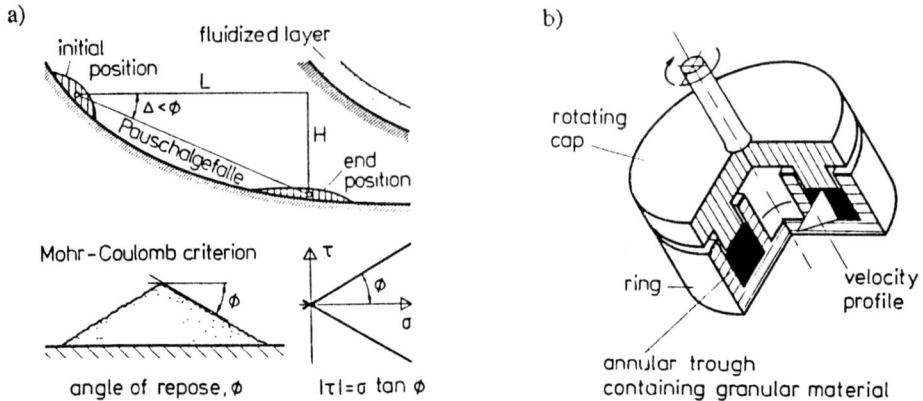

Fig. 11.4. (a) A mass of snow or rock in its initial and final rest positions. The inclination of the line connecting the centres of gravity is called "Pauschalgefälle"; pile of a granular mass defining the angle of repose ϕ: Mohr-Coulomb yield line relating normal stress N on an interior plane and shear stress S. (b) Annular shear cell: the granular material is contained in the annular trough in the lower half of the shear cell, and capped by an annular ring attached to the upper half of the shear cell. The cap is rotated with constant angular velocity, thus inducing a shear flow in the granular material.

mass of granular material starts to move from rest, however, the major part of the volume expansion takes place immediately at the start such that the subsequent motion may be nearly isochoric. This is indeed the case, and this property will be used later on.

Granular materials exhibit solid behaviour as well; indeed, they can form piles. For a given assemblage of granules, the free surface, when formed freely, is inclined at some maximum value that is peculiar to the material in question; the maximum angle is called the material's *angle of repose*, ϕ, see Figure 11.4a. If S and N are, respectively, the shear and normal traction acting on the element, then from a simple force balance we obtain

$$|S| = N \tan \phi. \tag{11.7}$$

This relation describes the limiting state of stress: If the inclination angle should be larger than the angle of repose, then granules will dribble along the free surface until the angle of repose is restored. Similar behaviour is conjectured to occur in the interior of a slowly deforming plastic material. In this latter case S is then the shear τ, N the normal traction σ, acting on any interior surface, and ϕ the material's *internal friction angle*. Typically, $\phi \approx 24°$ for spherical smooth particles (e.g., glass beads) and $\phi \approx 37°$ to $40°$ for rough, edgy particles (e.g. crunched marble chips). It is assumed in the above that the particles do not stick together because of adhesive forces; in other words, *cohesion* is ignored. This requires the particles to be dry, and not too small, so that electrostatic surface forces are negligible.

A law similar to (11.7) is also assumed to hold at the sliding base of a moving mass of granular material, i.e.,

$$S = \pm N \tan \delta, \tag{11.8}$$

where δ is the *bed friction angle*, clearly a property of both the granules and the bed. Relations (11.7) and (11.8) are generally referred to as the *Mohr-Coulomb yield criteria*.

Motivated by Bagnold (1954), who conducted the first laboratory experiments, a number of experiments have been performed using annular shear cells to determine the dependence of the shear and normal stresses upon the shear rate and particle concentration, see e.g., Savage (1979), Savage and McKeown (1983), Savage and Sayed (1984), Hanes and Inman (1985), Stadler (1986) and Buggisch and Stadler (1986). The results of such tests depend upon whether shear tests are performed at constant volume or constant normal stress, respectively, see Figure 11.4b:

- In both cases, the shear and normal stresses are related by a Mohr-Coulomb-type criterion with an internal angle of friction which is nearly constant: typically, for quasistatic processes, where particles remain in contact, $\phi = \phi_{stat}$ is slightly larger than for dynamic processes, where the internal stresses are primarily controlled by collisional properties; consequently, $\phi_{dyn} \lesssim \phi_{stat}$.

- When shear cell tests are conducted so as to maintain a *constant volume*, it is found that stresses vary with the square of the shear rate at high shear rates. Stresses also strongly increase with an increase in the solids concentration.

- When shear tests are performed with a *constant normal load*, and the layer of particles is allowed to expand freely, then it is found that the shear stress is nearly independent of shear rate. Stadler (1986), Buggisch and Stadler (1985) and Hungr and Morgenstern (1984a,b) have studied this behaviour, and Savage and Hutter (1989) discussed its relevance for the flow of granular materials down inclined chutes. It is indeed this behaviour which is applicable for the avalanching motion of a finite mass of granular materials; it justifies the use of the Mohr-Coulomb yield criterion (11.7) with constant internal friction angle ϕ.

During rapid flows, the particles of a granular material are in a highly agitated state, continually colliding with one another. In landslides and avalanches, the mechanism that supplies this energy is gravity; this energy is transferred into collisional energy and rubbing-like friction, and quickly dissipated. Savage (1989) quotes Erismann (1979), who has estimated the enormous energies involved in these events. "For instance, the prehistoric Flims landslide in Switzerland had a volume of about 1.2×10^6 m^3, and probably took place over a period of a couple of minutes. Its motion resulted in the dissipation of an amount of energy equal to the total present day energy consumption of the world for a period of 10 hours."

11.2.2 Large Travelled Distances, Size Effects

Landslides often spread out into very thin layers and flow on surfaces that are much less inclined than the angle of repose of the material. This is evident in avalanche events in which the two far ends of a landslide (or less frequently, but more logically, the centres of gravity in these two positions) are connected at rest (Figure 11.4a). The tangent of the inclination angle of this line, Δ, is denoted by the German term *"Pauschalgefälle"*. If (i), the landslide is regarded as a single mass, (ii), curvature effects of the bed are ignored, (iii), a Coulomb-type bed friction law (11.8) with constant bed friction angle δ, is employed and (iv), the loss in gravitational energy is assumed to be entirely consumed by basal friction, then

$$MgH = \int_0^s \tan\delta Mg \cos\zeta \, ds = \tan\delta Mg \int_0^L dx = \tan\delta MgL$$

or (11.9)

$$\frac{H}{L} := \tan\Delta = \tan\delta \Rightarrow \Delta = \delta$$

holds. Note that $\Delta < \phi$ by observation from historic and present rockfalls, sturzstroms and snow and ice avalanches on the Earth, Mars and the Moon. Thus, the bed friction angle δ, is smaller than the internal angle of friction ϕ; this fact is also the reason why the nose of a landslide or avalanche moves surprisingly long distances (below, a volume dependence will also be shown to exist), the reason being the fluidization of the granular material in at least a basal boundary layer. Ever since Albert Heim (1882, 1932) observed the Elm rockfall in Switzerland, hypotheses have been proposed to explain the fluidization. The essence of each of these hypotheses is to introduce some type of fluidization mechanism to produce the high mobility of the large volume rockfalls. Those involve upward flow of air (Kent, 1986), hovercraft action (Shreve, 1966, 1968), fluidization by high pressure steam (Goguel, 1978), mechanical fluidization aided by the presence of interstitial dust (Hsü, 1978), vibrational fluidization by large earthquakes (McSaveney, 1978), lubrication by a thin layer of molten granular material (molten rock, Erismann, 1979, 1986; molten snow at the bed of a flow avalanche, Hutter, 1991), accoustic fluidization (Melosh, 1986; Foda, 1994), and the presence of a thin layer of vigorously fluctuating particles beneath a densely packed overburden (Dent, 1986). Many of these hypotheses can be ruled out: e.g., there is not enough air or water on the Moon and Mars, and earthquakes almost never trigger the motion of an avalanche or rockfall. In flow avalanches of snow and extremely large landslides, melting may, however, contribute to lubricating the bed; even more likely is fluidization due to the rapid and vigorous collisions of the particles.

It has been mentioned before that, for very large landslides, the Pauschalgefälle is size-dependent. Savage (1989) quotes Heim (1932), Scheidegger (1975), Hsü (1975), Lucchitta (1978), Davies (1982), and Li Tianchi (1983) for having drawn attention to this fact, and presents Figure 11.5, which displays the Pauschalgefälle (in its first definition) as a function of the total volume for a number of field events. It can be seen that for rockfall volumes greater than about 0.5×10^6 m^3, there is a decrease in H/L with increasing volume, implying a reduction of the bed friction angle δ, with volume. Further, Savage demonstrates that this volume depencence is also present when the Pauschalgefälle is defined as the line from the initial to final centre of mass (Figure 11.5b). This is a scale effect, and remains a mistery.

Avalanche Dynamics

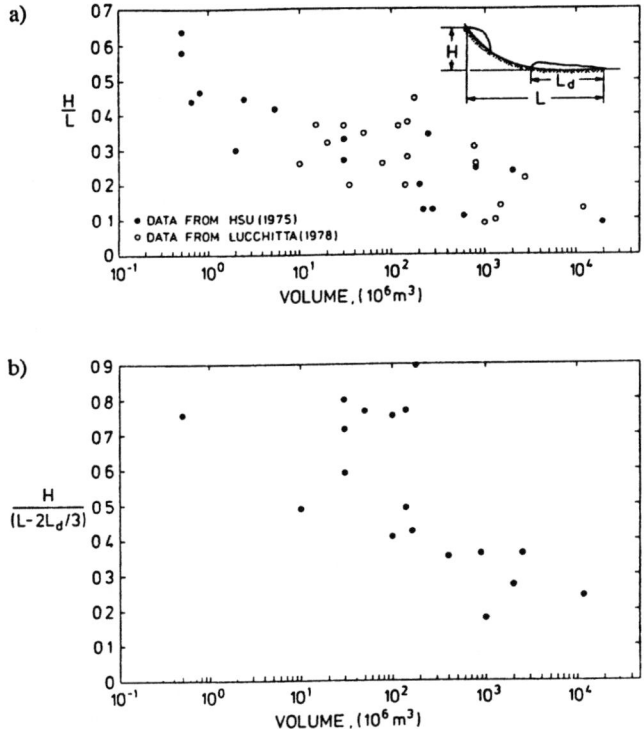

Fig. 11.5. (a) Variation of H/L with rockfall volume; based on data from Hsü (1975) and Lucchitta (1978), with additional information from Li Tianchi (1983). (b) Slope of line connecting estimated initial and final centre of mass positions (Pauschalgefälle), plotted against the rockfall volume (from Savage, 1989).

11.2.3 THREE DIFFERENT FLOW REGIMES

Let us ignore for the sake of simplicity the mass of the interstitial fluid and regard the bulk material to be made up of cohesionless solid particles alone. (This assumption is reasonable as long as the interaction force between the particles and the fluid (the mutual drag force) is negligible; this means, e. g., that the interstitial fluid cannot develop turbulent eddies large enough for the particles to become suspended in them and essentially carried away). During their motion, the particles will typically acquire a mean motion, relative to which individual particle velocities fluctuate as a result of collisions and particles riding one over the other. Three main mechanisms operate thereby in generating stresses due to deformations of the bulk material; these are as follows:

- Dry *Coulomb, rubbing friction*. The associated internal forces develop while particles are in contact with one another. These frictional processes must be dominant, when processes are quasistatic or when collisional activity is small. Natural bulk densities are generally large in this case (*Regime I*).

- Transport of momentum by *collisional interactions*. This mechanism is dominant when the particle contact is of short duration so that Coulomb rubbing friction cannot effectively develop. On the other hand, the bulk densities must be smaller than in Regime I, but cannot be so small that mean free paths are several particle diameters large (*Regime II*).

- In *Regime III*, the transport of momentum is by particle *translation*. Here, Coulomb-type rubbing friction is practically absent, collisions are infrequent and mean free paths are long.

In general, all three mechanisms are effective in the moderate to high density regime; there are limiting flow regimes, however, in which one or two of them play a dominant role. Regimes II and III are more important for industrial applications in powder technology, where enough fluctuation energy is provided from outside, where Regimes I and II have been applied for chute flow or steady hydraulic flow of a granular material down an inclined plane where particle concentrations are generally high (Savage, 1983); a reduced version of Savage's model forms the basis of the avalanche model presented below.

Before turning to it a few words should be spent on *kinetic theories*. In his pioneering paper, Bagnold (1954) looked at the situation in which nearly instantaneous collisions were the dominant processes of momentum and energy transfer (Regime II), and called it the *grain inertia regime;* he further showed that the shear and normal stresses were quadratically dependent upon shear rate. His findings were experimentally corroborated at high shear rates by Savage and McKeown (1983), Savage and Sayed (1984), Hanes and Inman (1985), Stadler (1986) and Buggisch and Stadler (1985). Constitutive relations that mimic this behaviour (McTigue, 1979; Savage, 1979; Jenkins and Cowin, 1979) were found, however, they admit no mathematical solution in a plane horizontal Poiseuille flow with gravity. At this point, the kinetic theories of the hardsphere-model-type that are used for the analysis of dense gases and fluids become relevant (e.g., Savage and Jeffrey (1981), Jenkins and Savage (1983), Haff (1983), Lun et al (1984), Jenkins and Richman (1985 a,b)). One essential feature in all these kinetic formulations is that, besides the balances of mass and momentum, a further equation for the collisional energy, sometimes called the *balance of granular temperature* must be added; this equation is completely analogous to the heat conduction equation in a fluid at the molecular level. The method also delivers expressions for the stress tensor, the flux of fluctuation energy, and the annihilation rate of fluctuation energy in terms of the properties of the collisions (particle diameter, coefficient of restitution). The advantage of the method is that it delivers a complete set of consistent model equations for a continuum in terms of only a few phenomenological parameters; on the other hand, the disadvantage is that it applies only to Regime II (e.g., in Jenkins and Savage, 1983), or Regimes II + III (e.g., Lun et al., 1984), but not Regime I. Furthermore, in an analysis of steady chute flows Hutter, Szidarovsky and Yakowitz (1986a,b) showed that the Jenkins-Savage model together with various physically reasonable boundary conditions at the free surface and at the bed, often yield an ill-posed boundary value problem. In addition, when it possesses solutions these exist in narrow bands of the parameter space which are physically of little relevance.The problem here is delicate, and connected with the singular behaviour of the governing equations. "Approximate" solution techniques may mask the singularities and reflect properties that are alien to the original equations (Richman and Marciniec, 1990). In a paper by Ahn, Brennen and Sabersky (1992), it is shown that such ill behaviour is avoided when the model of Lun et al. (1984) is used.

On the other hand, in a steady shear flow down an inclined plane only a relatively small basal layer is fluidized, with most of the bulk material above this boundary layer riding as a more or less passive mass on top. In other words, Coulomb-type internal rubbing forces are dominant in this case and thus a combination of Regimes I + II would be appropriate; however, a rigorous theoretical formulation for such a model does not yet exist. In lieu of this, quasi-static, rate-independent and the kinetic, rate dependent contributions, have been added, e.g., in the case of the stress tensor $\mathbf{t}$,

$$\mathbf{t} = \mathbf{t}_{stat} + \mathbf{t}_{dyn}. \tag{11.10}$$

For instance, if

$$\mathbf{t}_{stat} = -p(1 - \sqrt{2}\tan\phi(\mathbf{D}/\sqrt{\text{tr}\,\mathbf{D}^2})) \tag{11.11}$$

is used, then this part of the stress describes Coulomb yield. The dynamic stress part, $\mathbf{t}_{dyn}$, is either taken over from the Jenkins and Savage model (Savage, 1983), or from that of Lun et al. (1984) (see Johnson and Jackson, 1986), or an independent model (without the use of a balance of kinetic energy of the fluctuating motion) is proposed (Norem, Irgens and Schieldrop, 1987; Goddard, 1986). The boundary value problem for a steady plane shear flow is in this case well posed under conditions for which Hutter, Szidarovszky and Yakowitz (1986a,b) found ill-posedness when restricting the processes to Regime II alone, but, computations are rather formidable (see the papers by Johnson and Jackson, 1986; Johnson, Nott and Jackson, 1990).

It is for this reason that Savage and Hutter (1989) have proposed their yet further simplified model. Observations on laboratory chute flows of granular materials indicate that only a very thin basal boundary layer tends to be fluidized, while the major upper part of the moving and deforming mass is certainly in Regime I, where Coulomb-type rubbing friction is effective. Thus, despite the dynamic nature of the flow, a continuum subject to cohesionless Coulomb-type plastic yield is assumed. Instead of treating the fluidized layer as a granular material in Regime I + II, Savage and Hutter (1989) "collapse" the thin boundary layer to a boundary condition and suppose a Coulomb-type sliding law with bed friction angle δ. At first, δ is assumed constant, but later more general forms of the sliding law are permitted to accomodate for possibly overlooked processes within the fluidized layer.

At this point, a remark on the role of cohesion may be appropriate. Wet snow certainly exhibits cohesion. Its incorporation in a fluid model can be shown to lead in a plane gravity flow to a plug flow regime near the surface (the thickness of which is proportional to the value of the cohesion) while the layer below it is sheared, see Norem, Irgens and Schieldrop (1987). Such behaviour seems also to be observed in real avalanches through radar Doppler velocity measurements (Gubler, 1987). Hutter (1989) has shown, however, that a theoretical formulation of avalanche flow that accounts for plug flow and shear flow regimes is very complicated. On the other hand, since the model to be presented in Section 3 uses depth-integrated equations, the detailed constitutive postulates inside the body are not very critical.

11.3 One-Dimensional Model

The model proposed by Savage and Hutter (1989, 1991) is physically not so much different from the simple fluid or particle models of Voellmy (1955), Salm (1968) and Perla and Martinelli (1978), or Perla, Cheng and McClung (1980); it accounts, however, for the finiteness and deformation of the avalanching mass, and follows these through time. Furthermore, whereas the above authors employ their models to describe the dynamics of both dense flow and powder snow avalanches (by an adequate adjustment of a frictional coefficient), we wish to limit this model to flow avalanches, rockfalls and landslides.

11.3.1 Governing Equations

The material is treated as an incompressible, cohesionless continuum obeying a Mohr-Coulomb type yield criterion both in the interior and at the bed. The model equations are based upon the following physical statements:

- Balance laws of mass (in the form of the continuity equation) and linear momentum;

- kinematic equation and stress free boundary conditions at the free evolving surface;

- tangency condition of the velocity vector at the basal surface, and phenomenological postulate for the sliding mechanism in the form of a Coulomb-type friction law with bed friction angle δ, (see Eq. (11.8)). (Later on, this dry friction law will be generalized to also incorporate a viscous component);

- constitutive postulate for the material response inside the granular pile in form of Mohr-Coulomb yield, see Eq. (11.7).

For the ensuing analysis, the sketch in Figure 11.6 is relevant; in the figure, ξ, η are dimensional, and x, y dimensionless curvilinear coordinates, ζ the inclination angle, κ the curvature of the bed geometry and X, Y cartesian coordinates in the horizontal and vertical directions, respectively. Let p_{ij} be the pressure tensor; $p_{\eta\eta}$ denotes the normal stress perpendicular, $p_{\xi\xi}$ that parallel to the bed. At this surface, Eqs. (11.7) and (11.8) must be simultaneously satisfied. From this, using a standard Mohr's circle argument (Savage and Hutter, 1989), it can be shown that

Avalanche Dynamics

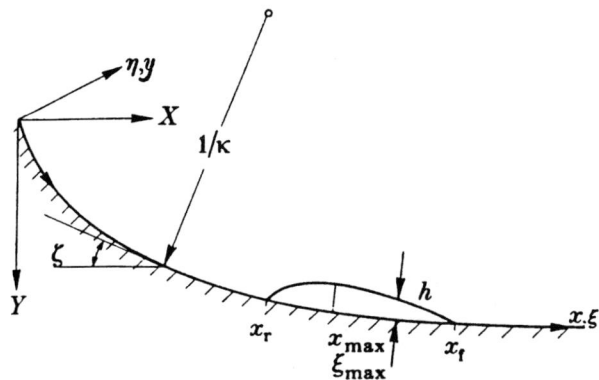

Fig. 11.6. Definition sketch of the curvilinear coordinate system and geometry of a finite mass of granular material moving down a rough curved rigid bed. Here, ξ and η are dimensional, x and y are dimensionless, and X and Y cartesian coordinates. Also, ζ is the bed inclination angle, and κ the curvature of the bed. Lastly, x_r, x_f and x_{max} denote the rear end, the leading edge, and the position of the maximum height within the avalanche.

$$P_{\xi\xi} = K_{act/pass} P_{\eta\eta}, \qquad (11.12)$$

where $K_{act/pass}$ can take active and passive values, K_{act} and K_{pass},

$$\left.\begin{matrix} K_{act} \\ K_{pass} \end{matrix}\right\} = 2\left(\frac{1 \pm \sqrt{1 - \cos^2\phi/\cos^2\delta}}{\cos^2\phi}\right) - 1, \quad \text{for} \quad \frac{\partial u}{\partial \xi} \begin{Bmatrix} > 0, \\ < 0. \end{Bmatrix} \qquad (11.13)$$

Savage and Hutter (1989, 1991) assume the pile of the granular material to be long and thin, the profile (perpendicular to the bed) of the streamwise velocity to be nearly uniform, the curvature of the bed to be moderate and $\delta < \phi$. On this basis, they write the equations in dimensionless form. Three different length scales, the longitudinal length scale, L, the depth scale, H, and a scale for the radius of curvature of the bed profile, R, will be used. With these, three different (independent) parameters arise, namely the aspect ratio

$$\varepsilon = \frac{H}{L} \ll 1, \qquad (11.14)$$

a dimensionless characteristic curvature

$$\lambda = \frac{L}{R}, \qquad (11.15)$$

and the bed friction angle δ. From actual situations, the following orders of magnitude are suggested and will be used:

$$\tan\delta = O(\varepsilon^\alpha), \quad \lambda = O(\varepsilon^\alpha), \quad 0 < \alpha < 1, \quad \text{typically } \alpha = \frac{1}{2}. \qquad (11.16)$$

Let the lengths, curvature, velocity, time and pressure tensor be scaled according to

$$(\xi, \eta, \kappa) = \left(L\tilde{\xi}, H\tilde{\eta}, \frac{\lambda\tilde{\kappa}}{L}\right),$$

$$(u, v, t) = \left(\sqrt{gL}\,\tilde{u},\ H\sqrt{gL}\frac{\tilde{v}}{L},\ \sqrt{\frac{L}{g}}\,\tilde{t}\right), \quad (11.17)$$

$$(p_{\xi\xi}, p_{\eta\eta}, p_{\xi\eta}) = \rho g H \cos\zeta_0(\tilde{p}_{\xi\xi}, \tilde{p}_{\eta\eta}, \tan\delta\,\tilde{p}_{\xi\eta}),$$

where quantities having a tilde are dimensionless, and are taken to be of order unity, and ζ_0 is a characteristic bed slope. Using (11.17), and performing an integration of the balance laws of mass and linear momentum over depth, thereby defining the depth-averaged dimensionless velocity by

$$\tilde{u} = \frac{1}{h}\int_0^h \tilde{u}\,d\eta, \quad (11.18)$$

Savage and Hutter (1991) derive the equations

$$\frac{\partial h}{\partial t} + \frac{\partial h\bar{u}}{\partial x} = 0, \quad (11.19)$$

$$\frac{d\bar{u}}{dt} = \frac{\partial \bar{u}}{\partial t} + \bar{u}\frac{\partial \bar{u}}{\partial x}$$
$$= \sin\zeta - \tan\delta\,\mathrm{sgn}(\bar{u})(\cos\zeta + \lambda\kappa\bar{u}^2) - \varepsilon K_{\mathrm{act/pass}}\cos\zeta\frac{\partial h}{\partial x}, \quad (11.20)$$

in which tildes have been omitted, and $\tilde{\xi}$ has been replaced by x for convenience. These equations involve approximations to the extent that terms of higher order than ε have been ignored. Equation (11.20) expresses a balance of acceleration (left-hand side) with the streamwise components of the forces (right-hand side). The first term on the right is the contribution of gravity; the second is the basal friction force, consisting of the contribution due to overburden and centrifugal forces, and the third term is a longitudinal pressure gradient due to depth variations of the pile. Furthermore, as long as δ and ϕ are constant, equations (11.19) and (11.20) are scale-invariant. This is in conformity with the inferences from the dimensional analysis in Section 11.3 (as the interstitial fluid is neglected and the particle diameter does not enter the above equations). Therefore the equations cannot reproduce a dependence of the Pauschalgefälle upon the total mass, as observed with landslides and rockfalls, and as illustrated in Section 11.2.2. It is also interesting to note that equations (11.19) and (11.20) contain the internal angle of friction ϕ, only indirectly through the earth pressure coefficient $K_{\mathrm{act/pass}}$, while the bed friction angle δ, appears explicitly (through $\tan\delta$) and implicitly (through $K_{\mathrm{act/pass}}$). This indicates a strong sensitivity of results based on these equations to variations in δ and a rather weak one on ϕ, see Greve and Hutter (1993). Provided the angle of friction δ, between the avalanche and the bed, as well as the basal

Avalanche Dynamics

geometry (in terms of the bed slope ζ, and the curvature κ), are known, the temporal evolution of h and $\bar{u}$ can be determined if an initial profile $h(x,0) = h_0(x)$ and depth-averaged velocity distributions $\bar{u}(x,0) = \bar{u}_0(x)$ are prescribed.

When either $\tan\delta$, λ or both are smaller than order ε^α, $0 < \alpha < 1$ and when $\bar{u}$ does not become too large, the term in equation (11.20) due to the centrifugal effects may be neglected. This does not mean that the bed curvature effects are ignored in this case, since ζ may still vary with position and thus incorporate some weak curvature effects implicitly.

It proved quite difficult to obtain accurate and reliable numerical solutions to the above equations (11.19) and (11.20) (some semi-analytic similarity solutions will be discussed in Section 11.3.3). For details, the reader may consult Savage and Hutter (1989). The gist of the method is that the finite difference net must deform with the pile (Lagrangian viewpoint); in addition, because an explicit algorithm is employed, numerical diffusion must be used to stabilize the system.

In closing, it should also be mentioned that equation (11.20) is based on a pure dry Coulomb-type bed friction law, eq. (11.8). As is clear from the discussion in Section 11.2.3, laws depend on more than just a constant bed friction angle could be introduced. For instance, Voellmy (1955) decomposes the shear traction S into

$$S = S_C + S_V, \tag{11.21}$$

where

$$S_C = -\operatorname{sgn}(\bar{u})|N|\tan\delta \tag{11.22a}$$

expresses Coulomb-type behaviour, while S_V is a viscous drag, i.e.,

$$S_V = -\operatorname{sgn}(\bar{u})\rho q(|\bar{u}|, |N|)|\bar{u}||\bar{u}|. \tag{11.22b}$$

Here, q is the dimensionless drag coefficient that may (but need not) depend on the moduli of the velocity vector and the stress normal to the basal surface. For a q independent of $|\bar{u}|$, eq. (11.22b) corresponds to a quadratic dependence of S_V on the velocity, another dependence is of course also possible. Avalanche specialists who are used to working with the Voellmy model usually define S_V in the form

$$S_V = \frac{\rho g}{\xi}|\bar{u}||\bar{u}|, \tag{11.23}$$

where $[\xi]$ has the dimensions of acceleration but is referred to as a viscosity. Incorporation of the viscous basal drag (11.22b) or (11.23) into the momentum equation and non-dimensionalizing as before shows that the following replacement in (11.20) is needed:

$$\tan\delta \Rightarrow \left(\tan\delta + q\frac{\bar{u}^2}{\varepsilon h}\right), \tag{11.24}$$

in which $q/(\varepsilon h)$ is assumed to be order unity. In laboratory chute experiments with granular materials, it has not been found necessary to incorporate the viscous drag. This may be different in three-dimensional granular flow, as claimed by Lang (1992), and in real snow avalanches, since observations indicate that the dynamic drag depends on the flow state (McClung and Schaerer, 1988).

Fig. 11.7. Series of photographs (Experiment no. 02 in Greve, 1991, taken at different dimensionless times) of a granular material (3 l of glass beads with 5 mm diameter on no. 120 SIA sandpaper) moving down a concave and convex curved bed.

Avalanche Dynamics

Fig. 11.7. *Continued.*

11.3.2 COMPARISON WITH EXPERIMENTS

Experiments were performed in a 100 mm wide chute with beds of different shapes: inclined *plane* (Huber, 1980; Savage and Hutter, 1989); *inclined plane connected via a circular bend with a horizontal plane* (Plüss, 1987; Savage and Hutter, 1991; Hutter, Koch, Plüss and Savage, 1993); *exponentially curved bed* (Koch, 1989; Hutter and Koch, 1991); and a *concave curved bed with a convex segment* (bump) (Greve, 1991; Greve and Hutter, 1993). The strip forming the bed was lined with drawing paper and no. 120 SIA sandpaper to vary the bed friction angle; it was mounted on a vertical plywood board that itself was coated by an off-white plastic folio. The front wall of the arrangement was made of clear, transparent plexiglass through which the moving granular pile could be photographed. A finite mass of granular material was confined by a plate, initially oriented normal to the bed, or later and better, vertically. The granular material was released from rest by quickly rotating the confining plate, thus permitting the granular material to move down the incline. The motion of the pile of granular material was photographed by a 50 mm camera with a high-speed motor drive capable of operating at nominally 6, 12.5, and 15 frames per second, respectively. The tests were performed with seven different sorts of granular materials: glass beads (3 and 5 mm nominal diameter), plastic particles (Vestolen), two different sorts of quartz sand (3 and 5 mm nominal diameter) and two different sorts of marble chips (3 and 5 mm nominal diameter). In each experiment, the internal angle of friction ϕ, and the bed friction angle δ, were determined. The effects of the confining walls (front and back walls) on the bed friction were also approximately determined by replacing δ by

$$\delta_{\text{eff}} = \delta_0(1 + \varepsilon k_{\text{wall}} h); \qquad (11.25)$$

here δ_0 is the measured bed friction angle, h the dimensionless avalanche height and k_{wall} a wall friction coefficient ($k_{\text{wall}} \approx 11°$). For details, see Hutter and Koch (1991) and Hutter, Koch, Plüss and Savage (1993).

Figure 11.7 shows a sequence of snapshots of the shape of a pile of granular material as it flows down a curved bed with a bump with details on the experimental conditions being listed in the figure caption. The advantage of this particular bed geometry with bump is that the granular avalanche will settle either as a single pile above or below the bump or in two piles, one above and the other below the bump, depending on the sorts and masses of the initial pile and depending on the bed lining.

In the experiments, which we compare with computational results below, the characteristic lengths with which the governing equations were non-dimensionalized are $L = H = R = 150$ mm; on this basis, the inclination angle $\zeta(x)$ takes the form

$$\zeta(x) = \zeta_0 \varepsilon^{-ax} + \zeta_1 \frac{\xi}{1+\xi^8} - \zeta_2 e^{-c(x+10/3)^2}, \qquad \xi = \frac{4}{15}(x-9),$$

$$a = 0.1, \quad c = 0.3, \quad \zeta_0 = 60°, \quad \zeta_1 = 31.4°, \quad \zeta_2 = 37°, \qquad (11.26)$$

$$\frac{dX}{dx} = \cos(\zeta(x)), \qquad \frac{dZ}{dx} = \sin(\zeta(x)).$$

Avalanche Dynamics

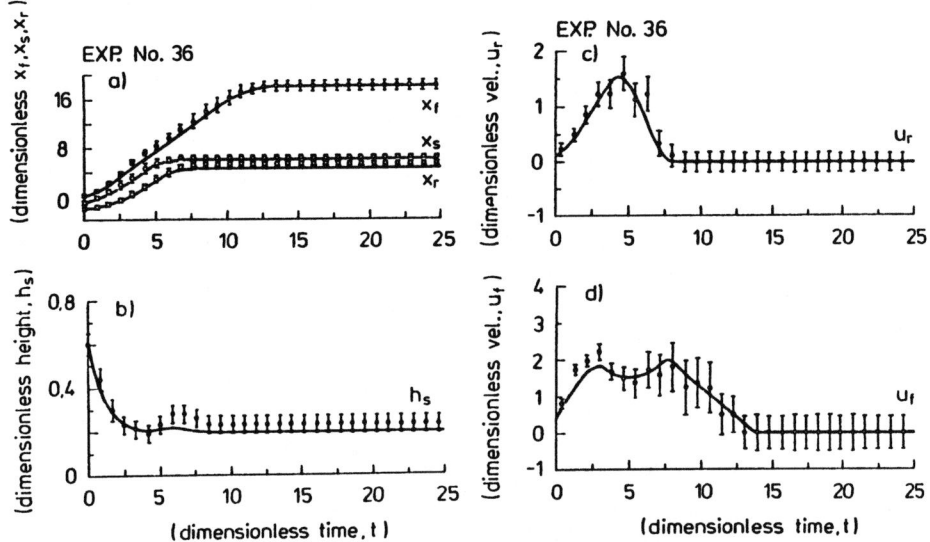

Fig. 11.8. Experimental measurements and theoretical predictions of, (a), the positions x_f (leading edge), x_r (trailing edge) and x_s (position of maximum height); (b), the maximum height h_s, (c) the rear velocity u_r, and (d) the front velocity u_f for experiment no. 36 as functions of dimensionless time t. Data points are shown with error bars and computational results are shown as solid lines. The model consists of 1.5 litre of 3 mm quartz sand on a bed lining made of drawing paper for which $\delta_0 = 30°$, $\phi = 44°$ and $k_{wall} = 12°$ (from Greve, 1991).

The profile $X(x)$, $Z(x)$ can easily be determined from $\zeta(x)$ through integration. Note that, the initial time is identified with the time at which the confining gate is rotated, and just about to detach from the granular mass (first photograph). This permits straightforward identification of the initial profile geometry and relatively easy evaluation of the initial velocity field (Greve and Hutter, 1993).

In Figures 11.8 and 11.9, we have plotted the dimensionless positions of the rear x_r, the maximum height x_s, and the front x_f (panel a), the dimensionless maximum height h_s (panel b), the dimensionless velocities at the rear u_r (panel c) and front u_f (panel d), all as functions of time. The details pertinent to the experiments are not given here. Solid lines are obtained from the numerical integration of equations (11.19) and (11.20), and the symbols represent experimental results inferred from the fast speed photographs including error bars. It can be seen that, except for a few isolated points, agreement between the theoretical curves and the experimental predictions is excellent. Most notable is the rapid evolution into a very thin layer. It is interesting to note and we emphasize that the two avalanches develop in this case into two piles such that x_r is the rear end of the pile above the bump, while x_f is the leading edge of the pile deposition below the bump. The final volumes above and below the bump are also reasonably well-predicted.

Finally, Figure 11.10 summarizes a detailed comparison for experiment no. 29 of the observed and the computed avalanche geometries for the consecutive times at which photographs were taken; agreement is excellent. Deviations between

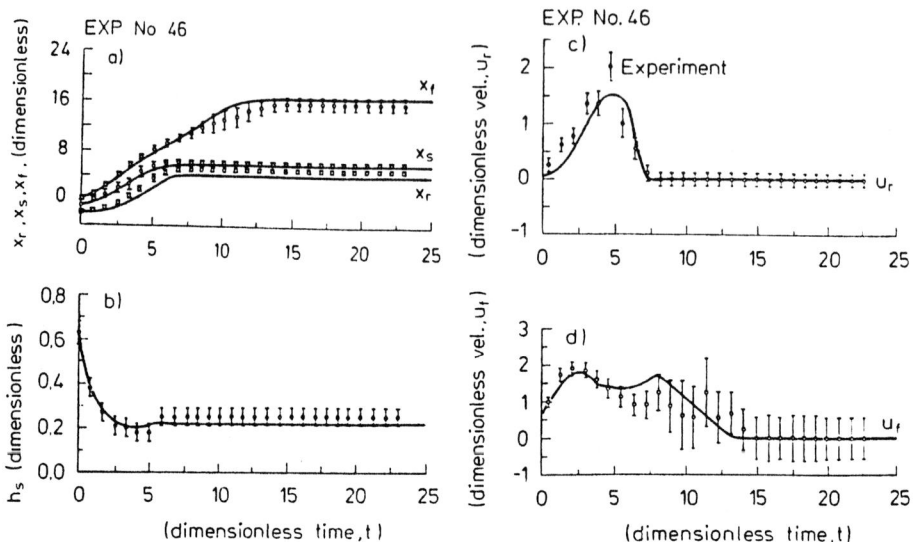

Fig. 11.9. Same as Figure 11.8, but now for experiment no. 46, i.e., 5 litre of 3 mm marble grains on a bed lined with drawing paper for which $\delta_o = 30°$, $\phi = 44°$ and $k_{wall} = 12°$ (from Greve, 1991).

the experimental and the computed avalanche geometries are generally within experimental errors; (in order not to overload the figures, errorbars have been omitted). Most astonishing is the fact that the complete avalanche geometry is correctly predicted, i.e., not only the positions of the leading and trailing edges are well predicted, but the entire mass distribution as well. These results are due to Greve (1991). Similarly convincing results were also obtained for other chute geometries, (Hutter and Koch 1991; Hutter, Koch, Plüss and Savage, 1993).

11.3.3 SIMILARITY SOLUTIONS

Similarity solutions of the governing equations (11.19) and (11.20) (with the possible replacement (11.24)) represent form, but not necessarily size, invariant solutions of these equations. Savage and Hutter (1989) have found such similarity solutions of the equations (11.19) and (11.20) for the case of the flow of a pile of granular material down a rough, inclined flat bed. Savage and Nohguchi (1988), Nohguchi, Hutter and Savage (1989), and Hutter and Nohguchi (1990) extended the solution procedure to the same plane flow configuration down rough *curved* beds having gradually varying slopes. These latter solutions incorporate in part the curvature effects of the bed, the position dependence of the bed friction angle δ (but holding the earth pressure coefficient constant), and the Voellmy type viscous drag (11.24). The emerging solutions are approximate solutions of (11.19), (11.20) and (11.22a and b), because first, the centrifugal term in (11.20) was ignored, second, the slope angle ζ, within the moving pile was replaced by a first-order Taylor series expansion about the centre of gravity and, third, it was supposed

Avalanche Dynamics

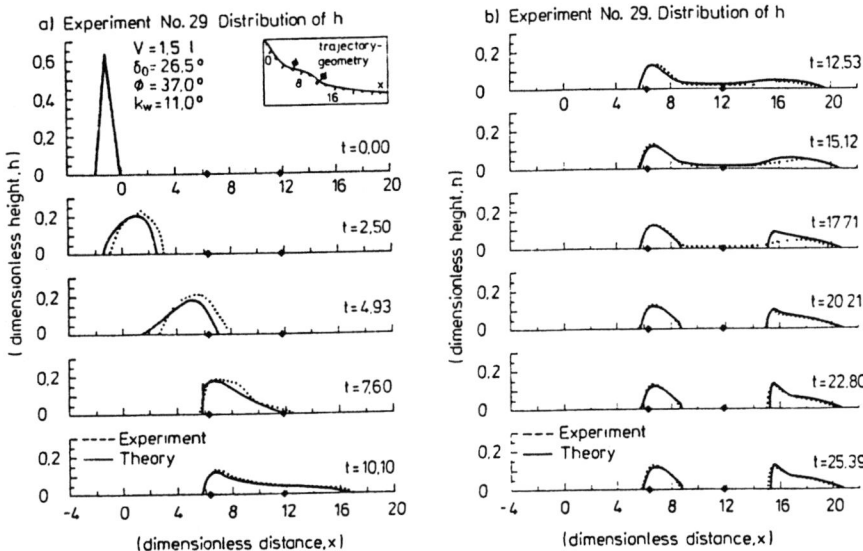

Fig. 11.10. Theoretical (solid lines) and experimental (dashed lines) dimensionless avalanche profiles $h(x, t)$, plotted in each panel for fixed dimensionless times (when photographs were taken) in dimensionless temporal steps $\Delta t \approx 2.5$ for experiment no. 29. In this experiment 1.5 litre of plastic beads (Vestolen) were used on a bed lined with drawing paper for which $\delta_0 = 26.5°$, $\phi = 37.0°$ and $k_{\text{wall}} = 11,0°$. The inset shows the geometry of the bed with the dimensionless length scale along the bed profile which is unrolled in the panels as the horizontal axis. The diamond symbols denote the segment of the bed with convex curvature.

that the Voellmy drag would affect the motion of the centre of gravity but not the spread of the moving and deforming pile (see Hutter and Nohguchi). Except for these approximations, the similarity solution in the form of a parabolic profile represents an exact solution of the governing equations. In fact, it can be shown that the motion of the entire parabolic cap can be deduced from a set of ordinary differential equations that describe (i), the motion of the centre of mass and (ii), the evolution of the semi-span, see Figure 11.11.

Let x_0 and u_0 be the arc length and the streamwise velocity, respectively, of the centre of mass of the parabolic cap, and X and Y the cartesian coordinates of this centre, as shown in Figure 11.11. Furthermore, let g represent the semi-length (or semi-spread) of the parabolic pile. Hutter and Nohguchi (1990) have shown that these variables can be obtained from a solution of the system of first-order

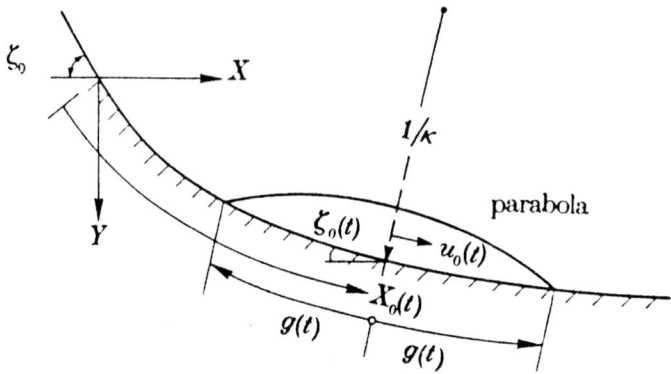

Fig. 11.11. Parabolic shape of the granular avalanche on a weakly curved bed. Shown are an arbitrary position of the centre of mass $x_0(t)$ with velocity $u_0(t)$, semi-length $g(t)$ and the cartesian coordinates X, Y.

ordinary differential equations:

$$\frac{dX}{dt} = u_0 \cos \zeta(x_0), \qquad \frac{dY}{dt} = u_0 \sin \zeta(x_0),$$

$$\frac{dx_0}{dt} = u_0, \qquad \frac{du_0}{dt} = \sin \zeta(x_0) - \mu_c \cos \zeta(x_0) - \frac{2q}{\varepsilon} \frac{\bar{u}^2}{V} g,$$

$$\frac{dg}{dt} = f, \qquad \frac{df}{dt} = \frac{3}{2} \varepsilon \bar{k} \cos \zeta(x_0) - \frac{\Delta \mu}{2} \cos \zeta(x_0)$$

$$- \lambda k(x_0) \cos \zeta(x_0) g,$$

(11.27)

in which the following definitions have been employed:

- $\mu_c = \tan \delta(x_0)$, $\Delta \mu = \frac{1}{2}(\mu_f - \mu_r)$; the bed friction angle is assumd to vary linearly between the front ($\mu_f = \tan \delta(x_f)$), and rear ($\mu_r = \tan \delta(x_r)$) ends and have a mean value μ_c. We think, this variation may account for certain ploughing effects.

- $\bar{k}$ is the value of the earth pressure coefficient at the centre of gravity: $\bar{k} = K_{\text{act}}(\delta(x_0))$, if $f > 0$ and $\bar{k} = K_{\text{pass}}(\delta(x_0))$, if $f < 0$.

- V denotes the total volume of the moving granular mass. Since snow entrainment is ignored, it is constant.

Equations (11.27) are derived from equations (11.19) and (11.20); the model based on these equations will be called the *finite mass Voellmy model*. They must be solved subject to the initial conditions on X, Y, x_0, u_0, g, and f at $t = 0$. For example, if the material is initially at rest, we have

$$X = Y = x_0 = u_0 = f(t_0) = 0, \qquad g(0) = d,$$

(11.28)

Avalanche Dynamics

corresponding to an initial semi-length of the parabolic cap of size d. Often $d = 1$; if this choice is taken, then $\varepsilon = H/L$ is the aspect ratio of the initial pile.

The advantage of the similarity equations and their (numerical) solutions is that the *qualitative* behaviour of the model equations can be studied. Hutter and Koch (1991) have amply shown that experiments cannot *quantitatively* be reproduced with them. The following inferences were found and can easily be deduced from eqs. (11.27):

- The existence of a rigid body motion (i.e. $df/dt = f = 0$) of the moving parabolic pile requires either a variable bed friction angle or a bed with gradually varying bed inclination angle or both. An avalanche on a plane bed with constant bed friction angle will extend forever. This follows simply from eq. $(11.27)_6$, since $df/dt > 0$ in this case.

- A *steady rigid body* motion $(du_o/dt = df/dt = f = 0)$ can (asymptotically) be reached only on an inclined planar bed provided that (see $(11.27)_4$)

$$(a)\ \tan \zeta = \mu_c, \quad \text{no viscous drag,}$$
$$(b)\ \sin \zeta - \mu_c \cos \zeta - \frac{2q}{\varepsilon} \frac{\bar{u}^2}{V} g = 0. \tag{11.29}$$

In this second case, to every inclination angle ζ there belongs an asymptotic avalanche spread $g = g_\infty$ and an asymptotic velocity $\bar{u} = \bar{u}_\infty$, to be evaluated from $(11.27)_6$ and $(11.29)_2$.

- The classical Voellmy (1955) model consists of equations that are formally identical with equations $(11.27)_{1,2,3,4}$ in which the coefficient of $\bar{u}^2$, namely $2qg/(\varepsilon V)$ is held constant. It follows that the classical Voellmy description matches only the finite mass Voellmy model (eqs. (11.27)) in the rigid body state. The transitional phases of the two models cannot correspond because in the finite mass Voellmy model, the approach to an equilibrium velocity depends on the evolution of the spread (which in turn depends critically on $\Delta \mu$, a parameter that does not appear in the Voellmy model).

- Along curved beds, rigid body motions of the moving pile are possible in accelerating or decelerating phases of the centre of mass motion. Computations also show that when the aspect ratio of the initial pile geometry is of the order of 10^{-2}, the motion along a concave curved bed is close to being rigid. This may be the reason why the classical Voellmy model (1955) is as successful as it is.

The above results are due to Savage and Nohguchi (1988) and Hutter and Nohguchi (1990).

11.4 Two-Dimensional Unconfined Flow

The above model has been extended to two-dimensional flow situations, i. e., situations in which the moving granular mass is not confined by sidewalls, but free to spread. Such dispersion perpendicular to the direction of principal motion takes place always, and should be incorporated in a realistic model of flow avalanches. Such studies have been performed, theoretically, computationally and experimentally and are completed (Greve, 1991; Lang, 1992; Hutter, Siegel, Savage and Noguchi, 1993; Greve, Koch and Hutter, 1993; Koch, Greve and Hutter, 1993) or near completion (Koch, 1994).

11.4.1 Equations

The theoretical developments are fairly routine: the shallowness assumption is used to derive a system of equations — balance laws of mass and linear momentum for an incompressible continuum — that are averaged over the avalanche depth, whereby a curvilinear coordinate system defined by the bed topography may have to be used. The now spatially two-dimensional equations take the form (Greve, Koch and Hutter, 1993)

$$\frac{\partial h}{\partial t} + \frac{\partial h\bar{u}}{\partial x} + \frac{\partial h\bar{v}}{\partial y} = 0,$$

$$\frac{\partial \bar{u}}{\partial t} = \frac{\partial \bar{u}}{\partial t} + \bar{u}\frac{\partial \bar{u}}{\partial x} + +\bar{v}\frac{\partial \bar{u}}{\partial y} = \sin\zeta - \tan\delta\,\text{sgn}(\bar{u})(\cos\zeta + \lambda\kappa\bar{u}^2)$$

$$- \varepsilon K_{x\,\text{act/pass}}\cos\zeta\frac{\partial h}{\partial x}, \qquad (11.30)$$

$$\frac{\partial \bar{v}}{\partial t} = \frac{\partial \bar{v}}{\partial t} + \bar{u}\frac{\partial \bar{v}}{\partial x} + +\bar{v}\frac{\partial \bar{v}}{\partial y} = -\tan\delta\,\text{sgn}(\bar{v})(\cos\zeta + \lambda\kappa\bar{u}^2)$$

$$- \varepsilon K_{y\,\text{act/pass}}^{(x\,\text{act/pass})}\cos\zeta\frac{\partial h}{\partial t},$$

in which the earth pressure coefficients $K_{x\,\text{act/pass}}$ and $K_{y\,\text{act/pass}}^{(x\,\text{act/pass})}$ are given by

$$K_{y\,\text{act/pass}}^{(x\,\text{act/pass})} = \begin{cases} K_{y\,\text{act}}^{(x\,\text{act})} & \text{for } \partial\bar{v}/\partial y > 0,\ \partial\bar{u}/\partial x > 0, \\ K_{y\,\text{pass}}^{(x\,\text{act})} & \text{for } \partial\bar{v}/\partial y < 0,\ \partial\bar{u}/\partial x > 0, \\ K_{y\,\text{act}}^{(x\,\text{pass})} & \text{for } \partial\bar{v}/\partial y > 0,\ \partial\bar{u}/\partial x < 0, \\ K_{y\,\text{pass}}^{(x\,\text{pass})} & \text{for } \partial\bar{v}/\partial y < 0,\ \partial\bar{u}/\partial x < 0 \end{cases} \qquad (11.31)$$

with

$$K_{x\,\text{act/pass}} = 2\left(\frac{1 \pm \sqrt{1 - \cos^2\phi/\cos^2\delta}}{\cos^2\phi}\right) - 1 \qquad (11.32)$$

Avalanche Dynamics

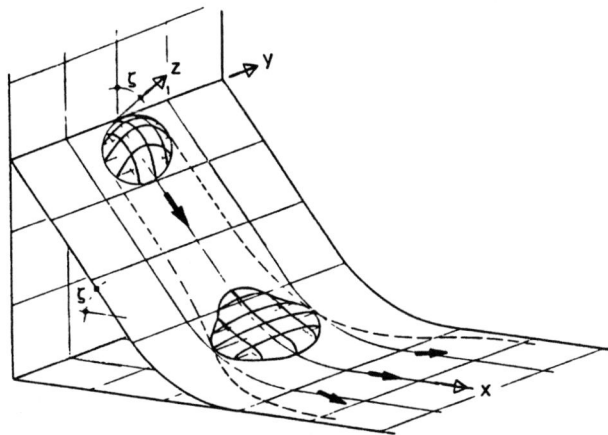

Fig. 11.12. Sketch of the curvilinear coordinate sytem: x is the arc length measured along the generatrix of the downhill direction, y is horizontal and perpendicular to the generatrices (the bed is not curved in the y-direction), z is perpendicular to the bed, and $\zeta(x)$ is the bed inclination angle in the direction of steepest descent.

and

$$K^{(x\ \text{act})}_{y\ \text{act/pass}} = \frac{1}{2}(K_{x\ \text{act}} + 1 \pm \sqrt{(K_{x\ \text{act}} - 1)^2 + 4\tan^2\delta}),$$
$$K^{(x\ \text{pass})}_{y\ \text{act/pass}} = \frac{1}{2}(K_{x\ \text{pass}} + 1 \pm \sqrt{(K_{x\ \text{pass}} - 1)^2 + 4\tan^2\delta}). \tag{11.33}$$

Here, upper, respectively lower signs correspond to the first, respectively second portion of the subscript in (11.32) and (11.33), i.e., the subscript "act" goes together with the minus signs and "pass" goes together with the plus sign. Figure 11.12 defines the dimensionless coordinates; h is the avalanche depth, now a function of x, y and t and $\bar{u}$, $\bar{v}$, are the downhill and sidewise averaged velocity components, respectively, also functions of x, y and t. These equations are spatially two-dimensional, and therefore much more difficult to integrate than the corresponding plane (chute) flow equations (11.19) and (11.20) (Koch, Greve and Hutter, 1993). For a dominant motion in the direction of steepest descent along the bed, a transverse averaging of the depth-averaged equations is suggested. To achieve it, the transverse distribution of the avalanche depth h, is assumed to be parabolic, and that of the streamwise velocity $\bar{u}$ to be uniform, i.e.,

$$h(x,y,t) = \frac{3}{2}\tilde{h}(x,t)\left(1 - \frac{y^2}{b^2(x,t)}\right), \qquad \bar{u}(x,y,t) = \tilde{u}(x,t), \tag{11.34}$$

where $b(x,t)$ is the semi-width of the moving pile of granular material. Koch (1994) has obtained the following evolution equations:

$$\bar{v}_R = \bar{v}(y_R, t) = \frac{db}{dt}, \qquad \frac{\partial(b\tilde{h})}{\partial t} + \frac{\partial}{\partial x}(b\tilde{h}\tilde{u}) = 0,$$

$$\frac{d\tilde{u}}{dt} = \frac{\partial \tilde{u}}{\partial t} + \tilde{u}\frac{\partial \tilde{u}}{\partial x} = \sin\zeta - \tan\delta\,\text{sgn}(\tilde{u})(\cos\zeta + \lambda\kappa\tilde{u}^2)$$
$$- \varepsilon K_{\text{act/pass}}\left(\frac{\partial \tilde{h}}{\partial x} + \frac{\tilde{h}}{b}\frac{\partial b}{\partial x}\right), \qquad (11.35)$$
$$\frac{d\bar{v}_R}{dt} = \frac{\partial \bar{v}_R}{\partial t} + \tilde{u}\frac{\partial \bar{v}_R}{\partial x} = -\tan\delta\,\text{sgn}(\tilde{u})(\cos\zeta + \lambda\kappa\tilde{u}^2)$$
$$+ 3\varepsilon K_{\text{act/pass}}\cos\zeta\frac{\tilde{h}}{b}.$$

These are now spatially one-dimensional equations; they can be integrated with the same methods with which eqs. (11.19) and (11.20) were numerically integrated.

Equations (11.30) were also scrutinized for the existence of similarity solutions. They exist only when further (mild) assumptions are made. These solutions would, for instance, develop from a circular pile having parabolic profile, whereby circles develop into ellipses. For the spreading of a granular avalanche down an inclined plane, such solutions have been constructed by Hutter, Siegel, Savage and Nohguchi (1993) when the bed friction is restricted to Coulomb-type resistance, as in (11.31). Extensions to a Voellmy-type viscous drag have also been worked out (Hutter and Greve, 1993).

11.4.2 Experiments and First Results

Laboratory experiments have been performed to investigate the three-dimensional motion of a mass of granular material. The mass is released from its spherical-cap rest position and then moves down an inclined plane, or an inclined plane merging via a curved segment into a horizontal plane which functions as runout zone (Lang, Leo and Hutter, 1989; Lang, 1992; Greve, 1991; Greve, Koch and Hutter, 1993; Koch, Greve and Hutter, 1993). The bed was covered with no. 120 SIA sandpaper, and various granular materials (the same that were used in the chute experiments) made up the avalanche pile. At first, high speed photography was used to follow the moving and deforming mass through time from a camera positioned such that the entire avalanche motion could be followed from initiation to runout. This led to sequences of birds-eye views of the avalanche, and permitted experi-mental detection of the outer contour, but not a detailed evolution of the pile geometry.

Figure 11.13a shows the flow of a finite mass of plastic beads down an inclined aluminium plane, with photographs taken perpendicular to the plane (and thus corresponding to the least optical distortion). As can be seen, the originally spherical avalanche mass develops into droplet-like shape as it moves down the inclined plane. Computations employing the doubly-averaged equations (11.35), initial geometry, measured internal angle of friction ϕ, as well as the bed friction angle δ, generate contour lines which match the somewhat fuzzy outer contour lines of the experimental droplet-like piles very well. This fuzziness is due to the relatively large bouncing of the particles close to the margin. Figure 11.13b displays a series of snapshots of a similar avalanche of quartz sand moving down an aluminium plane of 60° inclination; the other experimental conditions are the same as before. Evidently, the geometry of the moving pile is blunter than before, the aspect ratio

Avalanche Dynamics

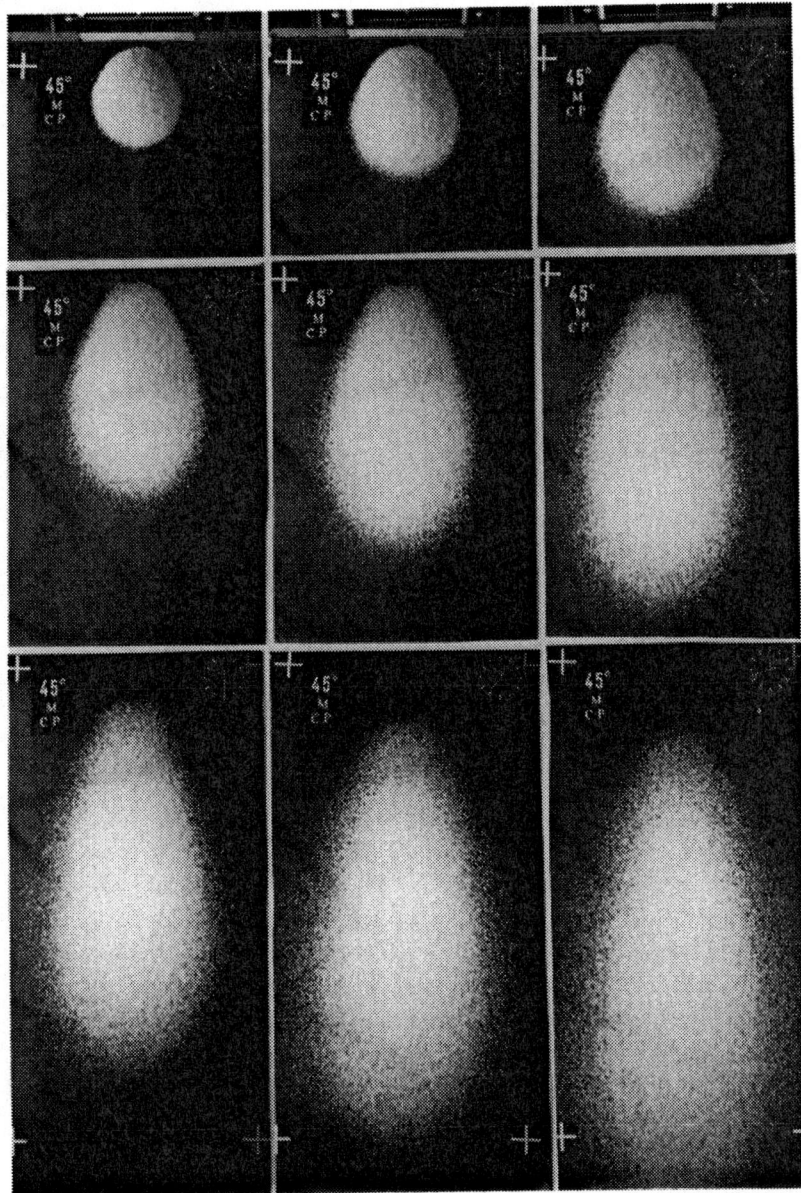

Fig. 11.13a. Sequence of photographs of a finite mass of plastic beads (Vestolen) moving down an inclined plane covered with sand paper SIA No. 120 and inclination angle 45°. The longer arm of the clock at the upper right corner is performing one revolution per second, so the camera is taking about 10 frames per second. Initially, the avalanche is confined by a spherical cap of diameter 520 mm, radius 208 mm, height 104 mm and volume 7656 cm^3. The cap is lifted by rotating it about a horizontal axis at its upper end, thus instantly freeing the granular mass. The mass develops into a droplet shape and, as time proceeds becomes elongated with a progressively smaller aspect ratio.

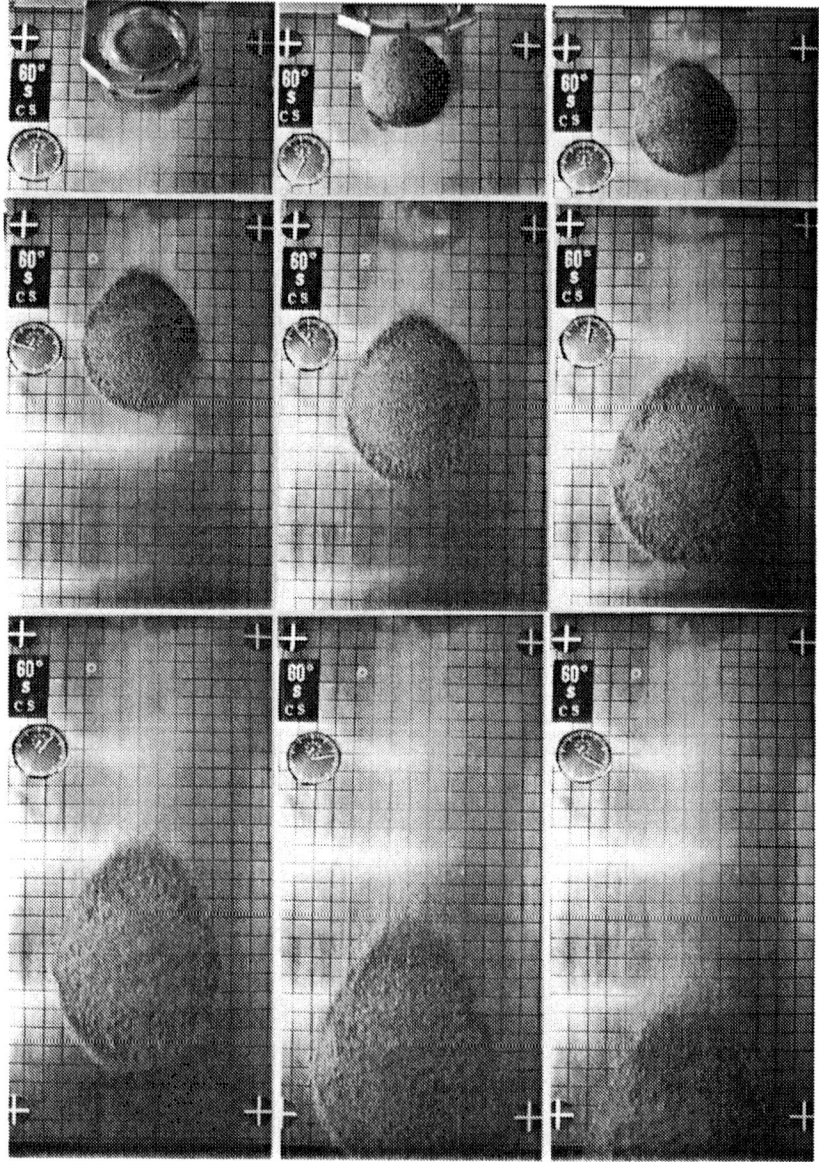

Fig. 11.13b. Same as in (a) but now for a finite mass of quartz sand with nominal diameter 3 mm moving along an aluminium plane with 60° inclination angle; other conditions are the same. Note that, because of the smaller bed friction angle than in Fig. 11.13a, the shape is blunter, and the aspect ratio bigger. Furthermore, the outer margin contour can better be identified because the particles' collisional agitation is smaller.

therefore larger. Moreover, because of the smaller bed friction angle, the collisional agitation of the particles is smaller, and identification of the margin therefore easier. Needless to say, theoretical predictions agree well with the experimental results, and comparison is convincing, see Koch (1994). Similar comparisons, using the twice averaged equations, were also made with experiments on avalanches moving along a curved bed; in this case there is poor agreement between experiments and theory. Greve (1991) concluded that the simple transverse distributions (11.31) cannot be maintained since the pile geometry experiences strong distortions in the course of the motion; integration of the original depth-averaged equations (11.30) is therefore compelling. On the other hand, Lang (1992) altered the basal drag, and added to the Coulomb-type drag a viscous drag, i.e., $S = S_C + S_V$, see Equations (11.20), (11.22a) and (11.22b) for the plane flow situation. As a result, agreement between theoretical prediction and experimental results was somewhat improved. It is concluded from these studies that the use of the transverse averaged equations is not justified, in general and the full equations must be used.

The ultimate test of the model equations is a comparison of theoretical predictions with observations from the motion of a finite mass of granular material moving down a curved inclined surface. Thus a 5 m long and 2 m wide slide was constructed, whose upper approximately 2 m long, inclined portion represents the plane acceleration zone. Below this lies a curved transition zone of about half a meter length in which the inclination angle decreases to zero, merging into a horizontal plane of about 2.50 m length that serves as a runout zone on which the moving pile comes to rest. The whole surface can be covered with different sheets, and different granular materials can be used to vary the internal ϕ and bed friction δ angles. At the top of the incline, a plexiglass hemisphere of 370 mm diameter can be filled with different granular materials; it defines the initial geometry of the avalanche. By suddenly removing this hemisphere, the motion of the avalanche is initiated. The motion upon release is recorded by high speed photography at a rate of approximately 10 frames per second. Figure 11.14 shows a sequence of snapshots from one such experiment. Using a special objective, each frame contains two simultaneous photographs with slightly shifted optical centres, thus allowing stereographic exploration. The frame in the upper left corner shows the hemispherical mass immediately after release, and that in the lower left corner shows the deposited mass in the runout zone. A close-up photograph of the deposited mass is shown in the lower right corner. Exploring these prints yields a sequence of curves marking "snapshots" of margin curves which also can be computationally traced by integrating the equations (11.30)–(11.33) numerically. A detailed description of the theory, its computational implementation, and comparison with experiments, is given in two papers by Greve, Koch and Hutter (1993) and Koch, Greve and Hutter (1993). Here, it suffices to discuss Figure 11.15. In it, the pile margin is represented by strokes that are indicative for a range of boundary positions (error bars), marking a monolayer particle margin. The numerical results are shown as two different contour lines; the outer one determines the numerical avalanche margin with vanishing height, the inner one, on the other hand, describes the contour with the height of one particle diameter. Given the rather coarse nature of the theory, the agreement is very good. The dashed lines in Figure 11.15 mark the

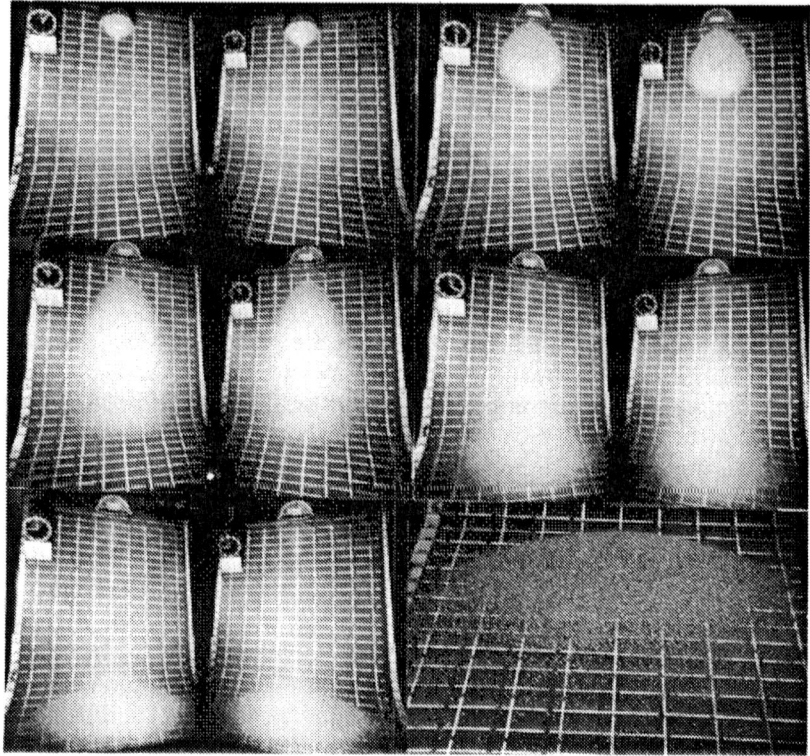

Fig. 11.14. Sequence of stereoscopic photographs of a finite mass of plastic beads (Vestolen) moving down an inclined curved slide ($\zeta = 45°$) coated with sand paper SIA No. 120. The longer arm of the clock at the upper left corner is performing one revolution per second. Initially, the avalanche is confined by a plexiglass hemisphere of diameter 370 mm and volume 13261 cm^3. The white lines generate a rectangular grid on the sliding surface of 100×200 mm^2. The experimental procedure is as described in the caption to Fig. 11.13a. The upper left frame shows the granular mass immediately at the start, that in the lower left at rest in the runout zone. The picture in the lower right displays a close-up of the deposited mass giving an impression how far individual grains spread. (Photos courtesy of R. Greve).

outer edge, where the surface is covered at least half with particles spread away from the avalanche body.

Note that quite a number of such comparisons have been made all with agreement comparable to or better than that in Figure 11.15.

Presently experiments are also being carried out on doubly curved inclines; we are looking for a comparison of theory and experiments in this still more general situation. The experimental technique is also presently being improved by implementing *stereophotography* in an attempt to follow the development of the entire three-dimensional geometry through time. The details of the method, which uses a single camera, are described in Greve (1991), but further work is needed before definite results can be reported.

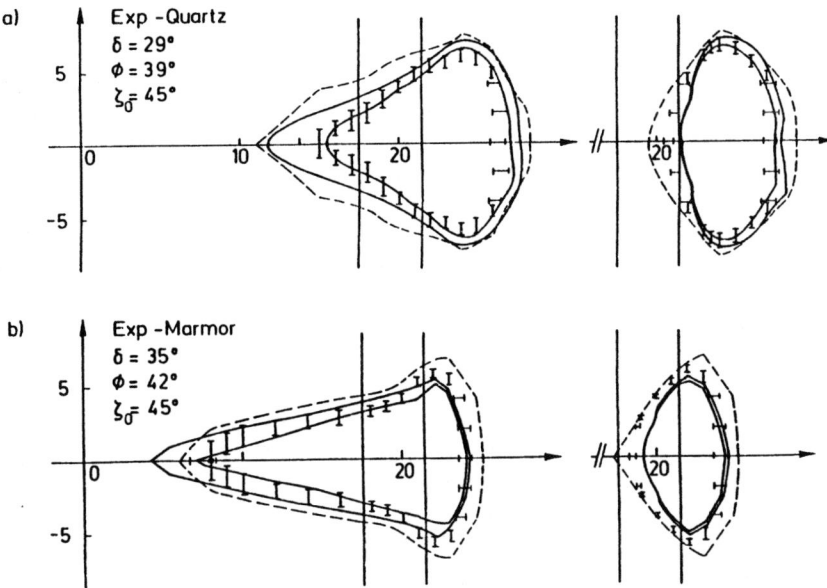

Fig. 11.15. Two "snapshots" of experimental and numerical avalanche positions for two experiments performed with quartz granules (a) and marble chips (b) performed on the experimental slide shown in Fig. 11.14. Bed internal friction and initial inclination angle are shown in the inset. The graphs to the right show the deposited masses at rest; hose on the left are snapshots of masses in motion. Solid lines represent computed margin positions (for $h = 0$ and $h = 3.5$ mm, corresponding to the particle size. The I symbols mark experimental positions of one-particle height, and the dashed lines indicate the boundaries where the basal surface is covered to approximately 50% by particles (adapted from Koch, Greve and Hutter, with alterations, 1993).

II. POWDER AVALANCHES

There are essentially two concepts according to which powder (snow) avalanches are treated. In the first, the simplest possible mixture model is used: The balance laws of mass and momentum for the mixture of the carrier fluid *plus* the suspended particles are complemented by a mass balance statement for the tracer component, here the particles. This additional equation takes the form of a buoyancy balance or a concentration equation and is formally identical to the energy equation of a single component fluid if the density difference between the density underflow and the ambient liquid is due to differences in the thermal states. This is the reason why turbulent density, gravity, turbidity and heavy gas currents share many common properties.

Global properties, rather than local ones, are emphasized, i. e., integrated or averaged field quantities. Turbulent boundary layer flows down inclined planes from steady sources and short finite avalanching masses from instantaneous sources, equivalent to the so-called "thermals", are analysed. Salt solutions, sand and baryta-suspensions are used in experiments, and photography, as well as visual observation, are the tools for data analysis. Entrainment of the ambient fluid and of snow from the snow cover into the current is partly accounted for; considerations are generally restricted, however, to the fully air-borne regime, so that snow deposition is ignored. The runout zone, that is practically of interest, has yet to be analysed this way and is even conceptually excluded in experiments when brine is used to generate the density difference. Theoretical formulations of this kind and associated laboratory experiments have been extensively studied in the last thirty years (e. g. Ellison and Turner, 1959; Hinze, 1960; Plapp and Mitchell, 1960; Middleton, 1966, Middleton and Hampton, 1976; Simpson, 1972, 1982, 1987; Escudier and Maxworthy, 1973; Tochon-Danguy and Hopfinger, 1975; Hopfinger and Tochon-Danguy, 1977; Britter and Simpson, 1978; Beghin, 1979; Chu, Pilkey and Pilkey, 1979; Hopfinger and Beghin, 1980; Britter and Linden, 1980; Beghin, Hopfinger and Britter, 1981; Parker, 1982; Parker, Fukushima and Pantin, 1986; Fukushima and Parker, 1990).

Our own understanding has been guided by the desire of the practitioners to obtain reliable information on amounts and location of snow depositions in the runout zone and dynamic pressures at any position within the avalanche. To achieve this, two phase mixture concepts of a particle suspension in a fluid are required, as is the determination of the *local* pressure and velocity fields. The governing equations are the local forms of the balance laws of mass and momenta for each constituent, with closure conditions for the peculiar Reynolds stresses, air and snow entrainment or detrainment and a postulate for the particle fluid interaction force. Experimentally, the distributions of the particle concentration, velocity and — ideally — the velocity of the carrier fluid ought to be measured in both the avalanche head and avalanche tail. Such studies were begun about ten years ago; related publications are, however, still relatively few: (e.g., Scheiwiller and Hutter, 1982; Scheiwiller, 1986; Scheiwiller, Hutter and Hermann, 1987; Hermann, Hermann and Hutter, 1987; Hermann and Scheiwiller, 1988; Hermann, 1990; Hermann and Hutter, 1991).

11.5 Density and Turbidity Current Concept

This section summarizes the research that has been conducted on flows of powder avalanches based on density and turbidity currents. We shall not so much review the early works if they have been superceded by more general model(s). Reviews that go into more detail in this regard, emphasize other aspects and contain further literature, are: Scheiwiller and Hutter (1982), Hopfinger (1983) and Tesche (1987). Two limiting cases will be studied: (i), the flow from a continuous source, leading to a theory of continuous turbidity currents and, (ii), the flow from an instantaneous source leading to a theory of thermals on inclined boundaries. In both cases, the driving force of the powder avalanche or turbidity current is the down slope component of the effective weight associated with the density difference between the density current and the ambient liquid. Snow or sand particles in snow avalanches or turbidity currents, are suspended by the turbulence and re-deposited if its intensity tapers off. The avalanche size tends to increase with time due to the entrainment of ambient fluid by turbulence. If the production of turbulence created by the motion becomes larger than the sum of the consumption of turbulence by

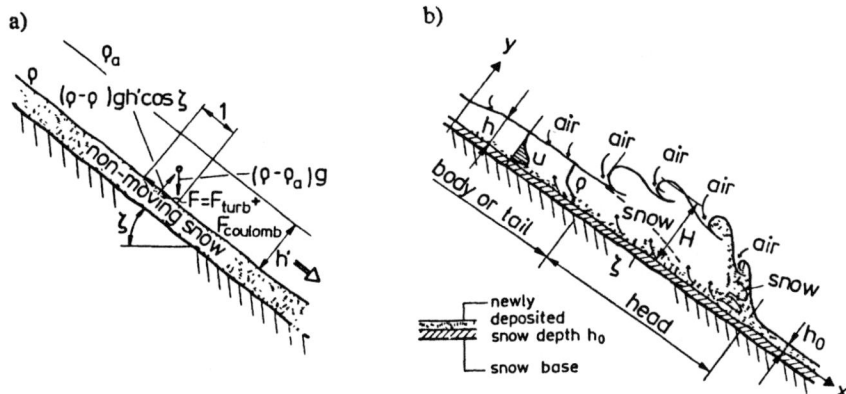

Fig. 11.16. (a) Sliding layer of snow of thickness h' on an inclined plane with slope angle ζ. The frictional force at the bed is composed of a turbulent viscous and a dry Coulomb component. (b) Sketch of a powder avalanche showing the head and the body or tail and indicating the snow and air entrainment mechanisms (after Hopfinger, 1983).

the suspension of heavy particles and the viscous dissipation within the avalanche, the powder avalanche or turbidity current can accelerate and grow to substantial size, to the extent that sufficient particle supply (snow, sand) is available from the bed. Otherwise, the powder snow avalanche or the turbidity current cannot maintain itself in motion. This is observed in nature in the runout zone when, owing to substantial snow deposition, a considerable amount of the turbulent energy has been lost, thus transforming the powder snow avalanche into a *snow fall*.

The above discussion suggests that the level of turbulence is an important physical mechanism that should somehow enter a theoretical formulation. This is only the case in the models proposed by Parker, Fukushima and Pantin (1986) and Fukushima and Parker (1990): Furthermore, because density differences between the ambient liquid and the density underflow may be large it is concluded that the Boussinesq approximation cannot necessarily be made (Tochon-Danguy and Hopfinger, 1975; Hopfinger and Tochon-Danguy, 1977; Hopfinger, 1983; Fukushima and Parker, 1990).

11.5.1 Long Gravity or Turbidity Currents

(a) *Voellmy model*. The simplest description is Voellmy's (1955) hydraulic model that ignores any microstructure and assumes the avalanche depth to be spatially and temporally constant. With reference to Figure 11.16, global momentum balance for an element of length 1 yields the equation

$$(\rho - \rho_a)gh' \sin\zeta - \mu(\rho - \rho_a)gh' \cos\zeta - \frac{\rho g}{x}u^2 = \rho h'\frac{du}{dt}; \qquad (11.36)$$

here, ρ and ρ_a are the densities of the suspension and the ambient liquid, respectively, h' is the avalanche depth, u the depth averaged velocity of the avalanche (neither the particle nor the fluid velocity, but supposedly a representative value for both), ζ is the inclination angle, $\mu = \tan\delta$ the dry Coulomb-type bed friction coefficient of a basal drag with values between 0.1 and 0.5, and ξ is a (reciprocal) friction coefficient for turbulent friction at the base (for avalanches having values between 1000 and 1800 m s^{-2} (see Schaerer, 1975 and Martinelli et al., 1980). The derivation of equation (11.36) shows that it is equally valid for a single mass; it possesses the general solution

$$u = u_{max}\text{Tanh}\left(\frac{u_{max}t}{\xi h'/g}\right),$$

$$u_{max}^2 = \xi h'\left(1 - \frac{\rho_a}{\rho}\right)(S - \mu\cos\zeta) \approx \xi h'(S - \mu\cos\zeta) \approx \xi h' S, \qquad (11.37)$$

$$S = \sin\zeta,$$

in which u_{max} is the steady asymptotic velocity; S is called the slope, and $\rho_a/\rho \ll 1$ has been ignored. Note that the dry Coulomb friction term is irrelevant for powder snow avalanches. Voellmy (1955) applied (11.37) to flow avalanches, and extended its application to powder snow avalanches by adjusting h' according to

$$h' = \frac{\rho_0}{\rho}(h + h_D), \qquad (11.38)$$

where ρ_0 is the mean density of the snow deposited along the avalanche track, h is the depth of the snow in its natural deposition and h_D (often = 0) the depth of the snow layer in front and below the avalanche. At the avalanche front, Voellmy assumed the pressure is hydrostatic and thus equates the static pressure at the base of the avalanche head to the dynamic stagnation pressure of the ambient fluid, i.e.,

$$(\rho_f - \rho_a)gh\frac{\rho_0}{\rho_f} = \rho_a\frac{u_f^2}{2}, \qquad (11.39)$$

where the index f signifies "front". Equation (11.39) in conjunction with (11.36) and the last of equations (11.37), implies $\rho_f = \rho_a\xi S/(2g)$, where $\rho_a/\rho_f \ll 1$ has been used; but this implies that the density of the avalanche head is constant along the track, which does not conform with observations. Obviously, the hydrostatic pressure assumption is too severe. Nevertheless, if this result is substituted into (11.39), the front velocity takes the form

$$u_f^2 = 2gh\frac{\rho_0}{\rho_f}\frac{1}{\rho_a}(\rho_f - \rho_a) = 2g\left(h\frac{\rho_0}{\rho_f}\right)\left(1 - \frac{\rho_a}{\rho_f}\right)\frac{\rho_f}{\rho_a} \approx 2gh'\frac{\rho_f}{\rho_a},$$

where again $\rho_a/\rho_f \ll 1$ has been assumed. Therefore, qualitatively, the avalanche front velocity does not depend upon the slope angle and, it grows with the square root of the avalanche height.

Voellmy also showed that, by requiring continuity of the stagnation pressure at the front, the velocity of the front is somewhat smaller than that of the tail, i.e.,

Avalanche Dynamics

$$\frac{\rho}{2}(u - u_f)^2 \leq \frac{\rho_a}{2} u_f^2 \Rightarrow u_f = u/(1 + \sqrt{\rho_a/\zeta}). \tag{11.40}$$

This is qualitatively corroborated also by observation. The model permits, using formulas (11.37) and (11.40), estimation of the maximum depth averaged velocities for a moving turbulent layer of thickness h' and density ρ. Local stagnation pressures cannot be inferred from it, however.

(b) *Tochon-Danguy and Hopfinger.* Contrary to Voellmy these authors based their model on rigorous mass and momentum balance statements in which the boundary layer and hydrostatic pressure assumptions, but not the Boussinesq approximation, are invoked. They extended Ellison and Turner's (1959) steady inclined-plume analysis to powder avalanches, which in most cases takes the form of an inclined starting plume consisting of a head and a body or tail, see Figure 11.16b (Tochon-Danguy and Hopfinger, 1975; Hopfinger and Tochon-Danguy, 1977; Hopfinger, 1983; Scheiwiller and Hutter, 1982). With these assumptions the two-dimensional local balance laws of mass and momentum in the avalanche body take the form (see Figure 11.16b for notation)

$$\frac{\partial u}{\partial x} + \frac{\partial v}{\partial y} = 0,$$

$$\rho \left\{ \frac{\partial u}{\partial t} + \frac{\partial u}{\partial x} u + \frac{\partial u}{\partial y} v \right\} - g \frac{d}{dx} \int_y^\infty (\rho - \rho_a \cos \zeta \, y') \tag{11.41}$$

$$-g(\rho - \rho_a) S + \rho \frac{\partial}{\partial y} (\overline{u'v'}) = 0.$$

The term involving the integral describes the longitudinal pressure gradient (and is obtained from the y-component of the force balance), and $\rho \overline{u'v'}$ is the xy-component of the turbulent Reynolds-stresses (the only component that survives if the boundary layer assumption is used). In addition, u and v represent the x- and y-components of the *barycentric* velocity of the mixture of fluid plus suspended particles, and the mass balance equation supposes incompressibility to be valid (thus part of the Boussinesq assumption is actually made, see Greenspan, 1968).

Tacitly, Hopfinger and Tochon-Danguy also use the mass balance statement

$$\frac{\partial c}{\partial t} + \nabla \cdot (c\mathbf{v}) + -\nabla \cdot \overline{(c'\mathbf{v}')} = -\nabla \cdot \mathbf{j}, \quad \mathbf{j} = c(\mathbf{v}_p - \mathbf{v}),$$

$$\mathbf{j} = c(\mathbf{v}_p - \mathbf{v}), \tag{11.42}$$

for the snow, where c is the volume fraction of the suspended particles, $\overline{(c'\mathbf{v}')}$ the turbulent mass flux, and $\mathbf{j}$ the diffusive flux of the particles relative to the barycentric motion. Ignoring the turbulent and diffusive mass fluxes on the right hand side of (11.42),[1] and they integrate eqs. (11.41) and (11.42) over the avalanche

---

[1] It is not difficult to show in this case that c in (11.42) may be replaced formally by $(\rho - \rho_a)/\rho_a$, the relative density difference between the mixture and the ambient liquid, in which case (11.39) becomes then the buoyancy balance. Then C in (11.43), as defined in (11.44), represents then a depth-averaged density difference $C = (\rho - \rho_a)/\rho_a$.

depth from $y = 0$ to $y = h\,(=\infty)$, and then obtain for *steady flow* in the avalanche tail the following global balances:

$$\frac{d}{dx}(ChU) = 0,$$

$$\frac{d}{dx}(Uh) + v_h = 0, \qquad (11.43)$$

$$\frac{d}{dx}(\xi_3 ChU^2 + \rho_a hU^2) + \frac{1}{2}\frac{d}{dx}(\xi_1 Cgh^2 \cos\zeta)$$
$$-\xi_2 CghS + \rho u'v'|_{z=0} = 0,$$

where ξ_1, ξ_2, ξ_3 are dimensionless shape factors, and U and C depth averaged velocities and particle concentrations, respectively, which satisfy the relations

$$Uh = \int_0^\infty u\,dy, \qquad \xi_3 ChU^2 + \rho_a hU^2 = \int_0^\infty \rho u^2\,dy,$$

$$U^2 h = \int_0^\infty u^2\,dy, \qquad \xi_2 Ch \doteq \int_0^\infty c\,dy, \qquad (11.44)$$

$$ChU = \int_0^\infty cu\,dy, \qquad \xi_1 Ch^2 = \int_0^\infty 2yc\,dy.$$

The first equation in (11.43), which is usually written as a buoyancy balance, shows that tracer mass along the avalanche body is conserved; consequently, snow entrainment or deposition at the bed, as well as diffusive loss of snow at the upper surface of the avalanche, are ignored. This deficiency is remedied in the work of Parker, Fukushima and Pantin (1986). Similarly, from the second relation in (11.43), which expresses total mass balance, it may be deduced that the volume flux is not conserved by the entrainment of air, v_h, into the avalanche body. Following Morton, Taylor, Turner (1956), Hopfinger and Tochon-Danguy set

$$v_h = -E_a U, \qquad (11.45)$$

where E_a is called *entrainment coefficient*. This represents one of the closure conditions for (11.43). The third of equations (11.43) is the integrated stationary momentum balance: its first two terms, defined in (11.44)$_4$, represent the longitudinal variations of the streamwise momentum fluxes of snow and air, respectively, its third term the change of longitudinal forces due to the hydrostatic pressure, its fourth term the driving component of the gravity force, and its last term the shear traction at the base. A viscous type drag relation of the form

$$-\rho(\overline{u'v'})|_{z=0} = C_D \rho_b U^2 \qquad (11.46)$$

is postulated for this last term, where ρ_b is the avalanche density at the base. It is not difficult to verify that the system (11.43), together with (11.45) and (11.46), possesses the following asymptotic solution:

Avalanche Dynamics

$$U^2 = \frac{\xi_2 C g h S}{\left(1 + \frac{\xi_1}{2}\mathbb{R}i\right)E_a + C_D \frac{\rho_b}{\rho_a}} = \text{constant}, \quad C = C_o \frac{1}{1 + \frac{E_a}{h_o}x}, \quad (11.47)$$

$$E_a = \frac{dh}{dx} = \text{constant} \Rightarrow h = E_a x + h_0,$$

where

$$\mathbb{R}i = \frac{1}{\mathbb{F}r} = \frac{C g h \cos\zeta}{U^2} = \text{constant} \quad (11.48)$$

is the *Richardson* (or *inverse densimetric Froude*) number, $g' = Cg$ the reduced gravity, and h_o and C_o the depth and mean concentration, respectively, at $x = 0$, the position of the (virtual) source. Notice also that the term involving ξ_3 does not enter (11.47) and (11.48). This term is due to the advective momentum transport, and thus affects the dynamics only when snow entrainment is included and/or transient effects are considered. According to Hopfinger (1983), $\xi_1 \approx 1$ and $0.6 \lesssim \xi_2 \lesssim 0.9$, while $C_D \simeq (1.3\text{–}3) \times 10^{-3}$. Note that, because $\mathbb{R}i \leq 0.05$, the Richardson number dependent term in (11.47)$_1$ may be ignored.

The velocity given in (11.47) is characteristic of the tail velocity; however, observations give the rate of advance for the leading edge. Following Middleton (1966), Ellison and Turner (1955) and Georgeson (1942), Tochon-Danguy (1977) obtains

$$U_f \propto \left(g_0' Q_o \frac{S}{E_a}\right)^{1/3}, \quad Q_o = (U_f h)_o, \quad (11.49)$$

via a dimensional analysis where g_0' is the reduced gravity evaluated at the virtual origin, and Q_o is the volume flow at the same position; on the other hand, experiments by Britter and Linden (1980) yield

$$U_f = (1.5 \pm 0.2)(g_0' Q_o)^{1/3}. \quad (11.50)$$

in the range $0° \leq z \leq 90°$ (i.e., no slope to vertical). For zero slope, Simpson and Britter (1978) find $U_f/(g_o U_o)^{1/3} = 1.26$, a value only slightly below the minimum value indicated in (11.50). The absence of the slope dependence in (5.15) does not conflict with (11.49), since this dependence is weak and approximately within the error band indicated in (11.50); see Figure 11.17a.

To be able to evaluate velocities from (11.47), values for the spatial growth rate of the density current must be known. Hopfinger (1983) quotes results from Ellison and Turner (1959) and Britter and Linden (1980), which he casts into the form

$$\text{ET}: \quad E_a = 9.5 \times 10^{-4}(\zeta + 5), \quad \text{BL}: \quad E_a = 4.0 \times 10^{-3}\zeta, \quad (11.51)$$

where ζ is given in radians. With the above numerical values, estimates for U are found to be between approximately 50 to 100 m s^{-1}, as expected. Tochon-Danguy

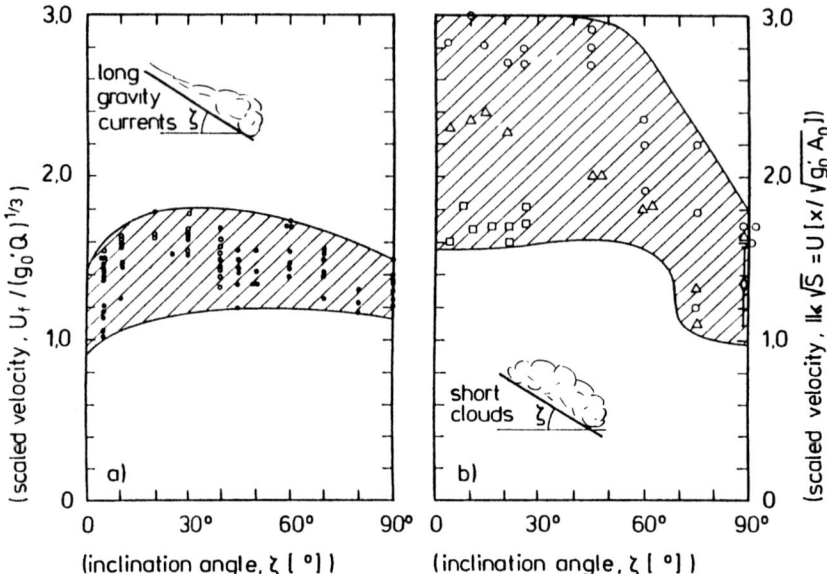

Fig. 11.17. (a) Non-dimensional front velocity $U_f/(g_0'Q)^{1/3}$, plotted against the slope ζ: ○ Tochon-Danguy (1977); ● Britter and Linder (1980); with data also from Georgeson (1942), Wood (1965), Tsang and Wood (1968). (b) Velocity coefficient $\mathbb{K}\sqrt{S}$, or non-dimensional front velocity $U(x/\sqrt{g_0'A_0})$ plotted as a function of slope angle ζ. ○ =, $A_0 = 8000$ mm^2; △, $A_0 = 2300$ mm^2; □, sand suspension $A_0 = 8000$ mm^2; ◇ Tsang's result for line thermals (after Beghin et al., 1981).

also studies short finite-mass gravity currents; results for these will be discussed in Section 11.5.2.

(c) *Parker, Fukushima and Pantin.* Real powder snow avalanches or turbidity currents differ from the simple, conservative density flow studied by Ellison and Turner (1959), Tochon-Danguy and Hopfinger (1975) and Hopfinger and Tochon-Danguy (1977) in that the source of the buoyancy difference, i.e., the mass of the suspended snow particles or sediment, is not conserved; the suspended particles are free to exchange with the particles forming the bed. Furthermore, as already mentioned earlier, the turbulence intensity is likely to affect the mean flow state, and transient features also need to be incorporated. Thus, the balance equations (11.43) need to be generalized and complemented by a balance law for the depth-averaged turbulent kinetic energy, and the closure conditions may need to be modified. Parker et al. (1986) present two such models, i.e., the three-equation and the four-equation models. The difference between the two is that the former does not include an equation of balance for the mean energy of the turbulence, whereas the latter does.

One of the basic differences between their work and that of Hopfinger and Tochon-Danguy, (1977) is that they do not ignore the turbulent mass flux and the diffusive flux of the particles relative to the barycentric motion; in fact, they

Avalanche Dynamics

employ the boundary layer approximation to the right-hand side of (11.42), i.e.

$$-\nabla \cdot (\overline{c'\mathbf{v}'} + \mathbf{j}) \Rightarrow -\frac{\mathrm{d}}{\mathrm{d}y}(F - v_s c), \qquad F = (\overline{c'v'}), \qquad j_y = -v_s c,$$
(11.52)

where v_s is the free fall velocity of heavy particles in an ambient fluid at rest. Thus it is supposed that no other diffusive flux than that due to excess gravity is at work. Parker et al. (1986) also focus on unsteady flow and impose the Boussinesq approximation. This makes their equations applicable to accelerating turbidity currents and laboratory avalanches; inferences from their model for real powder snow avalanches must be taken with some caution. Parker et al. (1977) average their equations by integrating (11.41) and the boundary layer approximation to (11.42) (see (11.52)) across the depth of the turbidity current, which yields

$$\frac{\partial(Ch)}{\partial t} + \frac{\partial(ChU)}{\partial x} = F_b - v_s c_b \cos\zeta = v_s(E_s - r_o C \cos\zeta),$$

$$\frac{\partial h}{\partial t} + \frac{\partial(hU)}{\partial x} = E_a U, \qquad (11.53)$$

$$\frac{\partial(Uh)}{\partial t} + \frac{\partial(U^2 h)}{\partial x} = -\frac{1}{2}gR\frac{\partial}{\partial x}(Ch^2 \cos\zeta) + gRChS - u_{*b}^2.$$

Here, C is interpreted as the depth-averaged particle volume fraction, and $R = (\hat{\rho}_s/\hat{\rho}_a - 1)$, where the hats indicate material densities. Equation $(11.53)_2$ corresponds to $(11.43)_2$ and incorporates the entrainment postulate for the fluid (11.45). Equation $(11.53)_1$ generalizes $(11.43)_1$ to time dependent processes; its right-hand side corresponds to the flux terms in (11.52), and the index 'b' refers to values at the bed. Note that $F_b = v_b E_s$, where E_s represents a dimensionless coefficient of bed sediment entrainment. Further, $C_b = r_o C$ holds for the concentration of the suspended particles immediately above the bed. Equation $(11.53)_3$ represents the streamwise momentum balance; the first term on the right corresponds to the longitudinal variation of the (hydrostatic) excess pressure force (the factor $\cos\zeta$ is absent in Parker et al.'s (1986) presentation because the slope in their study is, small); the second term is the driving force due to excess gravity and the third term the basal friction term: $u_{*b}^2 = -(\overline{u'v'})$.

Physically, the important addition in (11.53), as compared to (11.43), is the right hand side of $(11.53)_1$, which can also be written in the form $v_s(E_s - c_b \cos\zeta)$. This term is responsible for self-acceleration; indeed, if E_s exceeds $c_b \cos\zeta$, the current entrains more snow from below than it loses through deposition. As a result, it may become heavier, increasing the term $gRChS$ that constitutes the driving force of the avalanche in $(11.53)_3$; thus the avalanche will accelerate. The increase in mean speed, U, can be expected to increase the entrainment rate, $v_s E_s$, which in turn results in an increase in U, leading to a positive feedback. However, Parker et al. (1986) argue that "turbulent energy is expended in both, maintaining the existing load in suspension and entraining new sediment from the bed. An arbitrarily large rate of entrainment cannot be maintained, because the rate of expenditure of turbulent energy may exceed the supply from the mean flow. This would eventually cause the turbulence to collapse, the particles held in suspension to settle out, and the current to disappear". Evidently, a balance of the mean turbulent energy K is needed; and Parker et al. (1986) show that it takes the form

$$\frac{\partial}{\partial t}(Kh) + \frac{\partial}{\partial x}(KUh)$$
$$= Uu_{*b}^2 + \frac{1}{2}U^3 E_a - \varepsilon h - (Rgv_s Ch + \frac{1}{2}RgChUE_a + \frac{1}{2}Rghv_s(E_s - r_o C)),$$
(11.54)

where

$$\varepsilon h = \int_0^h \nu \operatorname{tr}((\operatorname{grad} \mathbf{v}')(\operatorname{grad} \mathbf{v}')^T) \, dy) \tag{11.55}$$

is the depth-integrated mean rate of dissipation of turbulent energy due to viscosity. The first two terms on the right-hand side of (11.54) quantify the rate of production of turbulent energy; the remaining terms quantify its annihilation. The three terms in parentheses represent the rate of turbulent energy expenditure due to working against the particle concentration gradient. Parker et al. (1986) also show that (within their general approximation) these three terms are equal to the depth-integrated buoyancy flux $\int_0^h RgF \, dy$, where F is defined in (11.52). The first of these can be interpreted as the work necessary to maintain a given suspension in equilibrium (Knapp, 1938; Bagnold, 1962), the second as the increase in the height of the centre of gravity, and thus of the potential energy of the suspension, due to entrainment of liquid, and the third as a rate of turbulent energy expenditure due to entrainment of particles from, or deposition to, the bed.

Equilibrium solutions of (11.53) and (11.54), assuming constant values for h, C, U and K, are given by

$$E_s = c_b \cos \zeta,$$
$$E_a = 0,$$
$$u_{*b}^2 = RgChS,$$
$$u_{*b}^2 U = RgChUS = \varepsilon h + Rgv_s Ch.$$
(11.56)

In this case, then, the amount of entrained particles at the bed equals that being deposited (i.e., (11.56)$_1$), entrainment of ambient fluid does not occur (i. e., (11.56)$_2$) and the driving force balances frictional resistance (i.e., (11.56)$_3$). Further, the work done by the frictional forces equals the power supplied by the action of the down-slope component of gravity on the suspended sediment, which in turn equals the power expended in viscous dissipation plus the work necessary to maintain the non-buoyant particles in suspension (i. e., (11.56)$_4$). This last result also implies (since $\varepsilon > 0$) that a necessary condition for an equilibrium, self-sustained turbidity flow of a constant thickness to exist is $US > v_s$ (Bagnold, 1962, Knapp, 1938). More generally, for dK/dt to be non-negative, the inequality

$$u_{*b}^2 U + \frac{1}{2}U^3 E_a > Rgh(v_s C + \frac{1}{2}E_a UC + \frac{1}{2}v_s(E_s - r_o C)) \tag{11.57}$$

must hold; it can readily be deduced from (11.54). If this inequality is violated for a sufficient amount of time, the turbulence must disappear. More appropriate for the cessation of the turbulence, however, is a threshold value for K, i.e., $K = K^*$, below which rapid sedimentation would occur. Such a condition has not been proposed so far.

Equations (11.53), complemented by appropriate closure conditions, comprise the three-equation model; equations (11.53) and (11.54), also complemented by (other) closure conditions, constitute the four-equation model. Parker et al. (1986) propose the following phenomenological conditions:

(i) Three equation model:

- $E_a = \dfrac{0.00153}{0.0204 + \mathbb{R}i}$, $\quad \mathbb{R}i = \dfrac{RgCh\cos\zeta}{U^2}$, $\tag{11.58}$

- $r_o = 1 + 31.5\mu^{-0.46}$, $\quad m = \dfrac{u_b^*}{v_s}$, $\quad$ or $\quad r_o = 1.8$, $\tag{11.59}$

- $E_s = \begin{cases} 0.3, & Z > Z_m, \\ 3 \times 10^{-12} Z^{10}\left(1 - \dfrac{Z_c}{Z}\right), & Z_c < Z < Z_m, \\ 0, & Z < Z_c, \end{cases}$ $\tag{11.60}$

- $u_{*b}^2 = c_D U^2$. $\tag{11.61}$

with $Z_c = \sqrt{\mathbb{R}_p}\,\mu$, $R_p = (\sqrt{RgD_s}\,D_s)/\nu$, $Z_c \approx 5.0$, $Z_m \approx 13$, where R_p represents the particle Reynolds number, ν the kinematic viscosity of water, and D_s the grain diameter.

According to (11.58), as $\mathbb{R}i \to 0$, $E_a \to 0.075$, which is appropriate for non-stratified flows; on the other hand, as $\mathbb{R}i \to \infty$ the formula of Egashira (1980), extensively supported by data for density-driven flows, is obtained. Formula (11.59) is supported by data obtained from Garcia (1985) and (11.60) was proposed by Akiyama and Fukushima (1985) on the basis of data for open-channel suspensions in flumes and rivers. The velocity scale $(RgD_s)^{1/2}$ arising in the definition of the particle Reynolds number R_p is the free fall velocity of a buoyant particle with diameter D_s after falling freely for a distance of $D_s/2$. Finally, (11.61) is the classical turbulent viscous drag relationship, with $c_D \approx (1.3-3) \times 10^{-3}$.

ii) Four-equation model:
In this case the closure conditions (11.58)–(11.60) for E_a, E_s, r_o are retained, so that E_s is still a function of u_{*b}, (11.61), however, is replaced, and a closure equation for the mean turbulent dissipation ϵ must also be postulated. Thus, the four equation model takes the form

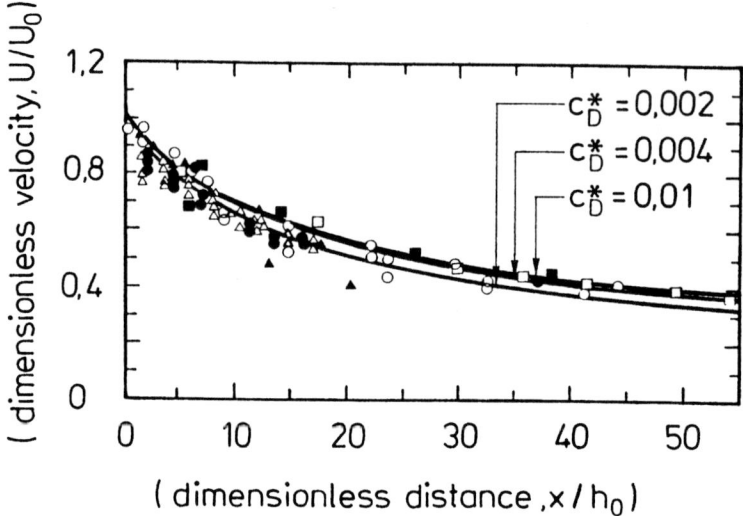

Fig. 11.18. Plot of U/U_o against x/h_o for a plane wall jet, showing the effect of varying c_{D*} on the predictions of the four-equation model. The points represent experimental data. o, Myers et al. (1961); •, Sigalla (1958); ▲, Schwarz and Cosart (1961); · · ·, Gartshore and Newman (1969); △, Rajaratnam (1976); □ Mathieu and Tailland (1965). The curves represent calculations based on the four-equation model, $a = 0.1$ (from Parker et al., 1986)).

- E_a, r_o, E_s as in (11.58), (11.59), (11.60),

- $u_{*b}^2 = \alpha K$,

- $\varepsilon = \beta \dfrac{K^{3/2}}{h}$, $\beta = \dfrac{\frac{1}{2} E_a(1 - \mathbb{R}i - 2(c_{D*}/\alpha)) + c_{D*}}{(c_{D*}/\alpha)^{3/2}}$,

(11.62)

where α and β are dimensionless parameters, $\mathbb{R}i$ the Richardson number defined in (11.58), and β has been selected so that the three- and four-equation models generate the same equilibrium solutions for U, h and C when $v_s = 0$ and c_D in (11.61) is replaced by c_{D*} (Parker et al., 1986). The expression for the turbulent dissipation ε is formally identical with that proposed by Kolmogorov (1941) and Heisenberg (1948). Two parameters must be prescribed, namely α and c_{D*}. Values of $\alpha = 0.1$ and $c_{D*} = 0.002$ match data on plane wall jets excellently (Rajaratnam, 1976; Mathieu and Tailland, 1965) see Figure 11.18 ($E_a = 0.05$ was used in lieu of $E_a = 0.075$). Note that depth-averaged velocities U react very insensitively to variations of a and c_{D*} in the intervals $0.05 < \alpha < 0.5$ and $0.000 < c_{D*} \leq 0.008$, respectively. The value $c_{D*} = 0$ must be excluded, as it implies $\beta \to \infty$.

In steady turbidity currents with $v_s = 0$ (i.e. neither entrainment nor deposition of particles at the bed), the equations reproduce Ellison and Turner's (1959) equilibrium flow solution with *constant* $E_a = dh/dx$, U, C and $\mathbb{R}i$ as given in (11.47) and (11.53) (in which $x_1 = x_2 = 1$, appropriate for the Boussinesq approxima-

Avalanche Dynamics

tion). No such equilibrium solution exists for turbidity currents and powder snow avalanches with $v_s \neq 0$. One can formally, however, determine values for U, C, h, Ri and K for which the right-hand sides of $(11.53)_{1,3}$ and (11.54) vanish, subject to the condition that the conditions (11.62) hold. A set of parameters U_I, C_I, h_I, Ri_I, K_I satisfying these conditions will be called a state at *Ignition*. It is uniquely determined, provided numerical values for S, c_{D*}, h_I/D_s, D_I (or R_I) and a are specified; in the subsequent analysis, it serves as a convenient set of *scaling factors* for the respective variables.

Consider next the steady state analogues of (11.53) and (11.54). A forward numerical integration technique may be used to calculate the downstream development of the turbidity current subject to prescribed upstream (initial) values of U_o, $h_o = h_I$, C_o and K_o. The results can be expressed in the form U/U_I, etc., where $U = U(x)$, where U_I is the ignition value of the depth-averaged velocity. Parker et al. (1986) construct the following phase- plane-type projections of the results: $(U/U_I, \psi/\psi_I)$, where $\psi = UCh$ is the *volumetric suspended sediment transport rate*, and $\psi_I = U_I C_I h_o$ its value at ignition, see Figure 11.19. Motions only exist in the supercritical regime, for which $Ri < 1 (\text{Fr} > 1)$, and so only the shaded area in Figure 11.19 is relevant. Any initial data are projected to a point in this graph, and the motion of the turbidity current maps onto a trajectory in this plane emanating from the initial point. All these trajectories merge into a very small converging band, yielding either larger or smaller velocities U, such that a current is either accelerating or decelerating depending upon whether it lies above or below the so-called autosuspension line (AGL). More generally, this line separates all possible initial conditions into those which result in *igniting fields* — the avalanche is stable and may grow — and subsiding fields — the avalanche will disappear. This is corroborated in Figure 11.20, where fields starting from points (a) and (b) in Figure 11.19 are growing, whereas those emanating from points (c) and (d) are subsiding. We might also mention (without proof) that in the three-equation model, the normalized volumetric transport rate ψ/ψ_I becomes exceedingly large at a finite distance, so that the inequality (11.57) is quickly violated, and the level of turbulence (a quantity that is not calculated in the three-equation model) is soon too low to sustain the motion. So, in the three-equation model, any long turbidity current along an inclined bed will disappear; this is the reason for abandoning this model.

The four-equation model also requires amendment, or further investigation, in particular with regard to the sedimentation of the suspended particles, when the level of turbulence is approaching a certain threshold value $K = K^*$, below which the motion essentially ceases to exist. It is this limiting situation that is of utmost interest to practitioners.

11.5.2 Short Gravity Currents. "Thermals" on Inclined Boundaries

Short, finite-length gravity currents develop when the buoyancy supply at an upstream position is not maintained. Such clouds or thermals on inclined boundaries were studied by Tochon-Danguy (1977), Hopfinger and Tochon-Danguy (1977),

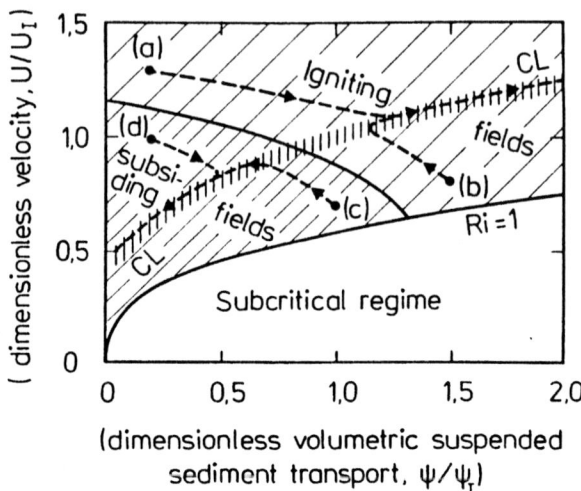

Fig. 11.19. Phase diagram computed from the four-equation model for the case $D_s = 0.1$ mm, $c_{D*} = 0.004$, $S = 0.05$ and $h_o/D_s = 2 \times 10^4$. Schematic plot even though obtained from concrete computations (from Parker et al., 1986).

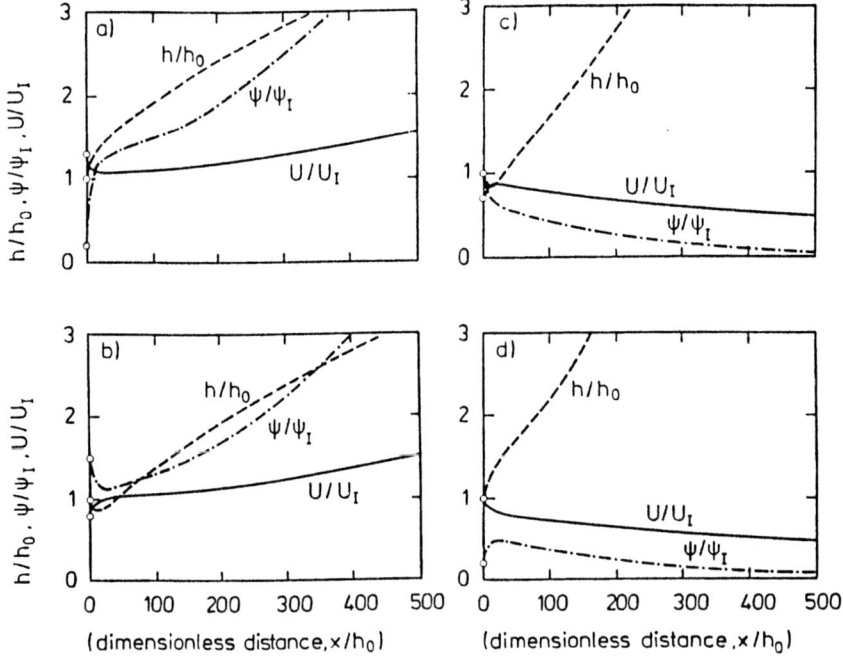

Fig. 11.20. Dimensionless height h/h_o volumetric suspended sediment transport rate ψ/ψ_I, and current U/U_I, plotted against dimensionless distance, a) from point (a) of Figure 11.19; b) from point (b) of Figure 11.19; c) from point (c) of Figure 11.19 and d) from point (d) of Figure 11.19 (from Parker et al., 1986).

Avalanche Dynamics

Beghin, Hopfinger and Britter (1981) and Fukushima and Parker (1990). They generalize work originally commenced by Escudier and Maxworthy (1973).

a) Hopfinger and Tochon-Danguy. These authors combine dimensional analysis, a buoyancy balance and observations to arrive at asymptotic relations for the moving finite cloud, Figure 11.21. Assuming that the speed of the avalanche front U_f depends on the mean density difference $\Delta\bar{\rho}$, the density of the ambient fluid ρ_a, the slope, S, and the cross sectional area of the (infinitely wide) avalanche A, i.e. $U_f = \text{fct}(\Delta\bar{\rho}, \rho_a, S, A)$, then simple dimensional analysis yields

$$U_f = C_1((\Delta\bar{\rho}/\rho_a)gS\sqrt{A})^{1/2}, \tag{11.63}$$

If it is further assumed that a layer of snow (with depth h_N and density ρ_N) is entrained, then the steady state excess mass balance is given by

$$\frac{d}{dt}(\Delta\bar{\rho}A) = \frac{d}{dx}(\Delta\bar{\rho}A)U_f = \Delta\rho_N h_N U_f. \tag{11.64}$$

This can immediately be integrated to yield $\Delta\bar{\rho}A = $ constant (if $h_N = 0$) and $\Delta\bar{\rho}A = \Delta\rho_N h_N x$ (if $h_N \neq 0$), where x is the distance from a virtual origin. If, moreover, the similarity assumption is made, then $A = \xi_A h^2$, with shape factor ξ_A. Finally, observations show h to grow linearly with the distance from a virtual origin, i.e., $h = \xi_h x$. Substituting all these results into (11.63) yields the following asymptotic results:

(i) *when snow entrainment is absent* ($h_N = 0$),

$$h \sim \xi_h x, \qquad U \sim \frac{c_1}{\xi_A^{1/4}\xi_h^{1/2}}\sqrt{\frac{\Delta\bar{\rho}_o}{\rho_a}gA_oS}\frac{1}{\sqrt{x}}, \qquad \Delta\bar{\rho} \sim \frac{\Delta\bar{\rho}_o h_o^2}{\xi_h^2}\frac{1}{x^2}, \tag{11.65}$$

where $\Delta\bar{\rho}_o = \bar{\rho}_o - \rho_a$ and A_o are the density difference and the cross sectional area at the virtual origin, respectively.

(ii) *With snow entrainment* ($h_N \neq 0$),

$$h = \bar{\xi}_h x; \qquad U \sim (c_1/(\xi_A^{1/4}\bar{\xi}_h^{1/2}))\sqrt{(\Delta\rho_N/\rho_a)gS}, \qquad \Delta\bar{\rho} \sim \frac{\Delta\rho_N h_N}{\xi_A \bar{\xi}_h^2}\frac{1}{x}. \tag{11.66}$$

It is assumed that the slope $\bar{\xi}_h$ differs from ξ_h.

Snow entrainment accelerates the flow since U is asymptotically constant when $h_N \neq 0$ but decays as $\xi^{-1/2}$ when $h_N = 0$. Although the laws (11.65) have been experimentally corroborated in the laboratory (Figure 11.21b which shows U_f rather than U, plotted against distance) the asymptotic relations (11.66) are recommended for real avalanches; these latter formulas are, however, not very useful since the location of the virtual origin must be known in advance.

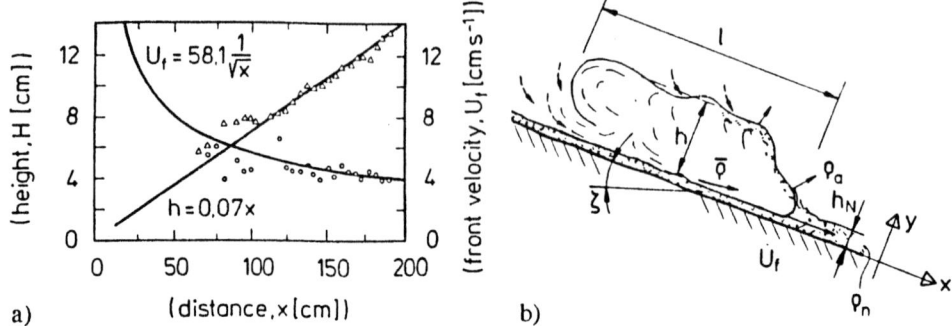

Fig. 11.21. (a) Sketch showing the characteristics of a "inclined thermal" with snow and air entrainment. (b) Height h and frontal velocity U_f of a short gravity current without entrainment from the bed as functions of the travelled distance x with $\zeta = 20°$, $A_o = 1670$ mm^2 (after Hopfinger and Tuchon-Danguy, 1977).

b) Beghin, Hopfinger and Britter. Beghin (1979) and Beghin, Hopfinger and Britter (1981) studied primarily laboratory avalanches of finite dimension in a plane flow situation. They ignore entrainment or sedimentation of snow, assume basal friction to be negligible, ignore any spatial variation in the density of the cloud (so that a mean density $\bar{\rho}$ characterizes the gravity current), and assume similarity i.e., if h, l and A are a typical depth, length and area of the cloud, respectively, then

$$K_1 = \frac{l}{h} = \text{const.} \quad \text{and} \quad K = \frac{A}{hl} = \text{const.} \tag{11.67}$$

are independent of position and time, but may still depend on ζ, the bed inclination angle of the bed. In addition, while the total amount of snow within the avalanche is kept constant, air entrainment through the upper surface is accounted for by a relation of the form

$$\frac{dA}{dt} = \bar{E}(\zeta) U \sqrt{(\bar{\rho}/\rho_a) hl}, \tag{11.68}$$

where U is the speed of the centre of gravity of the cloud along the incline and $\bar{E}(\zeta)$ an overall entrainment coefficient, assumed to be at most dependent on ζ. The analysis here goes beyond that of Hopfinger and Tochon-Danguy (1977) in which the following balance laws of mass and momentum for snow plus air and snow (in the latter case in form of a buoyancy equation) are formulated:

$$\begin{aligned}
\text{total mass:} \quad & \frac{dm}{dt} = \rho_a \frac{dA}{dt}, \quad m = \bar{\rho}A = \bar{\rho}Khl, \\
\text{buoyancy:} \quad & \frac{dB}{dt} = 0, \quad B = \Delta\bar{\rho}A = (\bar{\rho} - \rho_a)A, \\
\text{momentum:} \quad & \frac{d}{dt} = [((k_v \rho_a + \bar{\rho})AU] = BgS,
\end{aligned} \tag{11.69}$$

Avalanche Dynamics

where k_v is the coefficient of (virtual) added mass (Batchelor, 1974, p. 431, $k_v = 2h/l$ for an elliptic cylinder). Solutions to (11.69), subject to the similarity (11.67) and the entrainment assumption (11.68), are easily constructed (Scheiwiller and Hutter, 1982).

(i) *In the Boussinesq-regime* ($\Delta\bar{\rho} \ll \bar{\rho}$): In this case $\bar{\rho}$ and $(Kk_v\rho_a + \bar{\rho})$ on the left of $(11.69)_{1,3}$ may be regarded as constants, we obtain the relations

$$\frac{dh}{dx} = \frac{1}{2}\sqrt{\frac{\rho_a}{\bar{\rho}}} \frac{\bar{E}(\zeta)}{K\sqrt{K_1}} = \text{const.}, \qquad \frac{dl}{dx} = \frac{1}{2}\sqrt{\frac{\rho_a}{\bar{\rho}}} \frac{\sqrt{K_1}}{K}\bar{E}(\zeta) = \text{const.},$$

$$U^2 = U_0^2 \frac{x_0^4}{x^4} + \frac{2c}{3}\frac{1}{x}\left(1 - \left(\frac{x_0}{x}\right)^3\right), \qquad c = \frac{4K^2\Delta\bar{\rho}gSh_0l_0}{\rho_a\bar{E}_a(k_v + 1)}, \qquad (11.70)$$

where quantities with an index 0 are reference values at $x = x_0$, (the virtual origin of the cloud). Note that the velocity U reaches a maximum at $x_m = x_0(4 - 6x_0U_0^2/c)^{1/3}$. So, when the cloud is released with a velocity smaller than this, it will go through an accelerating phase just after release and then decelerate, otherwise the thermal decelerates throughout its motion. Asymptotically, as $x \to \infty$, the velocity approaches the final speed

$$U \sim \sqrt{\frac{2c}{3}\frac{1}{x}} = \sqrt{\frac{8K}{3\bar{E}(k_v + 1)}}\sqrt{S}\sqrt{\frac{\Delta\bar{\rho}}{\rho_a}gA_0}\frac{1}{\sqrt{x}} \qquad (11.71)$$

$$= \mathbb{K}\sqrt{S}\sqrt{\frac{\Delta\bar{\rho}}{\rho_a}gA_0}\frac{1}{\sqrt{x}} = \mathbb{K}\sqrt{S}\sqrt{g_0'A_0}\frac{1}{\sqrt{x}}, \qquad (11.72)$$

where $A_0 = Kh_0l_0$ is the initial volume of the cloud.

(ii) *In the non-Boussinesq regime*: here $\bar{\rho}$ is large in comparison to $k_v\rho_a$, which may now be ignored. Equations (11.69) become in this case

$$\frac{dm}{dt} = \frac{\bar{E}}{\sqrt{K}}M\sqrt{\frac{\rho_a}{m}}, \qquad \frac{dM}{dt} = BgS,$$
$$\frac{dB}{dt} = 0, \qquad M = \bar{\rho}AU, \qquad (11.73)$$

where M is the total momentum of the cloud and $U = M/m$. Straightforward integration of (11.73) yields $B = \text{const.}$, $M = (BgS)t + M_0$ and

$$m(t) = \left(\frac{3}{2}\frac{\bar{E}\sqrt{\rho_a}}{\sqrt{K}}\left(\frac{BgSt^2}{2} + M_0t\right) + m_0^{3/2}\right)^{2/3}, \qquad (11.74)$$

so that a simple computation yields

$$U(t) = \frac{M}{m} = \frac{BgSt + M_o}{\left(\frac{3}{2}\frac{\bar{E}\sqrt{\rho_a}}{\sqrt{K}}\left(\frac{BgSt^2}{2} + M_o t\right) + m_o^{3/2}\right)^{2/3}}$$

$$= \frac{2\sqrt{K}}{\bar{E}\sqrt{\rho_a}}\frac{d}{dt}\left(\frac{3}{2}\frac{\bar{E}\sqrt{\rho_a}}{\sqrt{K}}\left(\frac{BgSt^2}{2} + M_o t\right) + m_o^{3/2}\right)^{1/3}, \quad (11.75)$$

$$x(t) = \frac{2\sqrt{K}}{\bar{E}\sqrt{\rho_a}}\left(\frac{3}{2}\frac{\bar{E}\sqrt{\rho_a}}{\sqrt{K}}\left(\frac{BgSt^2}{2} + M_o t\right) + m_o^{3/2}\right)^{1/3}$$

with an appropriately chosen value for $x(0)$. With this choice, $(11.75)_2$ may be used in (11.74), and when the resulting equation is combined with the definition of m, (11.74) and the similarity assumption (11.67) imply

$$h = \frac{\bar{E}}{2K\sqrt{K_1}}\sqrt{\frac{\rho_a}{\bar{\rho}}}\,x, \qquad l = \frac{\bar{E}}{2K}\sqrt{\frac{\rho_a}{\bar{\rho}}}\,x. \qquad (11.76)$$

Because $\bar{\rho}$ may vary with x, however, neither h, nor l are linear functions of x. Squaring (11.76), and writing $\bar{\rho} = \rho_a + \Delta\bar{\rho}$ implies

$$(\rho_a h^2 + \Delta\bar{\rho}_o h_o^2)^{1/2} = \frac{\bar{E}\sqrt{\rho_a}}{2K\sqrt{K_1}}\,x, \qquad (\rho_a l^2 + \Delta\bar{\rho}_o l_o^2)^{1/2} = \frac{\bar{E}\sqrt{K_1}\sqrt{\rho_a}}{2K}\,x,$$

$$(11.77)$$

where h_o, l_o and $\Delta\bar{\rho}_o$ represent initial values for h, l and $\Delta\bar{\rho}$, respectively (evaluated at $x = x(0)$), and conservation of buoyancy has been employed. Furthermore, since B = constant,

$$\Delta\bar{\rho} = \Delta\bar{\rho}_o\left(\frac{h_o}{h}\right)^2, \qquad (11.78)$$

or upon substitution of $(11.77)_1$,

$$\frac{1}{\Delta\bar{\rho}} = \frac{1}{\delta\bar{\rho}_o}\frac{\bar{E}}{4K^2 K_1}\left(\frac{x}{h_o}\right)^2 - \frac{1}{\rho_a}. \qquad (11.79)$$

It follows that the density difference drops off asymptotically as x^{-2}.

In the Boussinesq regime, the predictions described above have been verified by laboratory experiments. In the theory, the variables x and U refer, however, to the position of mass centre measured from the virtual origin and to the mass centre velocity, respectively. A simple transformation to the position of the front x_f and its velocity U_f, which are observable, takes the form

$$x_f - x = \frac{1}{2}l, \qquad U = U_f\left(1 - \frac{1}{2}\frac{dl}{dx_f}\right). \qquad (11.80)$$

using these transformations, all equations can now be written in terms of x_f and U_f.

Avalanche Dynamics

Beghin, Hopfinger and Britter (1981) corroborate the constancy of dh/dx and dl/dx by plotting dimensionless $h/\sqrt{A_o}$ and $l/\sqrt{A_o}$ against $x_f/\sqrt{A_o}$, where A_o is the initial cross sectional area (volume per unit depth) of the cloud. They verified the assumed linear variation of h and l with x_f (Figure 11.22a,b) and demonstrated that dh/dx_f and dl/dx_f depend on z in a linear fashion, as indicated in Figure 11.22c, d, i.e.

$$\frac{dh_f}{dx_f} = 0.04 + 3.6 \times 10^{-3}\zeta \quad \text{and} \quad \frac{dh}{dx_f} = 0.26 + 4.4 \times 10^{-3}\zeta. \quad (11.81)$$

On the other hand, Figure 11.17b verifies relation (11.72) insofar as the parameter $\mathbb{K}\sqrt{S}$ is a function of ζ.

c) Fukushima and Parker. The simple cloud model of Beghin, Hopfinger and Britter (1981) is generalized by Fukushima and Parker (1990) in the following ways: (i), the total buoyancy of the avalanche need not be conserved, but may freely vary via erosion and deposition of snow, (ii), a balance law of mean turbulent kinetic energy is added to the balance laws of mass and momentum with accordingly adjusted closure conditions, (iii), the slope $S = \sin\zeta$ may vary in the flow direction, (iv), the effects of the shear tractions on both the free upper surface (wind drag) and the bed (frictional resistance) are incorporated; and (v), the Boussinesq assumption is not invoked. The model is very similar to the long gravity current model by Parker, Fukushima and Pantin (1986). However, since gross balances for the cloud as a whole are considered from the outset, equations involve phenomenological coefficients that account (i), for the influence of the shape of the cloud and (ii), the fact that model variables represent mean values over the volume of the cloud or its surface.

Fukushima and Parker (1990) assume that the shape of the avalanche is given by the slope angle ζ alone; therefore, if A, P_i, P_b denote the cross sectional area (= volume per unit width), the perimeter of the free surface and that along the bed, respectively, see Figure 11.23, then, according to this similarity hypothesis A, P_i and P_b may be expressed as

$$A = \xi_A(\zeta)h^2, \qquad P_i = \xi_i(\zeta)h, \qquad P_b = \xi_b(\zeta)h, \quad (11.82)$$

where h is the typical height. If P_b is identified with the characteristic length of the avalanche, then $\xi_b = K_1$, see (11.67).

Fukushima and Parker (1990) write balance laws of mass for the particles, the particles plus the fluid and the momentum balance for the particles plus the fluid in the form

$$\frac{d(CA)}{dt} = v_s(E_s - c_b \cos\zeta)P_b,$$

$$\frac{dA}{dt} = E_a U P_i, \quad (11.83)$$

$$\frac{d}{dt}((\bar{\rho} + k_v r_a)UA) = \Delta\bar{\rho}gAS - \tau_i P_i - \tau_b P_b.$$

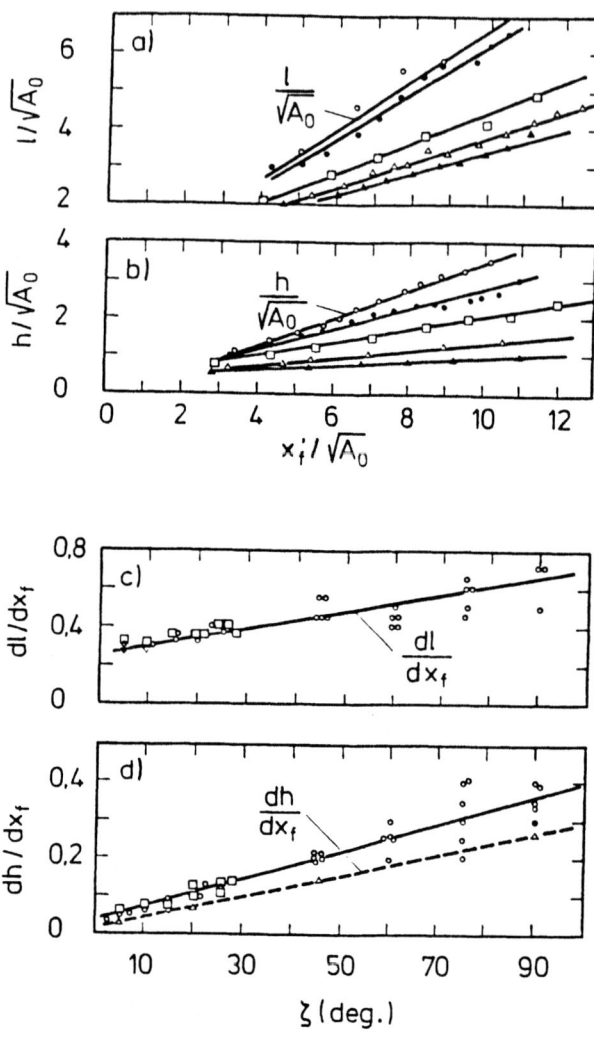

Fig. 11.22. (a), (b) Length (a) and height (b) of a plane thermal on an inclined plane as a function of distance from the position of release to the front x_f, normalized by $A_o^{1/2}$, where A_o is the initial cross sectional area △, 5°; ▲, 20°; □, 45°; •; 75°; o, 90°. (c) Growth rate in length as a function of slope angle ζ. The symbols o refer to the sand suspension thermal: nabla, $dh/dx_f \times (l/h)$; - - - $4.4 \times 10^{-3} + 0.26$. (d) Growth rate in length as a function of slope angle ζ. The symbols o refer to the sand suspension thermal: △ and - - - indicate the growth of a gravity current head (Britter and Linden, 1980); □, value of line thermal of Tsang (1971); — corresponds to $3.6 \times 10^{-3}\zeta + 0.04$. (All from Beghin, Hopfinger and Britter, 1981).

Here, C is the average volume fraction of particles in the avalanche, $\bar{\rho} = \hat{\rho}_a(1 + RC)$, $R = \hat{\rho}_s - \hat{\rho}_a/\hat{\rho}_a$. Further, $v_s E_s P_b$ is the entrained snow volume per unit time from the bed, v_s being a mean depositional velocity of the particles, and E_s is

Avalanche Dynamics

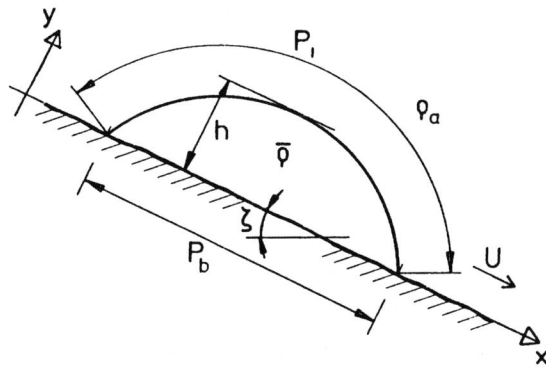

Fig. 11.23. Schematic diagram of a two-dimensional finite mass powder avalanche.

the entrainment coefficient of the snow. The term $v_s c_b \cos \zeta P_b$ represents the rate of particle deposition along the bed. When multiplied with the material density $\hat{\rho}_s$ of the particles, (11.83)$_1$ represents the particles mass balance. This equation is analogous to (11.53)$_1$. The second of equations (11.83) relates the time rate of change of the total volume to the volume of the entrained fluid. This corresponds to the global conservation law of mass for the particles plus the interstitial fluid only if (i), incompressibility is assumed and (ii), the contribution from snow entrainment or deposition is ignored.[2]

This violates the non-Boussinesq conditions, as stated by Fukushima and Parker (1990); a proper mass balance would be

$$\frac{d(\bar{\rho} A)}{dt} = \hat{\rho}_a E_a U P_i + \hat{\rho}_s v_s E_s P_b - \hat{\rho}_s v_s c_b \cos \zeta P_b, \tag{11.84}$$

Equation (11.83)$_2$ corresponds to (11.53)$_2$. The entrainment hypothesis of Fukushima and Parker (1990) differs, however, from that of Beghin et al. (1981), see (11.67), so that $\bar{E}(\zeta)$ and E_a differ from one another, in general. In the balance law of momentum, (11.83)$_3$, the virtual mass coefficient is assumed to depend on the shape of the avalanche, and thus on the slope angle ζ; t_i and t_b are streamwise shear stresses acting on the upper and the lower boundaries, respectively.

Equations (11.83), complemented by suitable closure conditions, constitute the three-equation model for the avalanche. In the four-equation model, a balance relation for the level of turbulence, K, is needed, this takes the form

---

[2] Actually, this second omission is not a serious one, since with $\rho = (1 - C)\hat{\rho}_a + C\hat{\rho}_s$, one has $\Delta \rho = \rho - \hat{\rho}_a = C(\hat{\rho}_s - \hat{\rho}_a) \simeq C\hat{\rho}_s$, and consequently

$$\frac{d}{dt}[\hat{\rho}_a(1 - C)A] = \hat{\rho}_a \frac{dA}{dt} - \hat{\rho}_a \frac{d(CA)}{dt} = \hat{\rho}_a E_a U P_i$$

holds for the mass balance of air. As long as $E_a U P_i \ll v_s(E_s - c_b \cos \xi) P_b$, the error implied by (11.83)$_2$ is negligibly small, since $C\hat{\rho}_a \ll \bar{\rho}$.

$$\frac{d}{dt}((\bar{\rho} + k_v r_a)KA)$$
$$= U(t_i P_i + t_b P_b) + \frac{1}{2}\rho_a U^2((1 + k_v)EUP_i$$
$$+ Rv_s(E_s - c_b \cos\zeta)P_b) - E - \rho_a RgCv_s A$$
$$- \bar{\rho}Rg \cos\zeta (\xi_\phi h)(\frac{1}{2}CE_a UP_i + v_s(E_s - c_b \cos\zeta)P_b). \tag{11.85}$$

Here, $\xi_\phi h$ denotes the y-coordinate of the centre of gravity. The physical interpretation of the terms in (11.85) is essentially the same as that of those in the layer-averaged balance of turbulence for continuous gravity currents, i.e., (11.54). The left hand side in (11.85) is the time rate of change of the level of turbulence for the entire cloud. On the right hand side, the first two terms account for the turbulent energy production caused by the shear stresses acting on the upper and lower boundaries, respectively. The next two terms represent the turbulent energy production caused by the loss of mean flow energy asssociated with the acceleration of newly entrained fluid from above, and particles from the bed, respectively. The remaining terms represent processes that annihilate turbulence energy. Of these, E is the mean rate of viscous dissipation. The next term expresses the work necessary to maintain the particles in suspension (Knapp, 1958; Bagnold, 1962). The last two terms express the fact that the potential energy of the avalanche changes by turbulence due to the entrainment mechanisms. Entrainment of fluid always increases the potential energy of the avalanche, and thus decreases its level of turbulence. The corresponding term due to snow entrainment and deposition, however, can have both signs. Obviously, for the turbulence not to disappear, the right hand side of (11.85) must be positive.

The three- and four-equation models must be complemented by appropriate *phenomenological statements*. Beghin, Hopfinger and Britter (1981), following Escudier and Maxworthy (1973), assume that the shape of the simple density cloud can be approximated by a half ellipse. The shape factors ξ_A, ξ_i and ξ_ϕ as well as the virtual mass coefficient k_v, are then given by

$$\xi_A = \frac{\pi}{4}\xi_b, \qquad \xi_i = \frac{\pi}{2\sqrt{2}}\sqrt{4\xi_b^{-2} + 1}\,\xi_b,$$
$$\xi_\phi = \tfrac{4}{3}\pi, \qquad k_v = 2\xi_b^{-1}. \tag{11.86}$$

Fukushima and Parker (1990) employ experimental data of Beghin et al. (1981), according to which ξ_b is given by

$$\xi_b = 8.47\zeta^{-1/3}, \tag{11.87}$$

with ζ in degrees. Based on experiments, they further suggest to employ the relations

Avalanche Dynamics

- $E_a = 0.1 \frac{\zeta}{\pi/2}$,

- E_s, ρ_o as stated in (11.59), (11.60)

- $\dfrac{\tau_i}{\rho_a} = \dfrac{\tau_b}{\rho_b} = u_*^2 = \begin{cases} c_D U^2 & \text{(three equation model)} \\ \alpha K & \text{(four equation model)} \end{cases}$

- $E = \beta \bar{\rho} K^{3/2} \dfrac{A}{h}$, where

$$\beta = \frac{1}{\xi_A}\left(\frac{\alpha}{c_D}\right)^{3/2}\left((\xi_i + \xi_b)c_D + \left(1 - \frac{c_D}{\alpha}\right)\frac{1}{2}(1 + k_v)\xi_i E_a\right)$$

$$- \xi_\phi \cot an\,\zeta \frac{\xi_i E_a}{2\xi_A}\left(\frac{3}{4}(1 + k_v)\xi_i E_a + (\xi_i + \xi_b)c_D\right).$$

(11.88)

Accordingly, the entrainment coefficient of the fluid grows linearly with the inclination angle ζ (Beghin et al. 1981, Escudier and Maxworthy 1973). On the other hand, E_s and r_o (= $c_b C$) are obtained from an analysis of continuous turbidity currents, Parker et al. (1986). The shear velocities τ_i/ρ_a and τ_b/ρ_b obey the usual drag force relationship (three-equation model), or are set proportional to the level of turbulence with constant of proportionality a whose value is $\alpha = 0.1$ (four-equation-model); actually, τ_i/ρ_a and τ_b/ρ_b should each be characterized by its own c_{Di}, c_{Db} and α_i, a_b, respectively; for lack of more detailed information, however, Fukushima and Parker (1990) set $c_{Di} = c_{Db} = c_D$, $\alpha_i = \alpha_b = \alpha$. Finally, the turbulent energy dissipation is given by a formula suggested by Kolmogorov (1941) and Heisenberg (1948), the coefficient β being determined by the following argument: for a stationary cloud ($v_s = 0$), for which snow entrainment and deposition are both zero on a constant slope, both the three- and four-equation models possess asymptotic solutions as $t \to \infty$. If it is required that both models predict exactly the same asymptotic solution for conservative currents, then β must be prescribed according to the last equation in (11.88). Since in that formula the various shape factors are functions of the slope factor, such α dependence is also expected for β.

As was the case for long, continuous turbidity currents, the *concept of ignition* for the prediction of accelerating or decelerating turbidity currents can be introduced, by setting the right hand sides of (11.83)$_{1,3}$ and (11.85) to zero and defining an igniting set $(U, C, K, h)_I$, in which $h = h_I$ is arbitrarily chosen as the lowest root of these equations. Integrating equations (11.83) and (11.85), subject to the closure conditions (11.88) and appropriate initial conditions, we may then plot U/U_I against B/B_I, where $B = CA$. Unpublished work by Fukushima and Parker which shows these results for continuous gravity currents are qualitatively very similar to those in Figure 11.19. Note that the plane $(U/U_I, B/B_I)$ separates into two subregions, divided by the so-called autosuspension line (AGL). If initial conditions are such that they lie on one side of AGL, then the avalanche accelerates on a converging band as in Figure 11.19. On the other hand, if initial conditions lie on the other side of AGL, the avalanche decelerates. Fukushima and Parker (unpublished) also show that the solutions $U(t)$, $h(t)$ and $C(t)$ of the three- and four-equation models do not differ much for an avalanche with finite mass, in contrast to corresponding results for continuous turbidity currents. Nevertheless the four-equation model should be used because it permits estimation of the evolution of the degree of turbulence, K, which might be used to define a minimum value of the level of turbulence $K = K^*$, below which the avalanche ceases to have sufficient energy to maintain its snow in suspension. Further, it would be advantageous to use (11.84) instead of (11.83)$_2$ as a snow mass balance. Fukushima and Parker (1990) have also demonstrated the usefulness of their approach by analysing a real avalanche event at Maseguchi (Japan).

11.5.3 OTHER MIXTURE MODELS AND CRITIQUE

The long and short gravity current models presented in Sections 11.5.1 and 11.5.2 provide insight into the gross, global behaviour of an avalanche under limiting situations (i.e., with or without entrainment from the bed, and for steady or asymptotic flow situations). By averaging equations over depth (for long gravity currents from a continuous source) or over the entire avalanche volume (for short, finite-mass clouds), relatively simple model equations were obtained that permitted estimation of the temporal and/or spatial averages of the avalanche front velocity, density deficiency, and size or depth. These quantities constitute very limited information to the practitioner, however, who is concerned, in general, with the amount and location of snow deposition as well as distribution of local velocities and dynamic pressures over arbitrary bottom geometries. A first step towards this end is the formulation and solution of a full boundary value problem involving the local equations that describe the diffusion of the heavy particles suspended in a turbulent flow.

Very little work has been done in this regard. Tesche's (and equivalently Brandstätter's, 1993) model is based on equations very similar to those used by Parker et al. (1986) and Fukushima and Parker (1990), but the corresponding turbulent closure conditions are more complex. Tesche (1987) uses the *balance laws of mass* and *momentum for the mixture as a whole* (i.e. the suspended particles plus the carrier fluid), as well as the *balance of mass for the non-buoyant particles* (the latter in the form (11.42)). These are complemented by an adaption of the two-equation turbulence model of Rodi (1985) that incorporates density stratification into the k-ε model. These equations are a turbulent kinetic energy transport equation for the specific averaged turbulent fluctuation energy k, and a dissipation transport equation for the specific dissipation of turbulent energy ε. The equations apply in the non-Boussinesq regime, and the diffusive mass flux, described by the vector $\mathbf{j}$ in (11.52), is given as in Parker et al. (1986), i.e., $\mathbf{j} = -v_s c \hat{\mathbf{z}}_s$, where v_s is the free fall velocity of the particles in the ambient fluid at rest, $\hat{\mathbf{z}}$ is the unit vector in the vertical direction, and c is the particle volume fraction. The turbulent Reynolds stresses and mass transport vectors are given by

$$\frac{\mathbf{t}^R}{\bar{\rho}} = -(\overline{\mathbf{v}' \otimes \mathbf{v}'}) = \frac{\nu_{\text{turb}}}{2}(\text{grad } \mathbf{v} + (\text{grad } \mathbf{v})^T) - \frac{2}{3}k\mathbf{1},$$
(11.89)

and

$$\mathbf{F} = (\overline{c'\mathbf{v}'}) = D_{\text{turb}} \text{ grad } c,$$

respectively. Here, the momentum, ν_{turb} and mass, D_{turb} diffusivities, are related to the turbulent energy k, and its dissipation ε, by

$$\nu_{\text{turb}} = D_{\text{turb}} = C_\mu \frac{k^2}{\varepsilon},$$
(11.90)

where the coefficient C_μ is expressed in terms of the stratification, as described explicitly by Gibson and Launder (1978). Tesche (1987) also reports some boundary conditions but they are incomplete, so that the formulation of a well posed boundary value problem is not achieved. In addition, he states neither how the free surface of the avalanche is defined nor how its evolution equation is obtained. He does state, however, that the free surface boundaries are treated as planes of symmetry for the velocity as well as the turbulence variables k and ε; this is unreasonable. Further, although he assumes that, on the free surface, there is no snow mass flux, he does not write it as

$$(\mathbf{F} + \mathbf{j}) \cdot \mathbf{n} = 0 \quad \Rightarrow \quad (D_{\text{turb}} \text{ grad } c - v_s c \hat{\mathbf{z}}_s) \cdot \mathbf{n} = 0,$$
(11.91)

where $\mathbf{n}$ is the exterior unit normal vector on the free surface. No explicit statement regarding the entrainment of the ambient fluid is made, a crucial variable. Similarly, no clear boundary conditions at the bed are prescribed, where a statement on entrainment and/or deposition of snow should be made. The details are possibly given in a manual to a commercial code (Tesche, 1986); for the present time explicit results are not very conclusive.

The avenue taken by Tesche is certainly a correct one that should be further pursued and modified. Incidentally, the imprecise definition of the free surface, is also characteristic for the turbidity current models of Section 11.5.1; the averaged quantities are defined in terms of integrals over the transverse coordinate y, from $y = 0$ to $y = \infty$ (not $y = h$, see (11.44)); the global balance law for mass (11.53)$_2$, however, explicitly assumes the upper boundary to be at $y = h$. Moreover, in the derivation of the depth-averaged equations, the boundary terms that arise at the free surface are evaluated more or less arbitrarily at $y = h$ or $y = \infty$. Observationally, however, the free boundary of an avalanche or turbidity current is clearly identified with the location where the particle concentration drops to zero. This transition is fairly abrupt, suggesting that the free surface is *material* for the suspended particle phase, across which the particle concentration suffers a finite non-zero jump. Consequently, to describe it, the particle phase velocity should be determined, and boundary conditions should correspondingly follow from jump conditions. Two-phase concepts are indispensible to achieve this goal.

11.6 Two-phase Flow Models

In their work, Scheiwiller and Hutter (1982), Scheiwiller (1986) and Scheiwiller, Hutter and Hermann (1987) view a powder snow avalanche as a two-phase mixture of air and snow under turbulent motion, and thus differentiate between the peculiar constituent mass and momentum densities. Given such a detailed description of the mass and velocity fields of both components, it does not appear to be meaningful to develop *global* equations in the sense of Parker et al. (1986); moreover, the practitioner's interest is in local velocities, concentrations, and pressures, making the global approach inadequate. Thus, the equations are based upon local forms of the balance laws of mass and momenta for each constituent under fully turbulent conditions with closure conditions for the peculiar Reynolds stresses, air and snow entrainment or snow sedimentation, and an assumption for the particle-fluid interaction force. For the fluid phase, k-ε closure concepts are used in the original form of Launder and Spalding (1974) and Spalding (1982) and therefore without specifically accounting for the special stratification effects as suggested by Rodi (1985). In future work, this particular point requires modification. The particle Reynolds stresses are set to zero because the characteristic size of the fluid eddies is much larger than the distances across which particle momentum transfer can take place. Finally, the particle fluid interaction force is set proportional to each constituent density and the square of the diffusion velocity. Simple, steady boundary layer flow along inclined planes has been analysed and used as a model for the tail behaviour of a long gravity current. The numerical method used to solve the relevant equations was a spectral-type functional expansion in the direction perpendicular to the bed combined with a finite-difference technique along the bed, a description of which is given by Scheiwiller (1986) and Scheiwiller, Hutter and Hermann (1987), with details obtainable from Scheiwiller (1985).

11.6.1 TREATMENT OF BOUNDARY CONDITIONS

It is not the place here to present a complete theoretical model; it seems, however, worthwhile to highlight the treatment of the boundary conditions for mass and momentum at the free surface and the bed, as well as the kinematic equation at these surfaces. To this end, let $F_f(\mathbf{x}, t) = 0$ and $F_b(\mathbf{x}, t) = 0$ be the equations of the free surface (f) and the bed (b), respectively, $\psi^\alpha(\mathbf{x}, t)$ the density per unit mass of a constituent physical quantity, and $\phi_\psi^\alpha(\mathbf{x}, t)$ its flux, for which the jump condition

$$[[\phi_\psi^\alpha \cdot \mathbf{n}]] - [[\rho^\alpha \psi^\alpha (\mathbf{v}^\alpha - \mathbf{u}) \cdot \mathbf{n}]] = 0 \tag{11.92}$$

holds on any singular surface with unit normal vector $\mathbf{n}$. In (11.92) $[[f]] = f^+ - f^-$ denotes the jump in the bracketed quantity from the negative to the positive side, the positive side being defined as the side into which $\mathbf{n}$ points; moreover, $\mathbf{v}^\alpha$ is the material velocity associated with the material quantity $\psi^\alpha(\mathbf{x}, t)$, and $\mathbf{u}$ the velocity of the singular surface.

Consider the free surface $F_f(\mathbf{x}, t) = 0$. If it is defined to be material relative to the particle phase, the time derivative of $F_f(\mathbf{x}, t) = 0$ and (10.1) yield

Avalanche Dynamics

$$\frac{\partial F_f}{\partial t} + (\operatorname{grad} F_f) \cdot v_s^- = 0, \quad \text{and} \quad [[\phi_\psi^s \cdot \mathbf{n}]] = 0. \tag{11.93}$$

with $\alpha = s$, $\mathbf{v}^\alpha = v_s^-$, $\mathbf{u} = v_s^-$. Equation of $(11.93)_1$ represents the kinematic equation describing the motion of the free surface. In the case of the balance of mass for the solid particles ($\phi_\psi^s = 0$), $(11.93)_2$ is trivially satisfied. For balance of momentum (i.e., $\phi_\psi^s = \mathbf{t}^s$, where $\mathbf{t}_s$ is the solid particle stress tensor) $(11.93)_2$ yields $(\mathbf{t}^s \cdot \mathbf{n})^- = \mathbf{0}$ (we use + for the atmospheric and − for the avalanche side of the singular surface), implying that the solid particles' traction at the free surface vanishes (since there are no particles outside the avalanche).

For the fluid phase ($\alpha = a$), balance of mass requires $\phi_\psi^\alpha = 0$ and $\psi^\alpha = \rho_a = \hat{\rho}_a c_a$; in this case (11.92) implies

$$\hat{\rho}_a c_a^- (v_a^- - v_s^-) \cdot \mathbf{n} = \hat{\rho}_a (v_a^+ - v_s^-) \cdot \mathbf{n}$$
$$\Rightarrow c_a^- (v_a^- - v_s^-) \cdot \mathbf{n} = -E_a^f, \quad c_a^- = (1 - c_s^-), \tag{11.94}$$

where $c_a^+ = 1$ and $\mathbf{u} = v_s^-$ have been used and $(v_s^- - v_a^+) \cdot \mathbf{n} = E_a^f$ defines the entrainment rate of air. With the aid of it, the kinematic equation $(11.93)_1$ can be rewritten as

$$\frac{\partial F_f}{\partial t} + (\operatorname{grad} F_f) \cdot v_a^- = \|\operatorname{grad} F_f\|(v_a^- - v_s^-) \cdot \frac{(\operatorname{grad} F_f}{\|\operatorname{grad} F_f\|}$$
$$= \|\operatorname{grad} F_f\|(v_a^- - v_s^-) \cdot \mathbf{n}$$
$$= \frac{\|\operatorname{grad} F_f\|}{(1 - c_s^-)} E_a^f. \tag{11.95}$$

Note that $v_s^- = v_a^- + (v_s^- - v_a^-)$. The relation (11.95) represents another form of the kinematic surface equation based on the motion of the fluid.

The jump condition of momentum of the fluid phase is obtained from (11.92) by setting $\phi_\psi^\alpha = \mathbf{t}^a$ (the fluid stress) and $\psi^\alpha = v_a$:

$$[[\mathbf{t}^a \mathbf{n}]] = [[v_a(\rho_a(v_a - v_s^-) \cdot \mathbf{n}]] = -[[v_a]]\hat{\rho}_a E_a^f. \tag{11.96}$$

It follows that the traction of the fluid on the avalanche side does not necessarily equal that on the ambient fluid side. Indeed, this would only be so if $E_a^f = 0$. Nevertheless, all theoretical studies except Scheiwiller, Hutter and Hermann (1987) apply such a condition, even when $E_a^f \neq 0$. This requires $[[v_a]]$, which corresponds to the change in momentum of the entrained fluid, to be negligible. If not, an additional closure condition on $[[v_a]]$ must be stipulated.

At the bed, snow can be entrained or deposited and air can be entrained to the extent that it is contained in the pore space of the snow cover; consequently, a similar approach to that above yields

$$\frac{\partial F_b}{\partial t} + (\operatorname{grad} F_b) \cdot v_a^- = -\frac{\|\operatorname{grad} F_b\|}{(1 - c_s^-)} E_a^b,$$
$$\frac{\partial F_b}{\partial t} + (\operatorname{grad} F_b) \cdot v_a^- = -\frac{\|\operatorname{grad} F_b\|}{c_s^-} E_s, \tag{11.97}$$
$$E_a^b = -c_a^{b+}(v_a^+ - \mathbf{u}) \cdot \mathbf{n}, \quad E_s = -c_s^{b+}(v_s^+ - \mathbf{u}) \cdot \mathbf{n},$$

corresponding to (11.93)–(11.95), where **u** denotes the velocity with which the bed is moving, and the (+)-sign refers to the bed. In the snow cover, $v_s^+ = v_a^+ = 0$ and so

$$E_a^b = \frac{\nu}{1-\nu} E_s = eE_s, \tag{11.98}$$

follows from (11.97)$_3$ with $n = c_a^{b+}$ the porosity and e the void ratio of the snow cover. E_s must still be prescribed by a constitutive relation, as e.g. given by Parker et al. (1986).

When applying simplified forms of equations (11.97), care must be taken; ignoring the advective terms (when the bed is slowly varying) requires $c_s^- = \nu$ for consistency; on the other hand, ignoring $\partial F_b/\partial t$ in comparison to the advective terms requires $(1 - c_s^-)(\mathbf{n} \cdot \mathbf{v}_a^-) = ec_s^-(\mathbf{n} \cdot \mathbf{v}_s^-)$. The second assumption is less reasonable, and so either (11.97)$_{1,2}$ and (11.98) are used without simplification, or else

$$cs^- \frac{\partial F_b}{\partial t} = E_s \quad \text{and} \quad c_s^- = \nu \tag{11.99}$$

are assumed as an approximation.

Note that, equations (11.97) involve the normal velocities $(\mathbf{n} \cdot \mathbf{v}_a^-)$ and $(\mathbf{n} \cdot \mathbf{v}_s^-)$ but not their tangential parts. Thus we are still free to impose conditions on these. No-slip of both phases, i.e., adherance of both phases to the bed implies

$$\mathbf{v}_s^- \times \mathbf{n} = 0, \quad \mathbf{v}_a^- \times \mathbf{n} = 0. \tag{11.100}$$

With these, the jumps of the constituent momenta need not be considered, since the basal tractions $\mathbf{t}_{s,a} \cdot \mathbf{n}$ will be determined through the solution of a boundary value problem.

11.6.2 EXPERIMENTAL AND COMPUTATIONAL RESULTS

A detailed description of the experimental procedure is described by Scheiwiller (1986) and Scheiwiller, Hutter and Hermann (1987). Experiments were performed in a 5 m long, 4 m high and 2 m wide water tank, with glass front to make the avalanching motion visible, and containing an inclined chute that could be extended by a kink element followed by a further straight chute element (Figure 11.24a). Water at rest forms the ambient fluid, and polystyrene particles were used to model the suspension. An ultrasonic measuring technique is used to determine the particle velocities through the Doppler shift, and their concentration through the attenuation of the reflected signal. This ultrasonic measuring technique is explained at greater depth by Hermann (1990). Avalanche head and tail can be measured with equal precision, since pointwise measurements are performed. Flow regimes can accordingly be adjusted to the autosuspension or sedimentation domains. Data obtained on the turbulent fluctuating quantities represent ensemble averages over seven repetitions of experiments performed under identical conditions. Measurements are taken at positions 1 and 2, 1.5 m and 2 m downstream of the inlet orifice

Avalanche Dynamics

(Figure 11.24b), respectively, and, in case of the bent chute, at positions A and B immediately behind the kink and 1 m downstream in the runout zone, respectively. The inclination angle a and the kink angle b were varied in the experiments. Position 1 was chosen sufficiently downstream of the inlet orifice to guarantee fully developed flow; indeed, the flow was in the autosuspension regime by the time it reached Position 1.

Consider results from the inclined plane chute experiments. Figure 11.25 displays downslope particle phase velocity profiles for two slope angles (306° and 456°) at positions 1 (left panel) and 2 (right panel) under steady conditions, while for 45° (panels on the left), it is less bottom concentrated. Comparing left and right panels, it is seen that the less steep avalanche decelerates considerably between the two cross sections, since maximum velocities are roughly half as large at 2 as they are at 1. On the steeper slope particles are accelerated between 1 and 2.

Further, profiles are blunter at 2 than at 1, and avalanche heights appear to be slightly larger at 2 than at 1. The corresponding particle concentration profiles are displayed in Figure 11.26. At 1, the particle concentration profiles at 30° and 45° do not markedly differ from one another. However, the downstream evolution of the density profiles corroborates the transition from a decelerating to an accelerating flow regime with increasing slope angle. On a slope of 30°, the suspension flow between 1 and 2 is concentrated toward the bottom of the avalanche, while on the slope of 45°, the opposite effect is observed: the particles are distributed more uniformly and the avalanche is thicker.

The spectral-type numerical integration scheme was based on a two function set, with the aid of which cross sectional distributions of particle velocity and concentration at position 1 were optimally fit to the measured profiles (left panels in Figures 11.25 and 11.26) no-slip between the particle and fluid phase was assumed for the forward integration to determine the respective profiles at position 2. The resulting particle fluid-difference velocity at profile 2 was then fed back to position 1 to see whether differences in the particle velocity and density profiles at 2 were negligible in the two cases. Such a procedure was necessary, since fluid velocity profiles themselves could not and still can not be measured; the computed velocity and density profiles at 2 (right panels in Figures 11.25 and 11.26), however, reproduce the observed features correctly.

Very extensive and detailed measurements of velocity and density in a laboratory powder snow avalanche moving in a bent chute indicate that convex kinks are locations of large sedimentation of particles (Hermann, 1990, Hermann and Scheiwiller, 1988). A comparison of measured density profiles at Positions A and 2 in the tail of a long gravity current from a steady source (Figure 11.24b) shows that densities are substantially reduced, particle distribution is more uniform, and the avalanche height is larger at A than at 2. Particle velocities are also affected; their reduction is about 40–60% between A and 2, they are also more uniform at A than at 2, and the avalanche height (inferred from these) is equally larger, again at A. This corresponds to a redistribution of momentum and mass, and must correspondingly also affect the dynamic pressure.

Figures 11.27–11.29 display the evolution of the velocity and density in a powder avalanche at position A (left) and position B (right) with time as the

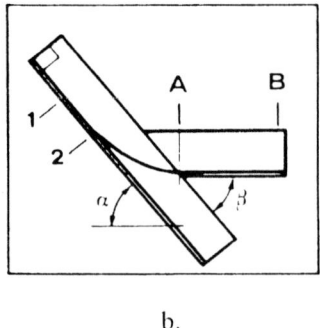

Fig. 11.24. (a) Water tank with front windows and submerged chute. (b) Explaination of the positions where measurements of particle velocities and densities were made.

Avalanche Dynamics

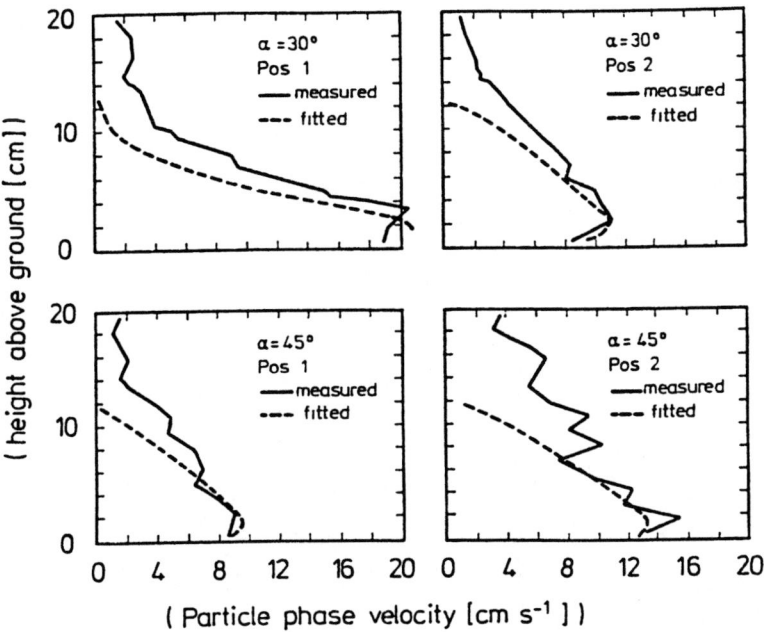

Fig. 11.25. Measured and calculated particle phase velocity profiles for different slope angles at positions 1 and 2.

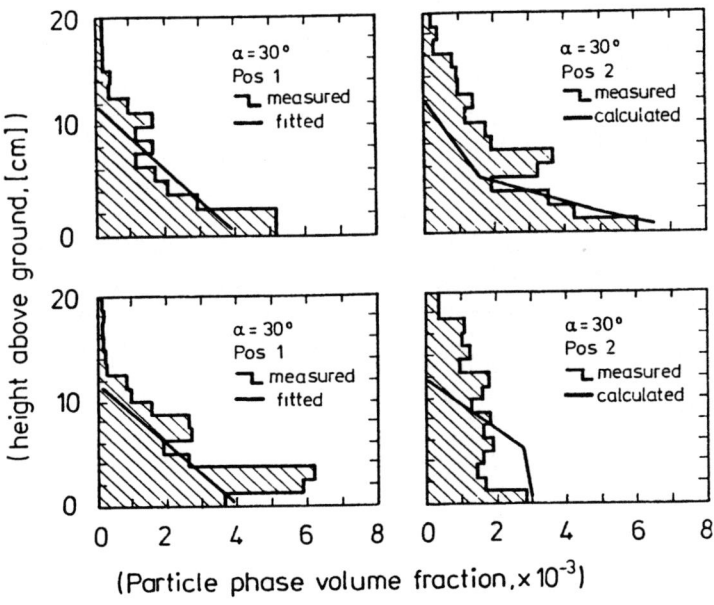

Fig. 11.26. Measured and calculated particle phase volume fraction profiles at positions 1 and 2.

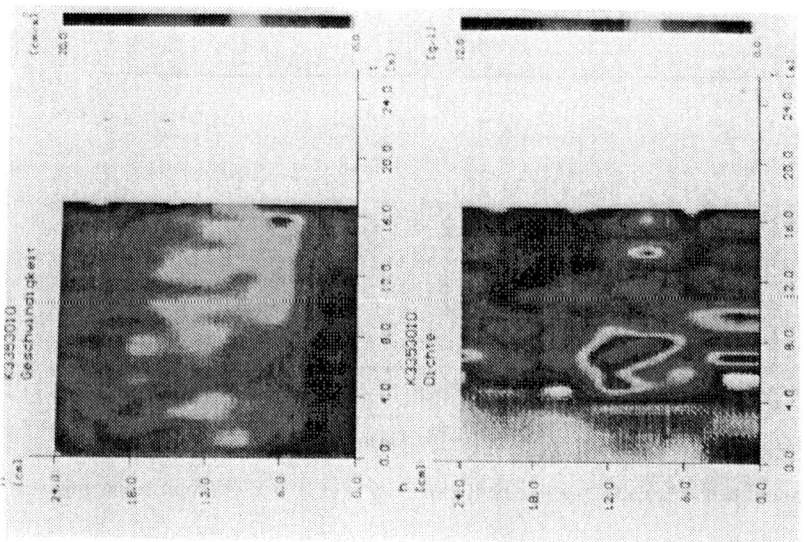

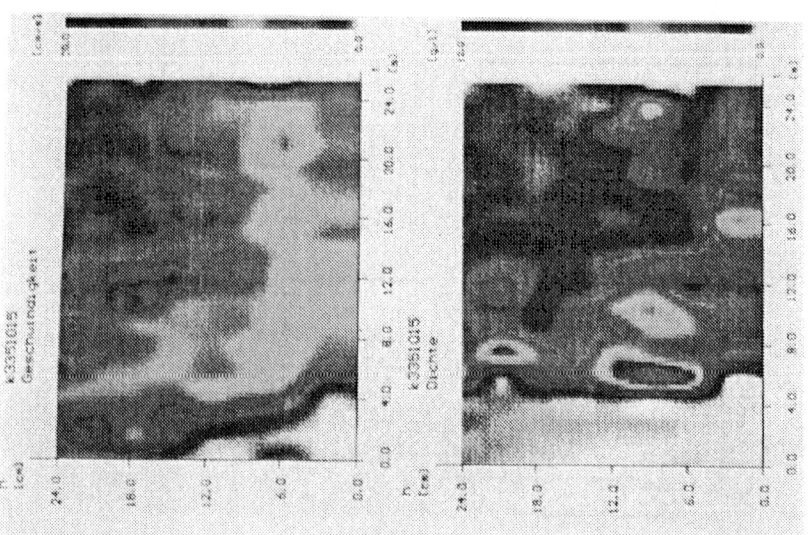

Fig. 11.27 Particle velocities (top) and densities (bottom) at position A (left) and B (right) in a bent chute (see Figure 11.22b). The horizontal coordinate gives the time in seconds, the vertical coordinate the height above ground in cm, and the colour the value of the streamwise velocity and the density in cm s^{-1} and g l^{-1}, respectively. The colour shading represents a linear interpolation. The angle of inclination of the chute is 35°, and the kink angle 55°.

Avalanche Dynamics

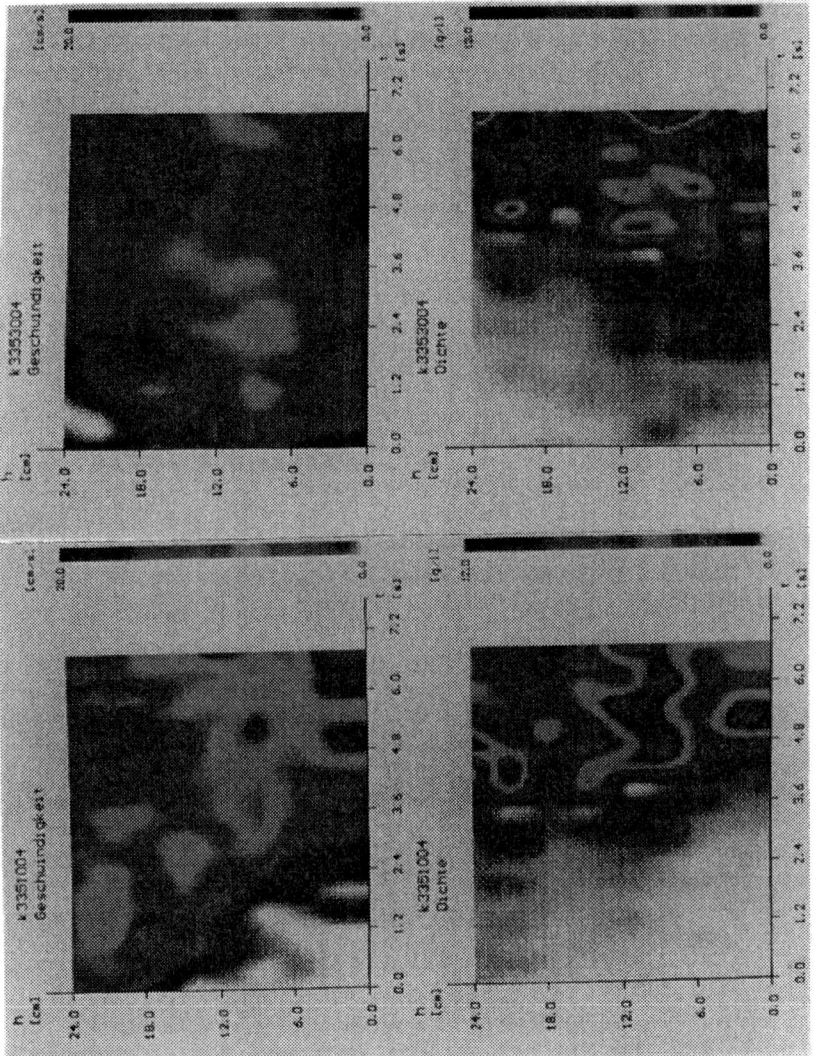

Fig. 11.28. Same as Figure 11.25, except that the measured frequency is greater, so that the data is more detailed, and covers the avalanche head.

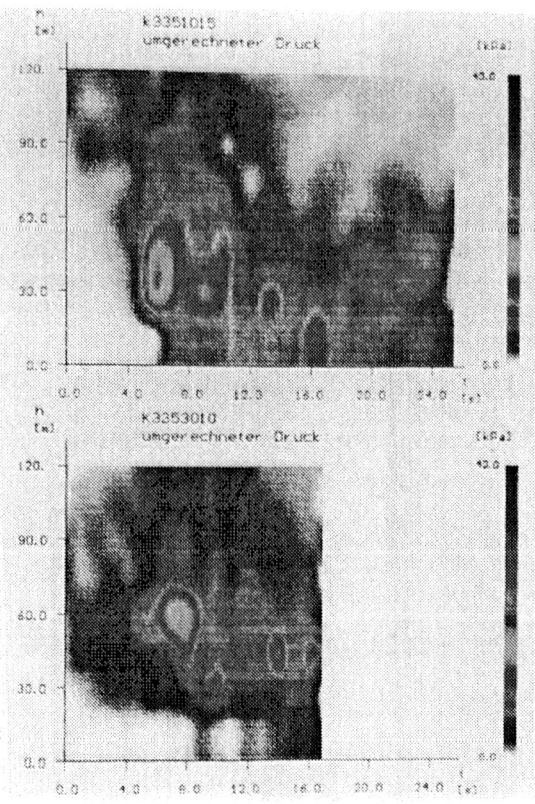

Fig. 11.29. Kinetic energy density at positions A (left) and B (right) for a chute with inclination angle 55°, as well as kink angles 75° (top) and 35° (bottom). The horizontal coordinates represent time in seconds, and the vertical coordinates height above ground. The kinetic energy is given by $k = \hat{\rho}_a(1 - c)(v_a^2/2) + \hat{\rho}_s c(v_s^2/2)$, where v_a is the fluid, and v_s the particle velocity, and we have assumed $v_a = v_s$. The graph is scaled to natural, and not laboratory, conditions, using a length scale of 1000 and density contrast $\Delta\rho/\rho = 10^{-3}$, requiring the densimetric Froude number to be invariant (from Hermann, 1990).

avalanche passes each position. The ordinate represents the height above ground from 0 to 24 cm, the abscissa gives time in seconds. The colour shading scheme represents variations in velocity and density, ranging from 0 (white) to 20 (dark purple) cms^{-1} for velocity, and 0 (white) to 12 (dark purple) gl^{-1} for density (i.e., 0 to 10^{-1} volume concentration). A comparison of the velocity and density profiles in Figure 11.27 implies the existence of a velocity precursor of up to four seconds in the ambient fluid. In the velocity plot, the areas of equal colour are large and their gradients in general small. Avalanche head and body are distinguished by the vertical extent of higher velocities and densities in regions above 12 cm. In contrast to the velocity, the density exhibits several distinct maxima. Comparing the velocity at the two positions shows that the largest decrease in velocity occurs in the lowermost 6 cm of the avalanche, with velocities being reduced by a factor of four. Whether this is due to *bottom friction* or a certain *"ski-jump effect"* is presently a matter of investigation. As for the density, it decreases substantially between positions A and B, in contrast to the velocity. Note that the reduction of particle concentration occurs mostly *between* the maxima and less at the maxima themselves. This can be seen in many other experiments as well.

Figure 11.28 is the same as the previous ones, with the exception that the recording intervals for measurements are very much shorter so that these only cover the avalanche front. The most striking feature is the difference between the velocity and density fronts. The visible edge of the avalanche coincides with the density front. All velocity data that are recorded before this front arrived at the measuring position must stem from precursory water movements, made visible by small impurities in the water. This zone is growing in the runout zone. The occurrance of strong winds prior to the powder snow avalanche is well known to foresters and mountaineers.

An avalanche variable of practical interest is the specific kinetic energy or dynamic pressure, which for the experiment of Figures 11.27 and 11.28 is shown in Figure 11.29 on the left, but now transformed via the laws of similitude to the natural (prototype) dimensions. By convention, 1.5 kPa corresponds to absolute endangered zones, and 0.5 kPa to harmless zones. Pressures in between are characterized as potentially dangerous. On the left side, at position A (top), the entire avalanche must be considered dangerous, at B (bottom), the bottom most 30 m are harmless. A similar transition is also shown in the panel on the right. The bottom zone experiences much lower dynamic pressures than zones higher up. This result has been corroborated in many other laboratory experiments. Despite this fact, such transformations from laboratory size to natural scales must be interpreted with caution because of the difficulties associated with the naive application of the laws of similitude.

The next step would be to verify these conclusions computationally, and to increase our knowledge of this previously unknown behaviour, in order to make inferences more reliable and codes more accurate. Lack of computational results on the velocity and density distribution of particles and interstitial fluid is indeed one of the missing links for making prediction of avalanche motion as well as estimation of their dynamic impact on objects quantitatively possible. Software for such computations is now becoming available and could be adapted to the problem at hand. Indeed, investments to make these adaptions could be less expensive than the money presently spent on avalanche protection measures.

The second missing link, at least as far as laboratory experiments are concerned, is the determination of the fluid velocity which differs from that of the particles. Its knowledge would make estimation of the fluid particle interaction force more reliable.

11.7 Concluding remarks

In this article I have reviewed recent work on avalanche dynamics, and concentrated on the discussion of two limiting cases: (i), the so called flow avalanches, and (ii), the powder snow avalanches. The former have been treated as moving masses of a granular fluid, the latter as turbulent air-borne density currents or turbulent two-phase flows. The review obviously draws on much of my own work, and emphasizes my own interests, so that unfortunately important contributions as well as many details had to be omitted. In the interest of space, to make up this abit, I now like to mention important additional work that would certainly not be omitted in a broader account dealing with the state of the art of avalanche dynamics.

First, other, earlier reviews of the subject should be mentioned. For example, Mellor (1978) and Perla (1980) both focussed on broader aspects of avalanche dynamics, as they treat avalanche release, motion and impact. Perla and Martinelli (1976) summarize the applied state of the art at that time in their Avalanche Handbook, and Hopfinger (1983) reviews the fundamental literature of 10 years ago. A further very broad literature review dealing with aspects from management policy to forecasting is given in a report by the Panel on Snow Avalanches of the United States (Anonymous, 1990), which contains more than 300 references. The Soviet research is compiled in the World Data Center for Glaciology, Report GD-16 (1984). Applied aspects are best drawn from the proceedings of specialized workshops and conferences (see e. g. Annals of Glaciology, 13; IAHS-Publication No. 162, (Salm and Gubler, 1987); Interpraevent Symposia and specialized institute reports (Gubler, 1991)). The spirit of the works contained in such literature differs in general from this one, as the focus is on immediate application of simple models. One purpose of my review has been to demonstrate that basic physics is indeed capable of explaining the prinicpal phenomena that are observed. Paired with modern computer methods, these rational models are likely superior to the earlier, simpler descriptions of avalanche dynamics; there is, however, still work to be done before they can be applied routinely in avalanche protection work.

Second, there are models others than that of granular media with which dense snow avalanches have been described. Lang et al. (1979), Lang and Dent (1979) and Dent and Lang (1980) model dense snow as a linearly viscous fluid, and solve the corresponding Navier-Stokes equations for free surface flows. Nonlinear viscous (turbulent) concepts are also the basis of the Saint Venant equations; these are the depth-integrated "shallow water equations" in which dissipation is often incorporated via a Chézy-Manning-Strickler formula. The French school is pursuing this avenue, see Martinet (1992) or Vila (1986). Other models of dense snow avalanches are based on Bingham-type materials (Dent and Lang, 1983). None of these, however, has been tested against laboratory experiments to the extent that the equations presented in this article have been tested.

Third, a word of caution is needed. In nature, pure flow avalanches, as well as pure powder avalanches almost never occur; the usual situation is a mixed-type avalanches, in which a dusty snow cloud rides on a flow-type core Future work will eventually have to concentrate on this mixed-type avalanche. Thus, the concepts developed separately for flow and powder snow avalanches will have to be combined with appropriate boundary conditions which link the two together. The development of such models is still in its infancy.

Acknowledgements

A first version of this paper was written while I was holding a visiting appointment at the Institute of Geography at ETH, Zürich (October 1991–March 1992). I thank Professor Ohmura for his support. The paper began as an invited survey lecture held at the First European Fluid Mechanics Conference, from September 15 to 20, 1991, Cambridge University, Cambridge, UK. An earlier draft of it was thoroughly reviewed in 1992 by Dr. D. Issler of the *Federal Institute of Snow and Avalanche Research* in Davos-Weissfluhjoch, Switzerland. This implied substantial revisions. These and extensions to it were incorporated into the text between January and September, 1993. I cannot sufficiently thank Dr. Issler for his substantial contributions from which the text has invaluably profited. I also thank F. Hermann for permitting me to reproduce Figures 27–29, and B. Svendsen, Ph. D. for thoroughly checking and improving the English. Mrs. Danner typed the various versions of the manuscript. Her help is equally acknowledged.

References

Ahn, H., Brennen, C. E. and Sabersky, R. H., 1992; Analysis of the fully developed chute flow of granular materials, J. App. Mech. 59 (109), 109 - 119

Anonymous, 1990; Soviet avalanche research, Avalanche bibliography update: 1977 - 1983, World Center for Glaciology (Snow and Ice), Report No GD-16

Akiyama, J and Fukushima, Y. 1985 Entrainment of noncohesive bed sediment into suspension. External Memo. No 175, St. Anthony Falls Hydraulic Laboratory, University of Minnesota, Minneapolis, USA.

Anonymous, 1990; Snow avalanche hazards and miligation in the United States, National Research Council, Panel on Snow Avalanches, National Academy Press, Washington, D. C., pp. I - X, 1 - 54

Bagnold, R. A., 1954; Experiments on a gravity free dispersion of large solid spheres in a Newtonian fluid under shear, Proc. R. Soc. London, Ser. A 225, 49 - 63.

Bagnold, R. A., 1962; Auto-suspension of transported sediment: turbidity currents, Proc. R. Soc. London Ser. A, 265, 315-319

Bailard, J., 1978; An experimental study of granular-fluid flow, Ph. D. Thesis, Univ. of Calif., San Diego, 172 pp.

Beghin, P., 1979; Etude des bouffées bidimensionnelles de densité en écoulement sur pente avec application aux avalanches de neige poudreuse, Thèse, Grenoble

Beghin, P., Hopfinger, E. J. and Britter, R. E., 1981; Gravitational convection from instantaneous sources on inclined boundaries, Journal of Fluid Mechanics, 107, 407 - 422

Brandstätter, W., 1993; Mehrdimensionale Simulation im Automobilbau, Klimatechnik und Alpinem Umweltschutz, Habilitationsschrift, Montanuniversität Leoben

Britter, R. E. and Linden, P. F., 1980; The motion of the front of a gravity current traveling down an incline, Journal of Fluid Mechanics, 99, 531 - 543

Britter, R., E. and Simpson, J. E., 1978; Experiments on the dynamics of a gravity current head, Journal of Fluid Mechanics, 88, 223 - 240

Buggisch, H. and Stadler, R., 1986; On the relation between shear rate and stresses in one-dimensional steady flow of moist bulk solids, Proc. World Congress Particle Technolgy, Part III, Mechanics of Pneumatic and Hydraulic Conveying and Mixing, Nürnberg, 16 - 18 April, 187 - 202.

Chu, F. H., Pilkey, W. D. and Pilkey, O. H., 1979; An analytical study of turbidity current steady flow, Mar. Geol., 33, 205 - 220

Davies, T. R. H., 1982; Spreading of Rock avalanche debries by mechanical fluidization, Rock Mech., 15 , 9-29.

Dent, J. D., 1986; Flow properties of granular materials large overburden loads, Acta Mecha-nica, 64, 111-122

Dent, J. D.and Lang, T. E., 1980; Modeling of snow flow, J. Glaciology, 26 (94), 1311-40

Dent, J. D.and Lang, T. E., 1983; A biviscous modified Bingham model of snow avalanche motion, Annals of Glaciology, 4, 42 - 46

Egashira, S., 1980; Basic research on the flow and mechanism of mixing of density-stratified fluids, Ph. D. thesis, Kyoto University, Japan

Ellison, T., H. and Turner, J., S., 1959; Turbulent entrainment in stratified flows, Journal of Fluid Mechanics, 6, 423 - 448

Erismann, T. H., 1979; Mechanisms of large landslides, Rock Mech., 12, 15-46.

Erismann, T. H., 1986; Flowing, Rolling, Bouncing, Sliding: Synopsis of basic mechanisms, Acta Mechanica, 64, 101-110

Escudier, M. P. and Maxworthy, T., 1973; On the formation of turbulent thermals, Journal of Fluid Mechanics, 61, 541 - 552

Foda, M. A., 1994; Landslides riding on basal pressure waves, Cont. Mech. and Thermodyn. 6, (to appear)

Fukushima, Y. and Parker, G., 1990; Powder snow avalanches: Theory and application, J. Glaciology, 36, (123), 229 - 237

Garcia, M. H., 1985; Experimental study of turbidity currents, M.S. Thesis, Dept. of Civil and Mineral Engineering, University of Minnesota, USA, 138 pp

Gartshore, I. S. and Newman, B. G., 1969; The turbulent wall jet in an arbitrary pressure gradient. Aeronaut. Q., 20, 25-56.

Georeson , E. H. M., 1942; The free streaming of gases in sloping galleries. Proc. R. Soc., Lond. A 180, 484-493.

Gibson, M. M. and Launder, B. E., 1978; Ground effects on pressure fluctuations in the atmospheric boundary layer, Journal of Fluid Mechanics, 86, 491 - 517

Goddard, J. D., 1986; Dissipative materials as constitutive models for granular media, Acta Mechanica, 63, 3 - 13.

Goguel, J., 1978; Scale dependent rock mechanisms, in Voigt, B., (ed.), Rockslides and avalanches, Vol 1 , Elsevier,167-180

Greve, R., 1991; Zur Ausbreitung einer Granulatlawine entlang gekrümmter Flächen - Laborexperimente und Modellrechnungen, Diplomarbeit, Fachbereich Mechanik, Technische Hochschule Darmstadt, Deutschland.

Greve, R. and Hutter K., 1993; The motion of a granular avalanche in a convex and concave curved chute: Experimenrts and theoretical predictions, Phil. Trans. R. Soc. London A 342, 573 - 6004

Greve, R., Koch, T. and Hutter K., 1993; Unconfined flow of granular avalanches along a partly curved surface. Part I: Theory. Proc. R. Soc. London A 445, 399-413

Gubler, H.-U., 1991; Proceedings of a workshop on avalanche dynamics, Mitt. des Eidg. Instituts für Schnee- und Lawinenforschung of avalanche dynamics.

Gubler, H.-U., 1987; Measurements and modelling of snow avalanche speeds, In: Avalanche Formation, Movement and Effects., (B. Salm and H.-U. Gubler eds.), IAHS, Publ. No. 162, 405 - 420

Hanes, D. M. and Inman, D. L., 1985; Observation of rapidly flowing granular-fluid mixtures, Journal of Fluid Mechanics, 150, 357 - 380.

Heim, A.,1882; Der Bergsturz von Elm, Deutsche Geol. Gesellsch. Zeitschrift, 34, 74-115

Heim, A., 1932; Bergsturz und Menschenleben, Beiblatt zur Vierteljahresschrift der Natf. Gesellschaft, Zürich, 20, 1-218

Heisenberg, W., 1948; Zur statistischen Theorie der Turbulenz, Z. für Physik, 124, pp. 628 - 657

Hermann, F., 1990; Experimente zur Dynamik von Staublawinen in der Auslaufzone, Mitteilung Nr. 107 der Versuchsanstalt für Wasserbau, Hydrologie und Glaziologie, ETH Zürich, 1 - 262

Hermann, F. und Hutter, K., 1991, Laboratory experiments on the dynamics of powder snow avalanches in the runout zone, J. Glaciology, 37, (126), 281-295.

Hermann, F. und Scheiwiller, T., 1988; Experiments on the deposition by laboratory powder snow avalanches, Mitteilung Nr. 94 der Versuchsanstalt für Wasserbau, Hydrologie und Glaziologie, ETH Zürich, 307 - 322

Hermann, F., Hermann, J. and Hutter, K., 1987; Laboratory experiments on the dynamics of powder snow avalanches. Avalanche Formation Movement and Effects (Proc. of the Davos Symposium, September 1986), IAHS Publ. No. 162, 431 - 440

Hinze, J., O., 1960; On the hydrodynamics of turbidity currents, Geol. Mijnb. 39e, 18 - 25

Hopfinger, E. J. und Tochon-Danguy, J. C., 1977; A model study of powder snow avalanches, Journal of Glaciology, 19, 81, 343 - 356

Hopfinger, E. J., 1983; Snow avalanche motion and related phenomena, Ann. Rev. of Fluid Mechanics, 15, 47 - 76

Hopfinger, E., J. and Beghin, P., 1980; Buoyant clouds appreciably heavier than the ambient fluid on sloping boundaries, Second Int'l Symp. on Stratified Flows, Trondheim, Norway, 1, 495 - 506

Hsü, K., 1975; On sturzstroms - Catastrophic debries streams generated by rockfalls, Geol. Soc. Am. Bull., 86, 129-140.

Hsü, K., 1978, Albert Heim; Observations on landslides and relevance to modern interpretations, in: Voigt, B., (ed.), Rockslides and avalanches, Vol 1 (Elsevier), 69-93.

Hungr, O. and Morgenstern, N. R., 1984a; Experiments on the flow behaviour of granular materials at high velocity in an open channel flow. Geotechnique, 34, 405-413.

Hungr, O. and Morgenstern, N. R., 1984b; High velocity ring shear tests on sand. Geotechnique, 34, 415-421.

Huber, A., 1980; Schwallwellen in Seen als Folge von Felsstürzen, Mitteilung No. 47 der Versuchsanstalt für Wasserbau, Hydrologie und Glaziologie an der ETH, 122 pp.

Hutter, K., 1989; A continuum model for finite mass avalanches having shear-flow and plug-flow regime. Internal report, Federal Institute of Snow and Avalanche Research, Weissfluhjoch, Davos.

Hutter, K, 1991; Two- and three-dimensional evolution of granular avalanche flow - theory and experiments revisited, Acta Mechanica, [Suppl.], 1 , 167-181.

Hutter, K. and Greve, R., 1993; Two-dimensional similarity solutions for finite mass granular avalanches with Coulomb and viscous-type frictional resistance, J. Glaciology (in press)

Hutter, K. and Koch, T., 1991; Motion of a granular avalanche in an exponentially curved chute: experiments and theoretical predictions. Phil. Trans. R. Soc. London, A 334, 93-138.

Hutter, K. and Nohguchi, Y.,1990; Similarity solutions for a Voellmy model of snow avalanches with finite mass. Acta Mechanica, 82, 99-127.

Hutter, K., Szidarovsky, F. and Yakowitz, S.,1986a; Plane steady shear flow of a cohesionless granular material down an inclined plane: a model for flow avalanches, Part I. Theory. Acta Mechanica, 63, 87-112

Hutter, K., Szidarovsky, F. and Yakowitz, S.,1986b; Plane steady shear flow of a cohesionless granular material down an inclined plane: a model for flow avalanches, Part II. Numerical results. Acta Mechanica, 65, 239-261

Hutter, K., Koch, T., Plüss, C. and Savage, S. B.,1993; Dynamics of avalanches of granular materials from initiation to runout, Part II. Laboratory experiments, Acta Mechanica, (in press)

Hutter, K., Siegel, M., Savage, S. B. and Nohguchi, Y.,1993; Two dimensional spreading of a granular avalanche down an inclined plane. Part I, Theory. Acta Mechanica, 100, 37 - 68

Jenkins J. T. and Cowin, S. C., 1979; Theories for flowing granular materials, in: Cowin, S. C., (ed.), Mechanics Applied to the Transport of Bulk Materials, ASME AMD-31,79-89

Jenkins, J. T. and Richman,W. M. 1985a; Grad?s 13-Moment system for a dense gas of inelastic spheres, Arch. Rat. Mech. Anal., 87, 355-377

Jenkins, J. T. and Richman,W. M. 1985b; Kinetic theory for plane flows of a dense gas of identical, rough, inelastic, circular disks, Phys. Fluids, 28, 3485-3494.

Jenkins, J. T. and Savage, S. B., 1983; A theory for rapid flow of identical smooth nearly elastic particles, J. Fluid Mech., 130, 187-202.

Johnson, P. C. and Jackson, R., 1986; Frictional-collisional constitutive relations for granular materials with application to plane shearing, J. Fluid Mech., 176, 67-93

Johnson, P. C. Nott, P. and Jackson, R., 1990; Frictional-collisional equations of motion for particulate flows and their application to chutes, J. Fluid Mech., 210, 501 - 535

Kent, P. E., 1986; The transport mechanisms of catastrophic rockfalls, J. Geol., 74, 79-83.

Knapp, R. T., 1938; Energy balance in streams carrying suspended load, Trans. A. G. U. 1, 501-505

Koch, T., 1989; Bewegung einer Granulatlawine entlang einer gekrümmten Bahn. Diplomarbeit, Technische Hochschule Darmstadt, 172pp

Koch, T., 1993; Bewegungf einer granularen Lawine auf einer geneigten und gekrümmten Fläche. Entwicklung und Anwendung eines theoretisch numerischen Verfahrens und dessen Überprüfung urch Laborexperimente. Doctoral dissertation, Technische Hochschule Darmstadt

Koch, T., Greve, R, and Hutter K., 1994; Unconfined flow of granular avalanches along a partly curved surface. Part II: Experiments and numerical computations, Proc. R. Soc. London , A 445, 415-435

Kolmogorov, A., 1941; The local structure of turbulence in incompressible viscous fluids for very large Reynolds numbers, Comptes rend. de l'Acad. des Sci. de l'URSS, 30, 301 - 305

Lang, R., 1992; An experimental and analytical study on gravity driven free surface flows of cohesionless granular media, Dr. rer. nat. Dissertation, Technische Hochschule Darmstadt.

Lang, R., Leo, B. and Hutter, K. 1989; Flow characteristics of an unconstrained non-cohesive granular medium down an inclined, curved surface: Preliminary experimental results, Annals of Glaciology, 13, 146-153

Lang, T. E., Dawson, K. L. and Martinelli, Jr. M., 1979; Application of numerical transient fluid dynamics to snow avalanche flow. Part I. Development of computer program AUALNCH, J. Glaciology, 22 (86), 117 - 126

Lang, T. E., Dawson, K. L. and Martinelli, Jr. M., 1979; Application of numerical transient fluid dynamics to snow avalanche flow. Part II. Avalanche modelling and parameter error evaluation, J. Glaciology, 22 (86), 107 - 115

Launder, B., E. and Spalding, D., B., 1974; The numerical computation of turbulent flows, Computational Methods in Applied Mechanics and Engineering, 3, 269 - 289

Li Tianchi, 1983; A mathematical model for predicting the extent of a major rockfall, Z. Geomorph., 27, 473-482.

Lucchitta, B. K. 1978; Large landslide on Mars. Geol. Soc. Amer. Bull., 89, 1601-1609

Lun, C. K. K., Savage, S. B., Jeffrey, D. J. and Chepurniy, N., 1984; Kinetic theories for granular flow: Inelastic particles in Couette flow and slightly inelastic particles in a general flow field. J. Fluid Mech., 140, 223-256

Martinelli, M., Jr., Lang, T. E. and Mears, A. I., 1980; Calculation of avalanche friction coefficients from field data. J. Glaciology, 26, (94) 109-119.

Mathieu, J. and Tailland, A., 1965; Jet pariétal, C. R. Acad. Sci. Paris , 261, 2282 - 2285

Mellor, M., 1978; Dynamics of snow avalanches, in: Rockslides and Avalanches, 1, Natural Phenomena (B. Voight, ed.), Elsevier Sci. Publ.Co, Amsterdam, 753 - 792

Martinet, G., 1992; Contribution à la modélisation numérique des avalanches de neige dense et des laves toorentielles, Th'se, Université Joseph Fourier I, Grenoble, pp. 1-218

McClung, D. and Schaerer, P. A., 1988; Determination of avalanche dynamics, friction coefficients from measured speeds, J. Glaciology, 20, (94) 109-120

McSaveney, M. J.1978; Sherman Glacier rock avalanche, Alaska, U.S.A, in: Voight, B., (ed.), Rockslides and Avalanches, Vol 1 (elsevier), 197-258.

McTigue, D. F., 1979; A nonlinear continuum theory for flowing granular materials. Ph. D. Dissertation, Department of Geology, Stanford University.

Melosh, J., 1986; The physics of very large landslides, Acta Mechanica, 64, 89-99

Middleton, G. V., 1966; Experiments on density and turbidity currents, Canadian Journal of Earth Sciences, 3, 523 - 546

Middleton, G. V. and Hampton, M. A., 1976; Subaqueous sediment transport and deposition by sediment gravity flows, in: Stanley, D. J. and Swift, D. J. P., (eds.), Marine sediment transport and environmental management (Wiley, New York), 197-218.

Morton, B. R., Taylor, G. I. and Turner, J. S., 1956; Turbulent gravitational convection from maintained and instantaneous sources, Proc. Roy. Soc. London, A 234, 1 - 23

Myers, G. E., Schauer, J. J. and Eustis, R. H., 1961; The plane turbulent wall jet. 1 Jet development and friction factor. Tech Rep. No 1 Department of Mech. Engng. Stanford University

Nohguchy, Y., Hutter, K. and Savage, S. B., 1989; Similarity solutions for granular avalanches of finite mass with variable bed friction, Continuum Mech. Thermodyn., 1, 239-265

Norem, H., Irgens, F. and Schieldrop, B., 1987; A continuum model for calculating snow avalanches, in: Salm, B. and Gubler, H., (eds.), Avalache Formation, Movement and Effects (IAHS Publ. No. 126), 363-379.

Parker, G., 1982; Conditions for the ignition of catastrophically erosive turbidity currents, Mar. Geol., 46, 307 - 327

Parker, G., Fukushima, Y. and Pantin, H. M., 1986; Self-accelerating turbidity currents, Journal of Fluid Mechanics, 171, 145 - 181

Perla, R., 1980; Avalanche release, motion and impact, in: Dynamics of Snow and Ice Masses (S. C. colbeck, ed.), Academic Press, New York, 397 - 462

Perla, R. and Martinelli, M., 1978; Avalanche Handbook, U.S. Department of Agriculture Forest Service, Agriculture Handbook, 489 pp.

Perla, R., Cheng, T. T. and McClung, D. M., 1980; A two parameter model of snow avalanche motion, Journal of Glaciology, 26, Nr. 94, 197 - 202

Plapp, J. E. and Mitchell, J. P., 1960; A hydrodynamic theory of turbidity currents, J. Geophys. Res., 65, 983 - 992

Plüss, C., 1987; Experiments on granular avalanches, Diplomarbeit, Abt XD, Eidg. Tech. Hochschule, Zürich, 113 pp.

Prandtl, L., Oswatitsch, K. und Wieghardt, K., 1984; Führer durch die Strömungslehre, Verlag Vieweg

Rajaratnam, N. 1976; Turbulent Jets, Elsevier

Reynolds, O., 1885; On the dilatancy of media composed of rigid particles in contact. Phil. Mag Ser. 5, 20, 469-481

Richman, M. W. and Marciniec, R. P., 1990; Gravity-driven granular flows of smooth, inelastic spheres down bumpy inclines, Transactions ASME, , 57, 1036 - 1043

Rodi, W., 1985; Calculation of stably stratified shear-layer flows with a buoyancy-extended k-ε turbulence model. In Turbulence and Diffusion in Stable Environments, edited by J. C. R. Hunt, Clarendon Press, Oxford

Salm, B., 1966; Contribution to avalanche dynamics, International Symposium on Scientific Aspects of Snow and Ice Avalanches (Proceedings of the Davos Symposium, 5 - 10 April 1965): IAHS Publ. No. 69, 199 - 214

Salm, B., 1968; On nonuniform, steady flow of avalanching snow, Union de Géodesie et Géophysique Internationale, Association Internationale d'Hydrologie Scientifique, Assembles générale de Berne, 25 Sept - 7 Oct 1967 (Commission de Neiges et Glaces), Rapports et discussions, 19 - 29, (Publication No. 79 de l'Association Internationale d'Hydrologic Scien-tifique)

Salm, B. and Gubler, H.-U., 1987; Avalanche formation, movement and effects, Proceedings of the Davos-symposium, 14 - 19 September 1986. IAHS-Publication No. 162, 1 - 686

Savage, S. B., 1979; Gravity flow of cohesionless granular materials in chutes and channels. J. Fluid Mech., 92, 53 - 96

Savage, S. B., 1983; Granular flows down rough inclines - Review and extension, in: Jenkins, J. T. and Satake, M., (eds.), Mechanics of Granular Materials: New Models and Constitutive Relations (Elsevier), 261-282

Savage, S. B., 1989; Flow of granular materials, in: Germain., P., Piau, M. and Caillerie, D. (eds), Theoretical and Applied Mechanics, Elsevier, 241-266

Savage, S. B. and Hutter, K., 1989: The motion of a finite mass of granular material down a rough incline. J. Fluid Mech., 199, 177-215

Savage, S. B. and Hutter, K., 1991; The dynamics of avalanches of granular materials from initiation to runout. Part I: Analysis. Acta Mechanica, 86, 201-223.

Savage, S. B. and Jeffrey, D. J., 1981; The stress in a granular flow at high shear rates. J. Fluid Mech., 110, 255-272.

Savage, S. B. and McKeown, S. 1983; Shear stresses developed during rapid shear of dense concentrations of large spherical particles between concentric rotating cylinders. J. Fluid Mech., 127, 453-472

Savage, S. B. and Nohguchi, Y., 1988; Similarity solutions for avalanches of granular materials down curved beds. Acta Mechanica, 75, 153-174.

Savage, S. B. and Sayed, 1984; Stresses developed by dry cohesionless granular materials sheared in an annular shear cell. J. Fluid Mech. 142, 391-430.

Schaerer, P. A., 1975; Friction coefficients and speed of flowing avalanches, Symposium Mecanique de la Neige, Actes. du Colloque de Grindelwald, Avril, 1974, IAHS-AISH Publ. 114, 425 - 32

Scheidegger, E., 1975; Physical aspects of natural catastrophes, Elsevier

Scheiwiller, T. and Hutter, K., 1982; Lawinendynamik: Übersicht über Experimente und theoretische Modelle von Flie?- und Staublawinen, Laboratory of Hydraulics, Hydrology and Glaciology, Report No. 58, ETH Zürich, Switzerland

Scheiwiller, T., Hutter, K. and Hermann, F., 1987, Dynamics of powder snow avalanches, Annales Geophysicae, Nr. 5B(6), 569 - 588

Scheiwiller, T., 1985; KANTAVAL, a set of PASCAL-computer programs for the calculation of plane steady turbulent gravity-driven dispersed two-phase flow. Int. Report of the Laboratory of Hydraulics, Hydrology and Glaciology, ETH Zürich

Scheiwiller, T., 1986; Dynamics of powder snow avalanches, Diss. ETH Nr. 7951, Mitteilung Nr. 81 der Versuchsanstalt für Wasserbau, Hydrologie und Glaziologie, ETH Zürich, 1- 115

Scheiwiller, T., Bucher, C., Hermann, F., 1985; Laboratory simulation of powder-snow avalanches, Int. Report No. 79 of the Laboratory of Hydraulics,, Hydrology and Glaciology, ETH Zürich

Schwarz, W. H. and Cosart, W. P. 1961; The two dimensional wall jet. J. Fluid Mech., 10, 481-495

Shreve, R. L., 1966; Sherman landslide, Alaska, Science, 154, 1639-1643.

Shreve, R. L., 1968: The Blackhawk landslide, Geol. Soc. Amer., Spec. paper, 108, 47 pp.

Sigalla, A. 1958; Measurements of skin friction in a plane turbulent wall jet. J. R. Aeronaut. Soc., 62, 873-877.

Simpson, J. E., 1972, Effects of the lower boundary on the head of a gravity current. Journal of Fluid Mechanics, 53, 759 - 768

Simpson, J. E., 1982; Gravity currents in the laboratory, atmosphere and ocean, Ann. Rev. Fluid Mech., 14, , 213 - 234

Simpson, J. E., 1987; Gravity currents in the environment and the laboratory, Ellis Horwood Ltd., Publ., Chichester

Spalding, D., B. 1982; Turbulence modelling: solved and unsolved problems, CHAM Ltd., Technical Report TR/58b, Appendix 2, London.

Stadler, R. 1986; Stationäres, schnelles Fliessen von dicht gepackten, trockenen und feuchten Schüttgütern, Dr.-ing Dissertation, Universität Karlsruhe, Deutschland.

Tesche, T. W., 1986; Sensitivity analysis of the AVALANCHE simulation model, Alpine Geophysics, Inc., Report No. AGI-86/010, Placeville, CA

Tesche, T. W., 1987; A three-dimensional dynamic model of turbulent avalanche flow. Paper presented at the International Snow Sciences Workshop, Lake Tahoe, California, October 22 - 25, 1986, pp. 111 - 137

Tochon-Danguy, J.-C. and Hopfinger, E. J., 1975; Simulation of the dynamics of powder avalanches, (Union Géodesique et Géophysique Internationale, Association Internationale des Sciences Hydrologiques, Commission des Neiges et Glaces), Mecanique de la neige, Actes du colloque de Grindelwald, Avril, 1974, 369 - 80 (IAHS-AISH, Publ. No. 114)

Tochon-Danguy, J.-C., 1977; Etude des courants de gravité sur forte pente avec application aux avalanches poudreuses, Thèse, Grenoble

Tsang, G and Wood, I. R., 1968; Motion of two-dimensional starting plume. J. Engng. Mech. Div. A.S.C.E. EM6, 1547-156

Vila, J.-P., 1986; Sur la théorie et l'approximation numérique de problèmes hyperboliques non linéaires, applications aux équations de Saint Venant et à la modélisation des avalanches de neige dense, Thèse, Université Paris VI, pp. 1 - 481

Voellmy, A., 1955; Über die Zerstörungskraft von Lawinen, Schweizerische Bauzeitung, Jahrg. 73, Hf 12, 159 - 62, (English translation: On the destructive force of avalanches, U.S. Department of Agriculture, Forest Service, Alta Avalanche Study Center Translation No. 2, 1964)

Wood I. R., 1965; Studies in unsteady self-preserving turbulent flows. University of New South Wales, Austral. Water Res. Lab. Rep. No. 81.

CHAPTER 12

Hydrological Disasters Associated with Volcanoes

Vincent E. Neall

ABSTRACT. Hydrological disasters associated with volcanoes display a wide variety of causal mechanisms. These can be broadly grouped into 8 principal categories. Those directly related to volcanic activity are steam (phreatic) explosions, eruptions through a crater lake, pyroclastic flows interacting with water, volcanic melting of snow and ice and volcanogenic tsunamis. Those indirectly related to volcanic activity include release of gases from a crater lake, non-volcanic initiated collapse of a crater lake and heavy rains on recently erupted materials. A review of each of these triggering mechanisms is described, together with the resulting disasters on nearby populations. In recorded history at least 116,000 persons have been killed by these processes, with property damage amounting to many billions of dollars. Any attempts to provide forewarning of volcanic/water interactions by volcanic monitoring, and meaningful responses by civilian authorities, will help to reduce the burgeoning number of casualties.

12.1 Introduction

Volcanoes are one of the most distinctive geological features on Earth. Not only do they mark the sutures where magma escapes to the surface to create new oceanic crust, but they also characterise subduction zones where oceanic crust is eventually destroyed. Further, they also appear in isolated positions away from plate boundaries. In oceanic settings there is ample opportunity for magma to interact with ocean water resulting in explosive hydrovolcanic eruptions, especially where submerged volcanic structures occur at shallow water depth. On land there is a wide array of opportunities for magma to react with groundwater, heavy rain, lakes, rivers, snow and ice. In this chapter, these interactions are discussed specifically with regard to known historical disasters where major loss of life has eventuated.

Many volcanic eruptions in the geological past which involved interactions with water have been of considerably greater magnitude than those described in this review. Where no information has been found about the impact on human populations these events have been omitted. Also to provide a limit to the scope of this review, 'disaster' has been taken to mean a 'sudden misfortune or calamity resulting in loss of life' based on Webster's New World Dictionary definition (Guralnik, 1970). Thus many lahars formed during historical eruptions that have not involved loss of life are excluded. A very few poorly known disasters have also been excluded for lack of recorded information on exactly what volcanic/hydrologic interactions occurred to create the disaster. Most volcanological terms used here are of standard usage. The frequently used term 'lahar' refers to a 'rapidly flowing mixture of rock debris and water (other than normal stream flow) from a volcano' (Smith and Fritz, 1989).

Hydrological disasters associated with volcanoes can be classified into 8 major categories, dependent upon the eruptive or non-eruptive triggering mechanism. These are:

A. Eruptive mechanisms.
1. Steam (phreatic) explosions.
2. Eruptions through a crater lake.
3. Pyroclastic flows interacting with water.
4. Volcanic melting of snow and ice.
5. Volcanogenic tsunamis.

B. Non-eruptive mechanisms.
6. Release of gases from a crater lake.
7. Non-volcanic initiated collapse of a crater lake.
8. Heavy rains on recently erupted materials.

12.2 Steam (Phreatic) Explosions

Disasters involving steam explosions fall into 2 main types. These are (i) phreatomagmatic, where magma encounters water resulting in combined magma/steam eruptions and (ii) nonmagmatic steam explosions. The extent to which magmatic eruptions involve a steam component is highly variable, most eruptions involving a small hydrological interaction. Those phreatomagmatic disasters where the steam component was only a minor contribution have had to be excluded from this review to maintain the context of a hydrological disaster. Those disasters caused by mainly phreatic explosions have been much easier to identify and have been included.

One obvious phreatomagmatic disaster where steam release played a major eruptive role occurred during the Mt Tarawera eruption in New Zealand on June 10, 1886. During the passage of the eruption, high velocity, steam-fluidised density currents exploded from Lake Rotomahana accompanying the eruption of basaltic scoria from nearby Mt Tarawera. The currents spread up to 6 km radially from source in an unpopulated region, outside which cold wet mud fell up to 60 km downwind. Where the mud was thickest, at least 116 lives were lost of the 153 total victims known to have died in the eruption. Nairn (1979) produced evidence that a continuous basalt dike encountered at its southwestern end the hydrothermal system beneath the Lake thus producing this highly explosive phreatomagmatic event.

Other examples of disastrous phreatomagmatic eruptions come from Volcano Island located in Lake Taal, in the Philippines. Here fatalities are recorded from these eruption types from September 24-27, 1716, August 11 to September 1749, May 13 to December 4, 1754 and on July 19, 1874. In January 1911 phreatomagmatic explosions formed ground-hugging surges that claimed about 1,500 lives mostly on Volcano Island (Oppenheimer 1991). Lives were lost again between September 28-30, 1965. Then again in 1969, eruptions created radial fissures into which the lake waters penetrated and phreatic explosions resulted with 150 lives lost (Rittman and Rittman, 1976).

The classic steam explosion-triggered disaster widely quoted in the literature occurred in 1888 at Bandai-san Volcano, on the Japanese island of Honshu, 220 km north of Tokyo. This event began without warning, with violent seismicity which began at about 7 am on July 15, 1888, lasting between 15 to 30 minutes. This was immediately followed by 15 to 20 explosions that were directed to the north. The initial explosion broke through the flank of the volcanic mountain, approximately 100 m above a place where steam had been observed, and all later explosions were generated from points close to the foot of the volcano (Williams and McBirney, 1979). The explosions were accompanied by a "rain of hot scalding ashes" (Glicken and Nakamura, 1988), now interpreted as a lateral blast. Then suddenly the northern flank of the 1819m high Bandai-san cone collapsed northwards forming a 1.5 km^3 debris avalanche that rushed 5 km down the slopes of the volcano at about 80 km/hr. It swept downward in surging waves that left behind arcuate ridges in the resultant unsorted debris. Near its margins it formed lahars along the major stream channels, burying 7 villages and nearby farmlands; in total it covered 34 km^2. When the explosions had ended a huge horseshoe-shaped amphitheater had been created within the volcanic edifice. A subsequent lahar was triggered by smaller debris avalanches of wet hydrothermally-altered materials from near the summit (Glicken and Nakamura, 1988). In all 461 people were killed and 70 persons injured (Blong, 1984).

All the products from the explosions seem to have been derived from the pre-existing Bandaisan edifice. There is a distinct lack of any juvenile magmatic material. Following the explosions a line of steam vents was observed to cross the floor of the new amphitheater suggesting that groundwater may have seeped down a major fissure and encountered hot rocks below. Thus the most commonly accepted triggering mechanism is that steam expanded violently producing the devastating blasts and this destabilised the northern flank of the volcano causing the debris avalanche.

A similarly triggered disaster occurred in earlier times in the Yatsugatake Volcanic Chain of Central Japan. Kawachi (1988) has established that this was the great disaster referred to in historical documents as the "Shinano-hukubu Earthquake". Exactly how many people were killed is unknown but the "river overflowed in six districts and castles and houses were completely destroyed. The bodies of victims, men and women, horses and cattle piled up into hill (*sic*)" (Kawachi, 1988). The disaster apparently originated by collapse of Mt. Inagodake on June 20, 888, and resulted in the deposition of the Ohtsukigawa Debris Avalanche, which totals about 0.35 km^3. There is a third example of a phreatic explosion disaster in Japan, recorded at Yake-dake. Here in 1585 an explosion generated a lahar that extended 4.5 km to the northwest of the summit burying an unknown number of victims in the village of Nakao (Kuno, 1962).

The steam explosion mechanism was also responsible for a similar disaster in West Java, Indonesia in 1772. Eyewitnesses reported sounds like "very heavy cannon shots" which accompanied steam explosions at Papandayan Volcano. This triggered collapse of the edifice creating a hot clay-rich volcanic debris avalanche. Lahars were then generated from watersaturated portions of the avalanche. Lithologies within the avalanche are mostly hydrothermally altered with a notable absence of any juvenile magmatic material; it was in fact very similar to the Bandai-san event. The resultant deposit totalled 0.14 km^3 and covers 18 km^2. About 40 villages were destroyed and a total of 2,957 people were killed from both the avalanche and accompanying lahars (Neumann van Padang, 1951, Glicken et al., 1987). Also in Indonesia, steam explosions were responsible for generating a mudflow at Butak Petarangan, in 1928 that resulted in casualties (Neumann van Padang, 1951).

12.3 Eruptions Through a Crater Lake

Mt. Kelut on the Indonesian island of Java is the type example for disasters caused by eruptions through a crater lake. In the last 10 centuries 29 lahars generated by this mechanism from Mt. Kelut have resulted in the loss of 15,400 lives.

Mt. Kelut is not a particularly high volcano (1731 m), and is surrounded by a dissected flank that grades to a lowland ring plain. The Brantas River begins to the south of the volcano and loops clockwise around it to flow northwards, thus draining a 200° sector of the radial rivers sourced on the mountain. Due to its low gradient and the plentiful volcaniclastic detritus, this river has built natural levees which border its main channel, elevating it several metres above the surrounding plains. Thus any extraordinary high flows created by expulsion of the crater lake waters readily inundate the adjacent lowlands.

The earliest historical eruption through the crater lake is recorded about the year 1000 A.D. Subsequent similar style eruptions where the casualties are unknown occurred in 1311, 1334, 1385, 1395, 1411, 1451, 1462, 1481 and 1548 A.D. In 1586 A.D. major loss of life resulted with over 10,000 persons killed. Later similar events are recorded in 1641, 1716, 1752, 1771, 1776, 1785, 1811, 1825, 1826, 1835, 1848, 1851, 1864 and 1901 A.D. (Alzwar, 1985; Suryo, 1985). In 1848 lahars were directed principally along the Gedok River to the west of the crater. However in 1875 heavy rains triggered a collapse that formed a breach in the southwest margin of the crater. The first reliable estimate of the volume of the crater lake dates from this time when it is estimated to have contained 78 million m^3 of water. The May 22-23, 1901 eruption showed for the first time the danger to the population in the southwest of the new configuration of the crater rim, leading to the first protective measures. A diversion dam was first constructed in the upper part of the Badak ravine above Blitar in 1905 to try and divert lahars from the most highly populated areas. By this stage the crater lake volume had dropped to about 40 million m^3.

During the subsequent eruption of May 20, 1919, all of the water was expelled from the crater. This created lahars up to 58 m deep in some ravines (Neumann van Padang, 1951), that travelled up to 38 km from source, inundating 131 km^2 of the Blitar district. These lahars destroyed more than a hundred villages, swept away the diversion dam in the Badak ravine and caused 5160 fatalities (Alzwar, 1985). It was the second most devastating eruption of Mt. Kelut (Van Bemmelen, 1949, Verstoppen, 1992).

Following this disaster the government initiated works to drain the bulk of the waters in the crater lake according to a plan proposed by H. Cool in 1907 to reduce the hazard from future lahars. Tunnelling began under the direction of an engineer named Von Steiger in September 1919 in the left wall of the Badak canyon at a height of 1111 m. The tunnel was to be about 955 m long and was intended to reach the bottom of the lake at 1114 m altitude (having a grade of 3%). The lake was dry at the time so tunnelling also began from the crater westwards. However, these efforts were unexpectedly terminated on December 6 and 7, 1920, when a lava plug rose from the crater close to the tunnel. Not to be thwarted by this unplanned event the tunnelling continued from outside the crater in an eastwards direction. By 1921 tunnelling had proved slow, particularly due to the high temperatures (up to 46°C) and high humidity. In November 1922, 735 m from the tunnel entrance, the volcanic debris of the inner crater wall was encountered and thereafter the tunnellers had to begin lining with concrete. Meanwhile the crater lake was reestablished and began to rise. By April 4-5, 1923, it was half-full with an estimated volume of 22 million m^3. Suddenly there was an inrush of water, mud and debris resulting in five of the workers being killed, so work was temporarily suspended (Van Bemmelen, 1949, Verstoppen, 1992).

This tragedy led to a new plan being undertaken by an engineer named Hettinga Tromp, whereby the lake level would be reduced in a step-wise procedure. The aim was to excavate seven tunnels and siphon the water through each tunnel progressively, thus reducing the lake level (Fig. 12.1). The first tunnel was driven under the lowest crater rim, which at the time was just above the lake level at 1185 m, preventing the lake rising any further. Then the tunnels were progressively completed at successively lower levels. The project was ultimately successful and the lake water was finally reduced to a height of 1129 m, with a volume of less than 2 million m^3. No major eruptive activity then followed for over 25 years.

Suddenly, on August 31, 1951, Mt Kelut erupted again and the small volume of crater lake waters quickly evaporated. No sizeable lahars were formed thus proving the effectiveness of the tunnel drainage system. However during this eruption the tunnels became blocked (Zen and Hadikusomo, 1965) and the crater had become deepened by about 70 m.

In 1954 an engineer named Yzendoorn was appointed to reactivate the tunnel system. Work on clearing the tunnels VIB, VIIA and VIIB was completed in 1955 when the crater lake level had just reached the floor of tunnel VIIB. This limited the volume of the lake to 23.5 million m^3 of water. A second stage of the plan was to drive a new tunnel 20 m below tunnel VIIB to further reduce the lake level. Two parallel galleries were begun from outside the crater extending inwards towards the lake. On reaching softer fragmental materials, presumably lining the inner crater

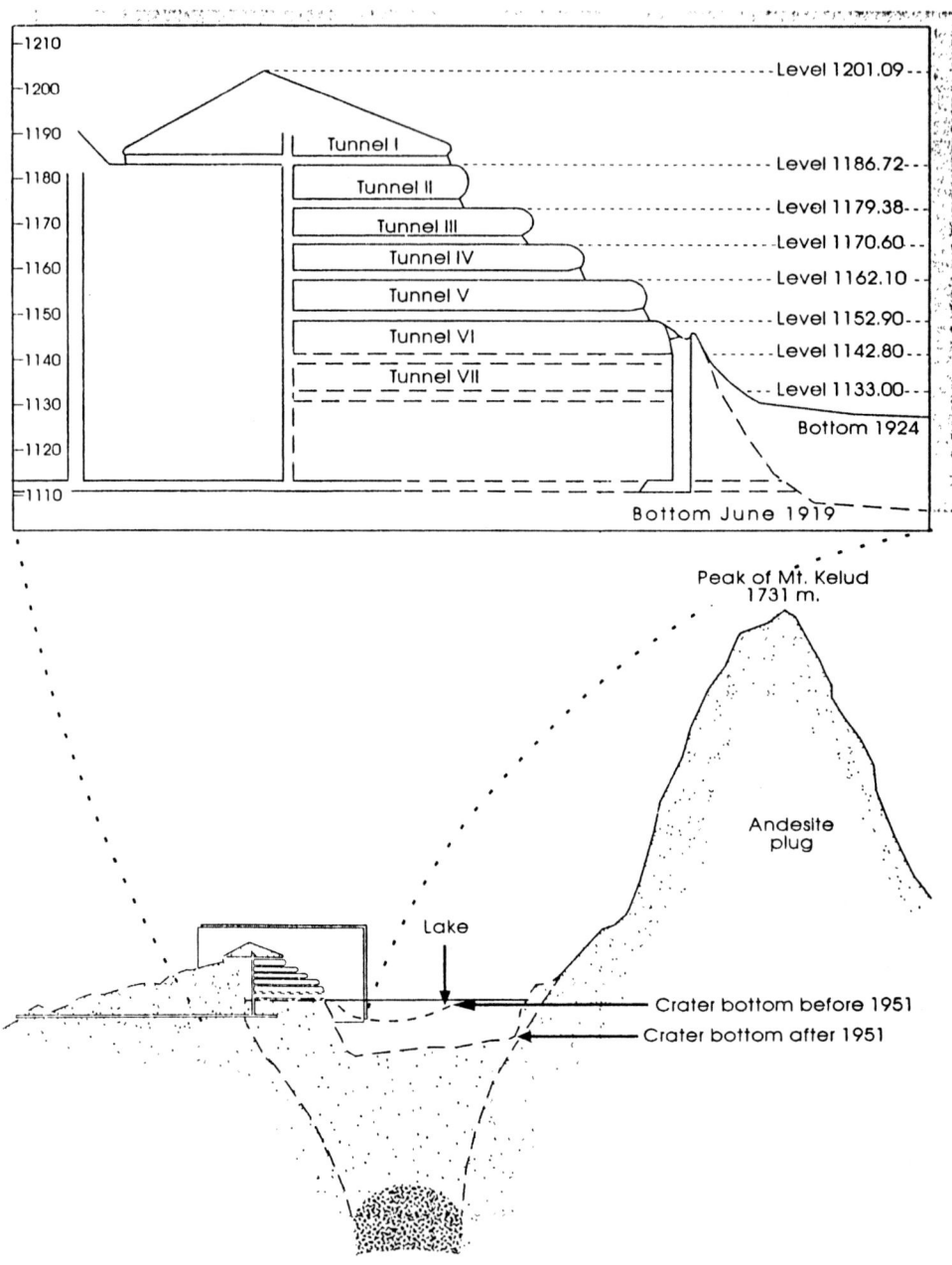

The Kelut Tunnel System

Fig. 12.1. Cross-section of the crater lake of Mt Kelut showing the network of tunnels used in lowering the lake level to reduce the formation of lahars (adapted from Zen and Hadikusumo, 1965).

wall, a system of adits were driven along a front. It was hoped that the lake water would seep through and reduce the lake level accordingly. The desired discharge of 200 l/s was never reached and on March 26, 1963, the project was terminated with a drainage rate of 70 l/s (Zen and Hadikusumo, 1965, Verstoppen, 1992). In 1965 rockfalls into the crater lake began to displace the water by elevating the lake floor, and these reduced the volume to 20.3 million m^3.

Unexpectedly, on April 26, 1966, a violent explosion expelled the 20 million m^3 of water creating voluminous lahars. The main lahar followed the Badak River and made its way into the Temas River. The reconstructed Badak diversion dam collapsed enabling lahars to enter the Blitar river catchment where further overflows from the Temas River added to its bulk. Many houses and a bridge were swept away, 30 persons were killed and 7 wounded. From eyewitness accounts the lahar lasted 3 hours. On the upper slopes it travelled at 12 m/s, whilst on the outskirts of Blitar it was gauged at 6.1 m/s. The Temas River lahar travelled 31 km destroying an irrigation dam, many villages and resulting in many casualties. Other lahars entered the Semut, Putih, Ngobo and Konto Rivers, branching and overflowing into many other catchments. Casualties from these lake-generated lahars totalled 208 with a further 73 wounded. The lahars inundated 45 km^2; 2000 ha of rice fields were destroyed and a similar area of arable upland was damaged. In total 732 houses were destoyed and a further 1897 damaged (Suryo, 1985).

By May 20, 1966, the crater lake was empty, the crater floor being 40 m above its former level. The southwest breach had been widened and the tunnel VIIB filled with volcaniclastic material and partially destroyed. In June 1966 a project was begun to construct a new tunnel 20 m below tunnel VIIB, connecting the crater with a previously unfinished tunnel. This was completed by the end of December 1967 when the inlet was 38m above the crater floor and the lake volume was limited to 4.3 million m^3.

A further eruption in February 1990 generated only limited lahars, reducing the death toll to 32 victims, which was markedly lower than prior to the tunnels being constructed (Verstoppen, 1992). Mt Kelut is likely to continue as a primary source of disastrous lahars whenever the crater lake accumulates a sufficient volume of water.

Another volcano in East Java with a crater lake is Kawah Idjen. In January 1817 an eruption began there that lasted for 33 days, during which the crater lake waters were expelled forming lahars to the southwest. The plain of Rogodjambi was inundated, several villages were destroyed and many people and animals killed (Neumann van Padang, 1960). This plain was also inundated by lahars from Raung Volcano in 1838. Neumann van Padang (1951) considered that prior to this eruption it was likely that a lake occupied the Raung crater causing the enormous floods and calamaties of 1638, 1730 and 1838. And in Sumatra, similar events occurred at Kaba Volcano in 1833 where the number of casualties is unknown (Neumann van Padang, 1951).

The famous eruptions of Mt Galunggung in West Java on October 8 and 12, 1822, were probably also of this type. Macdonald (1972) refers to an eruption through a crater lake forming the voluminous avalanche and lahars that inundated the plains to the southeast. Hot water and bluish mud were spread up to 38km from the mountain burying people in their houses. Lyell (1867, pp. 6-8) records "the face of the mountain was utterly changed, its summit being broken down, and one side, which had been covered with trees, became an enormous gulf in the form of a semi-circle." About 114 villages were destroyed and over 4,000 persons were killed (Lyell, 1867, Neumann van Padang, 1951). Some authors attribute all the casulaties to lahars (Haller et al., 1989) whilst others attribute casualties to both nuées ardentes and lahars (Sudradjat and Tilling, 1984). Following the 1982-83 eruptions, Galunggung has a new crater lake 70 m deep in 1990 and increasing at 5 m/yr. Its present volume is 7.5 million m^3 (Gerbe et al., 1990). Should another eruption occur or the edifice fail a similar scenario to Mt Kelut in 1919 could result.

At Awu in the Sangihe Islands of Indonesia, explosive eruptions through the crater lake have created enormous lahars. In 1711 there were 3177 victims and in the 18th century 5291 victims (Rittman and Rittman, 1976). Then again on August 12, 1966 about 30 million m^3 of crater lake waters were expelled initiating further destructive lahars (Hakikusumo 1966). A total of 39 people were killed, not all due to lahars (Blong 1984). But these disasters are not confined to Indonesia. At Kusatu-Sirane in Japan, an eruption beneath the waters of Lake Yu-gama formed a lahar in 1932 that travelled to the east killing 2 persons (Kuno, 1962).

Beyond the Pacific rim, another example occurred three days before the 1902 climatic eruption of Mt Pelée, Martinique, in the Caribbean. Towards the end of March 1902 increased fumarolic activity was observed in the crater of Mt Pelée, known as l'Etang Sec, and along the nearby Riviére Blanche that drains to the southeast. This led to phreatic explosions between April 23-25 and by April 26-27 a lake 200 m in diameter had formed. Four vigorous vents nearby were seen to have water cascading into the lake. There was no visible outlet but a fountain of warm water was observed in the headwall of the nearby Riviére Blanche canyon. Activity continued to accelerate and by May 2-3 light ash was falling in nearby Saint-Pierre. Continuous roaring and seismic tremor accompanied these eruptions (Tanguy, 1994). In what is a poorly understood triggering mechanism, a devastating lahar was generated shortly after noon on May 5, in the Rivière Blanche just 3 km north of St Pierre (Bullard, 1962). Either the south side of the 30 m high dam retaining the crater lake in L'Etang Sec failed or a phreatomagmatic eruption occurred which released the waters into the Rivière Blanche. Rushing down a gorge from 1000m altitude the torrent sped at about 120 km/hr as a deluge of boiling water, mud and boulders. In 3 minutes it had travelled to the mouth of the Rivière Blanche, where a rum distillery (the Usine Guérin factory) was destroyed killing 23 workmen (Tanguy, 1994). A 3-4 m tsunami was then generated by the lahar entering the sea. Within 3 minutes 3 successive lahars flowed down Rivière Blanche, in all totalling 5 million m^3, a volume corresponding to that of the lake (Chrétien and Brousse, 1989). Anderson and Flett (1903, p. 489) report that some of the other lahars accompanying the main 1902 eruptions of Mt Pelée were caused by emptying of a crater lake. However Tanguy (1994) records these as rain lahars (see Section 7).

Then just two days later on the nearby island of St Vincent, an eruption began at La Soufrière. The crater lake waters were violently expelled to form highly destructive lahars. On May 7, 1902, at about 10.30 am the crater was emitting a

Hydrological Disasters Associated with Volcanoes
403

continuous roar and the river beds draining the terrain to the west and the east (the Wallibou and the Rabaka Dry Rivers) had become raging torrents of boiling mud and water up to 15 m deep. By 2 pm this climaxed with the eruption of pyroclastic flows (and ? surges) that devastated one third of the island (Bullard, 1962). About 1,565 lives were lost, the bulk of them probably from the pyroclastic flows, but it is unclear how many deaths resulted from lahars (Robson and Tomblin, 1966).

12.4 Pyroclastic Flows Interacting with Water

One major disaster in Japan resulted from this triggering mechanism. On August 5, 1783, a pyroclastic flow rushed down the northern slope of Mt. Asama, and on reaching the river Agatuma-gawa produced a hot lahar (Kuno, 1962). According to Aramaki (1956) the pyroclastic flow dammed the river. The water level rose and the dam collapsed, generating a lahar that killed 1,377 persons and swept away 1,265 houses.

At Santa Maria, Guatemala, similar hot lahars were generated by devastating pyroclastic flows expelled from a lava dome on November 2, 1929 (Cotton 1944). "The glowing rocks and dust were shot into the rivers, heating the water and forming a mudflow, which spreading over the land, filled the valley with a deposit of hot mud" (Sapper and Termer, 1930 in Anderson, 1933). Mud covered a strip 10 km long, varying from 100 m to 1.5 km wide. Blocks up to 20 m^3 in size were spread like glacial erratics, while the stumps of trees were carbonised. Twenty-three persons were killed. First hand accounts of the devastation and a photograph are reproduced in The Volcano Letter (Jaggar, 1931). Also in Guatemala, at Fuego in 1971, hot ash flows and associated mudflows resulted in 10 casualties (Bonis, 1973). This mechanism also generates lahars from the numerous pyroclastic flows at Mt Merapi in Indonesia.

Some of the 1991 lahars at Mt Pinatubo were generated in this way, but since the bulk of them were triggered by heavy rains they have been collectively discussed in Section 7.

12.5 Volcanic Melting of Snow and Ice

The second worst volcanic disaster this century and the fourth most disastrous eruption in recorded history was initiated from volcanic melting of snow and ice. On the evening of November 13, 1985, eruptions from the 5389 m high summit crater of Nevado del Ruiz, Colombia, initiated the deadliest set of lahars in recorded history. They rushed down adjoining canyons to nearby villages and inundated the town of Armero and 22,000 of its inhabitants. This disaster could have been prevented and for this reason it has been the subject of much investigation to try and ascertain what went wrong.

Historical evidence shows that Nevado del Ruiz has erupted on at least 2 previous occasions causing widespread devastation. The first was recorded by Spaniards who on March 12, 1595, witnessed a pumice eruption. Lahars initiated from melting of snow and ice, descended catchments to the east of the volcano overflowing their channels and inundating adjoining lands (Simon, 1625 in: Voight, 1990). The second was on February 19, 1845, when a torrent of thick mud armoured with blocks of ice, debris, trees and sand emerged from the canyon of the Lagunilla River (Russell, 1985). It spread between 250–300 million tons of muddy debris across the present site of Armero to a depth of 8m, as well as destroying the town of Ambalema 30 km to the southeast, killing about 1000 persons (Voight, 1990).

The 1985 eruptions were relatively small on a global scale, the largest attaining a volcanic explosivity index (VEI) of 3. They initially comprised pyroclastic surges followed by pyroclastic flows that culminated in a small Plinian tephra. The total dense-rock-equivalent (DRE) volume erupted was only 1.9×10^7 m^3 (Calvache, 1990). Studies by Pierson et al. (1990), indicate that the pyroclastic surges were incapable of producing the meltwater that contributed to the sharp-crested lahars downstream. Rather it appears that of four recorded pyroclastic flows, two were mainly responsible for interacting vigorously with the snow and ice. Channels scoured into the ice were up to 100 m wide and 2–4 m deep. The eroded snow and ice became turbulently entrained into the pyroclastic flows dynamically increasing the area of contact between ice and hot rock. In the centres of these flows large quantities of meltwater were created transforming them into dilute slurries of water, lithic and pumiceous debris (Naranjo et al., 1986, Pierson et al., 1990). Mixed avalanches of varying proportions of snow, ice, liquid water, rock and colluvium were also initiated contributing slush, meltwater and rocky debris into the headwaters of lahar channels. These avalanches became wetter downstream due to melting from the entrained hot ejecta and the frictional heating during their 1 km descent (Pierson et al., 1990). The hot eruptive products mechanically mixed with about 10 km^2 of snow and ice (Thouret et al., 1987). About 16% (4.2 km^2) of the surface ice and snow area was lost, amounting to 0.06 km^3 or 9% of the total volume. About 43 million m^3 of liquid water was released, of which 11–12 million m^3 was incorporated into the lahars. The balance was included in other deposits, sublimated to steam or is still in storage on the volcano (Thouret, 1990).

The lahars were initiated within minutes of the onset of the eruptions. They were clearly formed by a complex sequence of diachronous events which were expressed in multiple-peaked flows close to source. Later lahars appear to have travelled more swiftly than earlier ones due to many channel roughness features being removed. This enabled later waves to overtake earlier ones.

From the head of each major catchment, the flows began eroding and entraining additional unconsolidated debris. Flows that spilled onto vegetated slopes quickly transformed into viscous, relatively slow-moving (4–6 m/s) debris flows. In the nearby canyons the peak flows were typically 20–30 m deep with a maximum of 44 m measured. These conditions lead to high boundary shear stresses, enabling the lahars to erode and incorporate unconsolidated materials together with their interstitial water. This markedly increased the total flow volume and peak flow rate, a process termed "bulking" (Costa, 1984; Scott, 1988).

By examination of remnant deposits along the main river channels it was possible for Pierson et al. (1990) to reconstruct the nature of the moving flows. The lahars clearly had watery tops, typical of low shear strength debris flow slurries, which showed considerable surface turbulence evident from irregular upper flow limits and splash marks. Beneath were debris flows carrying a considerable bedload component with evidence of high shear at the base. Reconstructed hydrographs show pulsing lahar flow which may have originated by (1) periodic self-damming and release, (2) inherent flow instability, (3) localised temporary damming by landslides and (4) restrictions of flow at bridges which finally failed. In addition, in one catchment, the Rio Lagunillas, a landslide had formed a 25 m deep lake the year before. This factor alone contributed a further 7.5×10^5 m^3 of water to the lahars above Armero.

In general, peak-flow mean velocities ranged from 5 to 15 m/s (maximum recorded was 17 m/s). Of considerable interest is that the Ruiz lahars moved slowly close to the volcano, but travelled relatively rapidly in distal channel reaches. This has been attributed to the persistent steep gradients in the distal reaches. Some flows took 6 hours to travel 70 km (to the west) whilst others took 45 minutes to travel the same distance (in the east). Some lahars displayed local supercritical flow with Froude numbers reaching 1.6, consistent with observed hydraulic jumping.

Maximum computed peak discharges for all lahars occurred at locations within 10–20 km of the crater. Farther downstream these peak discharges decreased due to flow attenuation caused by broader valleys. The maximum discharge is estimated at 48,000 m^3/s, at a point 9.6 km from source in the catchment above Armero (Pierson et al., 1990). This is roughly equivalent to the wave produced by the landslide-induced evulsion of the reservoir behind the Vaiont Dam in Italy in 1963 (Voight 1990). By the time this lahar burst from the canyon above Armero it had decreased to 27,000 m^3/s.

Pierson et al.'s (1990) calculations indicate that over the 104 km distance travelled by the lahars, net flow-volume increases (bulking) occurred by factors of 2 to 4. This realisation has a major social implication because lahars may far exceed their initial volume during their passage and thus extensively inundate inhabited regions over 100 km from a volcano.

Only at Armero, over 5 km vertical height below the summit, did lahars emerge onto an unconfined terrain. Here rapid deceleration resulted as flows shallowed and spread laterally, in this case into 3 lobes (Lowe et al., 1986). This is dramatically recorded by the account of one survivor who was carried along in the flow (Pierson et al., 1990, p. 60). Around Armero the lahar inundated 34 km^2 to an average depth of 1.5 m and a maximum of 8 m. The "one big beach of mud" had an estimated volume between 40 and 60 million m^3, and one week later it still comprised 35% by volume of water (Naranjo et al., 1986).

An estimated 21,559 persons were entombed by the lahars at Armero (SEAN 1985) and a further 1,090 in the vicinity of Chinchina to the west of Nevado del Ruiz (Russell, 1985). Over 5,000 persons were injured. About 3,400 hectares of agricultural land was buried along with 60% of the region's livestock and 30% of its grain sorghum and rice crops. About 50 schools and 2 hospitals were lost or damaged, 5,092 homes were destroyed and 58 industrial plants and 343 commercial establishments were damaged. The total financial loss was over a billion US dollars (Voight, 1990).

Another significant realization that can be drawn from this disaster was that such a small eruption (less than 5×10^6 m^3 of magma ejected) was able to generate up to 9×10^7 m^3 of lahar slurry from about 2×10^7 m^3 of meltwater (Pierson et al., 1990; Calvache, 1990).

Of major concern in retrospect is that this disaster could have been avoided. The Colombian government was unwilling to bear the economic or political costs of an early evacuation or a false alarm (Voight, 1990). Scientists foresaw the hazards and alerted the authorities but this was insufficiently precise for their purposes. A volcanic hazard map was prepared but it was not widely circulated. Even on the day of the disaster the Red Cross ordered a few hours in advance the evacuation of Armero, but it remains uncertain if this order was ever transmitted or why an alarm was not sounded (Hall, 1992). Pieced together in Voight's (1990) words "The catastrophe was not caused by technological ineffectiveness or defectiveness, nor by an overwhelming eruption, or by an improbable run of bad luck, but rather by cumulative human error — by misjudgement, indecision and bureaucratic shortsightedness ... Time proved to be a luxury squandered ... Armero could have produced no victims, and therein dwells its immense tragedy."

A second Andean volcanic giant also renowned for disastrous lahars triggered by melting of snow and ice is the 5897 m Mt Cotopaxi in Ecuador. Since the Spanish arrived in Ecuador in 1532, Cotopaxi has erupted about 30 times. Following a dormant interval of 208 years, the volcano reawakened in 1742 erupting in June, July and December. On all three of these occasions lahars triggered by melting of the 20 km^2 summit ice field have caused widespread damage to haciendas, livestock and people in the Latacunga Valley on the volcano's southwest side (Paradez, 1982 in Mothes, 1992). Further significant lahar-producing eruptions occurred in 1766 and 1768, inundating the main catchments. Then on June 26, 1877, the most disastrous historical eruption occurred.

The 1877 eruption is recorded by Whymper (1892) who quotes the inhabitants at Mulalo witnessing "molten lava pouring through the gaps and notches in the lip of the crater, bubbling and smoking, ... like the froth of a pot that suddenly boils over ... (when) out of the darkness a moaning noise arose, which grew into a roar, and a deluge of water, blocks of ice, mud and rock rushed down, sweeping away everything that lay in its course and leaving a desert in its rear." Enormous broad and deep channels were furrowed out in the ice and snow on the summit from which the great streams of water, locally referred to as avenidas were generated (Wolf, 1878 in Anderson, 1933). According to Mothes' (1992) estimates, lahar flows of 45,000–70,000 m^3/s would have reached Latacunga, 43 km from the crater in approximately 40–60 minutes. Initially travelling at 60–100 km/h, after 100 km they had slowed to 30 km/hr. In the three principal rivers draining the volcano, textile factories were destroyed, haciendas damaged and bridges swept away. A settlement in the Amazonian lowlands to the east was annihilated (Mothes, 1992). About 1000 people (Sigurdsson and Carey, 1986) and thousands of animals lost their lives, the lahars destroying "in one hour the work of many generations". Some people on galloping horses were even overcome by the advancing wave of one lahar (Sodiro, 1877, p. 13 in Mothes, 1992). The lahar that flowed down the volcano's north side is to the author's knowledge, the longest distance ever recorded for the effects of a lahar in historical times. Travelling 270 km from its source the northern lahar arrived at the Pacific port of Esmeraldas 18 hours after the eruption carrying cadavers and pieces of houses and furniture from the Chillos Valley.

Today a major concern must be the extent to which the local population has rebuilt homes, industries, schools and a hospital on the 1877 lahar deposits. Currently a population of 30,000 people live within the area of greatest lahar risk from Cotopaxi, and another 100,000 live in adjoining lower risk zones (Mothes, 1992). The last eruption of Cotopaxi was in 1914 and the travel time for lahars to these densely populated areas is between 30 minutes and 1 hour (Zupka, 1993). The United Nations Department for Humanitarian Affairs has sponsored a Volcanic Disaster Mitigation Project in Ecuador that has addressed these potential hazards and at other potentially dangerous volcanoes in Ecuador (Zupka, 1993).

A third Andean volcano, Villarrica in Chile, also typifies this type of disaster. On March 3, 1964, an explosion caused sudden melting of snow and ice at the summit, initiating a lahar that flowed down the northern and southern slopes of the volcano killing 25 persons (Katsui 1967). Then on December 29, 1971, another eruption fractured through the summit ice cover creating lahars to the north, west and south. Travelling at 100 k/h they carried tree trunks, ice and granite blocks up to 20m across. Four concrete bridges were swept away and 15 persons were lost before the lahars dissipated in two nearby lakes (Gonzalez-Ferran 1973).

Also in Chile, similarly generated lahars were created by eruptions at Mt Hudson on August 18, 1971. Here, melting at the head of Heumules Glacier mobilized pyroclastics into a lahar down the Rio de Los Heumules Valley destroying land, filling houses to the ceiling with mud and clasts and claiming three victims (Tobar, 1973).

The potential for future eruptions at Latin American volcanoes melting high elevation icefields and creating lahars is widely recognised (Sigurdsson and Carey, 1986). For example, many deaths are thought to have resulted from lahars generated by glacier melting on the 5245 m high volcano, Tolima in Colombia, but very little information is available (Rittman and Rittman, 1976). Vulnerable areas adjacent to these volcanoes should be zoned to prevent further urban encroachment.

A non-Andean disaster of a similar origin occurred at Tokachi-dake Volcano in Central Hokkaido, Japan in 1926. Prior to 1923 the volcano had shown fumarolic activity, but in 1926 activity in the central cone began to increase, culminating in an explosive eruption on May 24, 1926. The first explosion created lahars that swept over the western slope of the volcano. Four hours later a second explosion occurred from the summit destabilising the northwest flank. Between $2-4 \times 10^6$ m^3 of hot debris was mobilised which melted the snow on the mountainside and transformed into lahars. These rushed down the volcanic slopes at 50 m/s, pouring into two valleys to the west and northwest, and reaching the surrounding lowlands 25 minutes after the start of the eruption. The volume of the lahar deposits totalled 20×10^6 m^3 or ten times as much volume as the collapsed sector of the volcano. This is attributed to concomitant pumice and scoria pyroclastic flows but it may also have involved bulking en route. At about 25 km from source 29 km^2 of land was washed away, 5080 houses were ruined and 144 persons drowned (Murai, 1960).

During the May 18, 1980 eruption of Mt St Helens, possibly six of the casualties were killed by lahars in the Toutle Valley (Blong, 1984). Fairchild (1987) calculated that the forces generated by the pyroclastic surge were sufficient to trigger snow avalanches that initiated the South Fork Toutle River lahar. However Waitt (1989) has produced evidence for most of the lahars originating by swift snowmelt at the base of a hot and relatively dry turbulent pyroclastic surge. Melting snow and accumulating hot surge debris may have moved initially as thousands of small thin slushflows. These became self-sustaining as newly uncovered snow continued to partially melt from the surge above. The slushflows then grew over tens of seconds into sheetfloods that accelerated downslope at more than 100 km/h. Subsequently the most damaging lahar issued from the water-saturated (probably from slowly melted snow and ice) debris avalanche deposit and rushed down the North Fork Toutle River valley (Waitt, 1989).

An unusual mechanism whereby ice is melted by volcanic activity is unique to Iceland. Here, many volcanoes occur beneath the Vatnajökull ice cap and due to prolonged geothermal activity or to sudden explosive activity large quantities of water escape from their subglacial reservoirs. These glacial outbursts, named 'jökulhlaups' in Icelandic, have been responsible for creating the extensive sand plain (sandur) of the south Icelandic coastline. Their immense potential size is illustrated by two events, (i) the May 27, 1938 Skeidárarhlaup which reached a maximum discharge of 45,000 m^3/s when 7 km^3 of water escaped from the Skeidárarjökull glacier front and inundated about 1000 km^2 of sandur, and (ii) the 1918 Kötlulhlaup generated by the Katla subglacial volcano that flooded the Mýrdalssandur with a discharge between 100–200,000 m^3/s, about two or three times the discharge of the River Amazon. In the former, Skeidárarhlaups start so slowly that they are not likely to have killed people (Thorarinsson, pers. comm., 1976). In the latter case there was a community on the sandur in the Lágeyjarhverfi district which disappeared in 1918 (Thorarinsson, 1956). Earlier Kötluhlaups swept away farm houses and a church at the foot of Höfdabrekkuháls in 1660 and in 1721, but no people were killed at the time (Thorarinsson, pers. comm., 1976). However earlier Kötluhlaups in 1179 and 1311 probably did involve loss of life; the latter may have killed some tens of people (Thorarinsson, pers. comm., 1976).

Öraefajökull is another subglacial Icelandic volcano rising to 1850 m that has also been responsible for two major disasters in historic time. The first occurred in 1362 in one of the most violent and disastrous eruptions recorded in the history of Iceland. The catastrophic jökulhlaups that were generated completely destroyed two parishes resulting in at least 30 farms (maximum of 40) being abandoned for decades. Exactly how many people lost their lives is unrecorded but a reasonable estimate is between 50 and 100 killed (Thorarinsson, pers. comm., 1976). An eye-witness account of this event from a lone survivor is recorded by Sveinn Pálsson. Apparently a shepherd heard two crashing noises from the glacier and ran to the mountain above, where he sheltered in a cave. "Then came the third crash, and at the same moment the glacier exploded with enormous noise. Water and ice filled every ravine in the mountain and washed away all the people and the livestock in the settlement below or buried them all in deep mud, sand and glacial debris inside the farmhouses" (Bárdarson, 1971). When the farms were later revived, the district became known as Öraefi which means waste land. A second eruption in 1727 was not so violent but still caused great damage to the Öraefi district; three persons lost their lives clinging to the roof of a building that was swept away (Thorarinsson, 1958, p. 32) and two small farms were subsequently abandoned (Thorarinsson, 1956).

A review of all known global occurrences of historical eruptions perturbating snow and ice to trigger lahars and floods is provided by Major and Newhall (1989).

12.6 Volcanogenic Tsunamis

Historical records of all those killed directly by volcanic eruptions, show that about 25% have died as a result of tsunamis. The lethal effect of volcanically-induced (hereafter referred to as volcanogenic) tsunamis is the unexpected transfer of energy from what may be isolated, sparsely populated volcanoes to sea waves which can rapidly travel to densely populated shorelines. Latter (1981) has reviewed the historical record and recognises 92 proven examples of genuine tsunamis of volcanic origin which he attributes to 10 causal mechanisms. Earthquakes accompanying eruptions, pyroclastic flows, and submarine explosions each account for 20% of all volcanogenic tsunamis. Caldera collapse is responsible for about 10%, avalanches of cold rock and of hot materials about 7% each, with lahars, air waves and lava avalanching of minor extent.

Undoubtedly the most calamitous volcanogenic tsunami in recorded history occurred during the eruption of Krakatoa on August 27, 1883. The eruption entered its paroxysmal phase on August 26, with early small explosions generating small tsunamis that reached the Javan and Sumatran shores 40 km away shortly afterwards. This gave some coastal residents the opportunity to escape to higher ground inland. On the morning of August 27, two large waves swept along the Sunda Strait shorelines to be followed some time after 10 am local time by a gigantic wave or waves that inundated the nearby coasts (Self and Rampino, 1981). About 36,380 lives were lost (Judd 1888 in Simkin and Fiske, 1983, p. 303), mostly from drowning. The event is poorly recorded due to the few survivors and the poor visibility from falling tephra. However what is apparent is that at least one tsunami swept inland to heights of 40 m in the vicinity of Merak at the northwestern tip of Java (Wharton, 1888 in Simkin and Fiske 1983, p. 376). Photographs of the tsunami destruction are figured by Simkin and Fiske (1983, pp. 124–5) and in Tenison-Woods' words "It was a considerable time before the coastline could be visited and an approximate estimate of the damage done made. Even then the report was a series of negatives. Nothing was left. Not a house, scarcely a tree, not a road. All the divisions between the fields were obliterated, and the boundaries of properties destroyed. In fact, no one can tell where a house or a property stood." (Tenison-Woods, 1884 in Simkin and Fiske 1983, p. 113).

On the Sumatran coast the steamship Berouw was anchored in Telok Betong Harbour. Initially beached by an early wave it was later transported by the largest wave 2.5 km inland (3.3 km from its original anchorage) up the valley of the Koeripan River where it came to rest lying across the banks of the river (an engraving is figured in Simkin and Fiske, 1983, p. 129).

The origin of the tsunamis was likely to be caused by one of three mechanisms. The first was large-scale collapse of the northern part of Krakatau Island as part of a caldera-forming process (Verbeek, 1885 in Simkin and Fiske, 1983). The second is the discharge of subaerially generated pyrolastic flows. Self and Rampino (1981) have demonstrated the sequence and timing of the tsunamis were generated at the times of the explosions. They propose that column collapse produced pyroclastic flows that entered the sea within about 30 seconds of the explosions. About 12 km^3 of unwelded ignimbrite was violently emplaced into the sea for up to 15 km from source and this produced the catastrophic tsunami('s) responsible for the majority of casualties. Slumping and subsidence may have generated the later smaller tsunamis.

The third suggests a major debris avalanche entered the sea northwards from Krakatau. A hummocky submarine physiography has been cited as strong evidence in support of this theory (Camus et al., 1992). Sigurdsson et al. (1991) favour the pyroclastic flow origin because the erupted volumes closely approach the volume of the submarine caldera.

The second highest loss of life from volcanogenic tsunamis occurred on Kyushu, Japan in 1792. Known as the Shimabara Catastrophe in old Japanese manuscripts, the disaster began in the Unzen Volcano complex on the Shimabara Peninsula. On May 21, 1792, six months after the Unzen Volcano had commenced eruption in the west, a major collapse occurred on the coastal flank of the peak called Mayuyama, 4 km east of the previously active vents. There was no snow about and no heavy rains were recorded beforehand (K. Kobayashi, pers. comm., 1976). The collapse was caused either by a strong swarm of volcanic earthquakes (Ota, 1969, Latter, 1981) or a volcanic eruption (Furuya, 1974) or by saturation of the flank by hydrothermal waters preceding magma movement along an inclined plane (Ota, 1972,1973, Katayama, 1974). Katayama (1974) also considers a receding tide may have triggered the collapse, the weakening of Mayuyama's flank having been caused by saturation from escaping groundwater. Progressively shallowing earthquakes migrating eastward are considered to have ceased at the time of failure (Katayama 1974). At 8 pm on 21 May, two intense earthquakes were recorded when the flank then failed accompanied by loud rumblings. This created a 1km wide, 0.34 km^3 avalanche which swept into the Ariake Sea extending the shoreline 870 m seaward and forming between 200 and 300 new islets at the time (Anon, 1926). According to Aida (1975) the avalanche took 2–4 minutes to slide into the sea, reaching a maximum speed of 20 m/s. It appears that parts of the flow (probably on the margins) did not travel so rapidly and one resident was carried eastward about 1km together with his cottage, over a period of 3.5 hours (K. Kobayashi, pers. comm., 1976). Large quantities of water are recorded to have poured from the collapse scarp (Katayama, 1974). About 80% of the volume of the avalanche entered the sea and 3 ensuing tsunamis devastated much of the Shimabara coastline for a distance of 77km; 9745 persons were killed here and 707 wounded (Omori, 1907). However, the second tsunami which was about 10 m high (Omori, 1907, Aida, 1975) also traveled 20 km across the Ariake Sea inundating the shoreline of the Higo Province where a further 5100 persons lost their lives. In addition 343 people drowned in the Amakusa Islands (Omori, 1907).

Other similar disasters have also occurred elsewhere in the western Pacific. At Paluweh Island in Indonesia on August 4, 1928, a large landslide was initiated that entered the sea. Three waves, 5 to 10 m in height killed at least 160 people (Neumann van Padang, 1930 in Latter, 1981, p. 484). Another landslide entered the sea from the coastal volcano of Ili Werung, also in Indonesia, on July 18, 1979. The landslide was about 0.05 km^3 in volume and created a tsunami up to 9 m high which devastated four villages and caused more than 500 deaths at Lomblen Island (Latter, 1981, Siebert et al., 1987).

Clearly, steep-sided coastal volcanoes have been responsible for several highly destructive tsunamis. Despite isolated position or small volume they must be considered in any volcanic risk assessment where volcanogenic tsunamis could be generated.

One final form of volcanogenic tsunami is that recorded from lacustrine bodies of water. At Lake Taal in the Philippines, several eruptions generated from Volcano Island have created base surges that in turn formed tsunamis. Such waves were formed in at least five major eruptions, dated September 24, 1716, August 11, 1749, November 28, 1754, January 30, 1911 and September 30, 1965, all of which caused casualties around the lake shore (Latter, 1981).

12.7 Release of Gases from a Crater Lake

On August 21, 1986, at about 21:30 hrs local time, Lake Nyos in Cameroon suddenly released an estimated 240,000 tonnes of CO_2 (Giggenbach, 1990) which swept down adjoining valleys to distances of 25 km (SEAN 1986, Sigvaldason, 1989). Near to the lake survivors reported hearing a distant sound followed by a warm sensation and a smell of rotten eggs or gunpowder before they lost consciousness. Freeth (1990) however believes these reports may be false and were in fact illusionary or may have resulted from confusion with the taste of droplets in the aerosol. At least 1746 people were asphyxiated (Smolowe, 1986). The gas cloud uprooted trees and flattened vegetation. Over 3000 cattle were killed and the distribution of their dead bodies indicated the lethal CO_2-charged cloud had extended up to 120 m above the lake surface as it moved out of the crater over low spots in the rim. A drop in lake level of about 1m indicated a sudden loss of volume. Further, the surface waters to 10 m depth had turned from their normal clear blue colour to a dull red colour, attributed to the oxidation of iron hydroxide from the clear anoxic waters below (Kling, et al., 1987).

Lake Nyos occurs in a linear zone of crustal weakness known as the Cameroon Line that extends over 1400 km length and has the potential for future activity. The Lake occupies a maar or shallow explosion crater about 1.48 km^2 in area, which may have formed as recently as a few hundred years ago. It is 1925 m long, 1180 m wide and 208 m deep, and has a volume of 0.17 km^3. It is surrounded by steep walls on most sides, with a smooth flat floor and a natural spillway at its northwestern end (Walker, et al., 1992).

Three principal theories have been proposed to explain the unusual phenomenon experienced by the local residents. The first attributes the release to a large volume of hot volcanic gas emanating from beneath the lake causing a phreatic eruption (Tazieff, 1989, Barberi et al., 1989). This theory considers there is a deep-seated magma reservoir from which volatile components are exsolving as a gas phase. Carbon dioxide has a low solubility in magma, so is released in relatively large amounts during the early stages of degassing. Being readily soluble in water, the CO_2 probably dissolves in groundwater during its upward path to the lake bed. One observation in favour of a volcanic trigger was an elevated lake temperature, considerably above mean temperature soon after the event (Sigvaldason, 1989). However this can probably be explained by an addition of thermal spring waters in the bottom of the lake without invoking a volcanic eruption (Kling et al., 1989).

The second explanation considers the CO_2 gas to have slowly built up in solution within the lake and was released at or below ambient temperature, when stratification of the lake became disturbed (Kling et al., 1987, Kusakabe et al., 1989). Rapid exsolution of large amounts of CO_2 led to a burst of gas leaving the lake and resulting in major surface water waves. These waves rose to 25 m around the shoreline and on one promontory partly stripped vegetation and soil up to 80 m above the lake (SEAN, 1986, Sigvaldason, 1989).

Analysis of the deep waters in Lake Nyos show that below the stable surface layer the waters are about 20% saturated with CO_2 (Sigvaldason, 1989). This is due to the hydrostatic pressure enabling 20 times as much CO_2 to be dissolved as in surface waters. No typical volcanic gases have been detected. The CO_2 was at least 100 times more abundant relative to sulfur than in typical volcanic gases. However a $^3He/^4He$ ratio of 6, and ^{14}C and $\delta^{13}C$ data are consistent with a magmatic source. But no clear evidence of magma movement has been identified. Thus a subterranean non-volcanic source of CO_2 seems apparent, probably via CO_2-laden thermal spring water (Kling et al., 1989). The second explanation suggests the deep CO_2-saturated waters were disturbed, raising them to a point where a pressure reduction allowed the gas to escape from solution and bubble to the surface. Either by drag or further reduction of the weight of water above, the process quickly led to a runaway release of gases (Kerr, 1987).

The cause of the disturbance to the deep bottom waters is subject to conjecture. No earthquakes were reported in the area at the time and a 6 month seismic survey after the disaster revealed no significant crustal earthquakes beneath or close to the lake (Walker et al., 1992). Strong winds have been suggested as a mechanism that started water sloshing within the lake causing disturbance at depth. Heavy rain has also been suggested, leading to a cooling of surface waters that became denser than the bottom waters eventually causing convective overturn (Kling, 1987, Giggenbach, 1990). Another possible cause is rockslides into the lake. The seismic network established around the lake by British scientists after the disaster has detected many seismic events which can be explained as rockfalls and occasional landslides into the lake (Walker et al., 1992). Finally, it has also been suggested that an increased influx of high salinity water displaced the CO_2-saturated bottom waters upwards causing rapid exsolution of CO_2 bubbles (Kusakabe et al., 1989).

A third explanation occupied an intermediate position between the two other theories. This suggested that although the bulk of the gas had come from within the lake, its release was triggered by a gas release through the lake bed (Freeth and Kay, 1987). This suggestion has now been abandoned by its authors (Freeth, 1991).

Following a conference in Nancy, France in September, 1990, it seems a broad consensus of opinion favours the second explanation (Freeth et al., 1990). The undisturbed nature of the bottom sediment, the clear water at depth, only slightly elevated temperatures, the absence of acid gases and the low sulfur nature of the system point against a volcanic event having taken place (Kling et al., 1989). Nojiri et al. (1990) have also demonstrated that the lake is being fed by an influx of warm CO_2-enriched water. What is now clearly established is that Lake Nyos currently contains about 300 million m^3 of CO_2 and this gas is being added to the lake at a rate of 5 million m^3/yr (Freeth et al. 1990). Thus the bottom waters are becoming increasingly saturated with gas over time and there is an increasing risk of another CO_2 release causing a further disaster in the future. These revelations have shifted attention to the urgent need to reduce the CO_2 content of the deep bottom waters of the Lake. One method for controlled degassing would be to install pipes and remove gas-rich bottom waters (Freeth et al., 1990). Another might be to lower the lake level to reduce the hazard (Walker et al., 1992), but care would be needed to prevent triggering another fatal gas release. A third method suggested is to drill from the outside of the lake into the hydrothermal system beneath. This could intercept the CO_2-rich waters and reduce the CO_2 input into the lake. Concerns have also been expressed about the stability of the natural weir governing the lake level. If there should be signs of potential collapse, which could endanger populations downstream in Nigeria, then grouting the rock sill or lowering the lake level may be necessary (Sigvaldason, 1989).

Lake Nyos is not the only lake to show signs of rapid CO_2 degassing. Two years earlier on August 15, 1984, a gas outburst from Lake Monoun, 95 km to the south of Lake Nyos, killed 37 people (Sigurdsson et al., 1987). In fact, there are legends of three earlier cases of the lake exploding and causing mass deaths (Kerr, 1987). The similarity in timing between the two Cameroon disasters in the month of August, has led to the suggestion that these lakes become unstable at this time of year. Perhaps in the month of August the surface waters are more likely to mix with the anoxic deeper waters beginning a circulation in the lake that leads to convective overturn and CO_2 release.

12.8 Non-Volcanic Initiated Collapse of a Crater Lake

The best documented example of this type of volcanic/hydrologic disaster is the December 24, 1953 Tangiwai Disaster at Mt Ruapehu in New Zealand. Mt Ruapehu was in a period of volcanic quiescence at the time. It appears that part of the ice or ash barrier forming the lowest rim of the crater lake suddenly collapsed. An ice cave draining the lake rapidly enlarged to a 30 m high and 45 m wide tunnel at its entrance, allowing 340,000 m^3 of water to drain from the lake. This formed a lahar that filled the Whangaehu River channel about 32 m across to a depth of about 7 m. Travelling at an average velocity of 16 km/h the flow reached the Tangiwai railway bridge at about 10.20 pm coinciding with the crossing of the major night express from Wellington to Auckland. With an estimated peak discharge of 850 m^3/s, the flow caused the failure of the bridge piers (Healy, 1954). The railway engine was half way across when this happened and with its momentum plunged into the far western bank whilst 6 of the 9 passenger carriages fell into the river. One carriage was carried 2.4 km downstream where one of the bogies remains rusted on the river bank today. The casualties totalled 151 persons. Many of the survivors were covered in and had inhaled fine sulfurous silt which rescuers described as luminescent in the torch light.

By morning the destruction was apparent. Five tonne concrete blocks from the railway bridge had been carried 140m downstream and a 125 tonne pier had been shifted 64 m to a position where its centre of gravity was nearly 2 m higher than it was originally (Stilwell et al., 1954). This was New Zealand's worst rail disaster and it led to the installation of a lahar warning system upstream of the bridge which continues to operate today.

A similar type of event seems to have occurred on the extinct volcano, Agua, in Guatemala on September 29, 1541. It appears that a crater lake had either been in existence (Coleman, 1946) or had rapidly formed in the crater after days of torrential rain (Mooser et al., 1956). The crater wall suddenly collapsed in a torrent that created a huge ravine still preserved today (Cotton, 1944). Of historical importance was the devastation caused in Santiago de los Caballeros (now known as Ciudad Vieja), Guatemala's capital city at the time. Here over 700 Spaniards and 600 Indians were drowned leading to the capital being shifted to Antigua. Amongst the victims was Donna Beatrix de la Cueva, the widow of Pedro Alvarado, who was the first woman to head a government on the American continent (Kalijarvi, 1962).

Of the many lahars at Mt Kelut in Indonesia, the 1875 lahars were triggered by non-volcanic collapse of the crater lake.

Of major concern from this type of lahar event is the unexpected and unpredictable nature of its occurrence. It offers little warning for widespread evacuation of nearby civilians.

12.9 Heavy Rains on Recently Erupted Materials

After most eruptions, the slopes of stratovolcanoes are often mantled by voluminous loose pyroclasts. Heavy rain, particularly prevalent in tropical climates, has the capacity to remobilize these pyroclasts into lahars that transport volcaniclastic materials considerably farther away from the source area. Thus if recently active volcanoes are steep and high, rain-triggered lahars are an expected hazard in the post-eruptive period. In syn-eruptive and immediate post-eruptive time hot lahars can be generated from the accumulations of hot pyroclasts (primary lahars). With increasing time these deposits cool and cold lahars become prevalent, especially in the monsoon season of low latitudes (secondary lahars, rain lahars or murgangs). Such lahars may occur for years after an eruption.

One of the best documented recent examples of rain-triggered lahars comes from the aftermath of the June, 1991 eruptions of Mt Pinatubo in the Philippines. After a two-month preeruptive "seismic crisis" a series of paroxysmal Plinian eruptions occurred between June 12-16, 1991.

Pumiceous pyroclastic flows radiated outwards from the crater creating a hot pumice apron that infilled valleys to a total volume exceeding 6 km^3. An additional 0.2 km^3 of tephra was spread across the surrounding districts. On the volcano's slopes this led to a decrease in infiltration capacity and evapotranspiration. Thus when it rained there was an increase in the rate and magnitude of surface runoff. Both during the eruption and subsequently, typhoons were passing over the Philippines and their major impact was to remobilise huge volumes of the highly erodible pumice into lahars that inundated riparian strips on the surrounding lowlands. After a typhoon on July 18, 1991, 3 m-deep lahars spread up to 4 km across fan surfaces, burying homes, removing bridges and destroying valuable food-producing land. By July 26, 5 m-deep lahars had rushed through the streets of Concepcion, sweeping away several people. Then on August 20, heavy rains created lahars that caused 31 casualties. One pyroclastic flow dammed the pre-existing drainage, behind which a lake began to form. On September 7, the dam broke causing further lahars that destroyed 800 homes and killed seven. In total between June 12 and September 10 over 200 lahars were recorded. Of the 722 lives lost during the eruption and in the first 12 months after, 83 casualties were attributed to lahars (U.S. Dept. of Commerce 1992). After the first monsoon season 150 km^2 of land had been buried under several metres of coarse lahar deposits (Pierson, et al., 1992). During the 1992 monsoon season at least another 60 persons were killed by lahars, 50,000 fled their homes and over 1700 houses were destroyed (SEAN, 1992). Pierson et al. (1992) believe the first lahars were triggered from fine-grained ash-cloud deposits becoming water saturated and liquefying *en masse* from the slightest disturbance. This was accompanied by vigorous fumarolic activity and violent steam explosions due to rainwater seeping into the hot pyroclastic-flow deposits. In August, 5 to 10 explosions per day ejected ash to 7 km height and with accompanying rains generated more lahars. Most were triggered by intense rainfall but some were caused by breakouts from transient lakes created by debris dams. The magnitude of these lahars usually peaked at 1000 m^3/s but in late July some reached 5000 m^3/s. Many were up to 5 m deep and travelled at up to 11 m/s. Some reached temperatures of 70°C when derived directly from pyroclastic-flow deposits.

To provide an immediate response warning system, the Philippine Institute of Volcanology and Seismology installed telemetered rain gauges and flow sensors to relay information to the Pinatubo Volcano Observatory at the Clark Air Base. Lahar watchers were deployed and trip wires were installed to further warn residents of impending lahars in highly populated regions. With rapid post-eruptive aggradation in all major river channels on slopes of low gradient, a major problem has been one of reduced cross-sectional area for transportation of channel sediment. This increased the attendant risk of spillover from channels during high rainfall events. In an attempt to safeguard residents along the Pasik-Protrero River and at Porac, major excavation of the former river channel and construction of embankments with the lahar sediment has afforded some protection (Fig. 12.2). Inevitably these works will fill and become buried if maintenance does not continue. A solution needs to be found for the disposal of these immense volumes of sediment. One method is to create sediment-retention basins. This involves guiding lahar flows into settlement basins where the deposits can be later extracted for aggregate. Such

Fig. 12.2. Aerial view northwards up the Pasig-Protrero River; Mt Pinatubo in top left background. Lahars in 1991 spilled from the river to inundate homes and agricultural land (to right) and flow through Bacolor. Major embankments were constucted bordering the river channel (towards top left) to lower the lahar risk.

methods have proved successful in Indonesia. Unfortunately, the high population density around Mt Pinatubo has prevented this from happening, principally because no-one is prepared to forfeit their land for this purpose. The sediment will thus continue to accumulate and the problem will accentuate.

Pierson et al. (1992) estimate that 40–50% of the 1991 pyroclastics will be eroded before erosion rates return to normal. This implies between 1.2 and 3.6 × 10^9 m^3 of sediment could inundate the adjoining lowlands over the next decade; about 75% of this is expected to occur to the west of Mt Pinatubo.

Similar lahars have also been recorded on the island of Luzon in the Philippines at Mt Mayon. From 1616 to 1981, 29 lahar events were sufficiently devastating to be recorded in historical records; 21 events occurred during eruptions and 8 during the passage of typhoons in post-eruptive periods. More human fatalities have been associated with the latter (Umbal, 1986).

In 1766, 46 people were killed by lahars. On February 1, 1814, 720 persons lost their lives from lahars mainly in the towns of Cagsaua and Budiao (Blong, 1984). Two years after the 1871–73 eruptions, heavy rains triggered lahars that killed a further 1500 people. Major damage or loss of life due to lahars subsequently occurred in 1915, 1952, 1968 and 1981 (Alcaraz, 1968, Arguden and Rodolfo, 1990).

Early on the morning of September 23, 1984, a new eruptive phase began producing voluminous pyroclastic flows. Hot lahars were an immediate result, burying 47 hectares up to 4.2 m deep. Some lahars reached the coast at temperatures of about 50°C (Arguden and Rodolfo, 1990). Despite the eruption declining on September 25, sufficient debris had been produced to generate secondary lahars for years to come. Soon, two 5km long channels on the southeast margin became the main lahar conduits from the summit ravine. The southernmost of these is named the Mabinit channel. It terminates close to the highly populated cities of Legaspi and Daraga where a half of the 1 million inhabitants living within a 14 km radius of the summit dwell. By the end of 1984 an area of 3.9 km^2 had been inundated by 10×10^6 m^3 of sediment deposited by 41 lahar events. Lahar velocities were calculated between 2.0–5.6 m/s with peak discharges ranging from 71 to 2,584 m^3/s. About 8000 hectares of agricultural land and 158 houses were destroyed. Damage to infrastructural facilities amounted to about US$ 2M. Fortunately due to efficient evacuation measures only one life was lost (Umbal, 1986).

In 1985 a further 29 lahar events occurred. This dropped to nine events in the 1986 lahar season, two in 1987 and two in 1988 (Arguden and Rodolfo, 1990). Rodolfo (1986) has monitored the rain lahars over these monsoon seasons and mapped the resultant 1984–86 deposits. His results suggest rainfall in excess of 1mm/min lasting for more than 20 minutes is sufficient to trigger lahars, providing there has not been a prolonged dry period beforehand.

Methods currently employed to mitigate the impact of rain lahars in the Philippines have included evacuation of lahar prone areas, creation of protective embankments and the establishment of effective warning systems. Detecting the initiation of lahars from seismographs (Bautista, et al. 1986) offers a potential 15 to 30 minute warning for residents of high risk zones.

Rain lahars are also known from many other stratovolcanoes along the Pacific rim where disasters have been avoided by establishing efficient warning systems. For example, at Mt Merapi in Indonesia, viscous lava domes are often extruded and frequently collapse. The fallen materials are loose and unstable and frequently are remobilised into lahars during the monsoon season (Verstoppen, 1992). After 75 mm of rain within 35 minutes, a lahar may be generated and warning systems activated (van Bemmelen, 1949).

After the March 17, 1963 eruption of Mt Agung on Bali, Indonesia, heavy rains began to fall. Torrents of sediment-charged water swept downstream incorporating loose material from along their channels to devastate villages, bridges and roads. At river bends the lahars ploughed straight ahead. On March 21, one such lahar flowed into the village of Subagan where many had flocked into a mosque for protection. About 200 people lost their lives here and about 20 survived in the ceiling where a small portion of the building remained intact (Zen and Hadikusumo, 1964).

More recently in Indonesia in 1976, lahars were triggered by rains at Semeru following numerous ash eruptions. Forty persons died and 258 houses were destroyed in Sumber Wuluh village, by lahars following heavy rains (Sudradjat and Matahelumual, 1978).

Another disaster probably of this type occurred at Unzen Volcano, Japan following the eruptions of 1657. The following year 30 people lost their lives when 2 villages were swept away along Antoku-gawara (Anon 1926). More recently in 1974, at Sakurajima also on Kyushu, rains triggered lahars that killed 3 persons on June 17 and 5 persons on August 9 (Suwa and Ohura, 1976).

On the other side of the Pacific Ocean is Irazu Volcano in Costa Rica. Renewed volcanic activity began here in March 1963, leading to copious quantities of volcanic ash being spread throughout nearby areas. The ash destroyed the vegetation cover and then hardened to form an impervious crust which led to markedly increased runoff and flash floods. A combination of easily erodible, low shear-strength materials beneath the crust, high water discharges and high velocity flows of up to 10 m/s resulted in an exceptional eroding capacity of streams draining the volcano (Waldron, 1967). On December 9,, 1963, a high intensity cloudburst created a large lahar in the Reventado River which destroyed more than 300 houses and killed more than 20 persons at Cartago (Ulate and Corrales, 1966, Waldron, 1967). Over 1,000 persons escaped to safety due to a flood warning from the Civil Guard (Murata et al., 1966). Over 2.8 km^2 of land was devastated and damage was estimated at US\$ 3.5M. Economic losses were estimated at US\$ 15M.

The lahar flow reached a maximum height of 12 m but around Cartago spread laterally as a <2 m thick deposit. The flow was composed of about 35% sediment by volume and was capable of shifting boulders of an estimated 200 tonnes weight. It was calculated that the discharge of this lahar reached 407 cumecs, approximately 29 cumecs per km^2 of catchment. Control measures tackled the problem from two angles. First, soil conservation measures commenced that involved intensive planting, contour ditching of slopes and construction of check dams in the upper catchment to reduce surface runoff. Second, protective engineering works were constructed close to the populated area of Cartago. This involved the creation of artificial levees over a total length of 12.5km to contain the Reventado River. By March 1965 volcanic activity had ceased.

In nearby Guatemala, heavy rains triggered lahars from recently erupted ash and lapilli on Fuego in September 1963. One of these killed 7 persons on September 30 (Bohnenberger et al., 1971).

Immediately prior to the climactic eruption of Mt Pelée in 1902, thunderstorms lashed the area and mixed with ash on the upper slopes of the mountain. Between 2 and 5 am on May 8, the villages of Basse-Pointe, Macouba and Grand-Rivière were seriously damaged. But the worst lahar damage occurred to the west where about 400 people were killed in Le Prêcheur (Tanguy, 1994). A few hours later the 28,000 citizens of Saint-Pierre met their fate in the infamous nuée ardente.

It is worth reflecting on the point that the most ancient recorded lahar in historical literature was probably triggered by syn-eruptive rains on the flanks of Vesuvius in 79 A.D. Pouring down the western slopes as a "torrid, treacly river" it entombed the town of Herculaneum to a depth of 15–18 m. The debris cooled to a rock-like consistency upon which the current city of Resina is now sited. The effects of this lahar on the buildings was like a bulldozer action. Where structures withstood the onslaught the mud slowly filled the rooms without disturbing delicate items such as a cradle in a nursery, pots on a stove and a chicken prepared for lunch. The lahar was sufficiently hot to carbonise wood and scorch cloth and papyrus. Few bodies have been found at Herculaneum suggesting most inhabitants had sufficient warning to escape (Grant, 1971).

Over 15 centuries later, further lahars at Vesuvius created death and destruction. On December 16, 1631, strong explosions spread tephra onto inhabited villages. The next day two fissures opened on the southwest flank releasing lava that flowed to the sea. That evening it began to rain and mudflows invaded S. Giorgio a Cremano, Portici and Resina (formerly Herculaneum). By the eruption episode's termination in January 1632, nine towns had been wrecked by lahars triggered by rains on the recently deposited ash (Bullard, 1962). About 1750 lives were lost to lahars, constituting about one half of the total casualties of the eruption (Blong, 1984). As recently as May 1906, 2 persons were killed when rain on the porous upper flanks triggered lahars that inundated nearby cultivated and inhabited land (Perret, 1924).

12.10 Conclusion

In recorded history well over 116,000 people have lost their lives to the interplay between volcanic activity and water.

Due to inadequate historical records this figure is likely to be half the true record for the last five centuries. Almost half this total (45%) were victims of volcanogenic tsunamis. One fifth were casualties of eruptions through a crater lake and another fifth can be attributed to lahars generated by melting of snow and ice. For the last five centuries, 40% of the total casualties were recorded in the 19th century. This compares with nearly 30% for the 20th century and 20% for the 18th century. In contrast the 17th century showed a particularly low number of casualties compared with the preceding and succeeding centuries. Thus while hydrological disasters associated with volcanoes occur frequently, occasional large losses of life occur at irregularly intervals between centuries.

About 80% of the disaster casualties were unavoidable due to poor scientific knowledge at the time. Nowadays however, science has advanced to the stage where it can provide forewarning of impending volcanic activity and it may be possible to recognise in advance disastrous outcomes from the volcanic interaction with groundwater, lake water, sea water or heavy rains. It behoves the civil authorities in all countries to take positive action and respond to such warnings if loss of life and damage to property is to be minimized. Besides preparing action plans, authorities should support public education about volcanic hazards to better prepare civilians for their own knowledgable response in an emergency.

Acknowlegements

This review is dedicated to the memory of Dr S. Thorarinsson, Iceland and Professor K. Kobayashi, Japan, for their invaluable and willing assistance in providing me with further information about some of the disasters described. Their interest and helpfulness is most gratefully acknowledged. I also thank Dr J. Lecointre for kindly bringing to my attention a number of references that have clarified issues pertaining to this review.

References

Aida, I., 1975. Numerical experiments of the tsunami associated with the collapse of Mt. Mayuyama in 1792. Journal of the Seismological Society of Japan 28: 449-460.
Alcaraz, A., 1968. Mayon Volcano. Bulletin of Volcanic Eruptions 8: 8.
Alzwar, M., 1985. G. Kelut. Serials of Volcanology (Special edition) of the Volcanological Survey of Indonesia No.108. 60p.
Anderson, C.A., 1933. The Tuscan Formation in northern California with a discussion concerning the origin of breccias. University of California Publications; Bulletin of the Department of Geological Sciences 23: 215-276.
Anderson, T. and Flett, J.S., 1903. Report on the eruptions of the Soufrière in St. Vincent, and on a visit to Montagne Pelée in Martinique. Philosophical Transactions of the Royal Society of London, A, 200: 353-553.
Anon, 1926. Unzen Volcanoes. Guidebook Excursion E-1, 3, 4. Pan-Pacific Science Congress, Japan.
Aramaki, S., 1956. The 1783 activity of Asama Volcano Part I. Japanese Journal of Geology and Geography 27: 189 - 229.
Arguden, A.T. and Rodolfo, K.S., 1990. Sedimentologic and dynamic differences between hot and cold laharic debris flows of Mayon Volcano, Philippines. Geological Society of America Bulletin 102: 865-876.
Barberi, F., Chelini, W., Marinelli, G. and Martini, M., 1989. The gas cloud of Lake Nyos (Cameroon, 1986): results of the Italian technical mission. Journal of Volcanology and Geothermal Research 39: 125-134.
Bárdarson, H.R., 1971. Ice and Fire. H. Bárdarson. Reykjavík, Iceland.
Bautista, B.C., Bautista, L.P. and Garcia, D.C., 1986. Seismic monitoring: a useful tool for mudflow detection at Mayon Volcano, Albay, Philippines. Philippine Journal of Volcanology 3: 90-108.
Blong, R.J., 1984. Volcanic Hazards. Academic Press Australia, North Ryde. 424 p.
Bohnenberger, O.H., Bengoechea, A.J., Dondoli, C. and Castro, A.M., 1971. Report on Active Volcanoes in Central America during 1957-1965. Bulletin of Volcanic Eruptions 9: 19p.
Bonis, S.B., 1973, Description of volcanic eruptions - Guatemala. Bulletin of Volcanic Eruptions 11: 35.
Bullard, F.M., 1962. Volcanoes, in history, in theory, in eruption. University of Texas Press, Austin. 441p.
Calvache, V.M.L., 1990. Pyroclastic deposits of the November 13, 1985 eruption of Nevado del Ruiz, Colombia. Journal of Volcanology and Geothermal Research 41: 67-78.
Camus, G., Diament, M., Gloaguen, M., Provost, A. and Vincent P., 1992. Emplacement of a debris avalanche during the 1883 eruption of Krakatau (Sunda Straits, Indonesia). GeoJournal 28: 123-128.
Chrétien, S. and Brousse, R., 1989. Events preceding the great eruption of 8 May, 1902 at Mount Pelée, Martinique. Journal of Volcanology and Geothermal Research 38: 67-75.
Coleman, S.N., 1946. Volcanoes New and Old. The John Day Company, New York. 222 p.
Cotton, C.A., 1944. Volcanoes as landscape forms. Whitcombe and Tombs, Christchurch, N.Z. 416p.

Costa, J.E., 1984. Physical geomorphology of debris flows. pp. 268-317. In: Costa, J.E. and P.J. Fleisher (Editors), Developments and Applications of Geomorphology. Springer-Verlag, Berlin.
Fairchild, L.H., 1987. The importance of lahar initiation processes. In: J.E. Costa and G.F. Wieczorek (Editors), Debris Flows/Avalanches: Process, Recognition, and Mitigation. Geological Society of America, Reviews in Engineering Geology 7: 51-61.
Fletcher, H., 1992. Degassing Lake Nyos. Nature 355: 683.
Freeth, S.J. and Kay, R.L.F., 1987. The Lake Nyos gas disaster. Nature 325: 104-105.
Freeth, S.J., 1990. The anecdotal evidence, did it help or hinder investigation of the Lake Nyos gas disaster? Journal of Volcanology and Geothermal Research 42: 373-380.
Freeth, S.J., 1991. The Lake Nyos disaster: a steadily evolving consensus. Journal of African Earth Sciences 13: 553-555.
Freeth, S.J., Kling, G.W., Kusakabe, M., Maley, J., Tchoua, F.M. and Tietze, K., 1990. Conclusions from Lake Nyos disaster. Nature 348: 201.
Furuya, T., 1974. A geomorpholgical consideration of the Mayu-yama Great Landslide in 1792. Disaster Prevention Research Institute, Kyoto University Annuals, No.17B: 259264. (in Japanese with English abstract).
Gerbe, M-C., Morel, J-M. and Gourgaud, A., 1990. Évaluation des risques volcaniques au Galunggung (Java, Indonésie): apports de l'éruption de 1982-1983. Comptes Rendus de l'Académie des Sciences de Paris, 311, Serié II: 873-878.
Giggenbach, W.F., 1990. Water and gas chemistry of Lake Nyos and its bearing on the eruptive process. Journal of Volcanology and Geothermal Research 42: 337-362.
Glicken, H. and Nakamura, Y., 1988. Restudy of the 1888 eruption of Bandai Volcano, Japan. Proceedings of Kagoshima International Conference on Volcanoes: 392-395.
Glicken, H., Asmoro, P., Lubis, H., Frank, D. and Casadevall, T.J., 1987. The 1772 debris avalanche and eruption at Papandayan Volcano, Indonesia, and hazards from future similar events. Abstracts of International Symposium on How Volcanoes Work. January 19-25, 1987. Hilo, Hawaii.
Gonzalez-Ferran, O., 1973. Description of volcanic eruptions - Chile. Bulletin of Volcanic Eruptions 11: 41-42.
Grant, M., 1971. Cities of Vesuvius: Pompeii and Herculaneum. Penguin Books, Harmondsworth, England.
Guralnik, D.B., 1970. Webster's New World Dictionary. The World Publishing Company, New York and Cleveland.
Hadikusumo, D. 1966. Awu. Bulletin of Volcanic Eruptions 6: 8.
Hall, M.L., 1990. Chronology of the principal scientific and governmental actions leading up to the November 13, 1985 eruption of Nevado del Ruiz, Colombia. Journal of Volcanology and Geothermal Research 42: 101-115.
Hall, M.L., 1992. The 1985 Nevado del Ruiz eruption: scientific, social, and governmental response and interaction before the event. In G.J.H. McCall, D.J.C. Laming and S.C. Scott (Editors), Geohazards Natural and Man-made. Chapman and Hall, London. pp. 43-52.
Haller, D., Miller, D., Jakarta Domestic Radio Service, Agence France-Presse and United Press International, 1989. Galunggung. pp. 205-206. In: McClelland, L., T. Simkin, M. Summers, E. Nielsen and T.C. Stein, (Editors), Global Volcanism 1975-1985. Prentice-Hall, Englewood Cliffs, New Jersey and American Geophysical Union, Washington, DC. 655 p.
Healy, J., 1954. Origin of flood and Ruapehu lahars. In: Tangiwai Railway Disaster. Report of Board of Inquiry. Government Printer, Wellington, N.Z. pp.6-8 and 28-31.
Jaggar, T.A., 1931. Eruption of Santa Maria November 1929. The Volcano Letter 356: 1-4.

Kalijarvi, T.V., 1962. Central America. Land of Lords and Lizards. D. Van Norstrand Co. Inc., Princeton, NJ. 128p.
Katayama, N., 1974. Old records of natural phenomena concerning the "Shimabara Catastrophe". The Science Reports of the Shimabara Volcano Observatory, the Faculty of Science, Kyushu University 9: 1-45. (in Japanese with English abstract).
Katayama, N., 1974. Shimabara Taihen (Shimabara Catastrophe) and Higo Meiwaku (the nuisance to the Province of Higo) Kagaku (Science) 44: 566-570. (in Japanese).
Katsui, Y., 1967. Description of Volcanic Eruptions - Chile. Bulletin of Volcanic Eruptions 6: 8p.
Kawachi, S., 1988. Denial of the 887 "Shinano-Hokubu Earthquake (M= 7.4)" from the 888 Ohtsukigawa Debris Avalanche in the Yatsugatake Volcanic Chain, Central Japan. Proceedings of Kagoshima International Conference on Volcanoes: 471-474.
Kerr, R.A., 1987. Lake Nyos was rigged for disaster. Science 236: 169-174.
Kling, G.W., 1987. Seasonal mixing and catastrophic degassing in tropical lakes, Cameroon, West Africa. Science 237: 1022-1024.
Koenigsberg, E.J., Lockwood, J.P., Tuttle, M.L. and Wagner, G.N., 1987. The Lake Nyos gas disaster in Cameroon, West Africa. Science 236: 169-175.
Kling, G.W., Tuttle, M.L. and Evans, W.C., 1989. The evolution of thermal structure and water chemistry in Lake Nyos. Journal of Volcanology and Geothermal Research 39: 151-165.
Kuno, H., 1962. Catalogue of the active volcanoes of the world including solfatara fields. Part XI - Japan, Taiwan and Marianas. 332p.
Kusakabe, M., Ohsumi, T. and Aramaki, S., 1989. The Lake Nyos disaster: chemical and isotopic evidence in waters and dissolved gases from three Cameroonian crater lakes, Nyos, Monoun and Wum. Journal of Volcanology and Geothermal Research 39: 167-185.
Latter, J.H., 1981. Tsunamis of Volcanic Origin: Summary of causes, with particular reference to Krakatoa, 1883. Bulletin Volcanologique 44: 467-490.
Lowe, D.R., Williams, S.N., Leigh, H.H., Connor, C.B., Gemell, J.B. and Stoiber, R.E., 1986. Lahars initiated by the November 13, 1985 eruption of Nevado del Ruiz, Colombia. Nature 324: 51-53.
Lyell, C., 1867. Principles of geology. Vol.2. Murray, London. 659p.
Macdonald, G.A., 1972. Volcanoes. Prentice-Hall, Englewood Cliffs, New Jersey.
Major, J.J. and Newhall, C.G., 1989. Bulletin Volcanologique 52: 1-27.
Mooser, F., Meyer-Abich, H. and McBirney, A.R., 1958. Catalogue of the active volcanoes of the world including solfatara fields. Part VI - Central America. 146p.
Mothes, P.A., 1992. Lahars of Cotopaxi Volcano, Ecuador: hazard and risk evaluation. In G.J.H. McCall, D.J.C. Laming and S.C. Scott (Editors), Geohazards Natural and Man-made. Chapman and Hall, London. pp. 53-63.
Murai, I., 1960. 5. On the mudflows of the 1926 eruption of volcano Tokachi-dake, Central Hokkaido, Japan. Bulletin of the Earthquake Research Institute 38: 55-70.
Murata, K.J., Dondoli, C. and Saenz, R., 1966. The 1963-65 eruptions of Irazu Volcano, Costa Rica. Bulletin Volcanologique 29: 765-796.
Nairn, I.A. 1979. Rotomahana - Waimangu eruption, 1886: base surge and basalt magma. N.Z. Journal of Geology and Geophysics 22: 363-378.
Naranjo, J.L., Sigurdsson, H., Carey, S.N. and Fritz, W.G., 1986. Eruption of Nevado del Ruiz volcano, Colombia, 13 November, 1985: tephra fall and lahars. Science 233: 961-963.
Neumann van Padang, M., 1951. Catalogue of the active volcanoes of the world including solfatara fields. Part I - Indonesia. 271p.

Neumann van Padang, M., 1960. Measures taken by the authorities of the vulcanological survey to safeguard the population from the consequences of volcanic outbursts. Bulletin Volcanologique 23: 181-192.

Nojiri, Y., Kusakabe, M., Hirabayashi, J., Sato, H., Sano, Y., Sinohara, H., Njine, T. and Tanyileke, G., 1990. Gas discharge at Lake Nyos. Nature 346: 322-323.

Omori.F., 1907. Note on the Eruptions of the Unsen-dake in the 4th year of Kansei (1792). Bulletin of the Imperial Earthquake Investigation Committee, 1-3: 142-144.

Oppenheimer, C. 1991. People and volcanoes: Taal Island, Philippines. Geology Today 7: 19-23.

Ota, K., 1969. Study on the collapses in the Mayu-yama. 1. On the mechanism of collapse. The Science Reports of the Shimabara Volcano Observatory, the Faculty of Science, Kyushu University 5: 6-35. (in Japanese with English abstract).

Ota, K., 1972. Hot springs and their relationships to geologic structure and earthquakes in the Unzen Volcanic Region. Journal of the Japan Geothermal Energy Association Ser.No. 34, 9:76-81. (in Japanese with English abstract).

Ota, K., 1973. A study of hot springs on the Shimabara Peninsula. Science Reports of the Shimabara Volcano Observatory, the Faculty of Science, Kyushu University 8: 1-33 (in Japanese with English abstract).

Perret, P.A., 1924. The Vesuvius eruption of 1906 - study of a volcanic cycle. Carnegie Institution of Washington, Washington, D.C. 151p.

Pierson, T.C., Janda, R.J., Umbal, J.V. and Daag, A.S., 1992. Immediate and long-term hazards from lahars and excess sedimentation in rivers draining Mt. Pinatubo, Philippines. U.S. Geological Survey Water-Resources Investigation Report 92-4039.

Pierson, T.C., Janda, R.J., Thouret, J-C. and Borrero, C.A., 1990. Perturbation and melting of snow and ice by the 13 November 1985 eruption of Nevado del Ruiz, Colombia, and consequent mobilization, flow and deposition of lahars. Journal of Volcanology and Geothermal Research 41: 17-66.

Rittman, A. and Rittman, L., 1976. Volcanoes. Orbis Publishing Ltd., London.

Rodolfo, K.S., 1986. Mabinit lahar channel, Mayon Volcano, Philippines. Philippine Journal of Volcanology 3: 73-89.

Russell, G., 1985. Colombia's Mortal Agony. Time (NZ. ed.) 126(21): 12-18.

Scott, K.M., 1988. Origins, behaviour, and sedimentology of lahars and lahar-runout flows in the Toutle-Cowlitz River system. U.S. Geological Survey Professional Paper 1447A. 74p.

SEAN (Scientific Event Alert Network), 1985. Ruiz Volcano. SEAN Bulletin 10 (10): 2-4 and 25-35.

SEAN, 1986. Lake Nyos. SEAN Bulletin 11 (8): 2-5.

SEAN, 1992. Pinatubo SEAN Bulletin 17 (9): 8-10.

Self, S. and Rampino, M.R., 1981. The 1883 Eruption of Krakatau. Nature 294: 699-704.

Siebert, L., Glicken, H. and Ui, T., 1987. Volcanic hazards from Bezymianny- and Bandai-type eruptions. Bulletin of Volcanology 49: 435-459.

Sigurdsson, H. and Carey, S., 1986. Volcanic disasters in Latin America and the 13th November 1985 eruption of Nevado del Ruiz volcano in Colombia. Disasters 10: 205-216.

Sigurdsson, H., Carey, S., Mandeville, C. and Bronto, S., 1991. Pyroclastic flows of the 1883 Krakatau eruption. Eos 72: 377, 380-381.

Sigurdsson, H., Devine, J.D., Tchoua, F.M., Presser, T.S., Pringle, M.K.W. and Evans, W.C., 1987. Origin of the lethal gas burst from Lake Monoun, Cameroun. Journal of Volcanology and Geothermal Research 31: 1-16.

Sigvaldason, G.E., 1989. International conference on Lake Nyos disaster, Yaounde, Cameroon 16-20 March 1987: conclusions and recommendations. Journal of Volcanology and Geothermal Research 39: 97-107.
Simkin, T. and Fiske, R.S., 1983. Krakatau 1883 The volcanic eruption and its effects. Smithsonian Institution Press, Washington, D.C.
Smith, G.A. and Fritz, W.J., 1989. Volcanic influences on terrestrial sedimentation. Geology 17: 375-376.
Smolowe, J., 1986. The Lake of Death. Time (N.Z. ed.) 128 (10): 10-13.
Stilwell, W.F., Hopkins, H.J. and Appleton, W., 1954. Tangiwai Railway Disaster. Report of Board of Inquiry. Government Printer, Wellington, N.Z. 31p.
Sudradjat, A. and Matahelumual, J., 1978. Semeru. Bulletin of Volcanic Eruptions 16: 3132.
Sudradjat, A. and Tilling, R., 1984. Volcanic hazards in Indonesia: The 1982-83 Eruption of Galunggung. Episodes 7: 13-19.
Suryo, I., 1985. Report on the volcanic activity in Indonesia during the period 1964-1970. Bulletin of the Volcanological Survey of Indonesia No.106. 150p.
Suwa, A. and Ohura, E., 1976. Sukura-zima. Bulletin of Volcanic Eruptions 14: 40-41.
Tanguy, J.C. 1994. The 1902-1905 eruptions of Montagne Pelée, Martinique: anatomy and retrospection. Journal of Volcanology and Geothermal Research 60: 87-107.
Tazieff, H., 1989. Mechanisms of the Nyos carbon dioxide disaster and of so-called phreatic steam eruptions. Journal of Volcanology and Geothermal Research 39: 109-116.
Thorarinsson, S., 1956. The Thousand Years Struggle Against Ice and Fire. Museum of Natural History Department of Geology and Geography Miscellaneous Papers No.14. 52p.
Thorarinsson, S., 1958. The Öraefajökull Eruption of 1362. Acta naturalia islandica II, 2: 199.
Thouret, J.C., 1990. Effects of the November 13, 1985 eruption on the snow pack and ice cap of Nevado del Ruiz volcano, Colombia. Journal of Volcanology and Geothermal Research 41: 177-201.
Thouret, J-C., Janda, R.J., Pierson, T.C., Calvache, M.L. and Cendrero, A., 1987. L'eruption du 13 novembre 1985 au Nevado El Ruiz (Cordillère Centrale, Colombie): interactions entre le dynamisme éruptif, la fusion glaciaire et la génèse d'écoulements volcanoglaciaires. Comptes Rendus de l'Académie des Sciences de Paris, 305, Série II: 505-509.
Tobar, A.B., 1973. Description of volcanic eruptions - Chile. Bulletin of Volcanic Eruptions 11: 43-44.
U.S. Department of Commerce, 1992. The June 1991 Eruption of Mount Pinatubo, Philippines. National Geophysical Data Center Product Number 739-A11-007.
Ulate, C.A. and Corrales, M.F., 1966. Mud floods related to the Irazu Volcano Eruptions. Journal of the Hydraulics Division, Proceedings of the American Society of Civil Engineers 92, HY6: 117-129.
Umbal, J.V., 1986. Mayon lahars during and after the 1984 eruption. Philippine Journal of Volcanology 3: 38-59.
van Bemmelen, R.W., 1949. Geology of Indonesia. Vol.1. Government Printing Office, The Hague.
Verstoppen, H.Th., 1992. Volcanic hazards in Colombia and Indonesia: lahars and related phenomena. In G.J.H. McCall, D.J.C. Laming and S.C. Scott (Editors), Geohazards Natural and Man-made. Chapman and Hall, London. pp. 33-42.
Voight, B., 1990. The 1985 Nevado del Ruiz volcano catastrophe: anatomy and retrospection. Journal of Volcanology and Geothermal Research 44: 349-386.

Waitt, R.B., 1989. Swift snowmelt and floods (lahars) caused by great pyroclastic surge at Mount St Helens volcano, Washington, 18 May 1980. Bulletin of Volcanology 52: 138-157.
Waldron, H.H., 1967. Debris flow and erosion control problems caused by the ash eruptions of Irazu Volcano, Costa Rica. U.S. Geological Survey Bulletin 1241-1. 37p.
Walker, A.B., Redmayne, D.W. and Browitt, C.W.A., 1992. Seismic monitoring of Lake Nyos, Cameroon, following the gas release disaster of August 1986. In G.J.H. McCall, D.J.C. Laming and S.C. Scott (Editors), Geohazards Natural and Manmade. Chapman and Hall, London. pp. 65-79.
Whymper, E., 1892. Travels amongst the Great Andes of the Equator. John Lehmann, London.
Williams, H. and McBirney, A.R., 1979. Volcanology. Freeman, Cooper and Co., San Francisco, CA.
Zen, M.T. and Hadikusumo, D., 1964. Preliminary report on the 1963 eruption of Mt. Agung in Bali (Indonesia). Bulletin Volcanologique 27: 269-299.
Zen, M.T. and Hadikusumo, D., 1965. The future danger of Mt. Kelut (Eastern Java - Indonesia). Bulletin Volcanologique 28: 275-282.
Zupka, D., 1993. UN/DHA Action for volcanic disaster mitigation in Ecuador. Disasters 16: 16-17.

CHAPTER 13

Earthquakes

Akira Terakawa and Osamu Matsuo

ABSTRACT. Hydrological consequences of earthquakes are not directly recognized in the first stage of a disaster, but appear in the secondary disasters such as failures of dammed reservoirs and changes of groundwater conditions. The main features of the hydrological effects caused by the earthquakes are classified and their aspects are reported through several examples experienced in Japan.

13.1 Introduction

Earthquakes directly cause disasters in the form of morphological changes, but their hydrological consequences are not directly recognized in the first stage of a disaster. Table 13.1 shows major direct effects of earthquakes and the consequent hydrological effects. After the earthquake, the soil mass supplied by slope failures blockades river valleys and creates dammed reservoirs. They are liable to collapse, which may cause secondary disasters, such as a flooding of a downstream region by a surge wave. Fig. 13.1 shows a chain of hydrological consequences after slope failures caused by earthquakes. Table 13.2 shows major slope failures caused by the earthquakes in Japan. Morphological changes of river courses would affect the hydrological conditions of the region, which include a change of drainage area of surface water and groundwater, a disastrous collapse of slopes loosened by the earthquakes at a time of heavy storms and an increase of sediment discharge caused by the increase of sediment supply from the collapsed area.

In the following section, hydrological consequences are to be introduced through several examples experienced in Japan.

13.2 Example of Hydrologic Consequences of Earthquakes

13.2.1 THE ZENKOUJI EARTHQUAKE

The Zenkouji earthquake, which occurred in 1847 and is said to have caused death of more than 10,000 people and the loss of 20,000–30,000 households, is famous

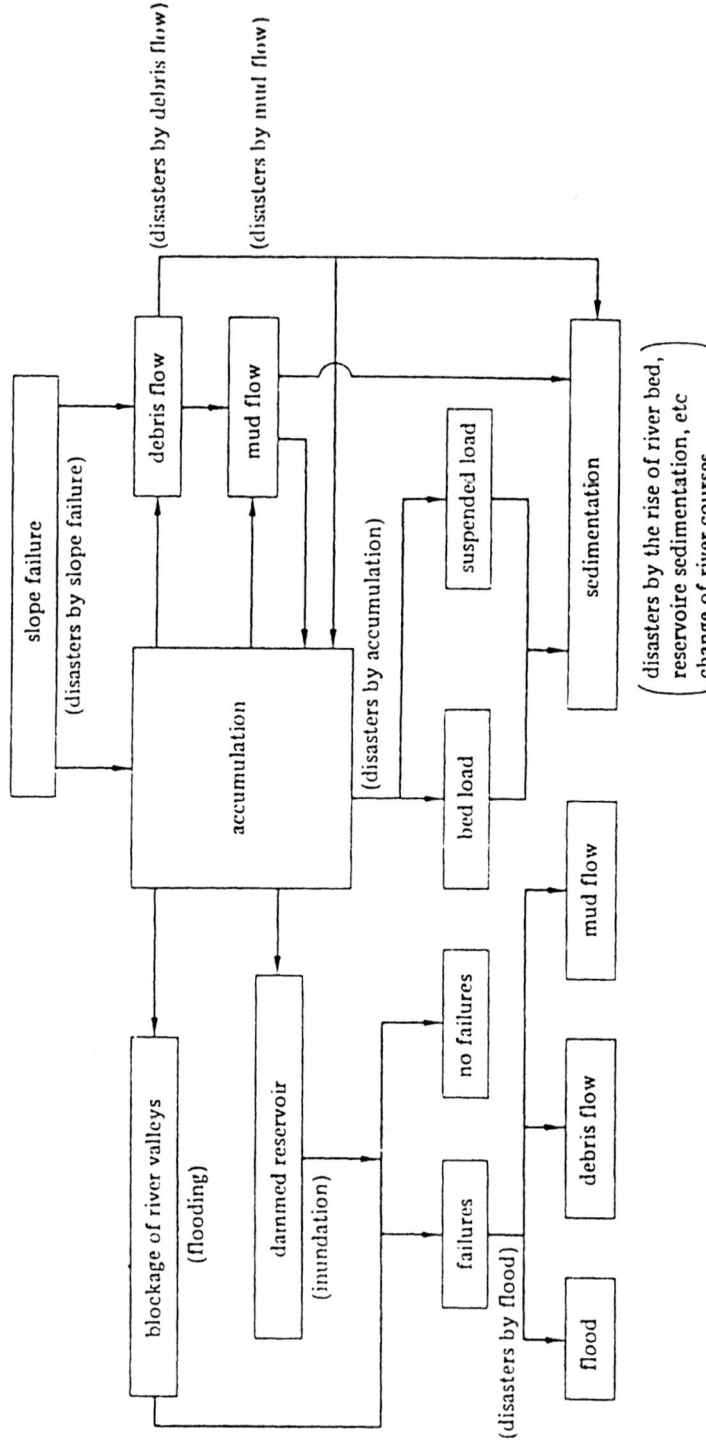

Fig. 13.1. Type of disasters with slope failure caused by earthquakes

TABLE 13.1
Hydrological aspects of earthquakes

Direct effects	Secondary effects	Effects on hydrological phenomena
Occurrence of faults	Discontinuity of soil layers	Change of drainage area
		Change of groundwater level
Slope failures	Blockage of river valleys	Propagation of surge in the valley and inundation
		Failure of dammed reservoir and flooding
	Sedimentation	Occurrence of debris flow
		Rise of river bed
Destruction of buildings	Failure of river banks	Flooding
	Dam failure	Occurrence of surge
Loosened slopes or cracks	Instability of slopes	Secondary slope failure
Soil liquefaction	Abnormality of groundwater condition	Change of groundwater level
Tsunami	Flooding in estuaries	

TABLE 13.2
Huge slope failures caused by earthquakes in Japan

Name of slope failure	Related river	Volume of sand mass (m^3)	Name of earthquakes	Magnitude	Date of occurrence
Oosawa	Huzi	7.5×10^7			19 Aug 1331
Kiunzan	Shou	2.5×10^7		7.9	18 Jan 1586
Ooya	Abe	1.2×10^8	Keichou	7.9	03 Feb 1605
Natale		4.0×10^7	Takata	6.6	20 May 1751
Kokuzouyama	Sai	$>10^7$	Zenkouji	7.4	08 May 1847
Ootonbiyama	Zyouganji	4.1×10^8	Hietu	6.9	09 Apr 1858
Ontake	Kiso	3.6×10^7	Naganokenseibu	6.8	14 Sep 1984

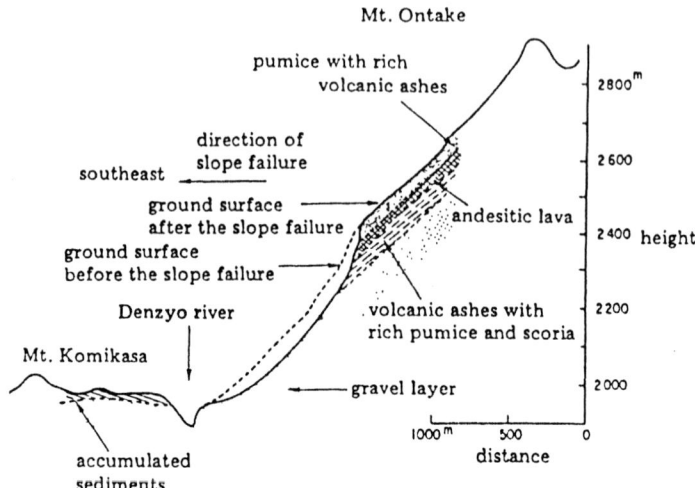

Fig. 13.2. Sectional figure of the slope failure on Mt. Ontake.

for the blockade of Sai River and the secondary disasters caused by the collapse of dammed reservoirs in the river valleys. More than 40,000 slope failures were counted. The largest one occurred on the slope of the Kouzou mountain and a huge amount of sediments accumulated in the Sai River, which is estimated to be about 700 m along its river course with a height of about 100 m and created a dammed reservoir more than 40 km of length. About 30 villages along Sai River were flooded. Three weeks after the earthquake, the dammed reservoir was collapsed by the flowing water caused by severe storms succeeding the earthquake. The stored water flowed into Sai River channel and killed more than 100 people.

13.2.2 THE NAGANOKEN SEIBU EARTHQUAKE

A large earthquake of M 6.8, with its center on the south slope of Mt. Ontake, occurred on September 14, 1984. The earthquake created many slope failures in the Ohtaki River basin. The total amount of sediments were estimated to be about 36 million m^3. The sediments moved down the tributary valleys in the form of debris flow. The survey revealed that the height of debris flow reached more than 100 m above the original river beds along Denzyou River, one tributary of the Ohtaki River. The sediments accumulated in the Nigori River, which has relatively mild river bed slope. The area of the accumulated sediments was estimated to be about 4.1 km long and about 450 m wide at a maximum point. Their depth reached 20–50 m above original river beds.

Fig. 13.2 shows cross sectional view of slope failure at the site of Mt. Ontake. The sediments were widely accumulated along the river courses. Therefore, it was thought that the accumulated river bed was not easily moved and secondary disasters by the collapse of dammed reservoirs might not be expected.

The Government dispatched a survey team which directed necessary preventive measures in order to mitigate secondary disasters. The survey team concluded

Earthquakes

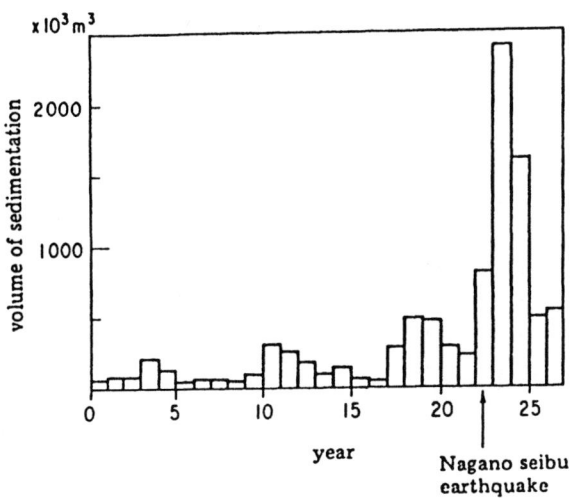

Fig. 13.3. Change of sediment discharge by the earthquake.

that secondary disasters would not be expected from the overstepping of dammed reservoirs, nevertheless discharging of waters from dammed reservoirs and preventive works against the movement of accumulated sediment should be undertaken quickly.

Fig. 13.3 clearly shows the change of sediment discharge after the earthquake in the Makio reservoir, which is located about 10 km downstream from the confluence point of Nigori River to Ohtaki River.

13.2.3 The Matsushiro Earthquake

Frequent earthquakes occurred at Matsushiro, Nagano Prefecture from August 3, 1965 till the end of March 1968. The total number of earthquakes reached about 680,000 including about 61,000 earthquakes recognized by man. The largest one was of M 5.2, which occurred on February 3, 1967. The depth of the earthquake center was estimated at 10 km under the ground surface. These frequent earthquakes did not cause direct damage, but secondary effects caused by the landslide occurred in the suburb of the town of Matsushiro.

The earthquakes caused many cracks in the surface of the ground. Fig. 13.4 shows a general description of surface abnormalities after the earthquakes. Many springs occurred in the plain, but groundwater discharge decreased in many wells in the mountainous area eastward of the Chikuma River.

Groundwater abnormality was recognized from May 1966. At the village of Tanaka, water welled up at many places, from north to south of the village, on May 20, 1966. Until September, welling up could be seen in the villages near Tanaka, discharging a significant amount of water. The quality of the groundwater was normal before the earthquake, but the concentration of CO_2 and Cl gradually increased. Finally, water from hot springs emerged in some places.

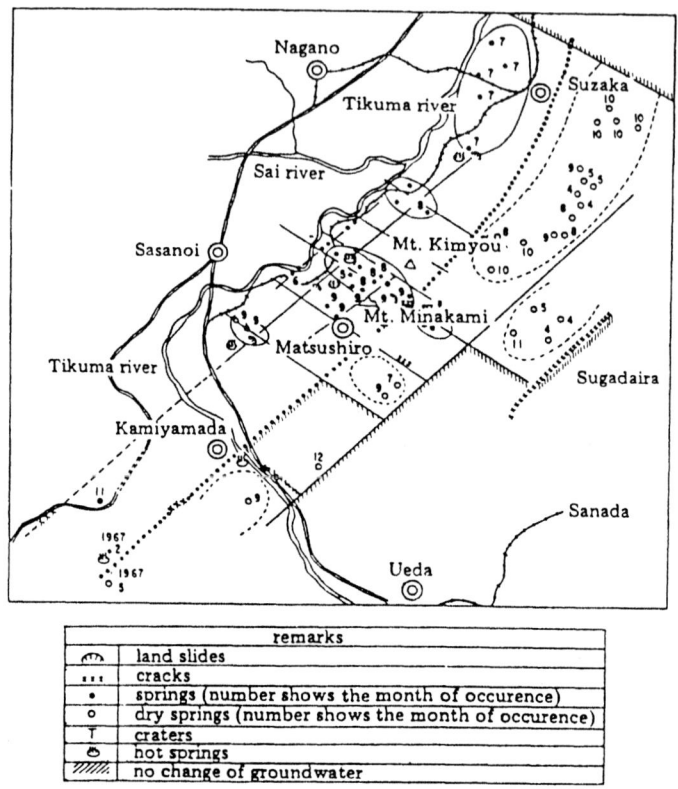

Fig. 13.4. Occurrence of groundwater abnormalities in the Matsushiro region (by Mr. Noboru Yamagishi, National Committee, 1969).

On the other hand, natural discharge from wells at Takimoto village gradually decreased from August 1966 and the water of Kiyotaki, which had had stable water discharge for many years (on the east slope of Mt. Kimyou), finally dried up. The same phenomena were observed eastward of Makiuchi village.

Moreover, the discharge from hot springs increased at Kagai. Fig. 13.5 shows a change of discharge and temperature at Kagai hot spring. At the end of August 1966, the discharge from hot springs rapidly increased. The period coincided with the period of rapid increase of discharge from newly welled up places and also the period of large movement of ground surface. At the same time landslides occurred successively in the district of Makiuchi.

Discharging of groundwater occurred in the area of cracked base rock, so it was reasoned that the landslide, as a secondary disaster of the earthquake, was triggered by the groundwater discharged from confined aquifer.

Earthquakes 433

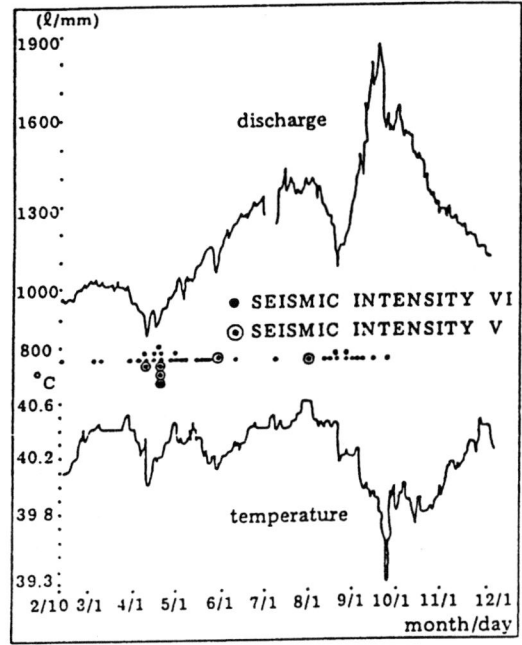

Fig. 13.5. Discharge and temperature of the Kagai hot spring (by Mr. Isao Kasuga, National Committee, 1969).

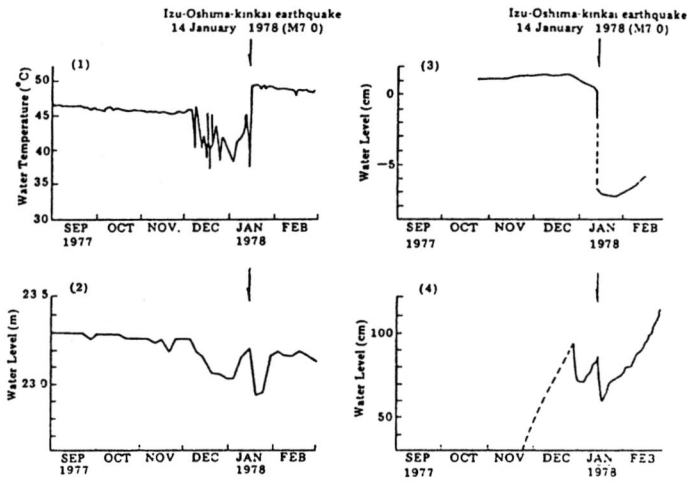

Fig. 13.6. Abnormal change of groundwater before and after the Izu-Oshima-kinkai earthquake (by Dr. Masakazu Otake, Japanese Association 1986); (1), (2) Nakaizu, (3) Hunahara, (4) Omaezaki.

13.2.4 THE IZU OSHIMA KINKAI EARTHQUAKE

The Izu Ohshima Kinkai Earthquake of M 7.0 occurred on January 14, 1978. Observed change of groundwater conditions (water level and water quality) are shown in Fig. 13.6, in which hydrological consequences of the earthquake are clearly recognized.

Observations revealed that radon content of groundwater gradually decreased at Nakaizu from about three months before the earthquake and drastically changed after the earthquake. The same tendency could be recognized in the temperature of hot springs at Nakaizu and groundwater levels at Nakaizu, Hunahara and Omaezaki.

References

Japanese Association for Disaster Prevention, 1986, The Forecast and Countermeasures Against Secondary Disasters (in Japanese).
National Committee for Countermeasures against Land Slides, 1969, Matsushiro Earthquake and the Land Slide (in Japanese).
Yoshino F, 1989, 'Hydrological Consequences of Earthquakes Experienced in Japan', Hydrology of Disasters, World Meteorological Organization, pp. 274–283.

List of Contributors

Dr. M.L. Anderson
Dept. of Civil and Environmental Engineering
University of California
Davis, CA 95618, U.S.A.

Dr. N.W. Arnell
Institute of Hydrology, Wallingford
Now at: Geography Department
University of Southampton
U.K.

Professor F. Ashkar
Department of Mathematics
University of Moncton
Moncton, N.B.
Canada E1A 3E9

Dr. M.A. Beran
Institute of Hydrology
Maclean Building
Crowmarsh Gifford
Wallingford, Oxfordshire OX10 8BB
U.K.

Dr. D.L. Fread
Hydrologic Research Laboratory
Office of Hydrology
National Weather Service, NOAA
U.S. Department of Commerce
1325 East-West Highway, Room 8348
Silver Spring, MD 20910, U.S.A.

Professor G. Gambolati
Instituto di Matematica Applicata
Universita degli Studi
via Belzoni 7, Padova
Italy

Professor K. Hutter
Tech. Hochschule Darmstadt
Institut fur Mechanik
Hochschulestrasse 1
D-64289 Darmstadt
Germany

Dr. P.A. Johnson
Dept. of Civil Engineering
University of Maryland
College Park, MD 20742-3021
U.S.A.

Professor M.L. Kavvas
Dept. of Civil and
  Environmental Engineering
University of California - Davis
Davis, CA 95216, U.S.A.

Mr. O. Matsuo
Public Works Research Institute
Ministry of Construction
1 Asahi, Tsukuba City, Ibaraki
Japan

Professor R.H. McCuen
Dept. of Civil Engineering
University of Maryland
College Park, MD 20742-3021
U.S.A.

Dr. V.E. Neall
Department of Soil Science
Massey University
Palmerston North
New Zealand

Dr. M. Putti
Dept. of Mathematical Methods
 for the Applied Sciences
University of Padua
via Belzoni 7, Padua, Italy

Professor V.P. Singh
Dept. of Civil & Environmental Engineering
Louisiana State University
Baton Rouge, LA 70803-6405, U.S.A.

Dr. Mohsen M.Sherif
Irrigation and Hydraulics Department
Faculty of Engineering
Cairo University
Giza, Egypt

Dr. P. Teatini
Dept. of Mathematical Methods for
 the Applied Sciences
University of Padua
via Belzoni 7
Padua, Italy

Dr. T.P. Gostelow
Engineering Geology Group
British Geological Survey
Keyworth, Nottingham NG12 5GG
U.K.

Mr. A. Terakawa
Public Works Research Institute
Ministry of Construction
1 Asahi, Tsukuba City, Ibaraki
Japan

Professor J. Wieringa
Department of Meterology
Wageningen Agricultural Univ.
6701 AP Wageningen
The Netherlands

Index

A
adsorption 289
advection-dispersion relation 284
Agua 414
Amakusa Islands 410
AMF series, pdf distribution 73
AMF model 72, 73
Angle of repose 326
annual flood series 65
annual maximum flood 65
arbitrary changes 45
arch dams 87, 89
Ariake Sea 410
Armero 402
at-site estimation 77
atmospheric system 129
Australia 56
avalanches 404
avenidas 406
Awu 402

B
backwater 90, 91, 96, 116
Bandai-san Volcano 397
BEED 90, 91
Belgium 51
blending height 24
Blitar 398
Boussinesq equation 211
Boussinesq number 324
BREACH 85, 87, 88, 89, 91, 92, 112
breach parameter(s) 94, 95
breach properties 94, 427
bridge(s) 108, 110
broad-crested weir 92, 106, 107
Buffalo Creek coal-waste dam 85
bulking 404
Butak Petarangan 398

C
caldera 409
canonical correlations analysis 78
carbon dioxide 42, 43
Cartago 418
CFCs 42
chemical explosion 1
chemical spills 2
chlorofluorocarbons 43
Ciudad Vieja 414
climate model simulations 47
climate system 41, 42
climate change 42
climate change scenarios 45
climatic change 1
cluster analysis 78
CO_2 exsolution 411
coastal aquifer 269
Cohesion 326, 331
collapse of a crater lake 395
collapse 409
column collapse 410
computational distance step(s) 102
computational time step(s) 101
concentration distribution 277
concentration gradient 278
cone of depression 276
confined aquifer 294
convective activity 49
conveyance factor 97
coupled apporoach 235
Courant number 296
crater lake 395
crater lake, gas releases 411
critical flow 111
critical slope 107
critical levels 45
critical threshold 45
cyclic flow 278
cyclone 49, 50
cyclones 1

D
dam breaching 1, 2
dam failure(s) 85, 87
dam-breach flood(s) 86, 87, 88, 96, 122
dam-break flood(s) 86, 87, 88, 96, 123
DAMBRK 116
dammed reservoirs 430
Darcy's law 273
dead storage 111

Debris slides 188
debris flow 161, 162, 163, 178
debris flow 430
debris basins 162
debris flow factors 162
debris 99
debris flow 13
debris avalanche 408
degassing 413
dendritic systems 111
density dependent 270
depressions 49
design floods, uncertainty in estimating 76
design flood 63
design flood, estimating 64
dispersion 276
dispersion zone 306
dispersivity 284
dispersivity, longitudinal 284
dispersivity, lateral 284
distribution candidates, functional relationships 72, 73
distribution candidates, descriptive abilities 75
distribution candidates, predictive abilities 75
distributions, comparing criteria 75
drawdown 277
drought risk 157
drought modelling 145
drought, definition 52
droughts 1, 2
droughts 42, 43, 44, 54, 59
dynamic routing 96

E

earth tides 270
earthen dam(s) 87, 89, 116
earthquakes 427
earthquakes 1, 3, 4, 9, 13, 14
earthquakes 270
El Niño-Southern Oscillation 50
energy balance models 146
Entrainment coefficient 356, 373
equilibrium scenarios 48
errors in river-flow data 69
eruptions 395
estimating design flood, parametric approach 64
estimating design flood, statistical methods 64
estimating design flood, data series 64

estimating design flood, statistical hypotheses 68
evaporation 270
expansion/contraction 97, 98
exposure correction 24
Extreme Value Type I 52
extreme value distribution 33
extreme wind 2, 19
extreme drought 127

F

famine 54
famine 1
feedback mechanisms 129
feedbacks 43, 44, 58
Fick's law 286
finite elements 251, 257
finite differences 259
finite element 295
finite difference 295
flood frequency analysis 71
flood risk assessment 52
flood data, seasonality in 67
flood quantiles 67
flood, T-year 68
floods 1, 2, 3, 5, 6, 7, 8, 9, 11, 14
floods 42, 43, 44, 49, 51, 59
floods, rain-generated 49
Flow avalanche 317, 320, 321, 324
flow pattern 278
fluid withdrawal 235
fluid density 290
formation factor 286
frequency 161, 169
freshwater 269
Friction angle 192
Froude number 323
Fuego 402

G

Galerkin technique 296
gamma 71
gamma distribution 73
gas release 413
Gaussian elimination 104, 111
general circulation models (GCMs) 47, 48, 51, 55, 58
general extreme value 71
general circulation models 146
generalized gamma distribution 75

Index

GEV distribution 75
Ghyben-Herzberg relation 271, 274
goodness of fit 77
Granular avalanche 325, 338, 346
gravity dams 88, 89
greenhouse effect 42, 43
groundwater conditions 433
groundwater abnormality 431
groundwater 269
growth 135
Grubbs-Beck test for outlier detection 70
Gumbel distribution 71
gustiness 20, 26

H

Höfdabrekkuháls 408
heat wave 2
heavy rains on volcanic material 395
Herculaneum 418
historical flood information 71
homogeneity 68
hot lahars 417
hurricanes 1, 2, 3, 5, 7, 8, 11, 13, 15, 21
Hydraulic conductivity 211, 217
hydraulic gradient 278
hydraulic jumps 111
hydraulic routing 96
hydrodynamic dispersion 280, 281, 284
hydrographs 404
hydrologic 96
hydrologic system 133
hydrological [processes] 44
hydrothermal waters 410
hydrothermally altered lithologies 398
hydrothermally-altered materials 397
hydrovulcanic eruptions 395

I

Iceland 408
ignimbrite 410
Ili Werung 411
immiscible fluids 270
incomplete means 76
independence of flood observation 68
index flood technique 77
initiation 135
Instability 220
interface 270, 271, 274
interface depth 275
Internal kinematics 188

Internal friction 326, 334, 346
intrusion length 273
IPCC 43, 44, 50
Irazu 418

J

jökulhlaups 408
jumps 69

K

Kötlulhlaup 408
Kaba Volcano 401
Katla Volcano 408
Kawah Idjen 401
Kelly Barnes Dam 85, 86
Kelvin-Helmholtz instability 320
Krakatoa 409
Kruskal-Wallis test 68
Kusatu-Sirane 402

L

Lágeyjarhverfi 408
L-moments 76
lahar 396
lahars, secondary 414
lahars, primary 414
Lake Monoun 413
Lake Rotomahana 396
Lake Taal 396
Lake Nyos 411
land subsidence 228
land slides 1, 3
land subsidence 1
landslide 431
landslide 410
landslide-formed dam(s) 86, 91
lateral dispersion 281
lateral flow(s) 99, 111
Laurel Run Dam 85, 86
Lawn Lake Dam 91, 94
Le Prêcheur 418
leaky 271
leaky aquifer 293
least squares 76
levee(s) 110, 111
level-pool routing 106, 107
light-tailed distribution 73
lognormal 71
Lomblen Island 411

longitudinal dispersivity 278
loop rating 97
LPIII distribution 75

M

Mýrdalssandur 408
Mach number 324
Madras aquifer 303
man-made disasters 1
Mann-Whitney test for jumps 69
Mann-Whitney test for non-homogeneity 68
Manning equation 98, 113, 122
Manning n 98, 113, 122
marked point process 65
maximum likelihood method 75
maximum entropy 76
mechanical dispersion 280, 281, 283
melting snow 408
meltwater 404
methane 42
method of characteristics 295
mitigation 161, 171
modelling errors 76
Mohr-Coulomb yield criteria 327, 332
molecular diffusion 280, 281, 283, 286
moment-ratio diagrams 77
moments, method of 75
moments, method of mixed 75
moments, generalized method of 75
monitoring network 270
monitoring networks 312
monsoon 417
Mt Tarawera 396
Mt Pelée 402
Mt Galunggung 401
Mt Merapi 402
Mt St Helens 407
Mt Ruapehu 414
Mt Pinatubo 402
Mt Inagodake 397
Mt Kelut 398
Mt Asama 403
Mt Cotopaxi 406
Mt Hudson 407
Mt Mayon 416
Mt Agung 417
mud flow 161
mud flow 99
multilayered aquifer 271
murgangs 415

N

natural disasters 1
natural recharge 278
nested meso-scale atmospheric models 48
Nevado del Ruiz 402
Newton-Raphson 104, 105
Nile Delta aquifer 308
nitrous oxide 42
non-magmatic explosions 396
non-Newtonian flow(s) 99, 104

O

observation wells 312
oceanic system 132
off-channel storage 112, 121
Ohtsukigawa Debris avalanche 397
Óracfajökull 408
outliers in flood data 70
overtopping 85, 91, 92, 94, 110

P

paleoclimates 45
Paluwh Island 410
Papandayan Volcano 398
Pareto (generalized) 71
partial duration series 65
peak discharges 405
peaks over threshold 65
Pearson Type 3 71
Peclet number 281
phreatic explosions 395
phreatic aquifers 271
phreatomagmatic explosions 396
piezometric head 270
piping 85, 89, 91, 92, 94
plotting flood sample values 77
point process 65
Poisson process 66
Poisson distribution 66
Pore water pressure 221
porous media 280
POT exceedances, pdf distribution 72
POT model 72, 73
potential wind speed 24, 30
Powder avalanche 317, 318, 320, 321, 322, 351, 353, 354, 358, 383
precautionary principle 60
prediction models 167
principal component analysis 78
probability weighted moments 75

Index

probability distributions 76
pumping 270
pyroclastic surges 404
pyroclastic flows 395

Q
quasigeostrophic theory 130

R
rail disaster 414
rain lahars 415
rainfall 164
rainfall 270
randomness, flood flows 68
rating curve(s) 108
Raung Volcano 401
recharge 270
recharge, natural 278
record augmentation procedure 79
recurrence interval 67
reducing lake level Mt Kelut 398
regional flood frequency estimation 77
regionalization technique 78
regression regionalization procedure 78
relative density 278
representative (-ity) 23, 25
reservoir water surface 89, 92
Resina 418
return period 33
return period 67
Reventado River 418
Reynolds number 323
Reynolds stress 355, 376
Richardson number 357
Rivière Blanche 402
robustness 60
roughness length 21, 27

S
Sacramento-San Jaoquin basin 51, 56
Sahel drought 58
Saint-Venant equations 91, 97, 112, 121
Sakurajima 417
saline water 269
saltwater 269
saltwater intrusion 286
saltwater intrusion 269
sampling errors 76
Sanata Maria, Guatemala 403

Santiago de los caballeros 414
Scaling factors 363
sea levels 42
seaward boundary 293
seawater 269
sediment transport 90, 91
sediment-retention basins 415
sediments 430
Seepage erosion 184
Seepage face 195, 196
seepage face 275
Seismic shock 221
Semeru 417
sensitivity analysis 45
Shape factors 356
sharp interface 270, 271, 278
sharp front 271
Shear stress 184
Shear strength 179, 184, 185, 216
Shimabara catastrophe 410
sinuosity factor(s) 97, 112
Skeidárhlaup 408
Skeidárjökull 408
Ski-jump effect 382
Slope stability 192
slope failures 430
snow avalanches 1
snow-melt floods 51
soil 163
soil moisture 55, 56
spatial analogues 45
spectral gap 20
spillway(s) 85, 86, 88, 90, 91, 92, 106, 109
St Vincent 402
stationarity of streamflow series 69
statistical distribution for fitting flood data sets 71
statistical distribution, choice of 71
steam explosions 395
steamship Berouw 409
stochastic process 66
storm 1
subcritical 102, 104
summer dryness 55
Sunda Strait 409
supercritical 102, 104
superfund sites 1
Surface erosion 219
surface roughness 24
sustainable safe discharge 277

Switzerland 51

T
Tangiwai disaster 413
temporal analogues 45
tephra 415
Teton Dam 85, 86, 91, 94, 116
thick-tailed distribution 73
thunderstorms 418
tidal waves 1
tidal effects 270
tidal effects 279
time series analysis of daily flow data 67
time of failure 86, 87, 95, 116
toe point 271
Tokachi-dake Volcano 407
Tolima 407
top tip point 271
tornadoes 1, 5
tortuosity 286
transient scenarios 48
Translational slide 188
Transmissivity 219
trends 69
tributary flow 116
tsunamis 1, 3
two-dimensional effect(s) 121
typhoons 415
typhoons 1, 3

U
uncoupled approach 236
ungauged sites 77
United Kingdom 56
Unzen Volcano 410
upconed interface 276
upconing of saltwater 276

V
Vatnajökull 408
Vesuvius 418
Villarica 407
Voellmy model 342, 343
volcanic debris avalanche 398
volcanic activity 395
volcanic melting 395
volcanic gases 395
Volcano Island 396
volcanoes 1
volcanogenic tsunamis 395
volume losses 123

W
Wald-Wolfowitz test 68
Wald-Wolfowitz test for trend 69
water reservoirs 57
water table 270
water resources 57
water use efficiency 59
weather type 48
wedge 270
Weibull distribution 71
Weibull wind distribution 31, 32
wild fires 4
wind spectrum 20
wind storms 2

Y
Yake-dake 398

Printed in the United States
21855LVS00002BB/11